Water Resources Engineering Risk Assessment

NATO ASI Series

Advanced Science Institutes Series

A series presenting the results of activities sponsored by the NATO Science Committee, which aims at the dissemination of advanced scientific and technological knowledge, with a view to strengthening links between scientific communities.

The Series is published by an international board of publishers in conjunction with the NATO Scientific Affairs Division

A Life Sciences B Physics	Plenum Publishing Corporation London and New York
C Mathematical and Physical Sciences D Behavioural and Social Sciences E Applied Sciences	Kluwer Academic Publishers Dordrecht, Boston and London
F Computer and Systems Sciences G Ecological Sciences H Cell Biology I Global Environmental Change	Springer-Verlag Berlin Heidelberg New York London Paris Tokyo Hong Kong Barcelona Budapest

NATO-PCO DATABASE

The electronic index to the NATO ASI Series provides full bibliographical references (with keywords and/or abstracts) to more than 30000 contributions from international scientists published in all sections of the NATO ASI Series. Access to the NATO-PCO DATABASE compiled by the NATO Publication Coordination Office is possible in two ways:

- via online FILE 128 (NATO-PCO DATABASE) hosted by ESRIN, Via Galileo Galilei, I-00044 Frascati, Italy.
- via CD-ROM "NATO-PCO DATABASE" with user-friendly retrieval software in English, French and German (© WTV GmbH and DATAWARE Technologies Inc. 1989).

The CD-ROM can be ordered through any member of the Board of Publishers or through NATO-PCO, Overijse, Belgium.

Series G: Ecological Sciences Vol. 29

Water Resources Engineering Risk Assessment

Edited by

Jacques Ganoulis

Dept. of Hydraulics and
Environmental Engineering
Aristotle University of Thessaloniki
54006 Thessaloniki
Greece

Springer-Verlag
Berlin Heidelberg New York London Paris Tokyo
Hong Kong Barcelona Budapest
Published in cooperation with NATO Scientific Affairs Division

Proceedings of the NATO Advanced Study Institute on Risk and Reliability in Water Resources and Environmental Engineering held at Porto Karras, Greece, May 18–28, 1991

ISBN-13: 978-3-642-76973-3 e-ISBN-13: 978-3-642-76971-9
DOI: 10.1007/978-3-642-76971-9

Softcover reprint of the hardcover 1st edition 1991

Typesetting: Camera-ready by authors
31/3140-543210 – Printed on acid-free paper

CONTENTS

PART 3 - RISK MANAGEMENT IN WATER RESOURCES AND ENVIRONMENTAL ENGINEERING

PART 4 - CASE STUDIES

Main presentations are listed in normal text; short presentations are in italics

PREFACE

Safety and performance of engineering projects have been always of primary concern for water resources and environmental engineers. The estimation of the T-year flood and the control of pollutant concentrations in a sewage treatment plant are two typical examples. In the design stage, a predominant factor for the reliability of the engineering project is the control of various uncertainties of the inputs and also the sample and model uncertainties.

Risk and reliability analysis provide the theoretical framework of methods and tools to evaluate and quantify these uncertainties. It is the aim of this book, based mostly on the lectures presented in the NATO/ASI, Porto Carras, Greece, May 18-28 1991, to summarize the up-to-date developments on risk and reliability analysis as applied to water resources and environmental engineering.

From Tuscon to Porto Carras

In May 1985 a NATO ASI was held in Tuscon, Arizona, USA for the purpose of classifying various concepts of engineering reliability and risk in water resources . The present NATO ASI, held in Porto Carras, Greece, six years later (May 1991) may be considered as a continuation and an extension of the initial endeavor. The main purpose was to provide a unified approach on risk and reliability in both water quantity and water quality problems reflecting in the same time concepts and techniques that have emerged during this time period.

In the edited volume of the 1985 meeting in Tuscon (**Duckstein & Plate,** 1987) a general systems engineering framework is provided for the engineering risk and reliability in water resources. Reliability investigations are presented in two groups: reliability in hydraulic structures and reliability in water supply systems. In the same volume, the last chapter is devoted to decision making under uncertainty and under multiple objectives.

In the present volume the structural reliability and standard techniques for reservoir management under "non-crisis" conditions are de-emphasized. Alternatively more emphasis has been placed on the quantification and management of risks related to a broad spectrum of problems: from the hydrologic estimation of the exceedance probability (**Bobée et al.**), the stochastic estimation of pollution risk in rivers (**Plate**), coastal waters (**Ganoulis**) and groundwater (**Bagtzoglou et al.**) to new techniques, such as the "envelope" approach for dynamic risk analysis (**Haimes et al.**) and the fuzzy set approach (**Duckstein & Bogardi**).

These techniques appear to be applicable to both scientific and decision making aspects of water resources and environmental engineering.

Organisation

The present volume includes the following topics:

1. Engineering risk and reliability related to hydrologic extremes under uncertainty.
2. Risk of pollution in surface and ground water problems.
3. Modeling of decisions under various uncertainties using Bayesian and related techniques.
4. Embedding of risk into multiobjective decision models and conflict analysis schemes.

Furthermore the following aspects are emphasized:

a. Water quality problems receive now wider attention whether in rivers, groundwater, estuaries or coastal waters, both from methodological and case study points of view.
b. Further new risks that are posed or examined include those related to global climatic change and oil pollution (**Stakhiv, Dracup & Kendall, Seip**). Concerning global climatic change the introduction of paleofloods and tree-ring information into the data base seems to be an interesting development (**Enzel et al., Ely et al., Dracup & Kendall**).
c. The results of applying more than one technique to the same problem are extensively compared, for example:
   - systems analysis versus water resources engineering (**Parent et al.**)
   - use of different stochastic simulation techniques to account for uncertainties in the analysis of seashore protection measures (**Stakhiv et al.**)
   - model versus sample and other uncertainties (**Bernier**)
   - remote sensing versus numerical models (**Caussade**)

With the above considerations, the present volume is organized in four parts

(1) Uncertainty analysis of hydrologic inputs
(2) Quantification of risks in water quantity and quality problems
(3) Risk management in water resources and environmental engineering
(4) Case studies

The ***first part*** deals with various aspects of input uncertainty such as the randomness or natural uncertainties and the sample uncertainties. Here the probability of exceedance (**Bobée et al., Rasmussen**) and the reliability of estimating hydrologic extremes such as droughts (**Correia et al.**) and floods (**Enzel et al., Ely et al.**) are considered together with the uncertainties in collecting groundwater quantity and quality data (**DeBacker**) and the use of remote sensing for oceanographic and coastal applications (**Gastellu-Etchegorry**).

The ***second part*** is devoted mainly to risk quantification i.e. the probability of the occurence of an adverse event. This part starts with a paper by **Parent** dealing with the stochastic dynamic hydrological systems. Using conceptual stochastic simulation and probabilistic modelling, risks of water pollution can be quantified in rivers (**Plate**), coastal waters (**Ganoulis**) and groundwater (**Bagtzoglou et al.**). The Bayes approach can be used not only in accounting sampling uncertainties but also modelling uncertainties (**Bernier**). The paper by **Duckstein & Bogardi** on the use of fuzzy sets and fuzzy arithmetic concludes this part and demonstrates that this relatively new technique is useful not only to quantify risks but also the expected losses corresponding to each action.

The ***thirt part*** of the present volume deals with risk management. Here the risks have been identified and, as much as possible, quantified. Various criteria have been defined to characterize risk, including performance indices related to effect of uncertainty. Some of these criteria may be probabilistic (mean and variance of a contamination mass in a polluted aquifer,**Woldt et al.**) or may be fuzzy (probability of ecological damage, **Stansbury et al.**). In any case, risk management provides the means to investigate the mitigation of the consequences of risks. For this purpose, trade-offs may be made at increasing by high levels between the various risk indications. For example, one risk may be traded off against another one at a certain level, and an overall risk index may be traded off against an overall cost at a higher level (**Woldt et al.**).A very important demonstration of how risk analysis can be useful to face new challenging problems such as the engineering implications from a potential sea level rise due to the climate change is given in the **Stakhiv et al.** paper.A multiobjective dynamic risk analysis is proposed by **Haimes et al.** to produce optimal flood warning threshold.Two papers on multicriterion reliability analysis are included in this part: the first by **Duckstein et al.** deals with a stochastic type estimation of risks in a hydrologic framework and the second by **Dahab & Lee** uses the fuzzy approach in the environmental engineering context.(short contribution). The operational control of denitrification in sewage treatment plants is the topic described in the short contribution of **Stamou.**

Some characteristic risk related case studies are presented in the ***fourth part*** . **Parent et al.** show the application of system analysis and hydrological modeling for the real time operation of the NESTE water system in France. Drought analyses related to climate changes in the Colorado river basin are applied by **Dracup & Kendall.**The three papers which conclude this part deal with problem of coastal pollution : the first by **Caussade** for the suspended sediment transport in the Gambia estuary, the second by **Samsunlu et al.** for the sewage coastal pollution in Istanbul and the third by **Seip** for the oil pollution of the sea in Norway.

Risk and reliability is a relatively new subject in water resources and environmental engineering. As a contribution to understanding and using this powerful tool, the present volume is aimed to serve as a textbook. This is the main reason to start with an introductory paper by **Ganoulis, Duckstein & Bogardi**. In this text the general framework for risk analysis is presented from the traditional engineering (conceptual modeling) rather than from the systems engineering point of view. Methods and tools from stochastic modeling and fuzzy sets are also presented in this paper.

It is with great pleasure that I would like to acknowledge here the sponsors of this meeting: mainly the NATO Scientific Affairs Division,the Aristotle University of Thessaloniki,the City of Thessaloniki, and the National Power Corporation of Greece (DEI).Particular thanks are addressed to Dr. **Luis Vega Da Cunha**, Director of the NATO/ASI program for his efficient help and assistance. I want to gratefully recognize here the friendly help of the organizing committee: **Bernard Bobée, Lucien Duckstein** and **Francisco Nunes Correia**. In particular I wish to thank: colleagues and friends **Lucien Duckstein, Istvan Bogardi** and **Erich Plate** for their help during the meeting and reviewing part of the papers that appear in this book; staff and graduate students from the Aristotle University of Thessaloniki **Marios Vafiadis, Yiannis Mylopoulos, Petros Anagnostopoulos, Nikos Theodossiou and Themis Mavroidis**; technical assistants **Gina Mavidou** and **Olympia Sortikou**, who also has typed parts of the final manuscript; and last but not least the oustanding lecturers and participants in this ASI. We have all worked hard in Porto Carras. I am sure that the beauty of the site, the local conditions and the proximity of the Mount Athos may have increased our intellectual effort and inspiration to further our knowledge of risk and reliability in water resources engineering by another small step.

Thessaloniki,Greece
July 1991

INTRODUCTION

RISK ANALYSIS OF WATER QUANTITY AND QUALITY PROBLEMS: THE ENGINEERING APPROACH

Jacques Ganoulis, Lucien Duckstein[1], Istvan Bogardi[2]
Department of Civil Engineering
Division of Hydraulics & Environmental Engineering
Aristotle University of Thessaloniki
54006 Thessaloniki
Greece

ABSTRACT

A unified approach to risk analysis in water quantity and quality problems is presented. This is done by overviewing the definitions and methodologies for risk and reliability analysis in water resources and environmental engineering, including the concepts of fuzzy sets.The various steps to be undertaken for a comprehensive application of engineering risk analysis to a specific problem are first analyzed. These are (1) hazard identification, (2) risk quantification, (3) consequences of risk, (4) perception of the consequences and (5) risk management. Then, methods and tools used in uncertainty analysis, stochastic simulation and the fuzzy set approach are summarized.

1.- INTRODUCTION

The aim of this paper is twofold

(1) to present a unified approach for engineering risk and reliability analysis in water quantity and quality problems

(2) to review the existing methodologies for engineering risk and reliability analysis, especially the stochastic and the fuzzy set techniques.

The approach followed herein is more from the traditional engineering point of view dealing with conceptual modelling and simulation, than from the systems engineering point of view where the system is considered as a black-box (Duckstein et al., 1987). Water resources engineers follow the traditional way to face the physical problems they have, which consists mainly of the following stages

---

[1]Department of Systems Engineering, University of Arizona, Tuscon, Arizona, 87521
[2] Department of Civil Engineering, University of Nebraska, Lincoln, Nebraska, 68588

NATO ASI Series, Vol. G 29
Water Resources Engineering Risk Assessment
Edited by J. Ganoulis

- Definition of the problem
- Formulation of a model
- Solution

Usually the problem is defined by a set of physical parameters and also the input and output variables. In groundwater flow the physical parameters are, for example, the porosity and the aquifer transmissivity. In coastal pollution they could be the eddy viscosity, the Coriolis coefficient and the turbulent dispersion tensor. In real situations examples of inputs or independent variables are the precipitation rate, the aquifer recharge and the pollutant loads discharging into the sea. The output or dependent variables could be the dissolved oxygen concentration, the velocity of the flow and the free surface variation .

According to the specific objectives of the engineering design and the existing standards, some of the output variables have to take on values belonging to a given interval or set. For example, the value of the dissolved oxygen concentration must exceed 5 ppm or the piezometric height must be below to a given value. "Crisis" conditions and losses are expected when the above conditions are not fulfilled. Generally speaking the engineering risk is a quantitative measure for accounting the expected losses in situations when an "incident" or a "failure" has occured.

In water resources and environmental engineering the main problem for evaluating the risk and reliability is that the physical parameters and the inputs of the system, because of the natural variability in space and time, show random deviations. To this natural randomness, one must add various other uncertainties due to the scarcity of the information concerning the inputs, the value of parameters (measurement and sampling uncertainties) and also the imperfection of the models (modelling uncertainties). It follows that the outputs are not deterministic, but they also show random variations. For dealing with randomness and uncertainties, risk analysis provides a general theoretical framework. This is based on the uncertainty analysis, the assessment of risk and the decision under risk conditions.

From the above considerations the organisation of this paper is the following. First a definition of the engineering risk and the various steps to follow for risk analysis are summarized. Then the methods for uncertainty analysis are reviewed from two point of views: the stochastic modelling and the fuzzy set approach. Risk assessment and risk management are further analysed before the conclusions are given.

2.- DEFINITION OF THE ENGINEERING AND HEALTH RISKS

In analysing a physical problem there is a characteristic variable, usually considered as an output which describes a specific situation. This characteristic variable is usually called ***"load"*** or ***"exposure"*** λ.

1. The exposure or load λ can be measured or calculated according to the methods outlined in section 3.1. In water resources problems, the exposure such as for instance the daily human nitrate consumption from drinking water, can be estimated with some level of uncertainty. This uncertainty can be encoded with probabilistic methods treating λ as a random variable $\underset{\sim}{\lambda}$ (section 3.1). Alternatively, fuzzy set methods can be used to represent uncertainty in the exposure by expressing λ as a fuzzy numbers $\hat{\lambda}$ (section 3.2).

2. The resistance (or capacity) ρ is to be expressed in the same unit as the exposure. The capacity of a water supply system is a typical engineering example. Or, for certain carcinogens a human threshold level, that is, resistance, can be defined. Exposure below this level would not cause cancer. The resistance is also uncertain in many cases. Then probabilistic methods or fuzzy set methods can be used to describe resistance either as a random variable $\underset{\sim}{\rho}$ or a fuzzy number $\hat{\rho}$.

3. Failure is defined as an event when exposure is larger than resistance $\lambda > \rho$. Due to the uncertainties in λ and ρ this event can be expressed as a probability $P(\underset{\sim}{\lambda} > \underset{\sim}{\rho})$, (see for example the case of a probabilistic approach in the dam-break problem (Ganoulis, 1987)) or as a difference between two fuzzy numbers: $(\hat{\lambda}\text{-}\hat{\rho})$. Often, a mixed probabilistic fuzzy set formulation can be used. For instance the exposure to a flood levee, the peak flood discharge is a common random variable in hydrology. On the other hand, the resistance of the levee may be expressed as a fuzzy number reflecting uncertainty in the levee soil material, construction and underground soil conditions. In this example, the levee failure event $P(\underset{\sim}{\lambda} > \hat{\rho})$ is a cumulative distribution function value obtained as a fuzzy number.

4. The consequence of a failure, in engineering risk analysis, is considered as economic loss, loss of human life or other adverse effects. This consequence $L(\lambda)$ is offen expressed as a deterministic function of the exposure . A typical example is an economic flood loss as influenced by the peak discharge above the level resistance level.

In health risk analysis such as cancer risk from nitrate exposure, the consequence of an exposure dose is analyzed by a dose-response relationship $DR(\lambda)$ which determines the probability of an event such as cancer development in an individual given an exposure dose λ (Zeise et al., 1987), above threshold value ρ (if any).

5. The perception of risk reflects the human attitude toward various kinds of risk. Risk perception may result in a higher or lower perceived risk as compared to the actual risk.The risk of automobile accidents is generally lower than the actual risk or, the perceived risk of airplanes is higher than the actual risk. It is of high importance to assess the perceived risk since the willingness to pay for risk reduction is controlled by the perceived and not by the actual risk.

The expression of risk depends on the way uncertainties in the elements of risk analysis are considered. The classical probabilistic formulation generally considers the expected value of risk using the probability density function g (λ) of exposure λ. In case of the engineering risk ER:

$$ER = \int_{\rho}^{\infty} L(\lambda)\ g(\lambda)\ d\lambda \qquad (2.1)$$

A typical water resources example consists of calculating the expected economic flood losses above resistance ρ. Similarly, in health risk analysis the so-called individual (health) risk, HR can be expressed again as an expected value, here expected probability.

$$HR = \int_{\rho}^{\infty} DR(\lambda)\ g(\lambda)\ d\lambda \qquad (2.2)$$

Here, ρ represents the threshold dose below which no health effect can be expected. In health risk analysis, the level of exposure is often fixed and the health risk is given by the corresponding dose-response function value. When combined with the size of the exposed population the individual health risk defined in equation (2.2) yields population risk. Population risk may be measured by the incidence of adverse health effect considered (e.g. cancer) in the exposed population per year.

If uncertainty in any element of risk analysis is encoded by fuzzy sets in form of fuzzy numbers, the corresponding risk figure will be also a fuzzy number. Two examples are given:

a) Engineering risk analysis: the exposure is probabilistic $\underset{\sim}{\lambda}$ the resistance is fuzzy $\hat{\rho}$ and the loss function is also fuzzy $\hat{L}$. Then the engineering risk can be calculated as a fuzzy number:

$$\hat{E}R = \int_{\hat{\rho}} \hat{L}(\lambda)\ g(\lambda)\ d\lambda \qquad (2.3)$$

The actual calculation is shown for example in Duckstein and Bogardi (this volume).

b) Health risk analysis: the exposure is fuzzy $\hat{\lambda}$, there is no threshold: $\rho=0$ and the dose-response relationship is fuzzy $\hat{D}R$. Then the individual health risk can be calculated as a fuzzy HR (Bogardi et al., 1988):

$$\hat{H}R = \hat{D}R\ (\hat{\lambda}) \qquad (2.4)$$

A systematic application of risk and reliability analysis can be performed by the following steps

(a) Risk assessment (analysis)
 (a.1) Hazard identification
 (a.2) Risk quantification
 (a.3) Consequences of failure
 (a.4) Perception of the consequences
(b) Risk management (control)

Uncertainty analysis provide the necessary tools and methods for the assessment of risk and especially the risk quantification.

3.- UNCERTAINTY ANALYSIS

3.1 The stochastic approach

Hydrological and water quality parameters are traditionally considered as random or stochastic variables.The connection between the stochastic and the deterministic approach has been analysed by Ganoulis & Morel-Seytoux (1985) in the case of aquifer systems. This is summarized here for the general case of a physical hydrological system.

As stated earlier, a physical system is characterized by a set of physical parameters $\{a_1, a_2, \dots a_k\}$. The system receives an input or excitation defined by a set of independent variables $\{X_1, X_2, \dots Xn\}$ and its reaction is characterized by a set of output or dependent variables $\{Y_1, Y_2, \dots Y_m\}$. According to the deterministic approach, there is a unique relation between the deterministic response $Y_{i,d}$ of the system and its excitation X_j in terms of the system parameters a_p. This means that a law can be found, usually as a result of integration of a partial differential equation, having the following functional form:

$$Y_{i,d} = f(X_j, a_p) \quad i = 1, \dots n \quad j = 1, \dots m \quad p = 1, \dots k \tag{3.1}$$

More precisely, $Y_{i,d}$ is ***a model deterministic response value*** $Y_{i,d}$. The actual (measured or true) value of Y_i can be written in the form:

$$Y_i = Y_{i,d} + \varepsilon_{i,d} \quad i = 1, \dots n \tag{3.2}$$

where $\varepsilon_{i,d}$ is the total error, having assumed a deterministic model, due to several causes such as: (a) the intrinsic randomness of the system, (b) the measurement errors due to the instrument precision and human actions, (c) the numerical errors, when relation (3.1) is computed, and (d) the structural model error.

According to the stochastic approach, the physical parameters of the aquifer a_p and the inputs X_j are considered as ***random*** or ***stochastic variables*** having some probability distribution functions. By using physical conservation laws (usually as partial differential equations) or empirical statistical analysis of the available data, the objective of the stochastic approach is to determine the probability distribution law of the dependent variable Y_i, in the form:

$$P\,[Y_i \leq y_i] = \phi\,(X_j, a_p, y_i) \quad i = 1, ...n \quad j = 1, ...m \quad p = 1, ...k \tag{3.3}$$

The stochastic value Yi,s can be expressed as the sum between the expected value <Yi,s> according to the probability law (3.3), and the stochastic deviation or error $\varepsilon_{i,s}$ in the form:

$$Y_{i,s} = < Y_{i,s} > + \, \varepsilon_{i,s} \quad i = 1, ...n \tag{3.4}$$

where generally

$$Y_{i,d} \neq < Y_{i,s} > \tag{3.5}$$

By comparing the relations (3.2) and (3.4) and setting $Y_i = Y_{i,s}$ one obtains:

$$\varepsilon_{i,d} = < Y_{i,s} > - \, Y_{i,d} + \varepsilon_{i,s} \tag{3.6}$$

Excluding any numerical or experimental or model structural error, the deviations $\varepsilon_{i,d}$ and $\varepsilon_{i,s}$ are due only to the physical uncertainties of the system. In this case, the stochastic approach furnishes a formal procedure for evaluating $\varepsilon_{i,s}$ or some measure of it, such as the standard deviation. This can be interpreted as an advantage of the stochastic approach over a deterministic analysis.

One important question in the stochastic modelling of hydrological systems is the change in the spatial heterogeneity scales (Ganoulis,1986).Furthermore various methods and tools have been extensively used in the past for stochastic simulation, such as

- Time series analysis, filtering, kriging
- Stochastic differential equations:
    - Spectral analysis

- Taylor series and perturbation analysis
- Monte-Carlo simulation

3.2. The fuzzy set approach

The concept of fuzzy sets

Fuzzy set theory (Zadeh, 1965; Zimmerman, 1985) is a mathematical method used to characterize and propagate uncertainty and imprecision in data and functional relationships. Fuzzy sets are especially useful when insufficient data exist to characterize uncertainty using standard statistical measures (e.g. mean, standard deviation and distribution type).

Fuzziness represents situations where membership in sets cannot be defined on a yes/no basis because the boundaries of sets are vague. The central concept of fuzzy set theory is the membership function which represents numerically the degree to which an element belongs to a set. In a classical set, a sharp or unambiguous distinction exists between the members and non-members of the set. In other words, the value of the membership function of each element in the classical set is either 1 for members (those that certainly belong to the set) or 0 for non-members (those that certainly do not). However, many of the categories we commonly employ, such as the classes of tall men, beautiful women, higly contaminated water, deep wells or numbers much greater than 1, do not exhibit this characteristic. That is, their boundaries are vague.

Since the transition from member to non-member appears gradual rather than abrupt, the fuzzy set introduces vagueness (with the aim of reducing complexity) by eliminating the sharp boundary dividing members of the class from non-members (Klir and Folger 1988). Thus, if an element is a member of a fuzzy set to some degree, the value of its membership function can be between 0 and 1. When the membership function of an element can only have values 0 or 1, the fuzzy set theory reduces to the classical set theory.

Fuzzy sets as extension of interval analysis

The simplest method of considering uncertainty, in model prediction is to perform an interval analysis. An uncertain parameter of a flow model, such as dispersion coefficient, can take on any value within such an interval. With more information on the uncertain parameter, the interval model can be "sharpened", that is, we determine the possibility that the parameter can take on certain value(s) within the interval. If the axioms and hypotheses of probability theory are verified, then the probabilistic procedure is simply an extension of interval analysis. However, if considerably weaker hypotheses than those of probability theory are then fuzzy set analysis which can be used with very few and weak prerequisite assumption can be selected as an extension of interval analysis.

The propagation of uncertainty characteristics measured as intervals is the basis of traditional sensitivity analysis. The arithmetic of intervals is straightforward and functions of interval numbers are easy to calculate (Dong and Wong, 1986). To utilize additional information on an uncertain model parameter, multi-intervals can be defined. For example, if we believe that a model parameter is definitely greater than 2 and less than 6, about 5 but possibly somewhere else in the interval 2 to 6, then the multi-intervals sketched in Figure 1 may be appropriate.

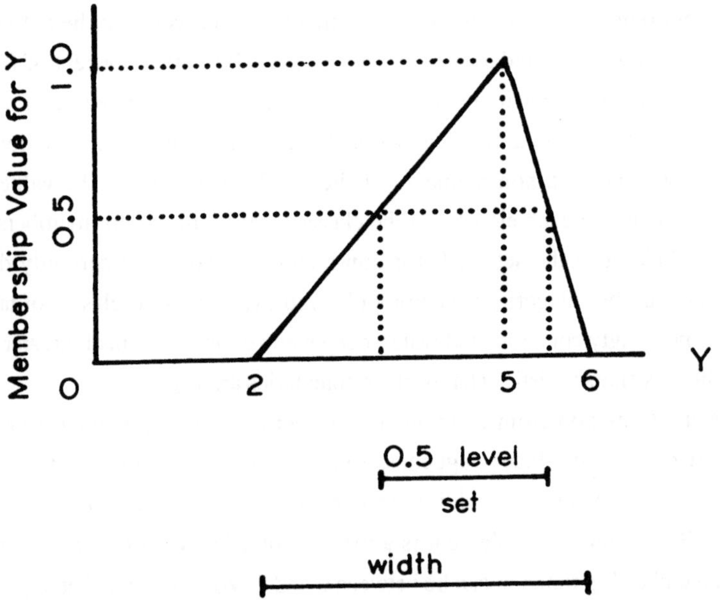

Figure 1 A membership function

Fuzzy numbers

A special class of fuzzy sets defined as fuzzy numbers describes the strength or degree of acceptance of the possible values of a model parameter. A fuzzy number has the following desirable properties:

1.- it is defined on the set of real numbers (rather than a set of linguistic variables)

2.- its membership function always reaches the maximum value of 1

3.- its membership function is unimodal consisting of an increasing and a decreasing part

4.- a complete arithmetic is available (e.g. Kaufmann and Gupta, 1985) to operate a rank or combine fuzzy numbers and furthermore, multidimensional functions of fuzzy numbers can be defined and transformed. This property is utilized in the fuzzy uncertainty analysis of model predictions.

Referring to Figure 1, the width of the membership function of a fuzzy number "about 5 but no less than 2 and no more than 6" is 6-2=4; this width is an internal length. If a so called credibility level h=0.5 is defined corresponding to a value for the strength of acceptance, then the level set h=0.5 is defined as the set [3.5, 5.5]. If h=0, the level set is [2,6].

4.- RISK ASSESSMENT

4.1.- Hazard identification

From the above, it appears that fuzzy sets and fuzzy numbers may be used to characterize input uncertainty whenever variables (and/or relationships) are not defined precisely. For example, the concentration of a pollutant may be described by the phrase "about 20 ppm but not less than 12 and not more than about 30". In general, an extreme loading (flood, dought, oil spill) or an abnormally low resistance (low DO high water temperature, weekened foundation) are hazards that may be imprecisely defined; in such cases, a fuzzy number representation may be used, specifying a range of real numbers X and the fuzzy set membership function m(x). This function m(x) should have an increasing left hand branch, a maximum of 1 and a decreasing right hand branch, as shown for example in Figure 1, or else, Figure 2 in Bogardi and Duckstein (in this volume).

A fuzzy number is thus defined as

$$\hat{A} = \{ (x, mA(x)) : x. \in .X;\ mA(x). \in .[0,1) \}$$

The closer mA(x) is to 1, the more "certain" one is about the value of x.

4.2.- Probability of an adverse event

If an event, on realization of a hazard is described by means of fuzzy logic, then the probability of that event may be calculated as a fuzzy number. A step example is given in Duckstein and Bogardi (this volume). Alternatively, the crisp (non-fuzzy) probability of a fuzzy event may be calculated . Either one of these two probabilities may be used to define performance indices of systems under consideration (Duckstein et al, 1988). In particular, the weighted fuzzy mean given below appears to have desirable characteristics:

$$\bar{A} = \frac{\int_{-\infty}^{\infty} mA(x)\, w(x)\, dx}{\int_{-\infty}^{\infty} mA(x)\, dx}$$

where w(x) = x mA(x) is a possible choice for the weighting function

4.3. Consequences of incidents or failures

The engineering, economic and environmental consequences may be expressed as fuzzy. For example, Stakhiv et al. (this volume) could have used fuzzy numbers to estimate the costs and benefits of sea level risk reduction. Note that fuzzy numbers and probability of occurrence of a fuzzy number can be conbined under the form of fuzzy.numbers, defined as follows:

$$Y = \{(\hat{A}, \underset{\sim}{x}) : \hat{A} \text{ is a fuzzy number, } \underset{\sim}{x} \text{ a random variable}\}$$

For example $\hat{A}$ may be an imprecisely defined storm surge above a stochastically defined high tide; in this case $Y=\hat{A}\ [+]\ \underset{\sim}{x}$ is the random/fuzzy loading on a sea level rise protection structure. The consequence of the loading L(d,y) were d is a decision is itself a hybrid number which can be estimated either in closed form or by Monte Carlo simulation.

As another example, let the present net worth B of a protection measure be a function of annualized construction, operation and maintenance costs $C(\hat{a})$ and averted losses $\underset{\sim}{\hat{K}}(\hat{a})$, where $\hat{a}$ is a discount factor.

$$\hat{B} = B(\hat{C}(\hat{a}), \underset{\sim}{\hat{K}}(\hat{a}))$$

Using fuzzy arithmetic, the membership function of the expected value of $\hat{B}$ can be calculated, even in the case of fuzzy probability of occurrence of adverse events. Fuzzy arithmetic is considerably simpler and more manageable than the algebra of random numbers, and functions of fuzzy numbers are calculated in a straightforward way using the extension principle, as illustrated for example in Duckstein and Bogardi (this volume).

In health risk analysis, the consequences of an exposure are evaluated by means of a dose-response relationship. Here, as in most cases where few imprecise observations of an output variable are available, fuzzy regression may by used (Bardossy et al, 1990, 1991a, Ramezani and Duckstein, 1991). Fuzzy regression, is used to calculate fuzzy number parameters $\hat{a}(1)$, ..., $\hat{a}(J)$ such that

$$\hat{Y} = \hat{a}(1) + \hat{a}(2)\, x(2) + \ldots + \hat{a}(J)\, x(J)$$

The data set is given as

$$(\hat{y}(t),\ x(2,t),\ \ldots x\ (J,t)),\ t=1,\ \ldots,\ T$$

when $\hat{y}(t)$ may be a fuzzy number. In the simplest case, the coefficients $\hat{a}(j)$ are represented as triangular fuzzy numbers (TFN) and are calculated by means of a linear programming problem. The objective functions to be minimized represent some measure of the combined widths of the a(j)'s, j=1, ..., J, for example the hypervolume under the J-dimensional membership function of the a(j)'s. The constraints of the linear programming problem express that at a given membership level the predicted value $\hat{y}(t)$ of y should embed the measured value $\hat{y}(t)$. Fuzzy regression may be applied if

(1) few data points are available thus, in Woldt et al (this volume) only 6 pairs of data points were used

(2) we know that a relationship exists; for example, runoff is related linearly to rainfall in a small watershed

(3) there are no outliers, it would be vain to attempt to deal with an outlier if only 6 obervations are available.

Fuzzy regression provides a measure of the goodness of the relationship between y and x(2), ...,x(J).

Next, if a large set of observations is available, but imprecision exists in the independent variables or the dependent variable or the relationship itself, or a combination thereof, then least squares regression may be used. For example, two ways of measuring earthquake magnitude, namely, body wave and surface wave magnitude, have been related by fuzzy least squares using more than 4000 pairs of observations (Bardossy et al. 1991b). In that paper, the normal equations for fuzzy regression are given and the pros and cons of that technique are discussed.

4.4. Perception of the consequences

Fuzzy numbers may represent value functions, prior probabilities, both together, or neither one. In the case of DO, a concentration x belongs to the set A of desirable concentration with a membership function value that can be used almost interchangeably with a value function ordinate: the perceptions of x may be expressed by either concept.

Similarly, consider a waste disposal site described by a set of economic and ecological indicators assessed by K experts; the elements of this set may be expressed as either value functions or fuzzy numbers defined on a discrete preference scale of, say, 1 to 7. Combining these assessments to come up with one number (per site) is a difficult problem; value functions used within a multiattribute approach, that necessitates many assumptions and trade offs, is just about the only choice ; in the other hand, fuzzy numbers may be combined with routine operations of set maximization and minimization, to yield a composite number with reasonable width representing the pooling of expert opinion. In general, then, fuzzy numbers may be used to quantify the consequences of adverse events for risk analysis.

Once these four steps of risk analysis have been performed, risk management may be undertaken. Note that in each step, fuzzy set analysis may be seen as a complement to statistical analysis. The extension principle is even applicable to Bayes' theorem.

5.- RISK MANAGEMENT

Here we distinguish between design or planning problems, and operation problems.

In ***planning problems***, a set of discrete alternatives (also called options, actions, schemes, decisions, ..) is defined and rated by a set of performance indices or figures of merit which usually may be deterministic stochastic or fuzzy and which usually includes non numerical criteria. The recommended multiobjective procedure (Tecle et al, 1988) is as follows

(1) define the form type objectives-specifications-criteria-preferences (weights, scales...)

(2) define the alternatives, which should be distinct-not just marginally different

(3) use the input uncertainty characterization/quantification to rate each alternative in terms of the selected criteria, including risk-related figures of merit. Use value functions to quantify the non-numerical criteria

(4) select at least one multicriterion decision making technique to rank the alternatives

(5) evaluate the ranking, for example by comparing the results of application of different techniques, as in Duckstein et al (this volume), and perform a sensitivity analysis.

The trade off between criteria, FM, or indicators may be done in a hierarchical manner : at first ecological indicators may be traded off, to yield a composite index of risk and then economic cost indicators, to yield a composite index of cost; as the second level, the two composit indices are then traded off.At each level, fuzzy numbers may be combined, compaired-using for example measure of distance between TFN'S.

In ***operation problems***, it is customary to include only numerical criteria, and use a sequential decision making scheme such as dynamic programming or water resources engineering (Parent, this volume).

Although the use of fuzzy sets in sequential decision-making has been formulated more than two decades ago in Bellman and Zadeh (1970) no realistic application has yet been put forth in hydrology and water resources engineering. The reason may be that a solid background in fuzzy set theory and in water resources engineering is necessary. Yet the calculations are order of magnitude simpler than in classical dynamic programming applied for example to reservoir operation or groundwater management.

It is quite natural to describe extreme events as fuzzy or hybrid numbers (especially droughts) Global change may then modify the fuzzy or hybrid numbers time series. Risk management is to be performed on such time series. Research is currently under way along this avenue of thought.

6.- CONCLUSIONS

A unified approach for risk and reliability analysis in water resources and environmental engineering has been presented.Mathematical modelling and simulation can be used to define "loads" and "resistance" of the physical system under risk conditions.This can be done by considering the hydrologic parameters,inputs and outputs as random variables or as fuzzy numbers.The steps to be followed in a systematic framework for the analysis of risks have been outlined and methods and tools for risk quantification and risk management have been presented.

Acknowledgement: The research leading to this paper has been supported in part by grant from the US NSF NO.BCS 9016462.

REFERENCES

Bellman, R. and L. Zadeh (1970) Decision making in a fuzzy environment, Management Science Vol.17.

Bardossy, A., I. Bogardi and L. Duckstein (1990) Fuzzy regression in hydrology, Water Resources Research , 26(7): 1497-1508, July.

Bardossy, A., I. Bogardi and L. Duckstein (1991a) Fuzzy non-linear regression analysis of dose-response relationships, in press, Eur.J.Oper.Res.

Bardossy, A., R. Magaman, L. Duckstein and I. Bogardi (1991b) Fuzzy least squares regression: theory and application , to appear in M. Fedrizzi and J. Kaczypryk (Eds.), Fuzzy Regression Models, Omnitech Press,Warsow.

Bogardi, I., L. Duckstein and A. Bardossy (1988) Uncertainty in environmental risk analysis, in: Risk Analysis and Management of Man-made and Natural Hazards, Y.Y. Haimes and E. Stakhiv (eds.), ASCE, New York, 156-174.

Dong, W.M. and F.S. Wong (1986) From uncertainty to approximate reasoning, Parts 1, 2, 3, Civil Engineering Systems, Sept., Dec. 1986, March 1987.

Duckstein, L., E.Plate and M.Benedini (1987) Water engineering reliability and risk: a system framework., in: Eng. Reliability and Risk in Water Resources, Duckstein L. & Plate E. (Eds.), E.M. Nijhoff, Dordrecht, The Netherlands, pp.201-228.

Duckstein, L., I. Bogardi and A. Bardossy (1988) A fuzzy reliability model of water supply during droughts, Fall AGU Meeting, San Francisco, December 1988. Available as paper 88-25, Systems & Industrial Engineering Department, University of Arizona, Tucson, Arizona 85721.

Ganoulis, J. and H. Morel-Seytoux (1985) Application of stochastic methods to the study of aquifer systems. Technical Documents in Hydrology ,UNESCO,Paris.

Ganoulis, J. (1986) Sur les échelles spatiales des hétérogenéités en milieu poreux, Hydrogéologie, No.2, 115, BRGM (France).

Ganoulis, J. (1987) Consequences of the failure of a water storage system, in: Eng. Reliability and Risk in Water Resources, Duckstein L.& Plate E. (Eds.), E.M. Nijhoff, Dordrecht, The Netherlands, pp.201-228.

Kaufmann, A. and M.M. Gupta (1985) Introduction to Fuzzy Arithmetic: Theory and Applications, Van Nostrand Rheinhold, New York, 1985.

Kaufmann, A. and M.M. Gupta (1988) Fuzzy Mathematical Models in Engineering and Management Science, North Holland, Amsterdam, 338p.

Klir, G.J. and T.A. Folger (1988) Fuzzy Sets, Uncertainty and Information, Prentice Hall, 355p.

Ramezani, R. and L. Duckstein (1991) Fuzzy regression analysis of the effect of university research on regional technologies, to appear in M. Fedrizzi and J. Kaczypryk (Eds.), Fuzzy Regression Models, Omnitech Press,Warsow.

Shrestha, B.P., K.R. Reddy and L. Duckstein (1990) Fuzzy reliability in hydraulics, Proceedings, The First International Symposium on Uncertainty Modeling & Analysis, University of Maryland, College Park, December.

Tecle A., M. Fogel and L. Duckstein (1988) Multicriterion selection of wastewater management alternatives, ASCE J. Water Res. Plan. Man , 114(4):383-398.

Zadeh, L.A. (1965) Fuzzy sets, Inform. Contr. 8, 338-353.

Zeise, L., R. Wilson and E.A.C. Crouch (1987) Dose-response relationships for carcinogens: a review, Environmental Health Perspectives, Vol.73, pp.259-308.

Zimmerman, H.J. (1985) Fuzzy Set Theory and its Application, Martinus Nijhoff,363 p.

Part 1.
UNCERTAINTY ANALYSIS
OF
HYDROLOGIC INPUTS

RISK ANALYSIS OF HYDROLOGICAL DATA USING GAMMA FAMILY AND DERIVED DISTRIBUTIONS

B. Bobée, F. Ashkar[1], R. Roy[2] and L. Perreault
INRS-Eau
2800 rue Einstein
Québec (Québec)
G1X 4N8 Canada

ABSTRACT

Risk can be estimated by the probability of exceedance of X_T (flow with return period T). An accurate determination of X_T with its confidence intervals is important in this context. We will consider estimation of X_T using several methods in the case of the Pearson type 3 and log Pearson type 3 distributions.

1. INTRODUCTION

If X_T is the event corresponding to a given return period T, in the case of flood for example, the risk of failure Π is such that:

$$\Pi = \Pr \{X \geq X_T\} = \frac{1}{T}$$

It can be shown that:

$$E(T) = \frac{1}{\Pi}$$

Fitting a statistical distribution to a sample $(x_1, \ldots, X_N)$ leads to the determination of the relationship:

---

1. University of Moncton, New-Brunswick
2. Hydro-Québec - Montreal

NATO ASI Series, Vol. G 29
Water Resources Engineering Risk Assessment
Edited by J. Ganoulis

$\tilde{X}_T(\Pi)$ or $\tilde{X}_T(T)$

where $\tilde{X}_T$ is the estimation of the true and unknown value X_T of return period T.

It can be shown that $\tilde{X}_T$ is approximately normally distributed with mean

X_T (the true value), and variance

Var $\tilde{X}_T$. Then,

$$U = \frac{\tilde{X}_T - X_T}{\sqrt{\text{var}\ \tilde{X}_T}}$$

follows a standardized normal distribution N(0,1) and the confidence interval at the level (1-α) % of the unknown value X_T is given by:

$$\tilde{X}_T - U_{\alpha/2}\sqrt{\text{var}\ \tilde{X}_T} \leq X_T \leq \tilde{X}_T + U_{\alpha/2}\sqrt{\text{var}\ \tilde{X}_T}$$

where $U_{\alpha/2}$ is the N(0,1) variate corresponding to the probability of exceedance α/2.

In this presentation, we will be concerned by the estimation of $\tilde{X}_T$ and var $\tilde{X}_T$, and then by the determination of the confidence interval of X_T using different methods of estimation of the Gamma family and derived distributions (Pearson Type 3, Log Pearson Type 3, Generalized Gamma). This important group of distributions shall be called the P-LP-GG group (Pearson Type 3 - log Pearson Type 3 - Generalized Gamma) in the present text. It will be seen in Section 2 how all members of this group are related to each other by simple mathematical relations, and how one can pass from simple one-parameter distributions (SP and G1) to two-parameter ones (G2, GG2), then to three-parameter ones (P, GG3, LP) and finally to the four-parameter GG4 distribution which is the most general

We shall see also in Section 2 how this system is derived, and how one group of distributions that has been very frequently employed in hydrological applications can be derived from one particular member of this system, namely the Pearson Type 3 distribution (3 parameters) or from one of its special cases, the one- or two- parameter gamma distributions (G1, G2).

A comprehensive presentation of all these distributions (mathematical and statistical properties, fitting techniques..) can be found in a book by Bobée and Ashkar (1991). This book is available with a PC software package "HFA" (HFA for "Hydrologic Frequency Analysis"), and presents a complete case study using a real-world hydrologic data series.

In the limited context of this paper, we shall present briefly the methods of fitting for the Pearson (P) and log - Pearson Type 3 (LP) distributions (Sections 3 and 4) since these methods are needed for the understanding and the demonstration of the PC package HFA and its applications.

2. THE PEARSON TYPE 3 - LOG PEARSON TYPE 3 - GENERALIZED GAMMA (P-LP-GG) GROUP OF DISTRIBUTIONS

2.1 Origin of the Pearson Type 3 (P) distribution and its interest in hydrology

In hydrology one often finds oneself faced with the problem of estimating the distribution function F_Z of a continuous univariate random variable Z from an observed sample $(z_1,\ldots,z_N)$. This kind of problem is not restricted to hydrology but is common to many other fields. As hydrological processes are bounded by physical limitations, the statistical distributions which should be used to represent them must conform to these limitations. The analysis of hydrologic data often shows that:

1. the empirical distribution $y = f(z)$ describing the data takes very different forms with possibly marked asymetry of skewness.

2. the empirical distribution $y = f(z)$ may possess a lower, upper, or both lower and upper bounds. For example, flood

flows can take only positive values and cannot exceed a physical upper bound (z_{max}) while keeping their physical meaning in the hydrogeographical context of the watershed. Due to problems in estimating this upper bound, however, the distribution of flood flows is often assumed to extend to $+\infty$. The estimation of the lower bound normally poses a less serious problem and can be set at $z = 0$ or at $z = m > 0$ depending on the hydrogeographical system of the watershed;

3. the empirical distribution $y = f(z)$ normally starts at $y = 0$, rises to a maximum and then falls back to $y = 0$ or asymptotically approaches $y = 0$ at an often different rate.

One way by which the above characterics can be met is by taking $y = f(z)$ to satisfy the Pearson differential equation:

$$(1/y)(dy/dz) = (a - z)/(b_0 + b_1 z + b_2 z^2) \tag{1}$$

where a, b_0, b_1, b_2 are parameters.
Obtainable by Eq. (1) are curves that are bell shaped, U shaped or J shaped. These curves can also be symmetrical, slightly skewed or very skewed. The system represented by Eq. (1) was originated by Pearson (1895).

The present text is dealing with the treatment of an important group of distributions derived from one particular member of this system which is the Pearson Type III distribution. This distribution along with the group of distributions derived from it, will be briefly introduced in 2.2 and presented in subsequent sections.

2.2 The Pearson Type 3 - Log-Pearson Type 3 - Generalized Gamma group of distributions (P-LP-GG Group)

For simplicity, the Pearson Type 3 distribution will be denoted in the remainder of the present text by the letter "P". The p.d.f. of this 3-parameter distribution is derived from the Pearson differential equation (1) by posing $b_2 = 0$. Then, it can be shown (Bobée and Ashkar, 1991) that the integration of equation (1) leads to:

$$y = f(z) = \frac{|\alpha|}{\Gamma(\lambda)} e^{-\alpha(z-m)} \{\alpha(z - m)\}^{\lambda-1} \tag{2}$$

where:

$\Gamma(.)$ is the gamma function defined as

$$\Gamma(\lambda) = \int_0^\infty e^{-x} x^{\lambda-1}\, dx \tag{3}$$

m is a location parameter, α is a scale parameter and λ is a shape parameter.

The domain of variation of z is such as:

$\alpha(z - m) \geq 0,$

. if $\alpha > 0$, $z \geq m$ and equation (2) corresponds to the P distribution with positive skewness (form A)

- if $\alpha < 0$, $z \leq m$ and equation (2) corresponds to the P distribution with negative skewness (form B).

In the remainder of the text, the distribution represented by Eq. (2) will be denoted by $P(z;\alpha,\lambda,m)$ or simply $P(\alpha,\lambda,m)$.

Two-parameter gamma (G2) distribution (Y) This is a special case of the Pearson Type 3 distribution $P(z;\alpha,\lambda,m)$ obtained by setting the location parameter m equal to zero or by a transformation of the form $Y = Z - m$ (see Figure 1).

One-parameter Gamma (G1) distribution (X) The G1 distribution is a special case of both the P and G2 distributions. It is obtained when the location parameter m of the $P(\alpha,\lambda,m)$ distribution is set equal to zero and the scale parameter α is set equal to +1. It can be obtained from the $G2(y;\alpha,\lambda)$ distribution by a transformation of the form $X = \alpha Y$ or from the $P(z;\alpha,\lambda,m)$ distribution by a transformation of the form $X = \alpha(Z-m)$ (Figure 1).

Standardized Pearson Type 3 (SP) distribution (K) The SP distribution is a P distribution with mean equal to zero and va-

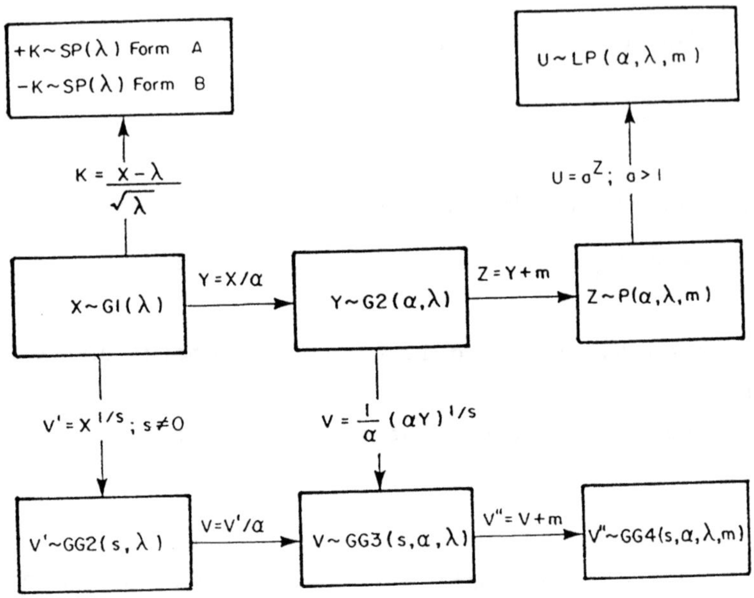

Figure 1. Diagram of functionnal relationships between members of the P-LP-GG group of distributions.

riance equal to unity ($\mu = 0$; $\sigma^2 = 1$). It depends only on one parameter, λ, which is a shape parameter. It can be of Form A (positive skewness) or Form B (negative skewness) depending on the skewness of the P distribution from which it is derived. The relation between the SP distribution (K) and the G1 distriburion (X) is $K = (X-\lambda)/\sqrt{\lambda}$ (Fig. 1).

Log-Pearson Type 3 (LP) distribution (U) The P and LP distributions are related to each other by a logarithmic transformation. A random variable U follows a log-Pearson Type 3 distribution $LP(u;\alpha,\lambda,m)$ if $Z = \log_a U$ follows a Pearson Type 3 distribution $P(z;\alpha,\lambda,m)$, where $a > 1$ is a general logarithmic base. The logarithmic bases most commonly used in practice are the decimal logarithm ($a = 10$) and the natural or base-e logarithm ($a = e \simeq 2.718281$).

Two- three- and four - parameter Generalized Gamma distributions (GG2, GG3 and GG4) (V) The $GG2(v';s,\lambda)$ and $GG3(v;s,\alpha,\lambda)$ distributions are derived from the $G1(x;\lambda)$ distribution by transformations of the form $V' = X^{1/s}$ and $V = X^{1/s}/\alpha$, respectively, where $s \neq 0$ is a parameter. The $GG3(v;s,\alpha,\lambda)$ distribution can also be obtained directly from the $G2(y;\alpha,\lambda)$ distribution (Form A) by a transformation of the form $V = (\alpha y)^{1/s}/\alpha$. The $GG4(v";s,\alpha,\lambda,m)$ distribution is obtained from the $GG3(v;s,\alpha,\lambda)$ distribution by adding a location parameter m using the transformation $V" = V + m$. Figure 1 gives the functional relationships between the variates of the different distributions in the P-LP-GG group.

P.d.f.'s of members of the P-LP-GG group of distributions

Since all members of the P-LP-GG group of distributions are functionally related to each other, their probability density functions (p.d.f.'s) are also related and can be derived.

3. PEARSON TYPE 3 DISTRIBUTION $P(\alpha,\lambda,m)$

3.1 Estimation by maximum likelihood (ML)

Given a sample $(z_1,\ldots, z_i,\ldots, z_N)$ the logarithm of the likelihood function, L, for the P distribution is given by:

$$\ln L = \lambda N \ln|\alpha| - N \ln\Gamma(\lambda) - \alpha \sum_{i=1}^{N} (z_i - m) + (\lambda-1) \sum_{i=1}^{N} \ln(z_i - m)$$

The maximum likelihood estimates $(\alpha_*, \lambda_*, m_*)$ are obtained by solving the three equations:

$$\frac{\partial \ln L}{\partial \alpha} = \frac{\lambda N}{\alpha} - \sum_{i=1}^{N} (z_i - m) = 0 \tag{4}$$

$$\frac{\partial \ln L}{\partial \lambda} \quad - N\,\psi(\lambda) + \sum_{i=1}^{N} \ln \alpha(z_i - m) = 0 \tag{5}$$

$$\frac{\partial \ln L}{\partial m} = \alpha N - (\lambda - 1) \sum_{i=1}^{N} \frac{1}{(z_i - m)} = 0 \tag{6}$$

where $\psi(\lambda)$ is the digamma function.

The solution of Eqs. (4) through (6) can be quite delicate because one has to solve a system of three non linear equations in three unknowns.

For a sample $(z_1, \ldots, z_i, \ldots, z_N)$ the method consists in computing from equations (4) and (6):

$$\lambda(m) = \frac{A}{A - B} \quad \text{and} \quad \alpha(m) = \frac{1}{N}\frac{A}{A - B}$$

with

$$A = \sum_{i=1}^{N} \frac{1}{(z_i - m)} \quad \text{and} \quad B = \frac{N^2}{\sum_{i=1}^{N} (z_i - m)}$$

For a fixed value $m = m_0$, $\lambda_0 = \lambda(m_0)$ and $\alpha_0 = \alpha(m_0)$ are calculated and then α_0, λ_0 and m_0 are substituted in Eq. (5) to obtain:

$$R_0 = - N \psi(\lambda_0) + \sum_{i=1}^{N} \ln \alpha_0 (z_i - m_0) \tag{7}$$

The value of m_0 is then varied until $R_0 = 0$, because then Eq. (5) will be satisfied. In practice, it is convenient to proceed by iteration, according to a method given by Bobée and DesGroseilliers (1985).

The calculation of $VarZ_T$ (variance of the event Z_T with return period T) is given, for $\lambda > 2$ in BOBEE and ASHKAR (1991).

3.2 Estimation by the method of moments

The method of moments consists in equating three moments of the population $P(\alpha, \lambda, m)$, which are functions of the parameters, to the corresponding moments of the sample. In practice, the mean $\bar{z}$, unbiased variance s^2 and coefficient of skewness C_s, of the sample, are used. We therefore obtain the 3 equations:

$$\bar{z} = m + \frac{\lambda}{\alpha} \tag{8}$$

$$s^2 = \frac{\lambda}{\alpha^2} \tag{9}$$

$$C_s = \frac{2}{\sqrt{\lambda}} \tag{10}$$

from which we deduce:

$$\tilde{\lambda} = \frac{4}{C_s^2} \tag{11}$$

$$\tilde{\alpha} = \frac{2}{s.C_s} \tag{12}$$

$$\tilde{m} = \bar{z} - \frac{2s}{C_s} \tag{13}$$

The estimates $\tilde{\alpha}$, $\tilde{\lambda}$ and $\tilde{m}$ are therefore expressed in a simple manner as function of $\bar{z}$, s^2 and C_s.

It can be shown (Bobée, 1973) that the variance of the estimated event $\tilde{z}_T$, with return period T, is given by the following relation:

$$\text{Var}(\tilde{z}_T) = \frac{\mu_2}{N} \{1 + \frac{K_T^2}{2} . (1 + \frac{3}{4} C_s^2) + K_T C_s + 6(1 + \frac{C_s^2}{4})(\frac{\partial K_T}{\partial C_s}) . (\frac{\partial K_T}{\partial C_s})(1 + 5\frac{(C_s)^2}{4} + \frac{K_T}{2} C_s \quad \} \tag{14}$$

where K_T is the standardized Pearson Type 3 variate with return period T and, C_s is the coefficient of skewness used in the estimation.

The method of moments is generally less efficient than the method of maximum likelihood as shown by Fisher (1921) for the P distribution. For the three parameters α, λ and m, this efficiency increases with λ and as λ increases indefinitely, C_s decreases and the method of moments becomes as efficient as the method of maximum likelihood. In order words, when the Pearson Type 3 distribution tends towards the normal distribution (C_s= 0), the two methods become equivalent.

4. LOG-PEARSON TYPE 3 DISTRIBUTION LP(α,λ,m)

4.1 Introduction

The first application of the LP distribution was done by Beard (1962) who suggested taking the logarithms of maximum annual flood flows and fitting a Pearson Type 3 distribution to these logarithmic values. A comparison of the LP distribution with several other distributions was then undertaken by the United States Water Resources Council which led to the recommendation of the systematic use of this distribution for representing maximum annual flood flows in the U.S. (WRC, 1976). The use of the LP distribution was later recommended in Australia by the Institute of Engineers (I.E.A., 1977). As a result of these recommendations, several theoretical studies have been undertaken on the LP distribution, and its application to the study of floods has been both extensive and widespread.

4.2 Mathematical properties

Here, we consider the most general form of LP distribution which is derived from the P distribution by a logarithmic transformation of arbitrary base a.

If $Z = \log_a U$ follows a P(α,λ,m) distribution then $U = a^Z$ follows a LP(α,λ,m) distribution. The p.d.f. of LP(α,λ,m) can therefore be obtained from that of P(α,λ,m). This p.d.f. (Bobée, 1975a) is given by:

$$f_{LP}(u;\alpha,\lambda,m) = \frac{|\alpha|}{\Gamma(\lambda)} e^{-\alpha(\log_a u-m)} \{\alpha(\mathrm{Log}_a u-m)\}^{\lambda-1} \cdot \frac{k}{u} \qquad (15)$$

where

$k = \log_a e = \frac{1}{\ln a}$

We shall suppose that $a > 1$ or $k > 0$.

As in the case of the P distribution, the parameters λ and m of the LP distribution are such that $\lambda > 0$ and $-\infty < m < +\infty$.

The domain of variation of the LP variable, u, is such that:

If $\alpha > 0$: $a^m \quad e^{m/k} \leq u < +\infty$

If $\alpha < 0$: $0 \leq u \leq a^m = e^{m/k} = u_{max}$

If natural logarithms are used in Eq. (15), the p.d.f. of the LP distribution becomes, (Bobée, 1975a):

$$f_{LP}(u;\alpha,\lambda,m) = \frac{|\alpha|}{\Gamma(\lambda)} \frac{e^{\alpha m}}{u^{1+\alpha}} \{\alpha(\ln u - m)\}^{\lambda-1} \tag{16}$$

which is easily obtained by replacing $\log_a u$ by $\ln u$ and putting $k = 1$ ($= \log_e e$) in Eq. (15). The mathematical properties of the LP(α,λ,m) distribution were studied in detail by Bobée (1975b). It can be shown (Bobée, 1975a) that the non central moment of order r of the LP(α,λ,m) distribution, exists only if

$$(1 - \frac{r}{\beta}) > 0 \tag{17}$$

This means that:

. if $\beta = \alpha k > 0$ (i.e. if $\alpha > 0$) the moments μ'_r(LP) exist only if $r < \beta$.

In order words, if $\beta > 0$ the LP moments are defined up to order r_0, where r_0 is the integer part of $\beta = \alpha k$. For example, if $\alpha = 2,5$ and a logarithmic transformation of base e ($k = 1$) is used, the non central moments of the LP distribution are defined up to order 2 inclusively, but moments of higher order do not exist.

. If $\beta = \alpha k < 0$ (i.e. if $\alpha < 0$) then by Eq. (17) r should be larger than β, which is always true, because $r > 0$ and $\beta < 0$.

In order words if $\beta < 0$ all the moments of the LP distribution exist.

When they are well defined, non central moments of the LP distribution are given by:

$$\mu'_r(LP) = \frac{e^{mr/k}}{(1 - \frac{r}{\beta})^{\lambda}} \qquad \text{with} \quad \beta = \alpha k \tag{18}$$

From these moments, the coefficients of variation, of skewness and of kurtosis are easily derived.

4.3 Estimation by maximum likelihood (ML)

From the p.d.f. of the LP distribution given in Eq. (16), the log-likelihood function can be determined:

$$\ln L = N\ln(|\alpha|.k) - N\ln\Gamma(\lambda) + \Sigma\alpha m - (1-\alpha k)\sum_{i=1}^{N} \ln u_i + (\lambda-1)\sum_{i=1}^{N} \ln\{\alpha(k\ln u_i - m)\} \tag{19}$$

The ML estimates α_*, λ_* and m_* are obtained by solving the system of equations:

$$\frac{\partial \ln L}{\partial \alpha} = \frac{N\lambda}{\alpha} - \sum_{i=1}^{N} (k\ln u_i - m) = 0 \tag{20}$$

$$\frac{\partial \ln L}{\partial \lambda} = N\,\psi(\lambda) + \sum_{i=1}^{N} \ln \alpha(k\ln u_i - m) = 0 \tag{21}$$

$$\frac{\partial \ln L}{\partial m} = N\alpha - (\lambda - 1)\,\Sigma \frac{1}{(k\ln u_i - m)} = 0 \tag{22}$$

where $\psi(\lambda)$ is the digamma function.

It can be shown by putting $z_i = k\ln u_i = \log_a u_i$ that Eqs. (20) through (22) are the same as Eqs. (4) through (6) which were obtained for the P distribution by maximum likelihood.

Therefore, a LP distribution fitted by ML to a sample $(u_1, \ldots, u_i, \ldots u_N)$ produces exactly the same estimates $\alpha*$, $\lambda*$, $m*$ as a P distribution fitted by ML to the transformed sample $z_i = \log_a u_i$.

Since $Z_T = \log_a U_T$ follows a P distribution, $Var(U_T)$ can be derived from $Var(Z_T)$ where $Var(Z_T)$ is the variance of the event Z_T obtained by fitting a P distribution to the sample in log space using maximum likelihood (Sec. 3.1). The basic relationships needed for calculating $Var(U_T)$ are:

$$U_T = a^{Z_T} = e^{Z_T/k}$$

$$Var(U_T) = \left(\frac{\partial U_T}{\partial Z_T}\right)^2 Var(Z_T)$$

$$Var(U_T) = \left(\frac{U_T}{k}\right)^2 Var(Z_T) \tag{23}$$

where $k = 1/\ln a$.

4.4 Indirect method of moments (WRC)

This method was proposed by the United States Water Resources Council (WRC, 1976) in which the LP distribution was recommended as a base technique for flood frequency estimation. The method has been described by Benson (1968). It consists in fitting a Pearson Type 3 (P) distribution to logarithmically transformed data. Given a sample $(u_1, \ldots, u_i, \ldots, u_N)$, the P distribution is fitted to the transformed values z_i such that:

$$z_i = \log_a u_i = k \ln u_i$$

where the fitting is done by equating the mean $\bar{z}$, the unbiased variance s^2 and the sample coefficient of skewness $(C_s)_1$ corrected for bias (WRC, 1976) to the corresponding moments of the population. This leads to the system of Eqs. (8) through (10) given in Section 3.2, which have to be solved for the parameters

α, λ and m. Hoshi and Leeyavanija (1986) consider a variation of this approach which uses a coefficient of skewness corrected for bias as proposed by Bobée and Robitaille (1975). The variance of $\tilde{Z}_T$ estimated by the indirect method of moments can be obtained using Eq. (14) of Section 3.2. As in the case of the method of maximum likelihood, $Var(\tilde{U}_T)$ can be derived from $Var(\tilde{Z}_T)$ using Eq. (23).

4.5 Direct method of moments

Bobée (1975a) proposed this method as a means for preserving the first three non-central moments of the LP variable, U. This method is able to reproduce the moments of the sample in real space rather than the moments of the logarithmically transformed data. The method consists in equating the first three non central moments of the LP distribution (μ_r') as calculated from Eq. (18) to their estimates m_r' obtained from the sample. Using natural logarithms, the resulting three equations which have to be solved for α, λ and m are:

$$\ln m_1' = \frac{m}{k} - \lambda \ln(1 - \frac{1}{\beta}) \tag{24}$$

$$\ln m_2' = \frac{2m}{k} - \lambda \ln(1 - \frac{2}{\beta}) \tag{25}$$

$$\ln m_3' = \frac{3m}{k} - \lambda \ln(1 - \frac{3}{\beta}) \tag{26}$$

where

$$m_r' = \frac{1}{N} \sum_{i=1}^{N} u_i^r$$

This method can be applied only if $\beta < 0$ or $\beta \geq 3$ because otherwise, the non central moments of the LP distribution do not exist (see eq. 17).

The previous system of equations can be written as:

$$\frac{\ln\ (1-1/\beta)^3\ /\ (1-3/\beta)}{\ln\{(1-1/\beta)^2\ /\ (1-2/\beta)\}} = \frac{\ln m_3' - 3\ \ln m_1'}{\ln m_2' - 2\ \ln m_1'} \tag{27}$$

$$\lambda = \frac{\ln m_2' - 2\ \ln m_1'}{\ln\{(1-1/\beta)^2\ /\ (1-2/\beta)\}} \tag{28}$$

$$m = k\ \ln m_1' + \lambda\ \ln(1-1/\beta) \tag{29}$$

Note that Eq. (27) is implicit in β; its right-hand member B is given by:

$$B = \frac{\ln m_3' - 3\ \ln m_1'}{\ln m_2' - 2\ \ln m_1'} \tag{30}$$

Where B can be calculated from the sample and used in Eq. 27) to solve for β.

When $|\beta|$ is large, Bobée (1975b) gave the following approximation formula for β:

$$\beta \simeq \frac{1 + \sqrt{(5B - 14)}}{B - 3} \tag{31}$$

Knowing β, using Eq. (27) or (31) the values of λ and m are determined, respectively, by Eqs. (28) and (29).

Calculation of the variance of U_T Bobée and Boucher (1981) calculated the variance of U_T estimated by the direct method of moments for the LP distribution and one of its special cases the LG (log-gamma) distribution.

4.6 Method of mixed moments

The direct and indirect methods of moments require the use of the non central moment of order 3, or coefficient of skewness, in their application. Numerous studies have shown that estimates of this third moment are generally biased, highly variable and sensitive to large values in the sample. Kirby (1974) also showed that in practice the coefficient of skewness of the sample is bounded by a function that only depends on sample size and is a biased estimate of the coefficient of skewness of the population. Wallis et al. (1974) studied this bias in C_s for a number of different populations.

One way of trying to correct the large sampling variance of the coefficient of skewness in the case of the indirect method of moments was to use the regional values of coefficient of skewness in the estimation (WRC, 1976). However, Landwehr et al. (1978) have shown that the use of regional skew maps, especially in log space, can lead to unreliable results.

Rao (1980) thus suggested the use of methods that do not depend on the coefficient of skewness in order to try to improve the estimates. This led to the proposal of the method of mixed moments which the author denoted by MM1. This method uses:

- the mean and variance of the observed sample assumed to be drawn from a LP distribution;
- The mean of the logarithmically transformed sample which follows a P distribution.

These moments of the sample in real and log space are used as estimates of the corresponding moments of the population. The method MM1 combines, therefore, the first two equations of the direct method of moments (Sec. 4.5) and the first equations of the indirect method of moments (Sec. 4.4). Rao (1983) later proposed a simplified approach for determining the parameters of the LP distribution using the method MM1.

It is possible to show, that the mean of the logarithmically transformed data is the log of the geometric mean of the observed data. Moreover, the geometric mean of the observed data can be considered, in real space, as a moment of order "quasi - zero" (Bobée and Ashkar, 1988a) because it corresponds to the moment of order r as r tends to 0. In subsequent discussions we shall denote the sample geometric mean by $m'_{\underline{0}}$ (with a dash under the zero) meaning that it is the moment of order quasi-zero.

The method MM1 calls, therefore, for the use of moments of order $\underline{0}$, 1 and 2 in real space and, strictly speaking, it cannot be called a method of "mixed" moment (Ashkar and Bobée, 1986).

4.7 Sundry averages method (SAM)

In general, it appears intuitively reasonable in a method of parameter estimation to use moments of lowest order possible. The reason for this is that the variance of these moments tends to be reduced and, as a result, $Var(U_T)$ could also be reduced.

This has led Bobée and Ashkar (1988b) to propose a method of moments based on:

- the arithmetic mean (moment of order 1)
- the harmonic mean (moment of order -1)
- the geometric mean (moment of order 0).

This method was called Sundry Averages Method (SAM). Because this method uses moments of order r = 1 and r = -1, condition (17) for the existence of these moments will be respected if

$$|\beta| > 1$$

When fitting an LP distribution with parameters α, λ and m to a sample $(u_1, \ldots\ldots, u_N)$, the three equations of the method SAM are:

Arithmetic mean:

$$\bar{u} = \frac{1}{N} \sum_{i=1}^{N} u_i = \mu_1'(LP) = \frac{e^{m/k}}{(1 - \frac{1}{\beta})^{\lambda}} \tag{32}$$

Harmonic mean:

$$\frac{1}{H} = \frac{1}{N} \sum_{i=1}^{N} 1/u_i = \mu_{-1}' \ (LP) = \frac{e^{-m/k}}{(1 + \frac{1}{\beta})^{\lambda}} \tag{33}$$

Geometric mean:

$$\log_a G = \frac{1}{N} \sum_{i=1}^{N} \log_a u_i = \frac{1}{N} \sum_{i=1}^{N} z_i = m + \frac{\lambda}{\alpha} \tag{34}$$

Equations (32) and (33) are obtained from relationship (18) by putting r = 1 and r = -1, respectively. Equation (34) is obtained, considering that $z_i = \log_a u_i$ follows a P distribution with parameters α, λ and m and that the mean of the P distribution $\mu_1'(P) = (m + \lambda/\alpha)$ is estimated by

$$\frac{1}{N} \sum_{i=1}^{N} z_i = \frac{1}{N} \sum_{i=1}^{N} \log_a u_i$$

After some computations, it is possible to show that the system of Equations (32) through (34) is equivalent to Equations (35) through (37):

$$f(\alpha) = \frac{\log_a(1 - \frac{1}{\alpha k}) + \frac{1}{\alpha}}{\log_a(1 - \frac{1}{\alpha^2 k^2})} - \frac{\log_a \bar{u} - \log_a G}{\log_a \bar{u} - \log_a H} = 0 \qquad (35)$$

$$\lambda = \frac{\log_a H - \log_a \bar{u}}{\log_a(1 - 1/\alpha^2 k^2)} \qquad (36)$$

$$m = \log_a G - \frac{\lambda}{\alpha} \qquad (37)$$

Equation (35) can be solved for α by a Newton - Raphson method. Once α has been estimated, Equations (36) and (37) can be used to calculate λ and m.

4.8 Generalized method of moments (GMM)

The "generalized" method of moments (GMM) as described by Bobée and Ashkar (1988a) for the LP distribution, uses three non central moments μ'_{r1}, μ'_{r2} and μ'_{r3} of arbitrary order r_1, r_2, r_3 and equates them to the corresponding moments m'_{r1}, m'_{r2} and m'_{r3} of the sample. Three equations are thus obtained which have to be solved for the parameters α, λ and m:

$$\mu_r'(LP) = m_r'(u) \qquad r = r_1, r_2, r_3 \qquad (38)$$

By Eq. (18) these can be written as:

$$\frac{e^{mr/k}}{(1 - \frac{r}{\beta})^{\lambda}} = \frac{1}{N} \sum_{i=1}^{N} u_i^r \qquad r = r_1, r_2, r_3 \qquad (39)$$

Note: when r = $\underline{0}$, $\mu_0'(LP)$ and $m_0'(u)$ are the geometric means of the population and of the sample, respectively.

This method which uses the system of Equations (39) is denoted by GMM (r_1, r_2, r_3), therefore:

- the direct method of moments presented in Section 4.5 is GMM (1, 2, 3);
- the method MM1 presented in Section 4.6 is GMM ($\underline{0}$, 1, 2);
- the method SAM presented in Section 4.7 (-1, $\underline{0}$, 1).

These three methods are therefore special cases of the generalized method of moments, and the calculation of Var(U_T) for each of these methods can be deduced from that of the GMM.

The detailed computation of Var(U_T) for the generalized method of moments is presented in Bobée and Ashkar (1991).

CONCLUSION

Flood flows data can generally be represented adequately by Pearson type 3 or Log-Pearson type 3 distributions.

In this paper we presented different methods of estimation of the parameters of these two distributions.

Then it is possible to obtain $\tilde{X}_T$ which is an estimation of the flood X_T of return period T, and to determine the var$\tilde{X}_T$ and the confidence intervals of X_T which is the design value associated with the risk $\Pi = \Pr\{X \geq X_T\}$.

REFERENCES

Ashkar F. and B. Bobée (1986) Comment on: Variance of the T - Year event in the Log-Pearson Type 3 distribution. Journal of Hydrology, 84(1): 181-187.

Beard L.R. (1962) Statistical Methods in Hydrology. Civil Works Investigation Project CW-151, US Army Corps of Engineers, Sacramento, California. 62 p. 2 appendices.

Benson M.A. (1968) Uniform flood-frequency estimating methods for federal agencies. Wat. Res. Res. 4(5): 891-908.

Bobée B. (1973) Sample error of T year events computed by fitting a Pearson Type 3 distribution. Wat. Res. Res. 9(5): 1264-1270.

Bobée B. (1975a) The Log-Pearson Type 3 distribution and its application in hydrology. Wat. Res. Res. 11(5): 681-689.

Bobée B. (1975b) Etude des propriétés mathématiques et statistiques des lois Pearson type 3 et log Pearson type 3. INRS-Eau, rapport scientifique no. 55, 175p.

Bobée B. and F. Ashkar (1988a) The generalized method of moments applied to the log-Pearson type 3 distribution. Journal of Hydraulic Engineering, ASCE, 114(8): 899-909.

Bobée B. and F. Ashkar (1988b) Sundry Averages Method for estimating parameters of the Log-Pearson type 3 distribution. INRS-Eau, rapport scientifique No. 251.

Bobée B. and F. Ashkar (1991) The gamma family and derived distributions applied in hydrology, 200 p. Water Resources Publications, Fort Collins, Colorado.

Bobée B. and P. Boucher (1981) Calcul de la variance d' un événement de période de retour T: cas des lois Log-Pearson type 3 et Log Gamma ajustées par la méthode des moments sur la série des valeurs observées, INRS-Eau, rapport scientifique no. 135, 17 p.

Bobée B. and L. Desgroseilliers (1985) Ajustement des distributions Pearson Type 3, Gamma, Gamma Généralisée et Log Pearson. INRS-Eau, rapport scientifique no. 105, 63 p.

Bobée B. and R. Robitaille (1975) Correction of bias in the estimation of the coefficient of skewness. Wat. Res. Res. 11(6): 851-854.

Fisher R.A. (1921) On the mathematical foundations of theoretical statistics. Roy. Soc. of London Phil. Trans. Ser. A, 222: 309-368.

Hazen A. (1924) Discussion on "Theoretical Frequency Curves" by Foster. Transactions ASCE 87:174-179.

Hoshi K. and U. Leeyavanija (1986) A new approach to parameter estimations of gamma type distributions. Journal of Hydrosciences and hydraulic engineering, 79-95.

I.E.A. (1977) Australian rainfall and runoff flood analysis and design. Institute of engineers, Australia Camberra, A.C.T., 149 p.

Kirby W. (1974) Algebraic boundedness of sample statistics. Wat. Res. Res. 10(2):220-222.

Landwehr J.M., Matalas N.C. and J.R. Wallis (1978) Some comparisons of flood statisties in real and log space. Wat. Res. Res. 14(5):902-920.

Pearson K. (1895) Contributions of the mathematical theory of evolution. Philosophical Transactions of the Royal Society of London Series A 186, 343-414.

Rao D.V. (1980) Log-Pearson type 3 distribution: method of mixed moments, ASCE, Journal of Hydraulics division, 106(6):999-1019.

Rao D.V. (1983) Estimating Log-Pearson parameters by mixed moments. ASCE, Journal of Hydraulics division, 109(8):1119-1132.

Wallis J.R., Matalas N.C. and J.R. Slack (1974) Just a moment. Wat. Res. Res., 10(2):211-219.

WRC (1976) Guidelines for determining flood flow frequency. US Water Resources Council hydrology Commettee, Washington, D.C.

RELIABILITY IN REGIONAL DROUGHT STUDIES

Francisco Nunes Correia
Maria Alzira Santos
Rui Raposo Rodrigues
Laboratorio Nacional de Engenharia Civil
Av. Brasil, 101
1799 Lisboa Codex, Portugal

ABSTRACT Droughts are essentially regional events and must be analyzed in explicit areal terms. Droughts are also long lasting events and this fact must be taken into consideration. Drought consequences can be devastating but never occur in short periods of time, as with other major natural disasters, namely floods, earthquakes and cyclones. Therefore, an engineering approach of drought problems must carefully model the areal and persistent nature of the phenomenon.
Contrary to other natural catastrophes, droughts are difficult to define, to identify and to characterize. They depend not only on hydrometeorological conditions and water availability, but also on very specific needs and demands of water in a given region. Their impacts must be equated in terms of natural environments and also in the context of engineering risk and reliability of man made water environments.
This paper presents a methodology that is very general and flexible, consisting by two different but complementary approaches. The first one takes into account the hydrometeorological features of droughts and makes their characterization through a number of random variables. The second approach gives a means of applying the concepts of risk, reliability, resiliency and vulnerability to the study of complex water resources systems.
The proposed methodology is tentatively and successfully applied to the river Sado basin located in southern Portugal. Results are presented and suggestions for further developments and more elaborated applications are made.

1.INTRODUCTION

Most major natural disasters such as floods, earthquakes and cyclones start in a comparatively sudden manner, have a relatively short duration and are restricted to local influence. Drought in contrast starts slowly, has long duration, often quite long, being of the creeping and pervasive nature and covering vast areas. While floods, cyclones and earthquakes are disasters associated with high extreme events, drought is the result of the low extreme as the "non-event", which is, in general terms, the non-occurrence of water. Drought is different in many ways from the other natural disasters. Basically, it evolves slowly and seldom causes dramatic losses of human life, except through famine in some regions of the world.
However, drought may have very significant impacts in terms of the environment and also on the adequate performance of water supply systems. With respect to water supply systems it should be noted that droughts are regional in nature.This implies that very often not only one source of water is affected but all sources of a complex water system may be affected, seriously decreasing the resilience of multi-source water supply systems.
In terms of environmental impacts, drought conditions may seriously affect the water-soil-vegetation relationship.Droughts damage agriculture production and natural soil cover, leading to a degradation of soil texture, triggering soil erosion and, ultimately, causing desertification. Drought may also have indirect environmental consequences by affecting the dilution capability of rivers with consequences in water quality, changing groundwater balances, causing saline intrusion, and affecting the processes of sediment transport and

NATO ASI Series, Vol. G 29
Water Resources Engineering Risk Assessment
Edited by J. Ganoulis

water-soil-vegetation relationship.Droughts damage agriculture production and natural soil cover, leading to a degradation of soil texture, triggering soil erosion and, ultimately, causing desertification. Drought may also have indirect environmental consequences by affecting the dilution capability of rivers with consequences in water quality, changing groundwater balances, causing saline intrusion, and affecting the processes of sediment transport and deposition in rivers, lakes and reservoirs.

The complexity of drought phenomena with their different causes and effects may be the reason why in the literature there are so many different definitions and approaches to their characterization. From the meteorological point of view drought is present when precipitation falls below a critical value, generally the mean value, during a certain period. From the hydrologic point of view drought occurs when there is a shortage of water including not only precipitation but also surface and subsurface water runoff and storage. From an environmental point of view drought is related to changes in the natural processes induced by long lasting water deficits. From a socioeconomic point of view, the definition of drought should consider not only water supply but also water demand in an explicit way. This means that a drought depends on the type of water use as well as on the density and distribution of users. More, a water deficit to be considered a drought has to last for a sufficiently long time and to be spread over an area of significant extension.

In general terms, one may say that a drought occurs when a significant water deficit that spreads both in time and space takes place. The definition of water deficit is quite important and it is different from case to case. A water deficit in pure meteorological or hydrological terms may have small importance in socioeconomic terms.

To define drought in general terms with reasonable precision is the first problem the researcher has to face. There follows the difficulty of how to characterize it. On the other hand, there is a need to make a distinction between ***point drought*** analysis,if only a given relevant point in space is considered, and ***regional drought*** analysis if the drought is referred to a region with a significant area.

Most definitions in the literature take droughts as point events caused by such hydrological phenomena as precipitation, evapotranspiration, river runoff, surface water stored in reservoirs, moisture and groundwater levels and water supply. Very frequently no measure of their consequences is made.

In a previous work (SANTOS, 1983), the author introduces a precise definition of regional drought and presents a methodology for drought characterization that grasps the long lasting and regional nature of the drought phenomenon allowing for its stochastic nature. This work is mainly oriented to a climatic characterization where such concepts as drought duration, deficit area, or drought intensity and/or severity are its main concern.From an engineering point of view there must be a need to consider other concepts related to risk of drought occurrences and drought consequences. These concepts are important because society is becoming more and more aware of social and economic losses caused by droughts and a rigorous characterization of this natural hazard is becoming more important.Making use of such concepts of systems analysis as risk, reliability, mission reliability, resilience and vulnerability, CORREIA et al (1987) presented a methodology to characterize regional droughts in the framework of an engineering risk analysis.

This paper summarizes both approaches from the theoretical point of view. A characterization of droughts in Southern Portugal is also presented using annual flows as drought determinants and taking into account significant loss functions.

2.REGIONAL DROUGHT STOCHASTIC CHARACTERIZATION

2.1.Basic Concepts

From an engineering viewpoint regional droughts are more important than point droughts because they are more directly related to the problems of water shortages and their impacts. Nevertheless, their study has been relatively neglected in the literature when compared with the attention given to point droughts.
In regional drought characterization, in addition to drought characteristics, some features of the affected region, namely its size and shape, must be considered.
To rigorously define regional drought some necessary variables have to be introduced. The first one is the ***determinant***, the phenomenon or number of phenomena which are assumed to determine or cause the drought. Among these phenomena, rainfall, evapotranspiration, river runoff, surface water stored in reservoirs, moisture and groundwater levels and water supply are the more commonly mentioned.
Assuming that a drought determinant is selected, the researcher has to decide on the discretization time unit. For drought studies this time unit generally is not less than one month and not larger than one year. In a previous paper (RODRIGUES et al. 1989) the problem of the appropriate time resolution for stochastic drought analysis is tackled.
The determinant factor and the discretization time unit are still not sufficient to define a water shortage. A threshold below which a water shortage exists must be considered. It is said to be the ***critical level*** or the ***truncation level*** and it may be a constant or a function of time, must be intimately associated with the determinant phenomenon and will depend on the type of drought under study and on the objectives that water users wish to attain. Also, if a regional drought is to be defined, other problems arise, namely whether a drought is considered to occur only when the whole area undergoes a shortage of water or whenever a significant area known as ***critical area***, is suffering from the drought effects. In this latter case, what is a significant area? Or, rather, should the affected area be discarded and a mean water shortage over the entire area be defined for drought purposes?
The definition and characterization of regional drought given below take into account these questions. But before a definition is given some other concepts must be introduced. Also, to make the definitions clearer, a mathematical notation must be presented.

Without loss of generality, a determinant will be represented by $\underset{\sim}{X}(t)$ with ~ meaning vector. The k-th component of this vector is represented by $X_k(t)$ (k = 1,..., n) and it measures the determinant phenomenon at station **k**, being **n** the total number of stations within the region under consideration.
The function f_k (t) measures the critical level at the same time and at the same station.
The basic concepts that are needed to measure the suffering area and the water shortage at a fixed time are the following:

- ***Instantaneous deficit area***, A_s (t): it is the drought-affected area, within the region R, at time t. It is given by

$$A_s(t) = \sum_{k=1}^{n} a_k \, I_{K(t)}(k)$$

where **k** is a station within the region, $K(t) = \{k \,/\, X_k(t) \le f(t)\}$ and $I_{K(t)}(k)$ is the indicator function; $\mathbf{a_k}$ is the effective area assumed for the station **k**. Generally, $\mathbf{a_k}$ is taken as the percentage of effective area to the total region area and A_s (t) takes values from 0 to 100, which gives the percentage of deficit area at time **t**;

- ***Instantaneous areal deficit***, D_S (t): it is the weighted sum of the deficits occurred in the region R. It is given by

$$D_S(t) = \sum_{k=1}^{n} a_k \left[f_k(t) - X_k(t) \right] I_K(t) \quad (k)$$

The term "instantaneous" used here is a general term and can be replaced with "monthly", "seasonal" or "annual" whenever the discretization time unit is the "month", the "season" or the "year", respectively.

2.2 Regional droughts

Regional drought can be defined as the time interval during which the instantaneous deficit area equals or exceeds a given critical area, $\mathbf{a_c}$.
The critical area $\mathbf{a_c}$ plays an important role and it must be previously established and based on some physical or other meaningful characteristics of the region under study.
Some drought characteristics aiming at giving a better understanding of the drought may now be introduced:

- ***Drought duration***, L, is the length of the time interval $[t_o, t_e]$, i. e.

$$L = t_e - t_o + 1$$

where $\mathbf{t_o}$ is an instant such that the deficit area $A_S(t_o) \geq a_c$ but $A_S(t_o - 1) < a_c$ and $\mathbf{t_e}$ is the first instant after $\mathbf{t_o}$ such that $A(t_e) \geq a_c$ but $A_S(t_e + 1) < a_c$;

- ***Deficit area***, A, is the average deficit area along the drought, that is, along the time interval $[t_o, t_e]$, i. e.

$$A = \frac{1}{T} \sum_{t=t_o}^{t_e} A_S(t)$$

Note that $As(t) \geq a_c$ for all $t \in [t_o, t_e]$;

- ***Total areal deficit***, D, is the sum of the instantaneous areal deficit $D_S(t)$ along the drought duration, i. e.

$$D = \sum_{t=t_o}^{t_e} D_S(t)$$

- ***Drought intensity***, I, is the ratio

$$I = \frac{D}{L}$$

In Figure 1 the concepts of instantaneous deficit area, instantaneous areal deficit and the meaning of a_k and f_k are presented. In Figure 2, adapted from SANTOS (1983), an example of the application of these concepts to drought characterization in Portugal, is displayed.

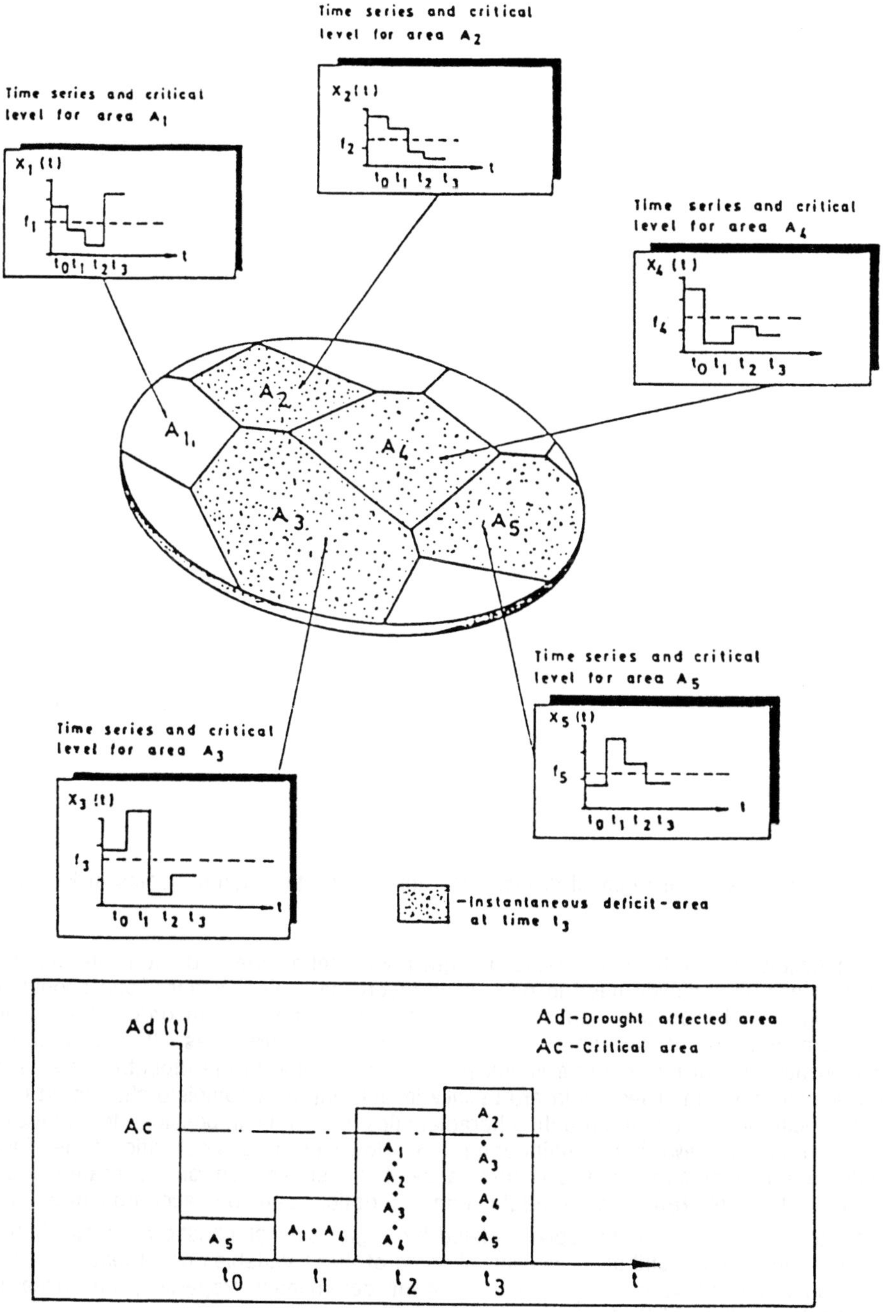

Fig. 1 - Illustration of areal concepts of droughts

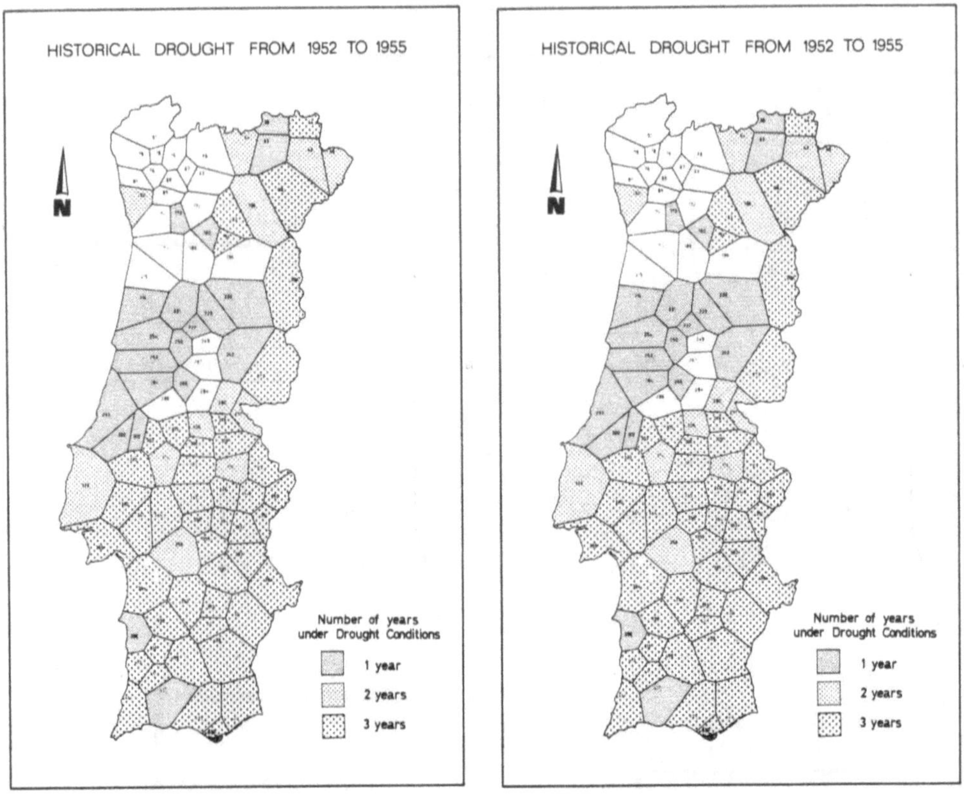

Fig. 2 - Application of regional drought characterization to drought studies in Portugal

For a complete comprehension of the drought phenomenon these drought characteristics which are random variables need to be studied theoretically. For instance, one should know their probability distributions and some characteristic parameters in order to be able to find return periods of droughts or probabilities of occurrences. If this theoretical characterization is not feasible one is limited to the estimation of the drought characteristics from the available data. These data are usually insufficient for a complete characterization of the statistical properties of drought characteristics and it is impossible to analyze the problem in the framework of engineering risk. However, if a stochastic model for the drought determinant can be established, a synthetic series can be generated and the statistical characterization of drought characteristics can be accomplished. If the determinant $X(t)$ can be reasonably modeled by a gaussian stochastic process, then it is possible to define the distribution function of some of these variables or, at least, to express their mean value and variance as functions of the determinant properties, the critical level and critical area (SANTOS, 1983).

2.3 Non-Areal Concept of Drought

The basic concepts for regional analysis of droughts presented in the previous sections are formulated in terms of affected areas. However, these concepts can be equated in more general terms and in a more practical manner with respect to its application in complex water supply systems.

Instead of several sub-areas, a given number of water sources can be considered. If these multiple sources are fully used in the scope of this system a unit weight can be associated to each one of them ($a_k=1$). A point drought occurs, if some drought determinant $X_k(t)$, in a given time interval t, is below a pre-defined critical level $f_k(t)$.

The concepts of instantaneous deficit area and instantaneous areal deficit can be generalized to the concepts of ***instantaneous deficit sources*** and ***instantaneous system deficit.*** These variables can be denoted also as $A_s(t)$ and $D_s(t)$, respectively. Accordingly, regional drought corresponds to ***system drought.*** A system drought occurs when the instantaneous deficit sources equals or exceeds a given critical value, a_c. The ***average deficit source*** can be defined as the average number of sources presenting a deficit during the drought, in analogy with the deficit area. The ***total system deficit*** is the summation of instantaneous system deficits along the drought duration, in analogy with the total areal deficit.

This generalization of the areal concepts for drought analysis is more flexible and, eventually more applicable to man made systems in which the source and its weight is more relevant than a hydrometeorological station and its surrounding area. This non-areal formulation is illustrated in Figure 3.

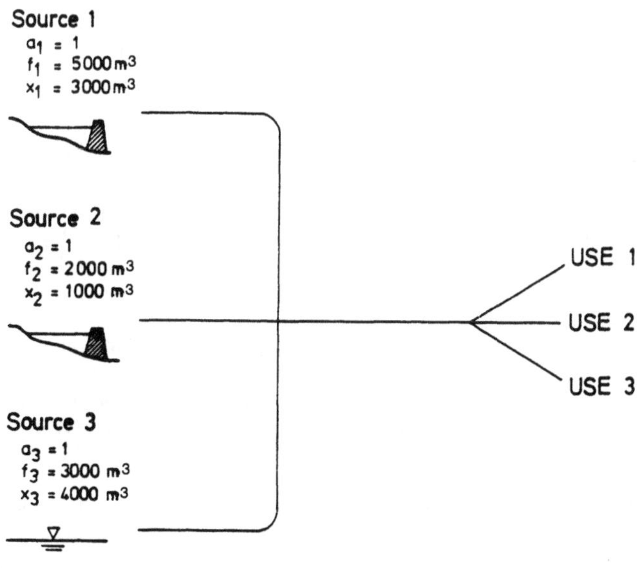

MINIMUM SUPPLY FOR A NON-DROUGHT SITUATION = 10 000 m³

Instantaneous Deficit Sources = 2

Instantaneous System Deficit = 3000 m³ (2000 x 1 + 1000 x 1)

Fig. 3 - Illustration of non-areal concepts of droughts

3. RISK ASSESSMENT OF DROUGHTS

3.1. Risk and Reliability

According to HASHIMOTO (1980) and HASHIMOTO et al. (1982) the ***reliability*** of a system can be described as the probability that a system is in a satisfactory state. ***Risk*** can be defined as the complementary probability. For practical purposes it is necessary to establish a criterion to decide whether the system is in a satisfactory or unsatisfactory state. Denoting the set of all possible satisfactory states by **S**, the set of all unsatisfactory states by **U**, the state of the system at time interval **t** by **X(t)**, the reliability of the system by α, and its risk by **ß**, the following relations hold:

$$\alpha = \text{Prob} \ [X(t) \in S]$$
$$\beta = \text{Prob} \ [X(t) \in U]$$
$$\alpha + \beta = 1$$

This definition of risk is somewhat strict because it does not include any consideration of severity. This concept, however, is considered in the definition of resiliency and vulnerability.

In the previous section, a rigorous definition of regional drought was provided. This definition can be used to identify the unsatisfactory states of the system. The system is taken as being the region under study, and the occurrence of a drought is equated with an unsatisfactory condition.

Drought studies are often based on monthly data as it is the case with the example presented in this paper. Under these circumstances, it is possible to assess risk and reliability in monthly terms. However, in many supply and firm yield studies, a yearly analysis is preferred. Therefore, it is adequate to define ***dry year*** as a year in which at least one dry month occurs and to characterize drought risk and reliability in annual terms. In this way, it is possible to define concepts such as return period of a given drought.

As described in the previous section a regional drought is defined by two distinct variables, namely a truncation level, usually corresponding to a given non-exceedance probability, and a percentage of drought affected area or number of supply sources in deficit conditions. Given a numeric value for one of these variables, it is possible to study the relationship between the other variable and return period. In the case study presented later, the relationship between truncation level (in terms of non-exceedance probability) and return period is established for several percentages of drought affected area.

DUCKSTEIN et al (1984) introduce the concept of mission reliability as being the probability that the system will not fail between the time when a non-null requirement arrives and the corresponding service is delivered. This concept is easily incorporated into the framework of regional drought characterization. In fact, it is very common that drought problems be relevant only during a given period of the year. This period will be called ***mission period.***

Mission reliability can be considered as the probability that a drought does not occur during the mission period and, likewise, ***mission risk*** is the probability that a drought does occur during that period. In this approach, a ***dry year*** can be defined as a year in which at least one dry month occurs during the mission period, and the study of annual mission risk can be easily done.

To study numerically the mission risk and reliability of a regional drought, it is sufficient to set the truncation level to zero for the period of the year in which drought is not relevant and different from zero for the mission period.

3.2. Resilience

Resilience is a measure of the recovery time of the system. An immediate formulation of resilience in the context of regional drought characterization is to equate it with the duration of the drought and to study this duration statistically.
This definition, however, may not be very significant from a physical point of view. In fact, if a region is under a severe drought, it does not recover immediately when the drought condition ends. If a large total deficit occurred during the drought period, it is reasonable to consider that recovery is accomplished when a certain fraction of the total deficit is compensated by an excess of water that replenishes, at least partially, the exhausted resources. This excess of water has to be computed with respect to a given truncation level that can be called the ***recovery level.*** The recovery level may be eventually equal to the drought defining truncation level. However, it is more realistic that the recovery level is higher than the truncation level. An adequate value for the recovery level is the mean monthly precipitation.
The fraction of the drought total deficit that needs to be compensated may be called the ***recovery rate.*** If the recovery level coincides with the drought defining truncation level and the recovery rate is zero, the drought duration can be equated with resilience. If the recovery level is higher than the truncation level and/or the recovery rate is greater than zero, a recovery time must be considered and the study of resilience must be based on it. Naturally, the recovery time is equal to or greater than the drought duration and corresponds to a more precise definition of resilience. More formally, a new function called ***excess*** at time **t**, can be defined:

$$e(t) = \begin{cases} X(t) - r(t) & r(t) \leq X(t) \\ 0 & r(t) > X(t) \end{cases}$$

where **r (t)** represents the recovery level. The cumulative excess at time **t'** after drought termination, $\mathbf{t_e}$, is given by:

$$E(t) = \sum_{t=t_0}^{t'} e(t)$$

For a recovery rate α_r, the recovery time $\mathbf{t_r}$, of a drought with total deficit **D**, can be defined as:

$$t_r = \min \left\{ (t-t_e) : \frac{E(t)}{D} \geq \alpha_r \right\}$$

For any given values of **r(t)** and α_r, it is possible to study the statistical properties of $\mathbf{t_r}$. This is illustrated in the case study.

3.3 Vulnerability

Vulnerability expresses the severity of failure in terms of its consequences. The concern is not how long the failure (*i.e.* drought) lasts, which is considered under resilience, but how costly it is.In the context of regional drought studies it is possible to consider several formulations for the vulnerability analysis.
A first approach consists in considering the total ***annual deficit*** as a good measure of the severity of droughts. In this case it is possible to conduct a vulnerability study based on the

statistical distribution of the total annual deficits. Proceeding in this way makes it possible to express a relationship between total annual deficits and return periods for a given drought definition (drought truncation level and drought affected area).
A second approach consists in focusing on ***drought intensity*** as the most relevant measure of drought consequences. To conduct a vulnerability study in annual terms, annual maxima of drought intensities can be used. It is possible, then, to establish a relationship between annual maximum drought intensity and return period.
A third and more general and attractive approach consists in establishing a ***loss function*** that may depend on any single drought characteristics or any combination of drought characteristics.

$$L = L(D, I, T, t_r)$$

In this expression **L** is the loss corresponding to a drought with a total deficit **D**, an intensity **I**, a duration **T**, and a recovery time $\mathbf{t_r}$.
If more than one drought occurs in one year, either the sum of losses or the largest loss can be selected to assess vulnerability in annual terms.
A simple but rather attractive example of a loss function is the following:

$$L_f = -\frac{1}{K} \ln \left(1 - \frac{D}{D_{max}}\right)$$

In this case only the total deficit **D** is considered. $\mathbf{D_{max}}$ represents a maximum value of total deficit corresponding to an infinite (catastrophic) loss **K** and is a parameter determining how fast the loss **L** approaches its asymptotic value at $\mathbf{D_{max}}$. The general shape of this particular univariate loss function is displayed on Figure 4.
In the case study presented in the following section, a loss function of this type is used to assess vulnerability in annual terms.

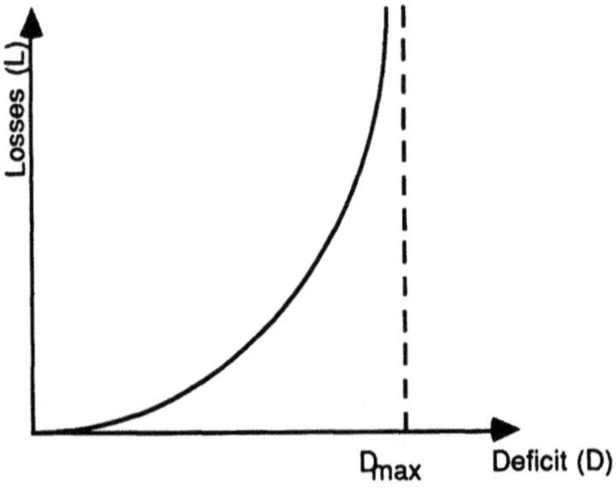

Fig. 4 - Example of drought loss as a function of areal deficit

4 . CASE STUDY

4.1. Basic Characteristics of the Sado River Basin

The proposed methodology was applied to the Sado river basin that drains an area of about 7,700 Km^2.
The Sado river basin is located in southern Portugal and has coincident boundaries with two international river basins from the Iberian peninsula : the Tagus river basin in the north and the Guadiana river basin in the east (Figure 5).
The climate within the watershed can be considered moderate with exceptions of a short wet coastal area in the SW and South headwaters, with annual precipitations higher than 800 mm, and a dry area in the East, where the annual precipitation is below 500 mm.
One of the main hydrological problems in the river basin is the within year partition of the rainfall amounts. In fact, the average number of rainy days in the several raingages covering the basin ranges from 60 (SE and E) to 100 (W, SW and N). These numbers are reduced, respectively, to 15 and 30 for daily rainfall amounts above 10 mm.
In order to introduce some regulation in the flow regimes determined by the intermittent rainfall amounts, a great number of reservoirs for irrigation purposes was built in the basin since 1949. Figure 6 locates the main reservoirs in the basin, shows the regulation volumes and the beginning year of activity for each reservoir.
The reasonably large size and different precipitation regimes coupled with flow data availability and the economic interests of the region makes the Sado river basin a very good case study for regional drought analysis.
In this study, data from four stream gages are used, covering approximately 50% of the whole drainage area as it can be seen in Figure 7. A common period of 26 years (1934/35 to 1959/60) was used to illustrate the regional drought analysis proposed in the first sections. Unfortunately, this record is not as long as it would be desirable to obtain a small degree of uncertainty in drought characterization. This situation could be overcome with the generation of a synthetic series of data as it was done in another study (CORREIA et Al. 1987). In this case, however, it was decided to use only the original record.
The empirical distribution functions of the annual flow series were plotted in Figure 8 in order to help deciding the values that should be used as truncation levels and its corresponding exceedance probability.

4.2. Risk Analysis of Regional Droughts

The methodology presented in sections 2 and 3 was applied to the annual flow series for regional drought analysis within the Sado river basin for different truncation levels and critical areas. Figure 9 illustrates the deficit volumes and associated recovery volumes that occurred in the study area from 1934/35 to 1959/60.
Although 26 years of annual data are clearly not sufficient to draw definitive conclusions on regional droughts in the Sado river basin, they can be useful for methodological purposes and for a preliminary characterization. As it was stated before, it would be interesting to have a longer record of data, as it always convenient when extreme events are analyzed. A methodology that can be used to compensate partially for the lack of observed data could be the generation of synthetic data, preserving as much as possible the stochastic properties of the observed values. This methodology was used with interesting results by (CORREIA et. al. 1987) in Ave river, in which 1,000 years of synthetic data were generated. It is interesting to note that results obtained in that study are not very different from those presented in this paper for Sado river, as far as general properties of regional drought events are concerned.

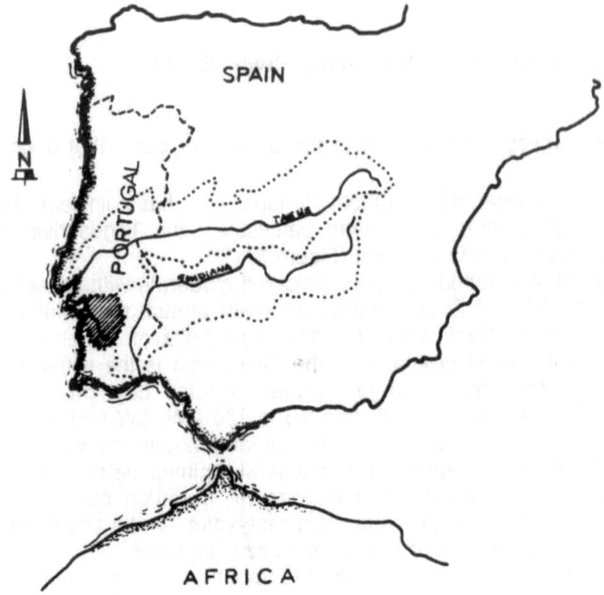

Fig. 5 - Location of the Sado river watershed

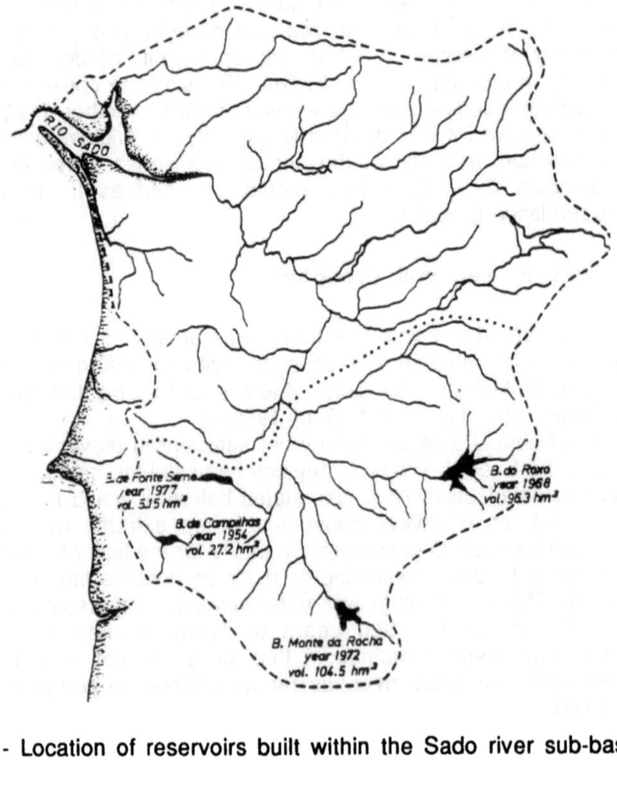

Fig. 6 - Location of reservoirs built within the Sado river sub-basin

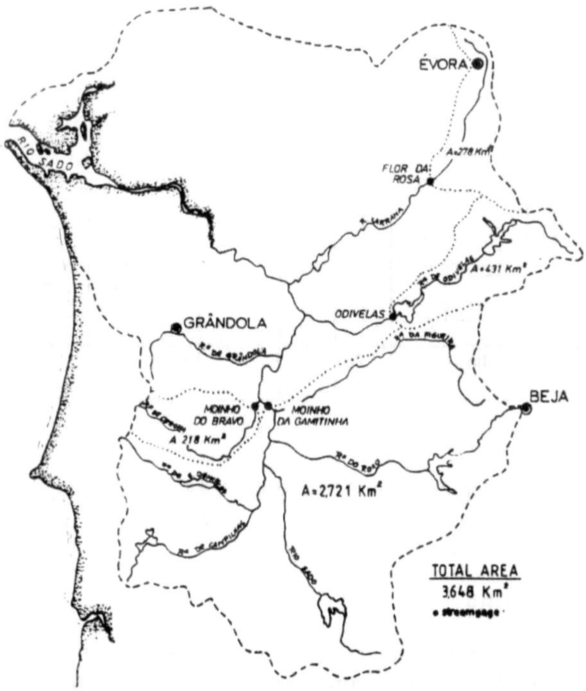

Fig. 7 - Location of stream gages used in the study

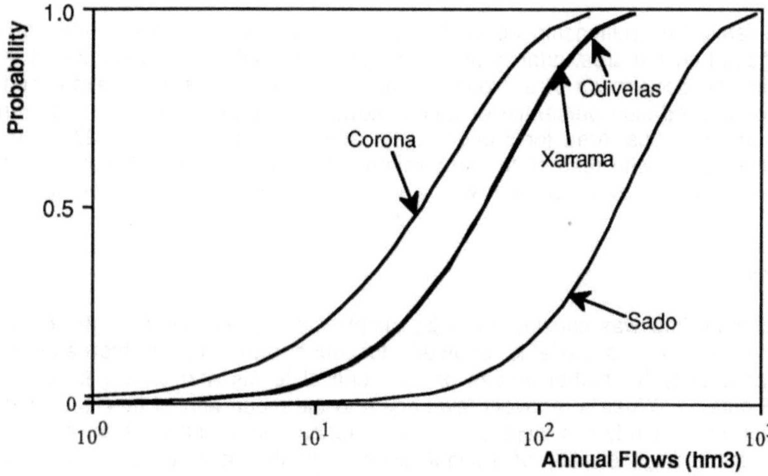

Fig. 8 - Annual flow empirical distribution functions

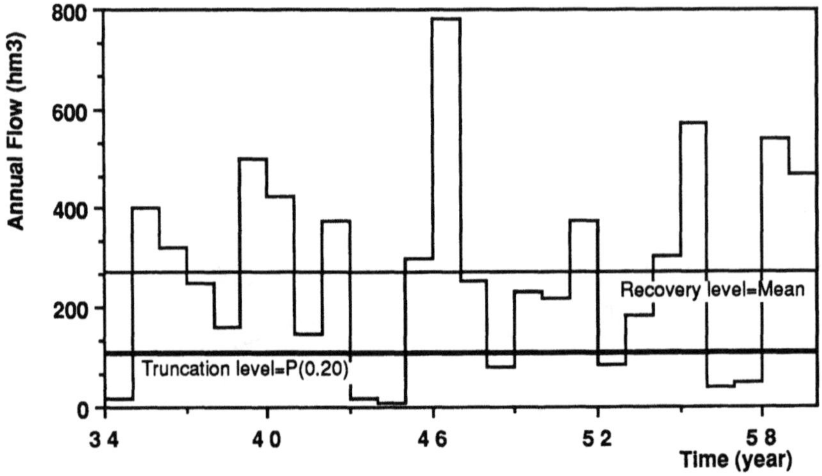

Fig. 9- Annual flow at the Sado stream gage

4.3. Engineering risk analysis

i) Risk and reliability

As described previously a regional drought is defined by two distinct variables namely a critical or truncation level, usually corresponding to a given non-exceedance probability, and a percentage of the drought affected area. Given a numeric value for one of these variables it is possible to study the relationship between the other variable and the return period, which is a concept directly associated to risk. In fact:

$$\text{risk} = P\ (x\ (t) \in U) = 1/T\ .$$

Figure 10 presents the relationship between the truncation level and the return period for two percentages of critical area, namely 50% and 100%, as well as an evaluation of the risk associated with the occurrence of a drought defined by a truncation level expressed as a determined non-exceedance percentile of annual flow. This figure shows the importance of the definition of the critical area for droughts with return periods from 7 to 27 years. This graph is not conclusive with respect to more severe droughts. It is known, however, that in Portuguese conditions a severe drought with a very large return period is likely to affect all the country.

ii) Resilience

When the recovery time was considered to be simply the drought duration, the distribution functions obtained for this variable showed, for each duration, an increase in non-exceedance probability for higher values of both critical levels and critical areas. This is displayed in Figure 11. When a recovery level equal to the mean annual flow is assumed for each sub-region under study this increase is less notorious, as it can be seen in Figure 12. When a more demanding definition of drought recovery is introduced, considering a recovery level greater than the critical level, larger recovery times and slower trends in distribution functions are obtained. These results can be observed in Figure 13. The study carried out showed that for a reasonable recovery flow, the mean annual flow, the associated recovery rate is always close to 100 %.

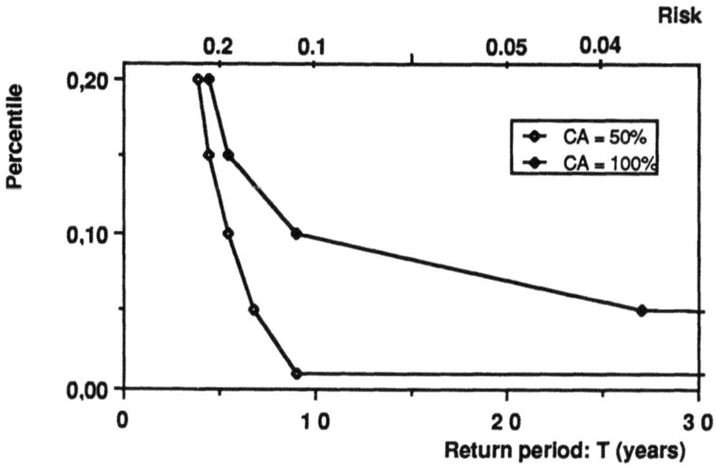

Fig. 10 - Drought risk for 50% and 100% critical areas

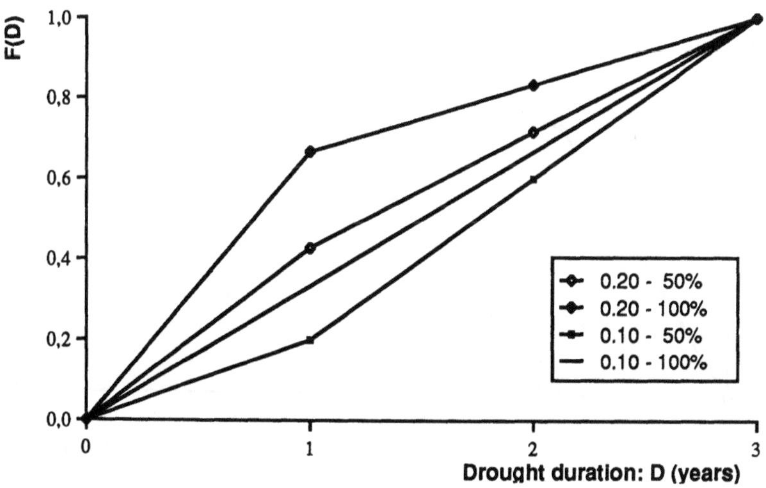

Fig. 11 - Probability distribution functions of drought duration for droughts defined by critical levels of 0.20 and 0.10 and critical areas of 50% and 100%

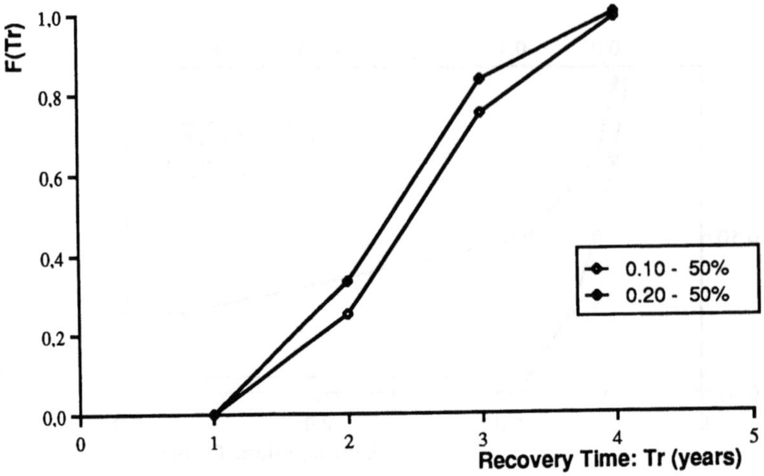

Fig. 12 - Probability distribution functions of drought recovery time for critical areas equal to 50% and critical levels of 0.20 and 0.10

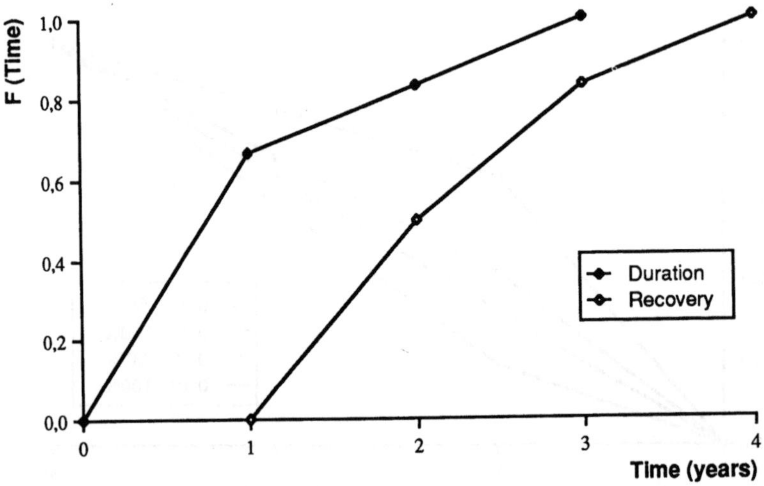

Fig. 13 - Comparison between the probability distribution functions of drought duration and drought recovery time for critical areas equal to 100% and a 0.20 critical level

iii) Vulnerability

As mentioned earlier vulnerability can be studied in terms of total areal deficit, maximum annual intensity and/or total annual loss.
Results for deficits and intensities are plotted on Figures 14 and 15 where, for the same return period, an increase in drought magnitude is related with an increase in the critical level. At the same time, that increase may correspond to a decrease in the critical area.

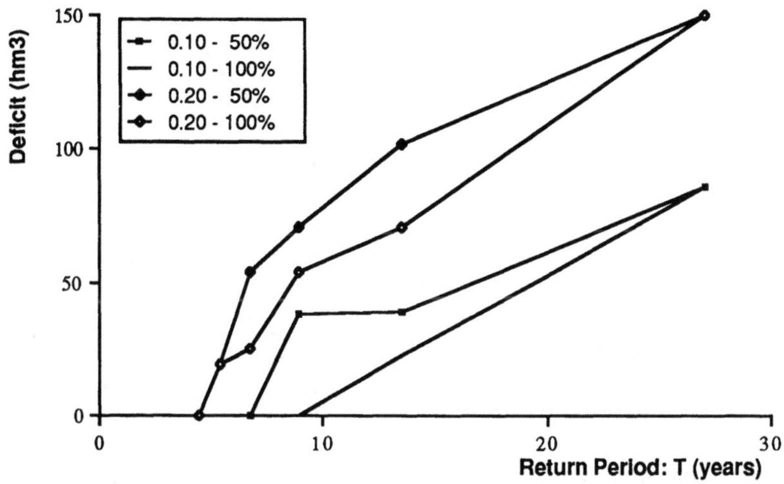

Fig. 14 - Drought vulnerability expressed in terms of total areal deficit for 0.20 and 0.10 critical levels and 50% and 100% critical areas

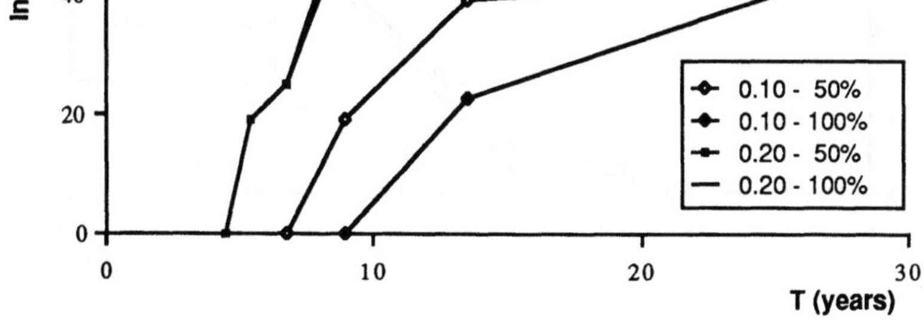

Fig. 15 - Drought vulnerability expressed in terms of maximum intensity for 0.20 and 0.10 critical levels and 50% and 100% critical areas

Figures 16 and 17 display the loss functions considered as a function of the total areal deficit by the expression referred to in section 3. The constant K, defining how fast the loss function approaches its asymptotic value, is set equal to 2. These figures show that the drought defining variables, namely the critical level and the critical area, have some impact on the way losses rise with return period .

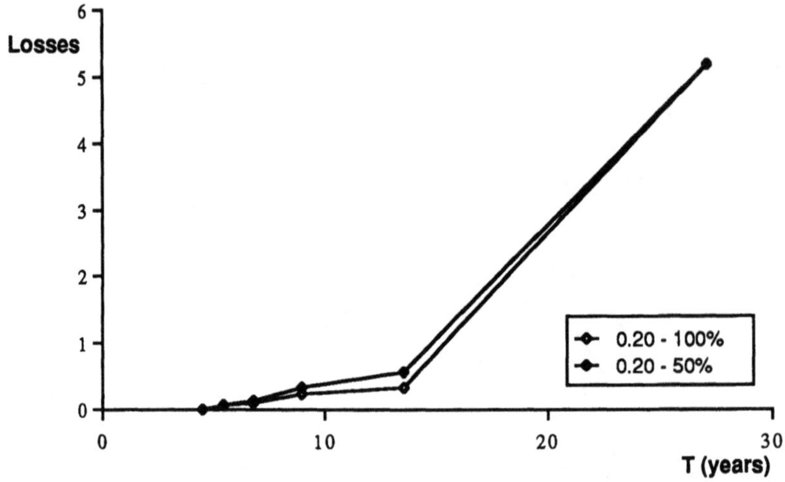

Fig. 16- Drought vulnerability expressed in terms of total areal loss for a 0.20 critical level, with critical areas equal to 50% and 100%.

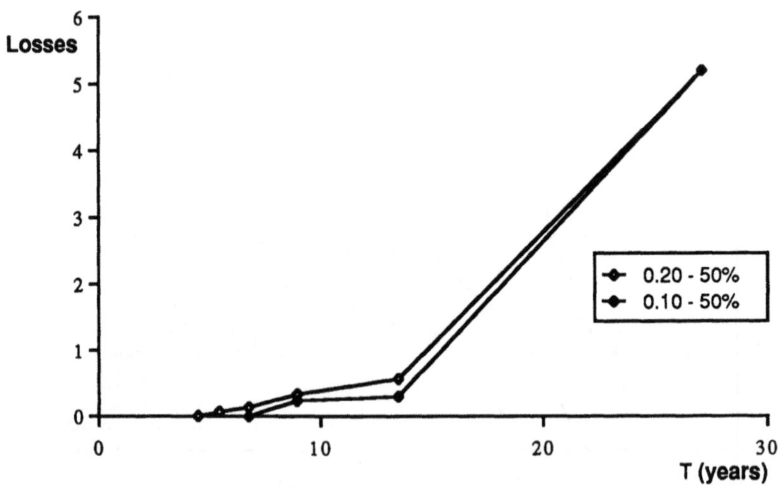

Fig. 17 - Drought vulnerability expressed in terms of total areal loss for a 50% critical area and 0.20 and 0.10 critical levels.

5. CONCLUDING REMARKS

An important characteristic of droughts which makes them different from other hydrological natural disasters is the fact that they may be a non-frequent phenomenon for a region. This explains why the public is usually inadequately prepared to face a drought. Unfortunately, people and decision-makers tend to react preferentially to ongoing or pressing problems. Infrequent problems, such as those caused by droughts, tend to be overlooked until they re-occur and attract the attention of everyone.

To face adequately the critical problems of droughts, it is essential to plan all the necessary adaptations, modifications and practices that could be activated in droughts.

To make such a plan possible, a precise knowledge of the main characteristics of historical droughts is essential. The methodology presented in this paper may contribute significantly to the required characterization.

Two different but complementary approaches aiming at characterizing regional droughts are shown. The first one takes into account the hydrometeorological aspects of droughts and quantifies their variability with time. Characteristics of droughts are described by a selected set of random variables.

The second approach gives a means of characterizing the risk of drought occurrences for droughts with pre-defined characteristics. The system analysis concepts of risk, reliability, resilience and vulnerability, are applied in the characterization of drought events. This approach is especially important for the decision-maker who need to evaluate drought situations in terms of engineering risk, and prepare adequate responses for several scenarios of risk and vulnerability of the systems under their responsibility.

The methodology proposed in this paper, and in previous reports by the same authors, is very general and broad. In fact, not only hydrometeorological droughts can be studied, but also drought phenomena in man made complex water systems can be considered. This "non-areal" formulation of droughts seems to be interesting and promising but requires more research work.

The proposed methodology has been successfully applied to several regions in Portugal. In this paper the Sado river basin is used as a case study. In this river basin, only 26 years of data were used which is a rather short record to study extreme events. Under these conditions the curves that were obtained relating drought characteristics and return period have a sharp-shaped.configuration .A synthetic generation of data could provide longer records and smoother configurations for the main functions representing drought characteristics, as it was done in previous studies (CORREIA et al. 1987) .

REFERENCES

CORREIA, F. N. ; SANTOS, M. A. ; RODRIGUES, R. R. (1987) - **Engineering Risk in Regional Drought Studies** in "Engineering Reliability and Risk in Water Resources".NATO ASI Series, The Netherlands, Martinus Nijhoff Publ. pp 61-86.

DUCKSTEIN, L.; PLATE, E. (1983) - **A System Framework for Hydrological Reliability with Application to the Design of Structures.** Proceedings of the Hamburg Symposium, International Association of Hydrologic Sciences.

DUCKSTEIN, L.; PLATE, E.; BENEDINI, M. (1984) - **Incidents and Failures in Reservoirs and Dams-A Mathematical System Framework.** Fall Meeting, American Geophysical Union.

HASHIMOTO, T. (1980) - **Robustness, Reliability, Resilience and Vulnerability Criteria for Planning Water Resources Systems.** PhD thesis, Ithaca, New York, USA, Cornell University.

HASHIMOTO, T. ; STEDINGER, J. ; LOUCKS, D. P. (1982) - **Reliability, Resilience and Vulnerability Criteria for Water Resource System Performance Evaluation**. Water Resources Research, vol. 18, nº.1, pp 14-26.

RODRIGUES, R. R.; SANTOS, M. A., CORREIA, F. N.(1989) - **Appropriate Time Resolution for Stochastic Drought Analysis.** Advanced Study Institute on Stochastic Hydrology and its use in Water Resources systems:Simulation and Optimization, Peñiscola, Valencia.

SANTOS, M. A. (1983) - **Regional Droughts: a Stochastic Characterization.** Journal of Hydrology, vol. 66, pp 183-211.

SANTOS, M. A.; CORREIA, F. N.; CUNHA, L. V. (1984) - **Drought Characterization and ImpactAssessment-Case Study of Portugal.** In Coping with Droughts, Edited by Yevjevich, V., Cunha, L. V. and Vlachos, E., Water Resources Publications, Littleton, Colorado.

SANTOS, M. A.; CORREIA, F. N.; RODRIGUES, R. R. (1988) - **Risk Assessment in Regional Droughts.** Workshop on Natural Disasters in European Mediterranean Countries, Colombela, Perugia.

RELIABILITY IN GROUNDWATER RECHARGE AND QUALITY PARAMETERS

L.W. De Backer
Unité de Génie Rural
Faculté des Sciences Agronomiques
Université Catholique de Louvain
Place Croix du Sud, 2 - Bte.2
B-1348 Louvain-la-Neuve, Belgium

ABSTRACT

The prevention of groundwater contamination is one of the most urgent tasks that a water system engineer must have in mind in all his projects. Rough approximations of the hydrological cycle have lead to the actual crise in global pollution due to local developments. Norms are the center element involved in the multicriteria decision process. Downward and upward movements of the water in the soils and the aquifer under unsaturated or satured conditions are discussed in the light of the modifications they infer.to the physico-chemical properties. Rainfall and evapotranspiration are the main estimates on which reliabilty efforts should be put . Regular field surveys of water quality in the springs help in the verification of the risks taken now for the long term.

1.INTRODUCTION

Groundwater contamination is the most dreadful health risk that humanity faces now at long term (Brookshire et al., 1986).This is particularly true in industrialized countries where the quality of groundwaters keeps on deteriorating (UNESCO, 1986).At the present rate of fertilizer use, for instance, nitrates are not expected to be held in the soil longer than 10 years and phosphates 100 years due to soil acidification by the rain (Behrent, 1988). The risk is threefold: (1) an unaffordable cleansing of the groundwater supplies due to very long residence time in aquifer and, downstream the water cycle, rivers polluted for centuries from their own springs. (2)Upward flow from water tables induce groundwater contaminants to pollute the soil. Desertification due to salinisation encroaches each year huge expanses of grass and arable lands. For instance, in Australia the felling of deep rooted trees has slowed down the evapotranspiration and caused a deadly rise of the water table.The risk of salt contamination of the root zone has been recognized even in regions under temperate climate (Mermoud et al., 1990, 1991).With strong evaporative demand, the soil can transmit higher fluxes for short periods. Therefore, the use of the steady-state flow analysis to describe the upward flux from the water table to the soil surface may lead, in particular cases, to under-estimate the salinisation risks.

NATO ASI Series, Vol. G 29
Water Resources Engineering Risk Assessment
Edited by J. Ganoulis

The salinisation risks of irrigated soils are so high that the area damaged (60 million hectares, i.e. one fourth of the 250 million hectares irrigated in the world) is equivalent to the estimate of needed new irrigation investments (Postel, 1989). According to the average estimate of world arable area of 30 million square kilometers, the earth could sustain some 8 billion peoples if the polluting conditions of the industrial society are prevented (Taube, 1985). Regular or continuous monitoring of water quantity and quality are hence more and more necessary because it permits observations of foreseable effects due to water contamination.(Duijvenbooden & Waegeningh,1987). (3) A third risk exists in the underground portion of the water cycle between the soil surface and the rivers : the modification of the physico-chemical properties of the soils, aquifers as well as groundwaters themselves. Injuring cations and anions react specifically with charged and uncharged surfaces of organic matter and minerals (Wiechmann, 1987) as well as with the fauna and flora of the soils. These phenomena occuring in the water cycle are yet poorly known and seldom taken into account in decision-making, if ever even in modeling (Jousma et al.,1989).They will illustrate this paper to show that we should develop a methodology relying on a global concept but toward local solutions which take into account the specificities of both the phenomena involved and the objectives wanted according to the time and space scales involved in the project (De Backer, 1989a). For the phenomena, all components related to hydrogeological conditions, soil type and land use must be considered (Duijvenbooden & Waegeninghet, 1987), measured (indicators) and controlled (indices) because they are very important in multiobjective evaluation (Hartman, 1987).

In the following chart (figure 1) a cyclic methodology for the conservation of natural resources for sustainable development in project management (De Backer, 1989b). It shows all the aspects of the exchange process involved with natural resources in our activities according to the objectives pursued. Risks to groundwater might appear in each of these aspects. These risks should be assessed especially with respect to those related to activities which involve directly the natural resources, namely the every day's, research and development activities. For those activities, norms are being implemented under the umbrella of the International Standards Organisation (ISO). When indicators are compared to the norms, they result in indices which are quite useful in Multicriterion Decision Making (Changkong and Haimes, 1982) to consider tradeoffs among options (Bogardi et al., 1984).

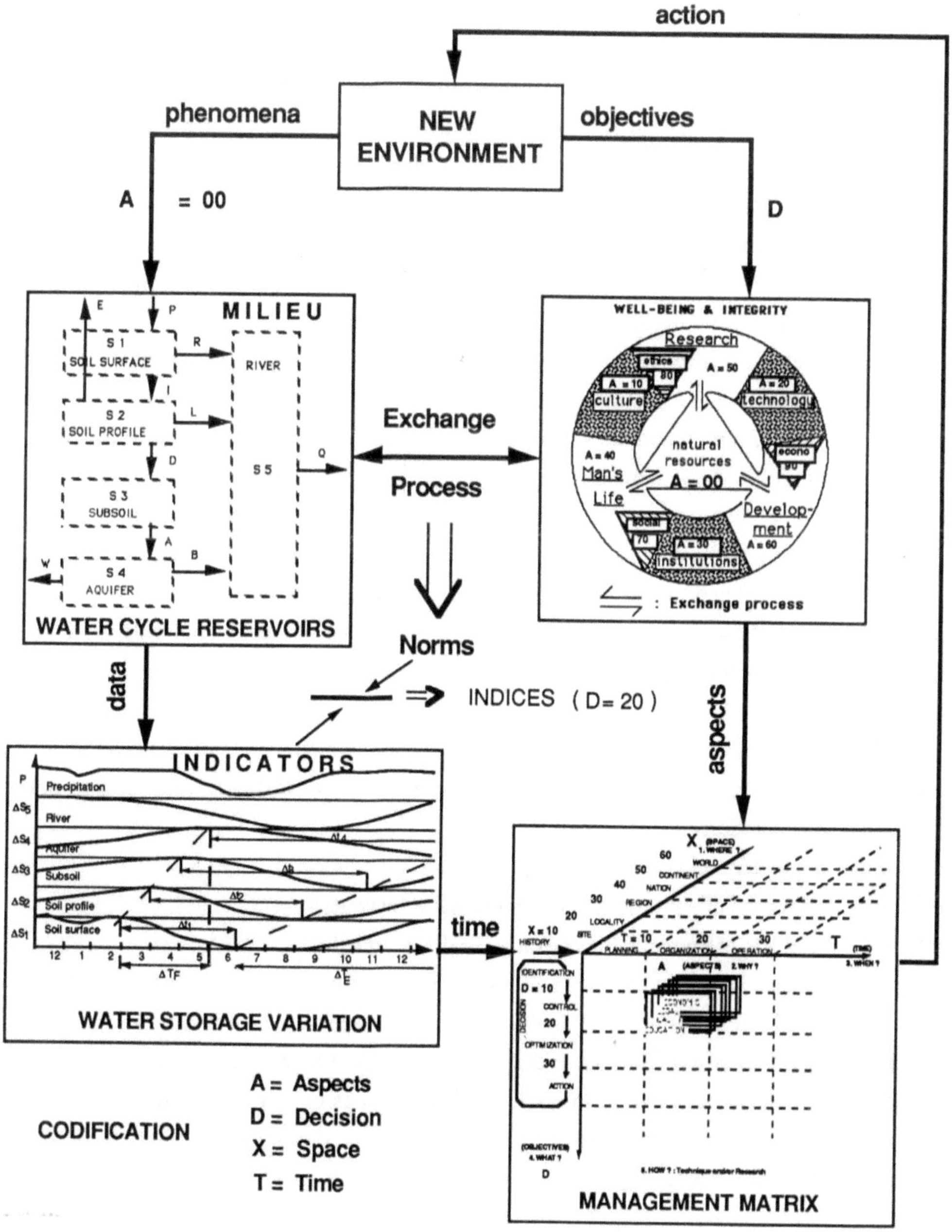

Fig. 1 A cyclic Methodology for the Conservation of Natural Resources for Sustainable Development in Project Management

LDB-UCL-89

2. GROUNDWATER RISK ASSESSMENT

The unsaturated zone is of special importance since it represents the first line of natural defence against groundwater pollution (Foster, 1987). Unsaturated zone water movement is normally slow and restricted to smaller pores with large specific surface, the chemical condition is normally aerobic and frequently alkaline. This results in considerable potential for a. interception, sorption and elimination of pathogenic bacteria and viruses,
b. attenuation of heavy metals, and other inorganic chemicals, through precipitation (as carbonates, sulphides or hydroxides), sorption or cation exchange,
c. sorption and biodegradation of many natural and synthetic hydrocarbon compounds.
It is therefore, essential that the unsaturated zone be fully considered in the evaluation of aquifer vulnerability. Should it be ignored, evaluations will be excessively conservative. However, its role can be complex and its ability to attenuate pollutants difficult to predict. ...In the case of persistent, mobile pollutants the unsaturated zone merely introduces a large time-lag before its arrival at the water-table, without any beneficial attenuation. In many other cases the degree of attenuation will be highly dependent upon the flow regime and residence time
From all these definitions and recommendations, it results that many uncertainties remain to assess groundwater pollution risk. To improve our assessment more measurements have to be made.
For example, the objectives of the Dutch national groundwater quality monitoring network are :
- to make an inventory of the present quality in relation to soil use, type of soil and hydrogeological conditions;
- to distinguish long term changes in quality;
- to provide data for the quality control and groundwater management;
- to indicate the extent of human influence on groundwater quality.

In the Netherlands where groundwater pollution is high, the network density is one well per 100 Km^2 and the basic programme of analysis is directed to diffuse sources like air pollution and agricultural use of soil. It consists of the following 17 parameters : electrical conductivity (EC), acidity (pH), Sodium (Na^+), Potassium (K^+), Magnesium (Mg^{++}), Calcium (Ca^{++}), Ammonium (NH_4^+), Bicarbonate (HCO_3^+), Chloride (Cl^-), Nitrate (NO_3^-), Sulfate (SO_4^{--}), Total Phosphate (t.P.), Total Organic Carbon (TOC), Aluminium (Al), Arsene (As), Nickel (Ni), and Zinc (Zn).
In the absence of monitoring wells, groundwater quality can be monitored at the springs. In the french speaking region of Belgium, the boyscouts measured simultaneously by colorimetry the nitrates content of 1,500 and 10,000 springs respectively in april of the year 1989 to test the sampling method and of the year 1991 to cover the whole territory of the region.

Data of the 1991 campaign which involved more than 4,000 boyscouts have not yet been compiled on Macmap. The results of "Operation Sources 89" have been given to the regional authorities to help them in water management. Their statistics indicate that barely 82 % of the springs are in conformity with the european guidance norm of 25 mg NO_3 per liter of water A study program is undertaken about the attenuation capacity of the altered limestone clays in order to determine the **appropriate local land use** and/or agricultural practices which will prevent the pollution of these aquifers. Strebel et al.(1989) have calculated the site- and time-specific nitrate leaching risk using a simulation model which describes the nitrate transport by convection, diffusion and dispersion.

3. UNCERTAINTIES IN MEASURING GROUNDWATER RECHARGE

Assumptions about the risk of modifications of the aquifer porosity due to recharge water quality changes are presented here.A first hypothesis is based on the observation of a slight but apparently irreversible reduction in water content. Calcium carbonate precipitates could so clog some pores. More significant water content variations would then be linked respectively to precipitation and redissolution of calcium carbonate when the $CaCO_3$ saturation index becomes negative (Gontier et al., 1990). A second hypothesis concerns the migration of dispersed fine particles. A third hypothesis could be air entrapment because of gas nucleation and dissolution as water tension increases and decreases due to flow unstability in stratified soils.

3.1. Rainfall

When no raingauge exists on the site where the water balance estimate has to be done, one has to rely on a closeby station. However, doubts might arise also about the data to choose when there are too many raingauges available.

Table 1 illustrates that case where precipitation data obtained with a lysimeter might seem overestimated although it records dew and light rains to which standard raingauges are insensitive. Closeby station (Ernage) is 10% lower and Lysi 25% higher than the annual mean. ASTR and GERU indicate a similar annual total while monthly data show some great discrepancies.

1989	mean	Ernage	ASTR	GERU	Lysi
month	mm	mm	mm	mm	mm
1	60	34	33	36	60
2	50	61	67	63	90
3	51	80	80	78	104
4	51	131	141	133	171
5	60	20	20	22	38
6	68	72	83	81	98
7	78	22	25	20	33
8	80	45	31	25	34
9	74	53	80	68	86
10	71	63	61	89	96
11	65	29	27	25	35
12	64	93	96	110	128
total	772	706	747	750	975

TABLE 1. Choice of rainfall data for the calculation of a water balance of the same area : mean is that of the closest national network station called Ernage ; ASTR, GERU and Lysi are a few meters apart : GERU is at ground level and Lysi weighs light rain and dew.

Table 2 illustrates the occurence of only 3 of the water cycle phenomena, namely rainfall (P), drainage (D) and evapotranspiration (ETR). They have been obtained on an hourly basis with the weighable lysimeter of Louvain-la-Neuve. Runoff and capillary rise are generally not measured in lysimetry. It is important to note that rainfall episodes represent barely 10% of the annual total and that they occur twice less often than dew or light rain which cannot be measured with raingauges.

Case	%	P	D	W	ETR	Comment
1	0,81	0	0	0	0	stationary
2	**10,9**	0	0	+	-	dew,wind effect
3	**9,6**	0	+	+	-	dew, light rain
4	**3,8**	+	+	+	0	P > D
5	0	+	+	0	0	permanent rate
6	0,03	+	0	0	0	none
7	**3,7**	+	0	+	0	none
8	0,3	0	+	0	-	D = dew
9	**35**	0	0	-	+	dry

10	1	+	+	-	-	P < D > W
		+	+	-	+	P < D < W
11	**35**	0	+	-	+	nul flux plane
		0	+	-	-	D > W
12	0,7	+	0	-	+	P < W

TABLE 2 . Percentage occurence of the 7520 hourly episodes recorded in 1988 :
P= precipitation, D= drainage, W= weight and ETR= actual evapotranspiration; + = occurence, 0 = no occurence and - = opposite occurence (dew can be considered as the opposite of ETR).

An instrument as precise as the frictionless weighable lysimeter can record data to the least error which is that of the electronic calibration, in our case ± 8.75 gm, i.e. a minimum precision cutoff of 17.5 gm seem desirable. It demonstrates that evapotranspiration (ETR) occurs during heavy rain.

As it could be expected, hourly averages of 6 minutes data overestimate significantly ETR when it is raining and to underestimate it in the case there is no rain. Hourly and 6' data agree fairly well when considering a precision cutoff value of 35 gm

Lysi rainfall values given in Table 1 are rough hourly data involving no cutoff. They may seem overestimated but one can say that they consist in more realistic liquid inputs (rain and dew) than those recorded with standard raingauges which are insensitive to light rain and dew.

3.2. Evapotranspiration

Much greater uncertainty remains in estimating evapotranspiration which consists in at least 75% of the episodes of 1988 (Table 2). This appears obvious looking at figure 2 where ETR is correlated to free water surface evaporation. This kind of relationship is done in engineering practice when local data are unavailable. Here it was necessary to estimate pan evaporation from a relationship with ETR when the two types of data are available. Although the correlation coefficient is satisfactory, there is a great risk of a 100% misestimation. The scatter around the relationship reaches easily ± 1 mm, the most frequent occuring value.

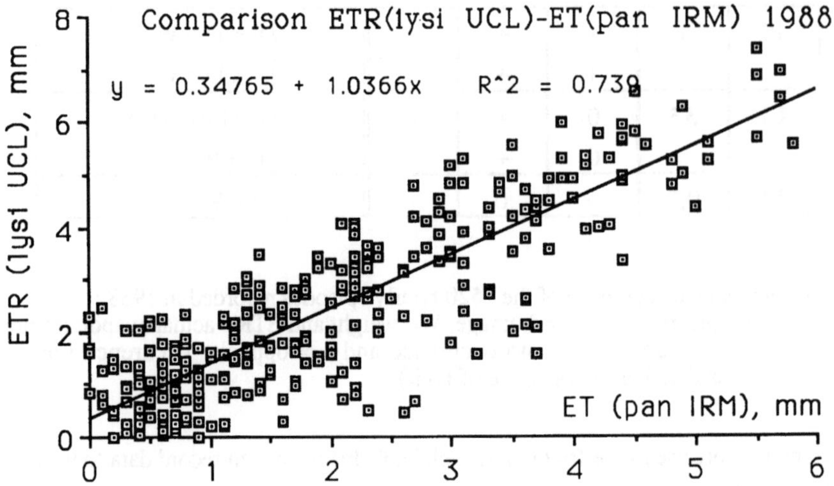

FIGURE 2. Correlation between evapotranspiration (ETR Lysi UCL) calculated from the data of the lysimeter of Louvain-la-Neuve and the evaporation of a free water surface (ET pan) measured by the Royal Institute of Meteorology (IRM), in Ernage, Belgium (R = 0.86)

Looking at the isolated ETR values of 10/08/90 (figure 3)) when there was no rain neither drainage one could have a similar first impression of uncertainty. However high these ETR values are used to compute daily water balance of the weighable lysimeter. These inputs must therefore be diminished by the amount of the dew liquid input. The substraction of dew from ETR results in an quite acceptable daily ETR of 3.5 mm whether it is calculated without cutoff or with cutoff not greater than 35 gm.

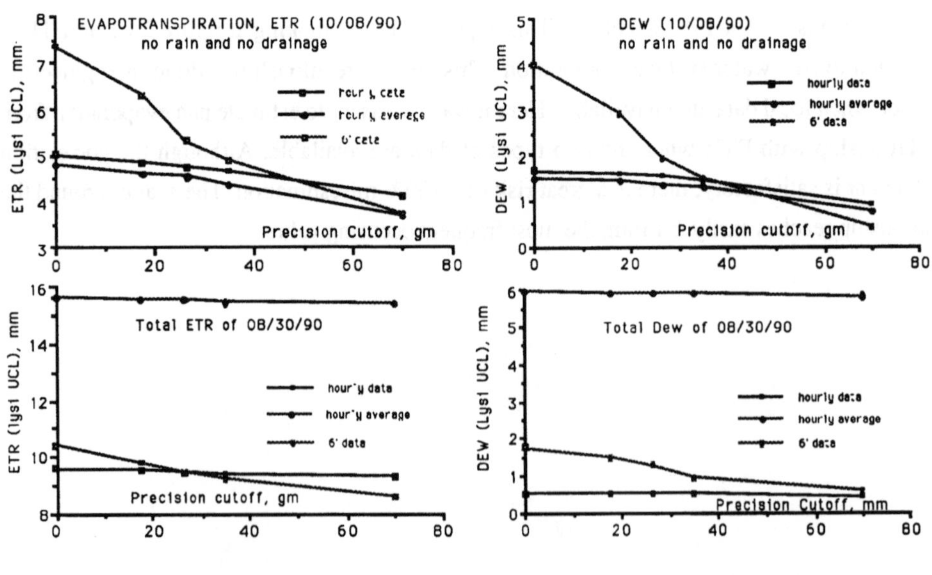

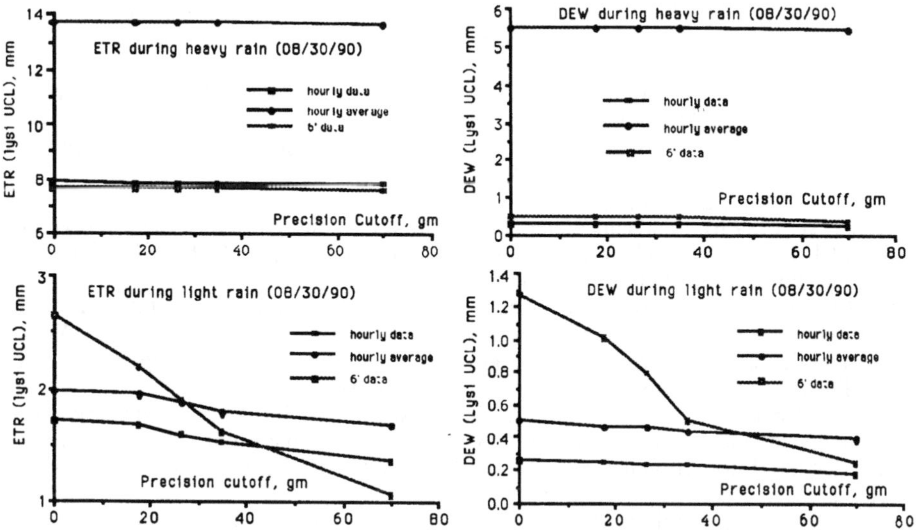

Figure 3. Effect of precision cutoff on the amount of ETR and Dew recorded with no drainage during morning light rain, afternoon storm and all day of 08/30/90 as well as with no rain and no drainage on 10/08/90 respectively for 6 minutes data, hourly average and hourly data.

As discussed in section 3.1, ETR occurs during light or heavy rain (an afternoon storm of more than 40 mm). ETR and dew are largely overestimated when using hourly averages for the water balance computation during rainfall. The analysis of figure 3 concludes also to a precision cutoff value between 17.5 and 35 gm. Indeed, a 26.5 gm cutoff didn't significantly affect the water balance calculations while a 70 gm did.

4. CONCLUSIONS

The risk which is to be taken into consideration in our engineering work and management will involve more and more new knowledges about the water cycle phenomena. Aquifer recharge evaluation is at stake when considering the actual modification of the water quality and its effects on aquifer properties changes.

REFERENCES

Behrendt H., 1988 : Changes in nonpoint nutrient loading into european freshwaters : trends and consequences since 1950 and not-impossible changes until 2080. Intern. Inst. for Appl. Syst. Anal., A-2361 Laxenburg, Austria, WP-88-026, 28 p.

Bogardi I.,L. Duckstein and A. Bardossy, 1984 : Tradeoff between cost and efficiency of pollution control, Proceedings of the VIth Int. Conf. on Multiple-Criteria Decision Making, Cleveland ,Ohio, June 1984.

Brookshire D.S., L.S. Eubanks and C.F. Sorg, 1986 : Existence values and Normative Economics : implications for valuing Water Resources, Water Resour. Res., Vol.22, 11, pp.1509-1518.

Changkong V. and Y.Y. Haimes, 1982 : Multiobjective Decision Making Theory and Methodology, 472 p., N.Y., Elsevier North Holland

De Backer L.W., 1989 a) : Background Concepts and Principles in Unsaturated Flow. in Hydrologic Modeling : Theory and Practice. Ed. Morel-Seytoux, Kluwer ac. Publ., Doordrecht,Boston, London; pp.3-25.

De Backer L.W.,1989 b): Land and Water Use : an iterative conservation process. Proceedings of the 11th International Congress on Agricultural Engineering, Dublin, Ed. Dodd and Grace,Pub. Balkema, Rotterdam pp. 241-253.

Foster S.S.D., 1987 : Fundamental Concepts in Aquifer Vulnerability, Pollution Risk and Protection Strategy. in Vulnerability of soil and groundwater to pollutants. Ed. W. van Duijvenbooden and H.G. van Waegeningh, TNO Committee on Hydrological Research. Proceedings and Information n° 38, The Hague, pp. 69-86.

Gontier E., A. Chabot, A. Navarre and L.W. De Backer, 1990 : Influence of Recharge Water on Groundwater and Aquifer Properties. in Groundwater and the Environment. Proceedings of the Second International Groundwater Conference IGW2-90, Kota Bahru, Ed. S. Awadalla, Univ. Kebangsaan Malaysia, pp.C22-C29.

Hartman L., 1987 : Methodological Guidelines for the Integrated Environmental Evaluation of Water Resources Development., UNESCO, Paris,152 p.

Jousma G., J. Bear, Y.Y. Haimes and F. Walter (Editors), 1989 : Groundwater contamination : Use of Models in Decision-making. Kluwer Acad. Publ., Dordrecht, 656 pp.

Mermoud A. et G. Gaillard, 1990 : Effets de la Formulation des Conditions aux Limites et de la Caractérisation hydrodynamique lors de la Simulation du Flux évaporatif isotherme en présence d'une nappe libre peu profonde. Proc. Intern. Conf. on Calibration and Reliability in Groundwater Modelling, The Hague, Vol. of poster papers, 241-249.

Mermoud A. et A. Musy, 1991 : Salinisation du Sol depuis une nappe peu profonde. Intern. Com. on Drainage and Irrigation (ICID). Proc. of the 42th meeting, Bejing, China (to be published).

Duijvenbooden, W. van , 1987 : Groundwater quality monitoring networks : Design and Results. in Vulnerability of soil and groundwater to pollutants. Ed. W. van Duijvenbooden and H.G. van Waegeningh, TNO Committee on Hydrological Research. Proceedings and Information n° 38, The Hague, pp.179-191.

Duijvenbooden, W. van and H.G. van Waegeningh, 1987 : Vulnerability of soil and groundwater to pollutants. TNO Committee on Hydrological Research. Proceedings and Information n° 38, The Hague, 1143 p.

Postel S., 1989 : Water for Agriculture : Facing the limits. Worldwatch Paper 93, Washington, 54 p.

Stiff H.A., 1951 : The interpretation of chemical water analysis by means of patterns. J. Petrol. Technol. 3, 15-17.

Strebel O., W.H.M. Duynisveld and J. Böttcher, 1989 : Nitrate Pollution of Groundwater in Western Europe. Agriculture, Ecosystems and Environment, 26 : 189-214.

Taube M.,1985 : Evolution of Matter and Energy, Springer-Verlag, N.Y.,289 p.

UNESCO,1986 : Pollution et Protection des Aquifères. Etudes et Rapports d'Hydrologie 30. Projet 8.3 du programme Hydrologique International, 434 p.

Weinberg G.M.,1975 : An Approach to General Systems Thinking, Wiley Intersciences, N.Y.

Wiechmann, H., 1987 : Criteria for Protection of Soil and Groundwater-Soil Aspects. in Vulnerability of soil and groundwater to pollutants. Ed. W. van Duijvenbooden and H.G. van Waegeningh, TNO Committee on Hydrological Research. Proceedings and Information n° 38, The Hague, pp. 111-124.

RELIABILITY IN REMOTE SENSING OF OCEANOGRAPHIC CHARACTERISTICS

J.P. Gastellu-Etchegorry
Centre d'Etude Spatiale des Rayonnements
Paul Sabatier University
31400 - Toulouse
France

Abstract: This paper introduces the notions of satellite platform, sensor and digital image processing, in the context of oceanographic and coastal applications with remotely sensed data. The most important interaction processes between water surfaces and electromagnetic waves are briefly reviewed. Major oceanographic parameters such as surface temperature and chlorophyll concentration that can be derived from satellite data are presented.

Keywords: Satellite , sensor, image processing, oceanography.

INTRODUCTION

Remote sensing, i.e. a technique for acquiring information about an object without being in physical contact with it, provides a means for synoptic and repetitive monitoring and mapping of earth resources. To a large extent, it consists of the measure of electromagnetic waves emitted and/or reflected from objects on the earth's surface. The object itself may be the immediate source of radiation, as in the case of emission of thermal radiation; or the surface where reflection processes take place, as in the case of reflected radiation. Remote sensing is said to be passive if the source of radiation is natural, e.g. measures of reflected sun/atmospheric radiations within the visible and infrared parts of the electromagnetic spectrum, and thermal emission in the infrared and microwave portions of the spectrum. Remote sensing is said to be active if the source of radiation is instrumental; e.g. radar remote sensing in the microwave part of the spectrum.

It must be kept in mind that satellite sensors do not measure directly caracteristics of Earth objects, but radiances that must be corrected for a

NATO ASI Series, Vol. G 29
Water Resources Engineering Risk Assessment
Edited by J. Ganoulis

number of atmospheric, radiometric, and geometric distortions; and transformed before deriving valuable information. The major hypothesis of remote sensing is that emitted and reflected radiations are modulated by a number of specific interaction processes, and consequently contain the "fingerprints" of the object. This allows one to infer the properties of the object from the characteristics of the measured radiation. In fact, specific portions of the spectrum, i.e. spectral bands, must be chosen with regard to the surfaces that must be studied, and also depending upon the conditions of observation.

The following is a brief introduction to the notions of platform, sensor and digital image processing. It is not a complete review of present platforms and sensors. The special emphasis on the SPOT satellite is intended to introduce the case study that is presented by B. Caussade (1991). This is followed by the presentation of some general digital processing approaches for satellite image analysis/interpretation (Martel and Flouzat, 1989; Martel, 1990). Finally, satellite remote sensing for oceanographic and coastal applications is introduced.

SATELLITE PLATFORMS AND SENSORS

Two major orbit types are available for satellite remote sensing missions: polar and geostationary orbits. The polar orbit, between 200 and 1000km altitude, takes the satellite close by the north and south poles. If the orbit is in the plane of the sun, it is sun synchronous. It means that the orbit remains in a constant plane relative to the sun while the Earth spins below. On each satellite orbit the Earth has rotated below the spacecraft about 30°, which makes possible the coverage of the Earth by successive orbits. On the other hand, geostationary satellites, also called weather satellites, are located at 35,900km altitude. They appear to be fixed above a well defined point on the equator. A major advantage is a very important revisit capability combined with a synoptic coverage.

The instruments on board the US satellites of the TIROS and Nimbus series series in the 60's, which offered a ground resolution of 1,000m, were the first to reveal the potential of space as a vantage point. Second generation remote sensing instruments, with a spatial resolution of 100m were developed during the seventies for the Landsat series of satellites. The images they returned enabled many users to become familiar with a new type of data and to develop the methods required for their interpretation. Spaceborne instruments with spatial resolutions of 30m and 10m were introduced in the 80's with the Landsat TM and SPOT HRV systems, respectively.

A wide variety of applications are available, depending on the mode of observation, e.g. the spectral range, type and viewing conditions of the sensor. Passive and active atmospheric sounders provide detailed profiles of atmospheric characteristics such as temperature and pressure. Large and medium scale multispectral imagers provide valuable cartographic and thematic information like repetitive global images of surface biomass cover and monitoring of anthropic and natural environmental changes. Spaceborn imaging spectrometers sensing from visible to far infrared wavelengths are being developed to acquire laboratory quality spectra, thus allowing direct identification of the surface or atmospheric composition. Active microwave sensors are for the survey, on a global basis, of the dynamics of the ocean: its topography, currents, near surface wind, polar ice,... Multispectral polarimetric imaging radars are intended to provide detailed maps of surface morphology and three dimensional maps of precipitation regions of the atmosphere.

SENSORS

Remote sensing observations are made by using a variety of detectors ranging from simple photographic films to complex devices such as spectrometers and multispectral scanners. Generally speaking, part of radiations emitted and/or reflected by an object are gathered by a collector, focused on a detector, and their properties measured.

Portion of the spectrum	Wavelength	Frequency	Platform
Visible	0.4 - 0.7 μm		NOAA, Landsat, SPOT
Near infrared	0.7 - 1.5 μm		NOAA, Landsat, SPOT
Middle infrared	1.5 - 3.0 μm		Landsat TM
Thermal infrared	8.5 -12.5 μm		NOAA, Meteosat
Microwave	10 - 300 mm	1 - 12.5 GHz	
X band	24 - 38 mm	8 - 12.5 GHz	ERS-1
C band	38 - 75 mm	4 - 8 GHz	Seasat
L band	150 - 300 mm	1 - 2 GHz	

Wavelengths and frequencies used in satellite remote sensing. Data acquisition in the visible and infrared is limited to atmospheric windows.

Precision photogrammetric technology was developed originally for aerial photographic missions. Such photographic camera systems once raised to the higher altitude of a satellite orbit allow a larger coverage, also called synoptic coverage, with a greatly improved geometric accuracy and a much better operational time repetitivity.

Scanning radiometers are the primary instrument for satellite remote sensing in the visible and infrared. These opto-mechanical systems consist mainly of a mirror to focus electromagnetic energy from the Earth onto one or several discrete detectors. The mirror oscillates to scan the surface below the satellite along a line perpendicular to the satellite track; i.e. the direction of the satellite orbit. The detector elements are placed behind filters that pass more or less broad portions of the spectrum in the visible, near infrared, and/or thermal infrared range. During the integration time, i.e. the time required by the scanner for crossing a scan line, incoming radiations are converted into voltages at the detectors. Each sample along the scan line corresponds to the residence time of the detectors in each instantaneous field of view (IFOV); it is referred to as a pixel. At the end of each scan the mirror views a set of calibration points commonly in the form of a grey wedge and other standard references. This allows calibration of the sensor signal (Harris, 1987). An image is constructed from a series of adjacent scan lines, each scan line comprising a line of pixels.

Major platforms with scanning radiometers comprise geostationary weather satellites such as Europe's Meteosat (1981-present), US GOES and Japan's GMS, but also sun synchronous satellites such as Nimbus, Landsat, and NOAA .

The Coastal Zone Colour Scanner (CZCS), on board the Nimbus 7 satellite (1978-1985), was designed specifically to measure the colour and temperature of the coastal zones of the ocean. It had a 1566km swath width and a 825m IFOV at nadir; nadir being the point directly below the satellite on the earth's surface. The first four bands are very narrow (0.02μm wide) and centered to enhance the discrimination of very subtle water reflectance differences. CZCS data were used to successfully detect chlorophyll (phytoplankton), inorganic suspended matter such as silt, temperature and gelbstoff (yellow substance) in the combination and concentrations typical of near-shore and coastal areas. These data allowed detection of ocean features such as pigment level variations in the range of 0.4 to 1mg.m^{-3}, and sometimes, because of the presence of particles and pigments, eddies and currents .

Band	Wavelength	Major applications
1	0.43 - 0.45 μm	Chlorophyll absorption (blue)
2	0.51 - 0.53 μm	Chlorophyll absorption (blue/green)
3	0.54 - 0.56 μm	Gelbstoff (yellow substance) (green)
4	0.66 - 0.68 μm	Chlorophyll concentration (red)
5	0.70 - 0.80 μm	Surface vegetation
6	10.5 - 12.5 μm	Surface temperature

Spectral wavebands of the Coastal Zone Colour Scanner. Bands 5 and 6 were used for masking clouds and land and for providing simultaneous thermal images. Band 4 was used for correcting atmospheric and surface effects.

The NOAA (National Oceanic and Atmospheric Administration) satellites originate from the Tiros series in the 60's. Since then the series is continuing. Thanks to their wide 2,400km swath each point on the Earth is covered twice a day; once during the day and once during the night. The Advanced Very High Resolution Radiometer (AVHRR), one of the major on-board sensors, operates in the visible (0.58-0.68μm), near infrared (0.725-1.1μm), short wave infrared (3.55-3.93μm), and thermal infrared (10.3-11.3μm and 11.5-12.5μm) portions of the spectrum. It is best suited to study areas which are relatively large. AVHRR data are used in meteorology and oceanography for the analysis of cloud patterns and sea surface temperatures. Sea surface temperatures as small as 0.2°C can be resolved. Moreover, monitoring of vegetation is now being done with this system.

The Landsat satellites, since 1972, have carried three principal sensor packages: the Return Beam Vidicon (RBV), the Multispectal Scanner (MSS), 185x185 km image with 79m pixel size, and the Thematic Mapper (TM). It is the TM that is now most widely used. It has a spatial resolution of 30m and operates in seven spectral bands. The primary emphasis of Landsat is for land applications. It has been widely used for geology, forestry, agriculture, and hydrology.

Band	Wavelength	Major applications
1	0.45 - 052 μm	Coastal water mapping - Soil/vegetation discrimination
2	0.52 - 0.60 μm	Green reflectance by healthy vegetation
3	0.63 - 0.69 μm	Chlorophyll absorption - plant species discrimination
4	0.76 - 0.90 μm	Biomass survey
5	1.55 - 1.75 μm	Vegetation moisture - snow/cloud discrimination
6	10.4 - 11.7 μm	Thermal mapping including plant stress
7	2.08 - 2.35 μm	Geological mapping

Wavebands of the Landsat Thematic Mapper (TM). Spatial resolution is 30m for bands 1-5 and 7, and 120m for band 6. Quantization range is one byte.

The primary limitation of scanning radiometers is linked to the short residence time of detectors in each IFOV for obtaining measurements with good spatial, spectral, and radiometric resolutions. One approach to increase the residence time of a detector in each IFOV is to use linear arrays of detector elements. In this configuration there is a dedicated pixel for each cross-track pixel. The French satellite SPOT (Chevrel et al., 1981) currently uses such line array detectors (see next paragraph).

Two other techniques are used to obtain a large number of wavebands. With the whiskbroom approach light from a pixel passes into a spectrometer where it is dispersed by wavelength and focused onto a line array. This approach is suited only to an airborne sensor that flies slowly enough for the readout time of the detector array to be a small fraction of the integration time (Vane and Goetz, 1988). This is the case of the Airborne Visible/Infrared Imaging Spectrometer (AVIRIS). The other approach uses two-dimensional area arrays of detectors, obviating the need for the optical scanning mechanism.

Radar remote sensing, specifically the Synthetic Aperture Radar (SAR), uses energy in the microwave portion of the spectrum. A major advantage of microwave use is their night and all weather capability. Microwaves are unattenuated by the atmosphere and clouds. Moreover, microwaves can be polarized, which adds another potential parameter for information extraction. The initial use of SAR was for oceanographic applications because of its ability to reveal surface and sub-surface ocean structures. It also has applications in ice and land studies, because of its all-weather capability. Imaging radars are at present in an experimental phase. Thus far only limited data from Seasat (1978) and the Shuttle Imaging Radars have been collected. ESA's ERS1 should be launched by mid 1991, whereas Japan's J-ERS1 and Canada's Radarsat should be launched in the near future.

THE SPOT SYSTEM

SPOT satellites have been orbiting the Earth since 1986. The orbital altitude and inclination of the orbital plane of these satellites were chosen so as to provide a circular sun-synchronous, near polar orbit (inclination: 98.8°) at an altitude of 832km. The orbital plane rotates around the Earth's axis, completing one revolution per year. The angle between the orbital plane and the sun direction is thus constant throughout the year. Thus, for any given region of the Earth the satellite always passes overhead at the same solar time, the actual time being determined solely by latitude. The SPOT satellite crosses the descending node, or point of intersection of its south-going track with the equator, at about 10.30 am, and at 10.30 am precisely on June, 15 of each year. The satellite's motion is also synchronized with the daily rotation of the Earth in such a way that the pattern of successive ground tracks is repeated exactly at 26 day intervals. It is thus possible to observe a given region of the Earth at regular intervals under the same viewing angles.

A payload and a standard multipurpose platform constitute the main parts of the SPOT satellite. The payload is comprised of the earth observation instruments, i.e. two identical high resolution visible (HRV) instruments, two magnetic tape recorders, and the mission telemetry package. The image data bit rate is 25Mbits per sec per instrument.

The HRV instruments can be activated independently. They are designed to operate in either of two modes, the visible and near infrared portions of the electromagnetic spectrum:

- a panchromatic mode for observation over a broad portion of the spectrum with a 10m pixel size (spatial resolution),
- a multispectral mode for observation in three narrower spectral bands, with a 20m pixel size (spatial resolution).

The detectors are of the CCD (Charge Coupled Device) type. Each array consists of 6000 detectors arranged linearly. This forms a "push-broom" scanner since it images a complete line of the ground scene in the cross-track direction in one "look" without any mechanical scanning. Each HRV has a 4°13' FOV (Field of View); i.e. a swathwidth of 60km (Fig. 1). When the two instruments operate in adjacent covering field, the ground coverage is 117km, and the two HRV overlap by 3km. Since the distance beween adjacent ground tracks at the equator is approximately 108km, complete Earth coverage may be obtained with this fixed setting of instrument fields. One of the key features of SPOT is off-nadir viewing; i.e. the instrument can look to one side or the other of the satellite track. Light from the scene being viewed enters the HRV instrument via a plane mirror that is steerable by ground control. The mirror allows one to adjust the viewing axis as required in the plane perpendicular to the orbit. This off-nadir viewing capability covers a range of -27° relative to the vertical, in 45 steps of 0.6° each. This allows the instrument to image any point within a strip extending 475km to either side of the satellite ground track. Because of the Earth's curvature, the maximum viewing angle at the ground is 33° from vertical. The width of the swath actually observed varies from 60km for nadir viewing upto 80km for extreme off-nadir viewing.

SPOT Scene	Multispectral mode	Panchromatic mode
Spectral bands	0.50 - 0.59 μm 0.61 - 0.68 μm 0.79 - 0.89 μm	0.51 - 0.73 μm
Pixel size (nadir)	20 m x 20 m	10 m x 10 m
Pixel coding format	3x8 bits	6 bits DPCM
Number of pixels per line	3 x (3000 upto 5200)	6000 upto 10400
Number of lines per scene	3 x (3000 upto 4900)	6000 upto 9800
Data volume	27 upto 76.5 Mb	36 upto 100 Mb
Scene coverage	60 km x (60 upto 85 km)	

DPCM (Digital Pulse Code Modulation) is a mode of data compression that does not degrade the radiometric accuracy of the image data (256 levels). On the SPOT 4 system there will be an additional short wave infrared band.

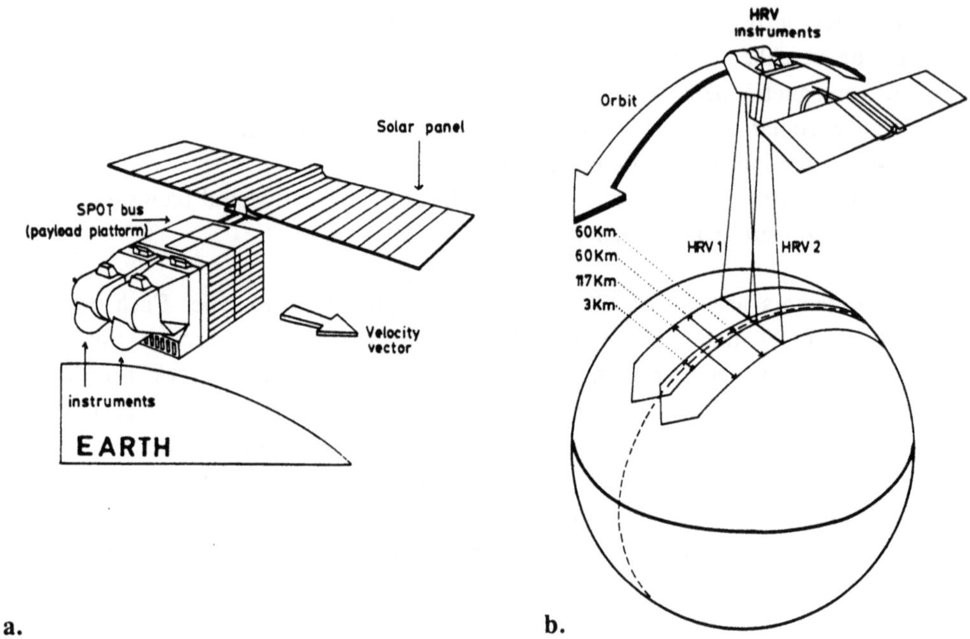

Figure 1: (a) The SPOT satellite. (b) Nadir viewing

Off-nadir (Fig. 2) viewing leads to both an increased revisit coverage and the acquisition of stereoscopic pairs of images of any Earth surface. Were the satellite instruments only capable of nadir viewing the revisit frequency would be 26 days. This interval is unacceptable for the observation of phenomena evolving on timescales ranging from several days to a few weeks, especially where cloud cover hinders the acquisition of usable data. The off-nadir capability of SPOT greatly improves the revisit coverage. Any area located on the equator can be targeted 7 times during the 26-day period separating two successive satellite passes over this area; i.e. 98 times in one year with an average revisit of 3.7 days. Areas at higher/lower latitudes are more frequently observed. At 45°, the average revisit is only 2.4 days with a time lag between 1 and 4 days.

Stereoscopic viewing is the other major possibility due to off-nadir capability. Images can be recorded at different viewing angles during successive satellite passes in the vicinity of the observed area. A stereoscopic pair comprising an image acquired at -24° (HRV looking east) and another acquired at +24° (HRV looking west), has a base-height ratio equal to one. This is particularly interesting for photogrammetry, geology, geomorphology, etc.

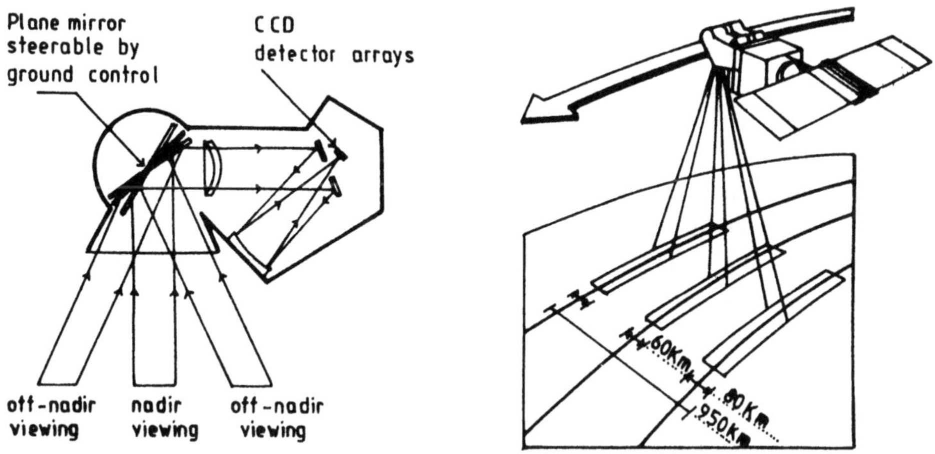

a. b.

Figure 2: (a) HRV instrument. (b) Extreme oblique viewings as compared to nadir.

SPOT data are widely used for cartography, agriculture, urbanism, forestry, coastal management, etc. applications. Highlights of this system are the good spatial resolution (especially the 10 m resolution), the revisit capability, and the option of stereoscopic viewing.

SATELLITE DATA PROCESSING

Basically satellite data are acquired as digital values. Only at a later stage are the digital values arranged in two dimensional image form and plotted as a picture. Consequently all initial processing for image corrections and enhancements are digital in nature. It must be noted that processing involves data volumes that are very large by current standards. For instance, a 170 x185km Landsat TM scene contains 273Mbytes of data; i.e. seven 1,600 bits per inch magnetic tapes. This induces important constraints on the equipment that are used for archiving and processing such large amounts of data.

Pre-processing

Generally speaking, the main objectives of computer processing of satellite images are to improve display of image data to the analyst, both from the geometric and color aspects, and to facilitate evaluation of the multispectral characteristics of the data. Data interpretations are often made from enhanced and/or classified images by an analyst trained in the field of concern; e.g. agriculture or oceanography. Digital image data, when first acquired from a satellite, contain a number of distortions that must be corrected before image interpretation. Such processing may be radiometric and/or geometric. Generally speaking, satellite data can be acquired raw or with a given pre-processing level. Basic pre-processing corresponds to the simple equalization of the radiometric response of detectors.

Geometric distortions originate from variations in the Earth's rotation, variations in the velocity, altitude, pitch, roll or yaw of the satellite platform. Geometric corrections allow one to registrate satellite data to a map projection, to other satellite data, or data from other sources. Corrections are conducted through modeling of the causes of distortion, and/or the use of Ground Control Points (GCP). Different precision levels and projection systems are possible. Geometric corrections comprise geometric transformation of pixel coordinates and resampling processes such as the nearest neighbour, bilinear interpolation, and cubic convolution.

Radiometric transformations are often applied for deriving physical parameters such as reflectance and temperature values. For visible and infrared remote sensing, such transformations may allow one to compensate for variable solar illumination and for atmospheric effects. These transformations must be used whenever it is intended to compare data acquired under different conditions. The sun elevation angle adjustment is made by dividing all brightness values in a scene by a constant that is the cosine of the incidence angle; surfaces are assumed to be Lambertian. This does not remove topographic effects and difference in solar illumination caused by changes in the azimuth of illumination. Adjustments for the atmosphere are difficult to determine because atmospheric effects have strong temporal, spatial, and spectral variations. Methods that rely uniquely on the satellite image at hand are often based on the lowest brightness level per waveband, on the presence of shadows, or on the presence of Earth objects that are partly illuminated and shadowed (Gastellu-Etchegorry, 1988). Usually, radiometric corrections rely on ancillary information about direct sun radiations and in-situ measures of temperatures.

Image enhancement

Image enhancement (Lillesand and Kiefer, 1987) is often a major step before image interpretation. This approach involves the following techniques:

- Contrast enhancement. This is done by assigning new brightness values to each pixel in the scene in order to enhance features of interest. Brightness tranformations may be linear or non linear. This results in the use of the total range of grey/color tones that are available with the display system. Brightness value parameters are usually determined by an analyst who determines what data are to be lost, and reassigned to maximum and minimum values on the output image. The analyst does this by examining image histograms or by interactively determining the truncation limits. Automatic truncation limits are sometimes determined by using a rule that a certain percentage of the data are to be truncated on either end of the histogram.

- Edge enhancement. Unenhanced satellite images can have subtle brightness variations that are difficult to detect. These variations may be related to variations in the illumination of topography or in turbidity levels on water surfaces. Edge enhancements are used to enhance radiometric patterns which have a certain spatial frequency in the image. For example, when a low frequency filter is employed, only major "edges/boundaries" are detected. A low frequency filter examines the average brightness value of a large number of pixels surrounding the pixel to be enhanced. A high frequency filter examines the average brightness value of a small number of pixels surrounding the pixel to be enhanced. The image enhancement techniques may have the effect of producing sharper images, but it may introduce artifacts into the data, such as the introduction of shadows adjacent to features which have abrupt changes in brightness values.

- Ratio enhancement. This technique is based on division of the brightness value in one band by the brightness value in another band, for each pixel within the scene. In fact, the numerator and denominator of the ratio may be linear combinations of brightness values of the same pixel in different bands; various forms of temporal ratioing may be also employed. The ratioed values are usually multiplied by a factor from a look-up table so that the contrast is optimized. Ratioing tends to reduce the effects due to topography and to emphasize the changes in brightness values between materials. Some band ratios have specific applications. Such the case is for the vegetation index, i.e. [(Near infrared - Red)/(Near infrared + Red)]; it gives high values for healthy green vegetation and low values for bare areas.

- Color enhancement. Various techniques exist for displaying several bands of a multispectral digital image on a single monitor image. They are often referred to as color composites. There are several common display modes (false color, natural color,..), depending on the "waveband - color screen" relationship. For example, SPOT derived color composites are obtained by assigning the near infrared band to the red colour, the red band to the green colour, and the green band to the blue colour. SPOT pseudo natural colour composites result from the creation of a simulated "blue" band from the 3 other bands.

- Principal component analysis. This approach involves finding mathematically a new set of pseudo bands which describe the variance of the original set of wavebands, but which are independent of each other and are caracterized by decreasing percentages of total variance. The major objective of principal component analysis is to reduce the original volume of data with a minimum loss of information, whenever variance can be regarded as an index of the level of information.

Image classification

Image classification (Fig. 3) is comprised of a dual approach: (1) multi-spectral analysis, i.e. the machine assisted delineation of multi-spectral patterns within the satellite image, and (2) multi-spectral classification, i.e. the identification of machine delineated multi-spectral patterns that may represent Earth features such as landscape cover patterns. There are two major classification processes (Lillesand and Kiefer, 1987); the supervised and unsupervised classifications.

Supervised classification is where the analyst "trains" the analytical processor by selecting samples of classes to be recognized. General rules guide the choice of samples. Whenever possible the sample pixels should be randomly chosen in a sufficient number. Ideally sample data have a multivariate normal distribution.

Figure 3: Classification of a SPOT image of Wonogiri reservoir, Central Java, Indonesia. Colors are for spectral classes. A number of classes correspond to different water turbidity levels. Since 1981 to 1986 water surface decreased by 35% due to silting-up (Gastellu-Etchegorry, 1988).

Depending on constraints such as computation time, accuracy required, and type of data different classification rules may be employed. Some are mentioned below.

- Minimum distance to means classifier: it is based on the calculation of the distance between the value of each pixel and each of the class means.
- Parallelepiped classification: once the multispectral limits of classes are computed, either interactively or automatically, the computer searches each pixel and determines which pixels have brightness values that fall within these extrema. The pixels which occur within these restricted ranges of brightness are classified; they may be identified by a specific colour code on imagery.
- Maximum likelihood classification: this approach relies on computation of the probability of each point in the feature space to belong to each class identified. The class to which a point belongs is the class for which it has the maximum probability. Probabilities are usually computed according to the normal law by considering the spectral mean and standard deviation of each class.

Unsupervised classification algorithms are employed whenever classes are not really identified. They rely on cluster analysis processes to determine "natural" groupings of multi-spectral data.

Texture analysis

Texture analysis relies on spatial information. Image texture varies from place to place and it is this variation that enables us to use texture as an indicator of image features. Texture indices are computed through convolution of the multispectral image with filters and may indicate the degree of homogeneity. They may be computed with filters that correspond to a more or less large neighborhood. Texture statistics and texture-derived pseudo-bands can be used for classification processes.

Techniques for satellite image interpretation are diverse, ranging from conventional photo-interpretation to the use of a host of electro-optical devices and computers. Image interpretation/analysis can be conducted under two general approaches, which are often combined: i.e. the physical and statistical approaches. Measurements can be used to derive physical parameters such as the reflectance and/or temperature, which may lead to the recognition and analysis of objects. On the other hand, one can attempt to remotely sense any materials under many conditions and hence derive, by empirical relationships, a basis for interpreting remote sensing observations of a poorly known area. A number of statistical techniques can be applied to quantify the empirical relations and to automate some of the analysis procedure using computers to "recognise" similar relationships. A major setback for this statistical approach is that we are dealing with nature, with a seemingly infinite number of variables. When one considers the wide variety of materials, natural settings, and climatic conditions, results obtained at one area under one set of conditions may significantly differ from those obtained under a different set of conditions.

OCEANOGRAPHIC SATELLITE REMOTE SENSING

Visible - infrared wavebands

Any sun radiation that interacts with water surfaces is the sum of direct sun radiations and atmosphere scattered radiations. Oceanographic applications of remote sensing take advantage of reflection, refraction, absorption, and scattering processes which take place at water surfaces and sub-surfaces. Consequently, ocean surface conditions play an important role. For instance, depending on sun elevation, there is little or no specular reflectances if water surfaces are rough, which occurs in presence of strong winds.

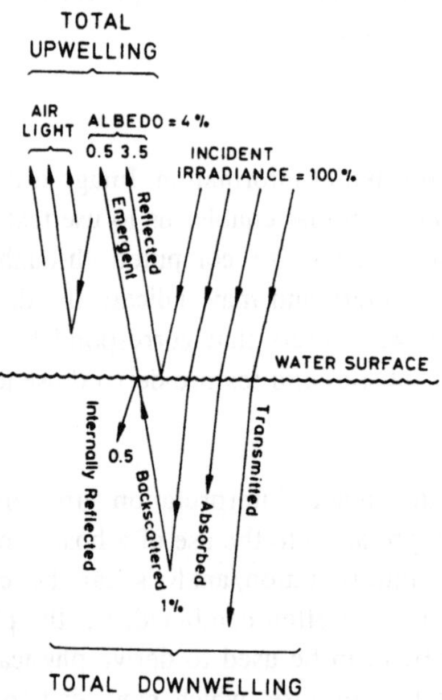

Figure 4: Schematic representation of the flux of the incident radiation in the air above the sea surface, at the surface itself, and beneath the surface (Szekielda, 1988).

Radiations that penetrate water bodies (Fig. 4) are refracted, and then absorbed or scattered by water particles. Most of the energy of radiations is transferred as heat to water molecules in the mix layer. Only a small portion of the incoming radiation is absorbed by pigments; part of it is re-emitted at other wavelengths (fluorescence). Finally, radiations that escape water surfaces make up the radiations that are remotely measured. Organic bodies (phytoplankton and yellow substances) induce strong absorptions at short wavelengths, whereas the importance of scattering processes increases with the concentration of mineral elements in suspension (Klemas and Hardinsky, 1987). In theory, the knowledge of the spectral signatures of water substances allows one to derive their concentrations if we have a sufficient number of spectral measurements.

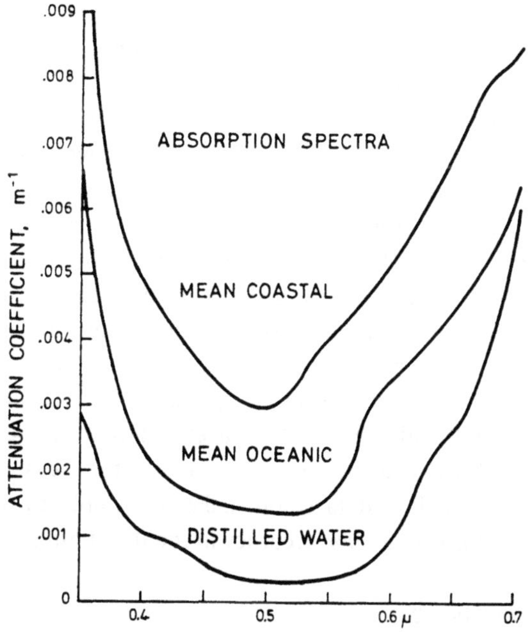

Figure 5: Absorption spectra of distilled, open ocean, and near coastal water (Szkielda, 1988).

Generally speaking, there is a strong dependence of absorption and diffusion processes, and consequently of water penetration depth, on wavelength, and on the type and volume density of material that is in suspension or dissolved in water (Fig. 5). Scattering processes are often expressed as λ^{-n}, n beween 0 and 4. Their intensity increases with material concentration. Absorption processes in the visible and near infrared are strongly related to the presence of photosynthetically active pigments such as chlorophyll and carotenoids. There are characteristic absorption bands in blue and red. This often explains the yellow/brown sea water. It must be noted that mixtures such as "algae + water" correspond to suspension elements and not chlorophyll suspensions. Moreover, planktons are not the only elements within water. The following table illustrates the spectral variations of penetration depth in the case of "pure water" over a perfectly reflecting body ($\rho = 100\%$) at 1m and 5m depth:

Depth	0.5-0.6μm	0.6-0.7μm	0.7-0.8μm	0.8-1.1μm
1m	$\rho = 86\%$	$\rho = 55\%$	$\rho = 11\%$	$\rho = 0.2\%$
5m	$\rho = 50\%$	$\rho = 7\%$	$\rho = 0\%$	$\rho = 0\%$

Spectral reflectance values of a perfectly reflecting object that is covered by 1m and 5m water depths. Bathymetry must be conducted with short wavelengths.

Because oceans are usually only sligthly reflective, it is very important for oceanic applications to take into account, and whenever possible to correct, atmospheric effects, i.e. aerosol and gas scattering and gas absorption such as ozone absorption in the upper atmosphere at short wavelengths. Aerosol scattering adds a signal which is highly variable in intensity, but has a smooth spectrum which can be reasonably approximated by a power law. For example, CZCS band 4 was used for determining this power law. Indeed, at 670nm the water reflectance becomes very small, and the remaining signal after Rayleigh correction, i.e. correction of Rayleigh scattering by atmospheric gases, can be used as a measure of the varying aerosol signal in the scene at this wavelength. Extrapolation to other wavebands, however, requires a knowledge of the power law spectrum, which can be provided with contiguous wavebands around 670nm. Errors exist wherever aerosol properties change, and where suspended material raises the water leaving radiance at the waveband that is used to map the aerosol contribution. Various evaluations showed that for areas with good clear water reference areas satellite data could make it possible to obtain pigment concentrations estimates with a 30% accuracy in the 0 to 1.5mg.m^{-3}.

Thermal infrared wavebands

As mentioned previously, measured radiances result from several components: atmospheric radiations, and atmospheric and sun radiations that have interacted with water surfaces. Atmospheric attenuation results mainly from absorption by gas molecules, especially water vapor, and aerosol scattering. Outside some portions of the spectrum, called atmospheric windows, atmospheric absorption is considered total. Atmospheric effects can induce errors larger than several Kelvin degrees for sea surface temperatures. Generally speaking, water emissivity is assumed to be independent of water temperature, wavelength, and water roughness as long as wind strength is low (Cassanet et al., 1981). It must be noted that temperatures that are measured correspond to the ocean top layer of less than 1mm depth. Thermal infrared remote sensing allows one to detect upwellings, eddies, cold and warm currents, and thermal fronts.

Microwaves

A major interest of microwave remote sensing is its all weather and night capability. Atmospheric attenuation can be neglected; water vapor absorption becomes significant for frequencies larger than 10GHz. . Water penetration of microwaves is much less than with visible radiations. With a 20°C temperature and a 3.6% salinity it is defined as 1cm at 1GHz (λ=30cm) and 0.1cm at 10GHz (Ulaby et al., 1986). Passive microwave radiometers allow one to derive surface temperature measures, salinity, wind surface, and atmospheric water vapor content. Because at these wavelengths, the energy of earth and sun radiations are very low, sensors tend to have large IFOVs. This is not necessarily the case of active microwave sensors that measure the backscattered radiations they emit, e.g. the altimeter leads to a quantitative determination of sea levels, the scatterometer provides wind intensity, through the evaluation of surface roughness (upto 50cm accuracy for wave heights, 1m/s for wind speed, and 20° for its direction), the imaging radar maps surface irregularities, currents, fronts,.etc.

CONCLUDING REMARKS

Previous remarks stressed the capability of satellite remote sensing for providing a synoptic and repetitive overview of oceanographic features and processes, both for physical and biological oceanography. The different satellite platforms and sensors give complementary information about water sub-surface and surface characteristics. For instance it is possible to characterize material in suspension, and to determine water movement, through the spatial distribution of concentrations. It must be noted that satellite data with high spatial resolution are required for coastal applications. Undoubtedly, the combination of conventionally gathered oceanographic observations and satellite data complement each other and should reveal a fuller perspective of oceanographic processes than either of them is individually capable of providing.

Acknowledgements: The author is indebted to G. Flouzat and C. Martel, from the Centre d'Etude des Rayonnements, Toulouse, France, and B. Caussade and S. Ouillon, from the Institut de Mécanique des Fluides, Toulouse, France, for providing valuable information for this paper.

References

Cassanet J., Thomas Y.F., Yoo H.R. and Zbinden R., 1981, Etude par télédétection de la température et de la turbidité des eaux côtières (Gironde, estuaire de la Loire, Manche occidentale et centrale). Rapport d'activité contrat CNEXO 80-6261, Ecole Normale Supérieure, Laboratoire de Géographie, Montrouge, 200 pages.

Caussade B., 1991, Numeric modeling and remote sensing contributions to identify estuarine pollution - Uncertainties and complementarity of these techniques applied to the case of Gambia, Risk & reliability in water resources & environmental engineering, Porto Carras, 18-28 may 1991.

Chevrel M., Courtois M. and Weill G., 1981, The SPOT satellite remote sensing mission, Photog. Eng. and Remote Sensing, Vol 47, No8.

Elachi C., 1987, Introduction to the Physics and Techniques of Remote Sensing, JohnWiley & Sons.

Gastellu-Etchegorry J.P., Remote Sensing with SPOT, An assessment of SPOT capability in Indonesia, Gama Press, Yogyakarta.

Harris R., 1987, Satellite remote sensing, an introduction, Routledge & Kegan Paul Ltd.

Klemas V. and Hardinsky M.A., 1987, Remote sensing of estuaries: an overview. Twenty first international symposium on remote sensing of environment, Ann Arbor, Michigan, October 26-30, 1987, 183-203.

Lillesand T.M. and Kiefer R.W., Remote sensing and image interpretation, John Wiley & Sons.

Martel C., 1990, Analyse géométrique et physique d'images d'observation de la terre. Eléments d'hydrodynamique côtière. PHd Université Paul Sabatier.

Martel C. and Flouzat G., 1989, Etude du panache des matières en suspension à l'estuaire de la Gambie : méthodologie de segmentationpour des images représentant des phénomènes à faible gradient, S.F.P.T., Bul. 114, 4-6.

Szekielda K.H., 1988, Satellite monitoring of the Earth, John Wiley & Sons.

Ulaby F.T., Moore R.K. and Fung A.K., 1986, Microwve remote sensing Vol 3, (Dedham: artech house), 2162 pages.

Vane G. and Goetz A.F.H., 1988, Terrestrial imaging spectroscopy, Remote Sensing of Environment, 24:1-29.

ON THE SAMPLING DISTRIBUTION OF EXCEEDANCE PROBABILITIES

Peter Funder Rasmussen
Institute of Hydrodynamics and Hydraulic Engineering
Technical University of Denmark
DK-2800 Lyngby

ABSTRACT

Using a partial duration series model as an example the impact of parameter uncertainty on the true exceedance probability is examined . Analytical expressions for the probability density function and the mean value of the true exceedance probability are derived. The true exceedance probability is positively biased implying that the traditional design method on the average tends to be too optimistic.

INTRODUCTION

In flood frequency analysis we are concerned with the relationship between water discharges of certain magnitudes and the corresponding exceedance probabilities. Hydrologists frequently encounter the situation where design standards require that the probability by which the annual maximum discharge exceeds the design flood should be no greater than, say, p. By means of a statistical analysis of observed discharge data we may attempt to determine the design flood, z_p, which fulfils the above requirement. Our problem is that we do not know the true statistical nature of floods, since, in practice, the statistical analysis must be based on relatively short data series. This introduces model errors and sampling errors. Consequently, we do neither know the true exceedance probability of our design flood estimate, $\hat{z}_p$ (nor may we live long enough to find out). Most design procedures emphasize the optimality of the design flood estimator, $\hat{Z}_p$, for instance with respect to biasedness and variance. It is, however, also relevant to consider the statistical properties of the true exceedance probability. Possible bias of the exceedance probability should in some way or other be reflected in the design procedure, since as nature is unknown we can only work in expected terms. In the following we shall discuss the distribution and bias of the true exceedance probability corresponding to certain design flood estimators. The emphasis is on the impact of parameter uncertainty, whereas it is assumed that the correct model can be

NATO ASI Series, Vol. G 29
Water Resources Engineering Risk Assessment
Edited by J. Ganoulis

specified. Reference is given to Rasmussen and Rosbjerg (1989) and Rasmussen (1991).

SAMPLING PROPERTIES OF THE EXCEEDANCE PROBABILITY

Our point of origin is the partial duration series approach, which today is a well-established method for flood frequency analysis. If we assume that all exceedances of the threshold discharge are identically exponentially distibuted with the mean value α, and that they are independent of each other and of the time at which they occur, the distribution of the annual maximum flood is

$$F_Z(z) = \exp(-\lambda\, e^{-\frac{z}{\alpha}}) \qquad z > 0 \tag{1}$$

where λ denotes the average number of exceedances per year. This cumulative distribution function (CDF) to some degree resembles the Gumbel distribution. Note, however, that it is bounded below at $z=0$ where it has a discrete probability mass. As we are interested in the upper, right tail, the truncation has no practical implications. Moreover, the CDF given by (1) is exact, contrary to the Gumbel distribution which is an asymptotic distribution. The level which is exceeded with probability p is

$$p = 1 - F_Z(z_p) \quad \leftrightarrow \quad z_p = F_Z^{-1}(1-p) = \alpha \ln\left(-\frac{\lambda}{\ln(1-p)}\right) \tag{2}$$

If y_p is introduced as the (1-p)-quantile in the standardized Gumbel distribution, we can express z_p simply as

$$z_p = \alpha\,(y_p + \ln\lambda) \tag{3}$$

where $y_p=-\ln(-\ln(1-p))$. In practice, it is necessary to replace the population parameters α and λ with their estimated values in order to determine the relationship between level and exceedance probability. Let N denote the number of exceedances observed in a period of t years. The maximum likelihood estimator of α is the average exceedance magnitude, i.e. $\hat{\alpha}=S_N/N$, where S_N denotes the sum of the observed exceedances. The parameter, λ, is estimated as $\hat{\lambda}=N/t$. Insertion in (3) gives

$$\hat{z}_p = \frac{S_N}{N}\left(y_p + \ln\frac{N}{t}\right) \tag{4}$$

Clearly, the exceedance probability of $\hat{z}_p$ given by (4) is not p as desired, since the population parameters have been replaced by their estimates. If we actually knew the population parameters, we could determine the exceedance probability of $\hat{z}_p$ as

$$p_{tr} = 1 - F_Z(\hat{z}_p) = 1 - \exp\left(-\lambda \exp\left\{-\frac{1}{\alpha}\frac{S_N}{N}\ln(y_p + \ln\frac{N}{t})\right\}\right) \qquad 0<p_{tr}<1-e^{-\lambda} \tag{5}$$

The upper restriction of p_{tr} corresponds to the situation where z_p has been estimated to zero, which happens with zero probability. Note that $1\text{-}e^{-\lambda}$ is the probability that the threshold discharge is exceeded in a single year. In practice, we are prevented from calculating p_{tr} because we do not know α and λ. We may, however, deduce its sampling properties as shown below. In (5), S_N and N were interpreted as the result of one single experiment (i.e. data record). Henceforth we shall consider S_N and N as random variables. If we assume that exceedances occur according to a Poisson-process, then the variable N is Poisson-distributed with mean λt. For a fixed N the variable S_N, being a sum of N exponentially distributed random variables, is Gamma-distributed with parameters N and α (since the shape parameter is an integer, this distribution is sometimes referred to as the Erlang distribution). The knowledge of the distributions of S_N and N allows us to determine the distribution of p_{tr} (as a function of the population parameters)

$$\begin{aligned} F_{p_{tr}}(\delta) &= P\{p_{tr} \le \delta\} \\ &= \sum_{n=1}^{\infty} P\left\{1-\exp\left(-\lambda \exp\left[-\frac{1}{\alpha}\frac{S_n}{n}(y_p+\ln\frac{n}{t})\right]\right) \le \delta \ \cap \ N=n\right\} \\ &= \sum_{n=1}^{\infty}\left[1-P\left\{S_n < \alpha n\,\frac{y_\delta+\ln\lambda}{y_p+\ln(n/t)}\right\}\right] P_c\{N=n\} \\ &= 1 - \sum_{n=1}^{\infty} F_{S_n}\left(\alpha n\,\frac{y_\delta+\ln\lambda}{y_p+\ln(n/t)}\right)\frac{(\lambda t)^n}{n!\,(e^{\lambda t}-1)} \end{aligned} \tag{6}$$

where $y_\delta = -\ln(-\ln(1-\delta))$. P_c denotes the truncated Poisson distribution (i.e. the case that N=0 is omitted), since in practice at least one exceedance is needed to perform a statistical analysis. We are especially interested in the probability density function (PDF) of p_{tr}, which is obtained by differentiation of the CDF

$$f_{p_{tr}}(\delta) = F'_{p_{tr}}(\delta) = -\sum_{n=1}^{\infty} f_{S_n}\left(\alpha n \frac{y_\delta + \ln\lambda}{y_p + \ln(n/t)}\right) \alpha n \frac{y'_\delta}{y_p + \ln(n/t)} \frac{(\lambda t)^n}{n!\,(e^{\lambda t}-1)} \qquad (7)$$

where y'_δ denotes the derivative of y_δ with respect to δ. It is readily verified that $y'_\delta = [(1-\delta)\ln(1-\delta)]^{-1}$. Upon introduction of the Gamma density we eventually get the PDF of p_{tr}

$$f_{p_{tr}}(\delta) = -\frac{1}{e^{\lambda t}-1} \sum_{n=1}^{\infty} \frac{\exp\left(-n \frac{y_\delta + \ln\lambda}{y_p + \ln(n/t)}\right) (\lambda t)^n}{(n-1)!\,n!\,(y_\delta + \ln\lambda)\,(1-\delta)\ln(1-\delta)} \left[n \frac{y_\delta + \ln\lambda}{y_p + \ln(n/t)}\right]^n \qquad (8)$$

$$0 < \delta < 1 - e^{-\lambda}$$

Despite the complexity of this expression the PDF of p_{tr} can easily be computed by means of appropriate computer support. Note that similar derivations can be used to obtain the PDF of $\hat{Z}_p$ in cases where the properties of this estimator is of interest. The parameter α disappeared in (8), which seems reasonable since it is only a scaling

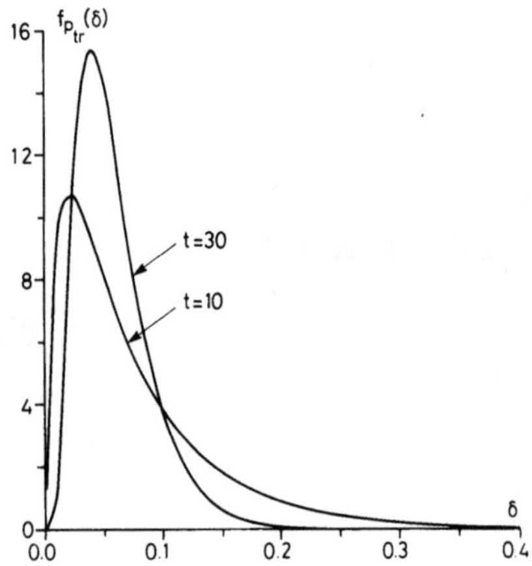

Figure 1
PDF of p_{tr}. $p=0.05$, $\lambda=2$.

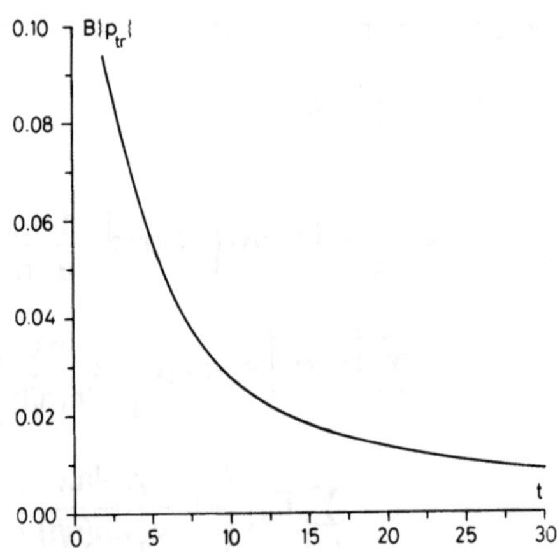

Figure 2
Bias of p_{tr}. $p=0.05$, $\lambda=2$.

factor. Examples of (8) are given in Figure 1. The shape of the PDF clearly indicates that major mistakes can be made if one assumes that the exceedance probability of the design flood estimate equals the design exceedance probability, *p*. It can be shown that the design flood estimator is approximately normally distributed with the mean z_p. This implies that there is about 50 percent probability that p_{tr} is greater than p and 50 percent probability that it is less than p. However, due to the non-linear relationship between level and exceedance probability, the distribution of p_{tr} is skewed. The skewness is particularly pronounced for small values of λt. We are especially interested in examining the mean value of p_{tr} corresponding to the estimator, $\hat{Z}_p$. An analytical expression for $E\{p_{tr}\}$ can be obtained by solving

$$
\begin{aligned}
E\{p_{tr}\} &= \int_0^{1-e^{-\lambda}} \delta f_{p_{tr}}(\delta)\,d\delta \\
&= \int_0^{1-e^{-\lambda}} \delta \sum_{n=1}^{\infty} -f_{S_n}\left(\alpha n \frac{y_\delta + \ln\lambda}{y_p + \ln(n/t)}\right) \alpha n \frac{y_\delta'}{y_p + \ln(n/t)} P_c\{N=n\}\,d\delta
\end{aligned}
\tag{9}
$$

We introduce the substitution

$$
\begin{aligned}
u &= n \frac{y_\delta + \ln\lambda}{y_p + \ln(n/t)} \quad \leftrightarrow \quad \delta = -\sum_{k=1}^{\infty} \frac{(-\lambda)^k}{k!} \exp\left(-u \frac{k}{n} \left\{y_p + \ln\frac{n}{t}\right\}\right) \\
\frac{du}{d\delta} &= n \frac{y_\delta'}{y_p + \ln(n/t)} \quad < 0
\end{aligned}
\tag{10}
$$

which upon insertion gives

$$
\begin{aligned}
E\{p_{tr}\} &= -\int_0^{\infty} \sum_{k=1}^{\infty} \frac{(-\lambda)^k}{k!} \exp\left(-u \frac{k}{n} \{y_p + \ln(n/t)\}\right) \sum_{n=1}^{\infty} \frac{e^{-u} u^{n-1}}{(n-1)!} P_c\{N=n\}\,du \\
&= -\frac{1}{e^{\lambda t}-1} \sum_{n=1}^{\infty} \sum_{k=1}^{\infty} \frac{(-\lambda)^k \; (\lambda t)^n}{k!\, n! \left[\frac{k}{n}\{y_p + \ln(n/t)\} + 1\right]^n}
\end{aligned}
\tag{11}
$$

The last expression can be found by reversing the order of integration and summation. The expected value of p_{tr} is illustrated in Figure 2. Clearly, p_{tr} is positively biased, and particularly when viewed at a relative scale the bias is quite large.

DISCUSSION

How are we to interpret the bias of p_{tr}? If a large number of hydraulic structures are designed to resist an annual flood with exceedance probability p, then more than 100p% of the structures are likely to fail each year due to the positive bias of the true exceedance probability. It seems that what the hydrologist is looking for is an estimator, $\hat{z}_p^*$, with the property that $P\{Z > \hat{z}_p^*\} = p$, where we consider the randomness of both the largest annual flood, Z, and the estimator, $\hat{z}_p^*$. This is somewhat in contrast to the classical approach, where we use the estimator $\hat{z}_p$ for z_p, the latter being defined by $P\{Z > z_p\} = p$. The probability that Z exceeds $\hat{z}_p$ is equal to the expected value of the true exceedance probability. One way to obtain a design flood estimator with corresponding unbiased probability is to use equation (11) to determine the value of y_p which corresponds to the requirement

$$E\{p_{tr}\} = p \tag{12}$$

Using this value of y_p in equation (4) instead of the previously defined value results in a probability-unbiased design flood estimator. Equation (11) cannot be solved analytically for y_p, so one has to use an iterative method to determine y_p. In practice, it is necessary to use the estimated value of λ in (11), but the formula is relatively robust to variations in $\hat{\lambda}$, so that we obtain an *approximate* probability-unbiasedness. It is readily acknowledged that the above procedure leads to an expected overdesign, since y_p determined by (11) always exceeds the classical y_p. This is, however, a necessary consequence if we pursue an unbiased exceedance probability.

REFERENCES

Rasmussen, P.F. (1991) The partial duration series approach to flood frequency analysis. Series paper no. 55. Institute of Hydrodynamics and Hydraulic Engineering, Technical University of Denmark

Rasmussen, P.F. and D. Rosbjerg (1989) Risk estimation in partial duration series. Water Resour. Res., 25(11), 2319-2330

THE CONTRIBUTION OF LONG-TERM RECORDS OF HYDROLOGIC EXTREMES TO RISK ANALYSES

Yehouda Enzel[1], Lisa L. Ely[1], P. Kyle House[1], Victor R. Baker[1], Lucien Duckstein[2], and Jean Weber[3]
[1]Department of Geosciences
[2]Department of System and Industrial Engineering
[3]Department of Statistics
University of Arizona
Tucson, Arizona 85721
USA

ABSTRACT

We discuss the advantages of lengthening the records of extreme floods and droughts by geological and botanical data and propose a methodology to incorporate such data into risk analyses. By themselves, such data provide information on the most extreme events that actually occurred in a drainage basin or region during the last several centuries to millennia. Through the adaptation of existing methodologies, these records serve as prior information within the Bayesian framework and substantially improve the estimated parameters of the probability distribution functions of extreme phenomena.

INTRODUCTION

Recent experience with high-risk projects reveals that risk is partly a matter of perception. Current public perception of consequences and actual probabilities of occurrence may be far more important for project success than long-held concepts of economic consequences and distributions based on theory alone (Baker et al., 1990). Various geological and botanical methods can provide precise data on the magnitude and frequency of hydrological phenomena, such as extreme floods and severe droughts. These data may be used to illustrate actual consequences and risks to be expected from future hypothesized events. Because of their accuracy, these data can also be used in conjunction with engineering practices to assess the risk from, and the probability of, future droughts or floods. This approach provides quantitative information on the actual occurrence and effects of extreme events over hundreds

NATO ASI Series, Vol. G 29
Water Resources Engineering Risk Assessment
Edited by J. Ganoulis

to thousands of years, a time scale usually not available from the shorter, observational hydrologic record.

In this paper we emphasize the contribution of a variety of geologic, geomorphologic and botanical studies to lengthening the gaged or observational hydrologic records and outline a methodology to apply such data toward engineering and risk assessment. Results of the above studies can also be incorporated directly into risk analysis by providing critical information about the magnitude of the largest floods that have occurred in a specific basin in the last several thousand years or by identifying the duration of the longest and most severe regional droughts.

GEOLOGICAL AND BOTANICAL DATA SOURCES

The main sources for data on extreme hydrological events during the last several thousand years include documented historical observations, lake sediments and lake-level fluctuations, paleoflood deposits, tree-rings, flood-scarred trees (Table 1), and in some unique cases continental ice cores. For example, in China, where the consequences of the hydrologic extremes are a direct risk to the lives and/or livelihood of millions of people, historical records and paleoflood information provide a source of high-quality hydrologic data and are routinely used in the design of large dams (Sauer, 1987; Zhan, 1987). These data have diverse levels of resolution and precision; therefore, only a probabilistic approach can incorporate them into a unified and usable data set.

Because specific meteorologic and climatic conditions are the related cause for these records, an effort to use them together and to extract from them the long-term information for the purpose of risk reduction is justified (Baker et al., 1990). For example, it is not only good science to confirm a design number (e.g., probable maximum flood, or 100-year flood) with data on event magnitudes that can be documented to have actually occurred in a region, it is also a scientifically responsible approach to good engineering and to good public relations (Baker et al., 1988). The use of these multiple data sources must be based in an understanding of the characteristics of each recorder (flood deposits, tree rings, etc.). For example, in the same region lake level rise, high river discharge, and increased tree-ring width probably result from the same climatic cause. However, each record has its own advantages and limitations, thus demanding an interdisciplinary approach involving close cooperation between scientists and engineers.

Table 1: Temporal resolution, abundance, and characteristics of various types of geological and biological data sources for extending hydrological records.

Data Source	Temporal Resolution (yrs)	Time Scale (yrs BP)	Abundance[1]	Type of Data (Examples)
Historical Records	1-5	0-3000	+	Floods above threshold Single events
Flood Deposits	1 to 100	0-10,000	++	Largest flood in record Single events
Tree Rings	annual	0-2000	++	Time series of annual flow Length of droughts
Scarred Trees	annual	0-500	+	Occurrence of floods Minimum magnitude
Lacustrine, Varved	annual	0-10,000	+	Time series, single floods, years and episodes with drought
Lacustrine, Other	100-1000	0-10,000	++	Increased or decreased water depth, episodes of increased flooding and droughts
Fluvial Deposits	1-1000	0-10,000	+++	Single floods, episode of regional flooding, rising water table

[1] +++ -abundant, + -less abundant

__Paleofloods__. Flood deposits record the magnitude and timing of individual extreme floods that occurred in specific river systems or in regions (Ely et al., this volume). Such paleoflood data provide information concerning the stationarity in flood magnitude series, the relationships of extreme flooding to climatic variations, and the magnitude of the largest flood in a given basin over the last few hundreds or thousands of years. Paleofloods, historical floods, and gaged records are all very similar types of data and consist of the timing and magnitude of individual floods. Tools have been developed to combine them into one data set for flood frequency analysis and this has contributed to flood risk assessment (Ely et al., this volume). Efforts should be made to adapt and develop other techniques for using this type of compatible flood data and routinely applying it in risk analyses. A new challenge to both geoscientists and engineers is the integration of diverse data to improve risk analyses.

As in other environments, the short-term risk of floods in the arid and semiarid regions of the world is the potential damage to life and property. In such regions, the absence of large floods bears much greater risk, as they are usually the only source of recharge for the groundwater aquifers (e.g. Enzel et al., 1989). Existing long-term data sets on magnitude, frequency, and persistence of hydrologic extremes and their relationships to climatic changes could be used to estimate the risk to the water supply in such basins.

Lacustrine Deposits. Playas and lakes in closed basins record the timing and magnitudes of annual flow, floods, and droughts (or lack of flow) by changes in the chemical, mineralogical, and sedimentological characteristics of their deposits and by evidence of water-level fluctuations (Street-Perrott and Harrison, 1985). These environments are excellent recorders of basin and regional hydrologic conditions in arid and semiarid environments. Where these deposits are preserved, lakes provide essential data on past events. In ungaged basins, lakes and reservoirs may be the only source of information even for the modern period (Anderson, 1986; Hereford, 1987). With the understanding of the hydrological processes that produced such deposition, the chronology and/or magnitude of hydrologic events can be established. Most lakes record longer episodes (10 to 10^2 years) of increased flooding or drying (Enzel et al., 1989), resolving problems related to causality of floods, and increasing the understanding of the long-term natural variability and stationarity of hydroclimatic conditions. Studies of varved lacustrine deposits result in a annual time series directly related to hydrological processes such as annual or seasonal inflow and droughts (Anderson, 1986). Furthermore, specific lacustrine settings record high- and low-frequency variations in ground-water levels and precipitation.

Tree Rings. Yearly tree-ring width variations are the basis for reconstructions of temperature, precipitation, and runoff (e.g. Fritts, 1976). Multivariate statistics are used to define the relationships between gaged hydrologic data and associated tree growth and to construct a transfer function for the several hundred-year time series of the tree-ring width variation (Stockton, 1975; Fritts, 1976; Meko, 1981; Hughes et al., 1982). The reconstructed time series of annual and seasonal precipitation and runoff are useful in estimating risks of low water supply in water resources management, as trees are especially sensitive to droughts. Michaelson and Haston (1988) use tree-ring data to determine the probability of occurrences of severe droughts for the purpose of managing water storage in reservoirs in southern California. Tree-ring widths do record, but are less sensitive to, excesses of precipitation and runoff. In some areas (see below) it is possible to identify the probability of a specific number of floods occurring above a chosen censoring level during a year with high annual discharge. Such relationships are then used to identify the probability of occurrence of floods for the length of the reconstructed runoff time series and to provide a prior probability distribution function (Fig.1).

Recently, a promising new technique has been developed that can directly determine the timing and magnitudes of individual large paleofloods through the analysis of flood-scarred trees. Trees in the valley-bottom environments of high-gradient, mountain streams are vulnerable to scarring by the impact of floating debris during the large floods. Identification of the specific annual ring that was damaged, in conjunction with the height of the scar, is used to generate a time series of the large floods during the life span of the tree (McCord, 1990). This type of data is similar in its characteristics to other types of flood data and can be incorporated into data sets of floods and paleofloods or can provide additional information on the prior probability distribution function (pdf) of the floods in a basin.

A PROPOSED METHODOLOGY

Case studies in which gaged and paleoflood data are integrated for flood frequency analyses are discussed by Ely et al (this volume). Here we outline a Bayesian procedure for incorporating annual flow time series generated from tree-ring analyses. The following approach may be used to combine partial duration series data of paleofloods, historical, and gaged hydrological records together with additional data, such as the time series of total annual discharge derived from tree-ring analyses.

Gaged, historical, and paleofloods (Q) are fitted to a probability distribution $f_Q(q \mid \theta)$, such as a lognormal pdf. Let Q(h) be the flood exeedance above level h from either a partial duration series or paleoflood data. Let y(h)= Log Q(h), then y = y(h) follows a normal pdf with mean μ and variance σ^2 (The argument h will be dropped without loss of generality):

$$y \sim N(\mu, \sigma; y) = N(\theta; y) \qquad (1)$$

If the observations of y form a random sample $(y_1,...,y_I)$, then the likelihood of (μ,σ) is by definition (Davis et al., 1972):

$$L(\theta) = L(\mu,\sigma) \mid y_1,...,y_I = \prod_{i=1}^{I} N(\mu,\sigma;y_i). \qquad (2)$$

This corresponds to the specific case of the L(θ) illustrated in Fig. 1.

The tree-ring analysis provides annual flow volumes; to incorporate these volumes into the annual peak series, the number J(h) of flow exceedance above level h must be identified. Preliminary results of the analysis of data for central Arizona rivers show a probabilistic relationship between the following quantities:

(1) number of exeedances J(h) and peak flows (during a given year) and

(2) number of exeedances J(h) and annual flow volume V.

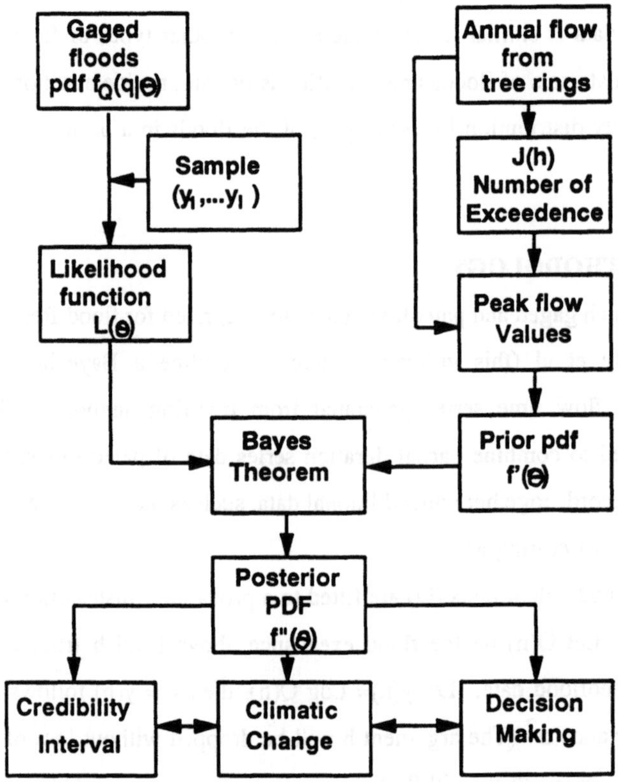

Figure 1: An example of a procedure to incorporate long time series of annual flow reconstructed from tree rings into flood frequency analyses.

Rearranging these two relationships yields the right branch of the diagram in Fig. 1, resulting in a prior pdf, f'(θ) of peak flows estimated from tree-ring records. Intermediate steps to estimate f'(θ) may include: a) specifying the form of f'(θ) and then estimating the so-called hyper parameters of f'(θ); b) using the empirical distribution of J(h) and of peak flows; and c) determining an adequate value of the exceedance discharge h.

Once L(θ) and f'(θ) have been estimated, Bayes theorem is used to obtain a posterior pdf f''(θ) of θ. In the log-normal example, f' and f'' are both normal gamma distributions (Davis and Dvorechik, 1971)

$$NG''(\mu,\sigma \mid y_1,\ldots,y_I) = K^{-1}L(\mu,\sigma \mid y_1,\ldots y_I)NG'(\mu,\sigma), \qquad (3)$$

where K is the so-called marginal distribution (Berger, 1985). As a result, a tighter bound on $\theta=(\mu,\sigma)$ can be determined. The confidence interval based on the posterior pdf f'' is called the credibility interval (Duckstein et al., 1983, 1987), and is an actual probability statement (Berger, 1985). Another possibility in using these types of data, or in a first step of the above procedure, is to consider only annual peak floods above h (in such a case, some years will experience 0 floods if a larger h is chosen).

CONCLUSIONS

Long-term records of hydrologic extremes produced by geoscientists can significantly improve risk analyses. Bayesian procedures incorporate geological and biological data into risk analyses, mainly as a prior information and prior pdf to improve the posterior pdf. From a geologic perspective such a procedure is highly desired, as it make use of the information that records the actual occurrence of past extreme events and the experience of the basin. Cooperation between geoscientists, engineers and applied mathematicians to exploit this vast source of information is a stimulating scientific endeavor and provides mutual benefits.

ACKNOWLEDGMENTS

The research was supported in part by Engineering Directorate, Natural and Manmade Hazards Mitigation Program, National Science Foundation, Grant BCS-8901430 to VR Baker and Grant BCS-9016462 to L Duckstein.

This publication is Contribution No. 7 of the Arizona Laboratory for Paleohydrological and Hydroclimatological Analysis (ALPHA).

REFERENCE

Anderson RY (1986) The varve microcosm: propagator of cyclic bedding. Paleoceanography 1:373-382.

Baker VR, Kochel RC, Patton PC (eds) (1988) Flood Geomorphology. John Wiley, New York, 503 p.

Baker VR, Ely LL, O'Connor JE (1990) Paleoflood hydrology and design decision for high-risk projects. In Chang HH, Hill RH (eds) Hydraulic Engineering, Proceedings of the 1990 National Conference, San Diego, Am. Soc. Civil Eng. New York, 433-438.

Berger JO (1985) Statistical Decision Theory and Bayesian Analysis. Springer-Verlag, New York 617 p.

Davis DR Dvoranchik WM (1971) Evaluation of the worth of additional data. Water Resources Bulletin 7:700-707.

Davis DR, Kisiel CC, Duckstein L (1972) Bayesian decision theory applied to design in hydrology. Water Resources Research 8:33-41.

Duckstein L, Davis D, Plate EJ (1983) Use of climatological information in river basin planning under risk. Proceedings, Int. Sym. on Statistical Clim., Lisbon, Portugal. Sept. 26-30 P. 9.3.1-9.3.10.

Duckstein L, Bobee B, Bogardi I (1987) Bayesian forecasting of hydrological variables under changing climatology. Proceedings, Int. Ass. Hyd. Sci. Vancouver B.C.

Ely LL, Enzel Y, O'Connor JE, Baker VR (this volume) Paleoflood records and risk assessment: Examples from the Colorado River Basin.

Enzel Y, Cayan RD, Anderson RY, and Wells SG (1989) Atmospheric circulation during Holocene lake stands in the Mojave Desert: evidence of a regional climatic change. Nature 341, 44-48.

Fritts HC (1976) Tree Ring and Climate. New York, Academic Press 567 p.

Hereford R (1987) Sediment-yield history of a small basin in southern Utah, 1937-1976: Implications for land management and geomorphology. Geology 15:954-957.

Hughes MK, ed. (1982) Climate from Tree-Rings. Cambridge University Press 223 p.

McCord VAS (1990) Augmenting flood frequency estimations using flood-scarred trees. PhD Dissertation, Dept. of Geosciences, University of Arizona.

Meko DM (1981) Applications of Box-Jenkins methods of time-series analysis to reconstructions of drought from tree rings. PhD Dissertation, Univ. of Arizona 149 p.

Michaelson J and Haston L (1988) Final report for Water Resources Center Project W-707, University of California, Santa Barbara.

Sauer VB (1987) Flood-frequency analysis with historical data in China. In Singh VP (ed) Hydrologic Frequency Modeling. D. Reidel, Dordrechet, p. 173-182.

Stockton CM (1975) Long-term streamflow records reconstructed from tree-rings. Univ. of Arizona, Laboratory of Tree-Ring Research Paper 5.

Street-Perrott AF, and Harrison SP (1985) Lake levels and climatic reconstruction. in Hecht, DA, ed., Paleoclimate Analysis and Modeling. New York, John Wiley & Sons, 291-340.

Zhan Daojiang (1987) Heavy rainstorm floods in China and their estimation. In Singh VP (ed) Application of Frequency and Risk in Water Resources. Reidel, Dordrechet, 34-44.

PALEOFLOOD RECORDS AND RISK ASSESSMENT: EXAMPLES FROM THE COLORADO RIVER BASIN

Lisa L. Ely, Yehouda Enzel, Jim E. O'Connor, and Victor R. Baker
Department of Geosciences
University of Arizona
Tucson, AZ 85721
USA

ABSTRACT

Geological evidence of extraordinary floods provides quantitative information on magnitude and frequency that increases the reliability of risk analyses for extreme floods. The long-term paleoflood records can be incorporated into flood frequency analyses, identify the maximum expected flood, and document the stationarity of flood series. Case studies from the Colorado River Basin are presented to illustrate the applicability of these data and procedures.

INTRODUCTION

Traditional analyses for assessing the risk of extreme floods are based either on extrapolations from gaged discharge records in which the largest events are greatly underrepresented or on hypothetical models with little connection to real experience. Geological and historical evidence extend the records of large floods on a river by hundreds to thousands of years, providing long-term data on the magnitude and frequency of rare, catastrophic events and increasing the reliability of risk analyses (Baker, 1987).

Fine-grained slackwater flood deposits demarcate the peak stages of the associated floods, and can be used to reconstruct paleoflood discharges. These flood deposits selectively preserve evidence of only the largest floods, which is precisely the information lacking from relatively short gaged discharge records. Recent advances in geochronology, hydraulic flow modeling, and statistical techniques for incorporating the paleoflood data into frequency analyses have greatly improved the accuracy and utility of this type of paleoflood information.

Paleoflood and historical information can be a key element for improving the assessment of flood hazards and flood frequency on rivers in a wide variety of environments (e.g. Kochel and Baker, 1982; Baker et al., 1983; Stedinger et al., 1988; Webb et al., 1988; Ely et al., 1991). In China, where gaged records are scarce, historical records of floods covering over

NATO ASI Series, Vol. G 29
Water Resources Engineering Risk Assessment
Edited by J. Ganoulis

1000 years have been used in determining the design floods for structures, including major dams (Hua, 1987). One emphasis in recent paleoflood studies is the relationship between extreme floods and climatic change, which involves the question of climatically related stationarity of flood series. The climatic implications of paleofloods have been established for the southwestern U.S., Australia, Peru, and India (Wohl, 1988; Enzel et al., 1989; Ely and Baker, 1990; Wells, 1990).

METHODS OF DATA COLLECTION

Fine-grained slackwater flood deposits provide the most complete paleoflood information. During large floods in stable bedrock canyons, fine-grained sand and silt fall rapidly out of suspension in areas of markedly reduced flow velocity, such as back-flooded tributaries and eddies at channel irregularities (Baker and Kochel, 1988). Many sites preserve multiple deposits, each representing a separate flood. Individual flood deposits are delineated by abrupt changes in grain size or sediment type, incipient soil development, or buried vegetation. Evidence of the largest events is selectively preserved; deposits from smaller floods lie closer to the active river channel and are more likely to be removed by subsequent erosion (Ely and Baker, 1990). This natural censoring threshold imposed by the depositional environment makes the geological data ideal for flood frequency analysis, as one can reasonably assume that within a given time range, all events over a particular discharge are preserved.

The ages of the floods are determined through radiocarbon dating of associated organic or archaeological materials. These dates establish the flood chronology and aid in correlating deposits among sites along the river. For flood frequency analysis, the number of floods within the total period of record is more important than the precise age of each event (Stedinger et al., 1988a).

The deposits provide minimum water-surface elevations of the associated floods. Paleoflood discharges are calculated by comparing the heights of the flood deposits with water-surface profiles generated by a step-backwater hydraulic model (Fig. 1). Under optimum conditions the heights of the deposits closely approximate the actual peak flood stages, as demonstrated by flume studies, comparison with gaged discharges, and field observations of the close relationship between high water marks and associated flood deposits (Ely and Baker, 1985; Partridge and Baker, 1987; Baker and Kochel, 1988).

FLOOD CHRONOLOGIES

Paleoflood studies on the Colorado, Verde, and Salt Rivers in the southwestern U.S. have focused on estimating the magnitude and frequency of extreme floods for flood hazard assessment. The Colorado Basin is a major source of water and hydroelectric power for a large part of the western U.S. and contains numerous dams. Fifteen floods with minimum peak discharges comparable to the maximum historic peak have occurred within the last 4500 years on the Colorado River in the Grand Canyon (Ely et al., 1991; O'Connor et al., in prep). The largest paleoflood in the 4500-year record had a peak discharge between 13,600 and 14,200 cubic meters per second (m^3s^{-1}). The remaining paleofloods yielded minimum peak discharges between 5700 and 8900 m^3s^{-1}). The youngest deposit in this record (8900 m^3s^{-1}) was identified at several locations along the study reach and may correspond to a large historic flood in 1884 (Fig. 1).

The Verde and Salt Rivers in central Arizona converge approximately 20 km upstream of the large urban center of Phoenix, Arizona. Each river contains a series of flood-control dams that were recently under consideration for renovation or replacement. Paleoflood studies on these rivers (Ely and Baker, 1985; Partridge and Baker, 1987) provided information on the occurrence of large floods that could impact the dams. The largest floods in the last 1000-

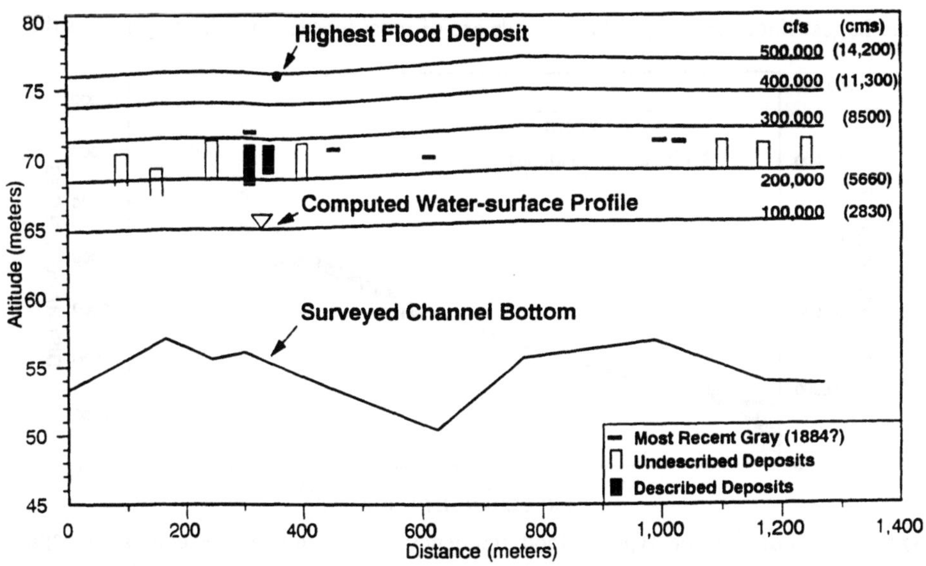

Figure 1. Comparison of flood deposit heights on the Colorado River with water surface profiles produced by a step-backwater model, resulting in the discharge reconstructions for the paleofloods (O'Connor et al., in prep.).

2000 years were 5000-5400 m^3s^{-1} on the Verde River and 4100-4600 m^3s^{-1} on the Salt River. Each river also contained evidence of one to three paleofloods with minimum discharges between 3000 and 3800 m^3s^{-1}. The youngest of these may correspond to a historic flood in 1891.

FLOOD FREQUENCY ANALYSIS

Paleoflood and historic flood data consist of isolated peak discharges that exceeded a designated threshold of observation. Depending on the type and quality of the data, the peaks are described either as precise discharges, as falling within a range of values, or simply as above or below some threshold. The maximum likelihood method efficiently incorporates this type of data into statistical flood frequency analyses in which flood flows are assumed to come from a given parametric distribution (Fig 2). The maximum likelihood method estimates the parameters of a distribution by maximizing the probability of the observed flood events (Stedinger and Cohn, 1986; Stedinger et al., 1988a,b).

The addition of the paleoflood data can greatly improve the estimates of the parameters of a chosen flood frequency distribution. Even more significantly, this type of information identifies which probability distributions are most consistent with the frequency distribution of the largest floods (Fig. 2). These floods are of most interest in risk analysis, but they are

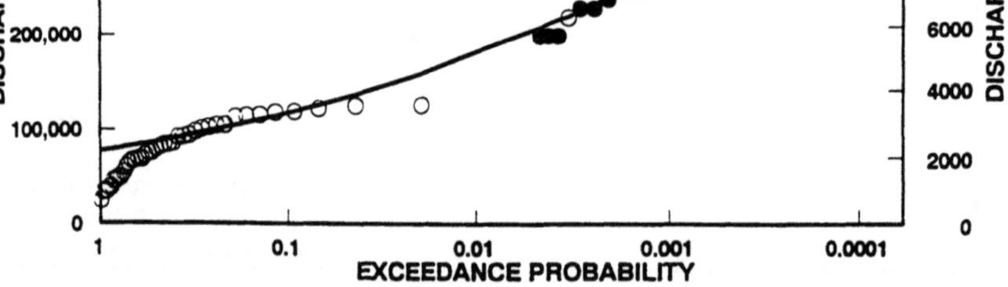

Figure 2. Log-Pearson type 3 flood frequency distribution of the combined gaged and paleoflood data from the Colorado River near Lees Ferry (O'Connor et al., in prep.). The distribution parameters were estimated using maximum likelihood techniques (Stedinger et al., 1988a), based on only those floods with discharges greater than 100,000 cfs (cubic feet per second) (2830 m^3s^{-1}).

also the events for which there is usually the least information (Stedinger et al., 1988a). For example, Stedinger et al. (1988a) found a single frequency distribution could not simultaneously describe the entire range of discharges represented by the gaged and paleoflood discharges on the Verde and Salt Rivers. This problem was overcome by fitting the flood frequency distributions to only the larger floods, thus eliminating misleading information from smaller floods that departed from this distribution.

On the Verde and Salt Rivers, 30-35% differences occur between the 100-year flood magnitudes estimated using the gaged record alone and those estimated using the combined gaged and paleoflood data (Table 1). These discrepancies illustrate the inadequacy of the gaged records on these rivers to characterize the frequency of extreme events. In contrast, a similar comparison on the Colorado River shows only a slight difference between these two 100-year estimates. In this case, the gaged record alone accurately depicts the frequency distribution up to the 0.01 exceedance probability (100-year flood), but the paleoflood record greatly increases the number data points describing the upper tail of the frequency distribution (Fig. 2).

Table 1. Estimates of extreme floods on the Colorado, Verde, and Salt Rivers.

	Discharge (m^3s^{-1})		
	Colorado River	Verde River	Salt River
Maximum Peak Discharge			
Paleoflood record[1]	13600-14200	5000-5400	4100-4600
Maximum historical peak (estimate)	8500	4250	4250
Gaged record	6230	2690	3310
Flood Frequency Analyses			
100-year flood, log-Pearson 3			
Gaged record[2]	5370	4500	4670
Combined gaged and paleoflood records[3]	5100	3100	3100
Probable Maximum Flood[4]	19700	21640	19200

1) Colorado River (O'Connor et al., in prep.); Verde River (Ely and Baker, 1985); Salt River (Partridge and Baker, 1987).
2) Gaged record flood frequency analyses from Anderson and White (1979).
3) Maximum likelihood analyses: Colorado River (O'Connor et al., in prep.); Verde and Salt Rivers (Stedinger et al., 1988b).
4) U.S. Bureau of Reclamation official PMF values.

MAXIMUM EXPECTED FLOOD

Another valuable result of the paleoflood studies is the information provided on the magnitude of the largest flood in the extended record. This information could be more important than the frequency analysis in flood risk evaluations and in the choice of a design flood for structures along a river. The dams on the rivers in these examples are designed to accommodate the Probable Maximum Flood (PMF), determined by modeling the hydrological response to a hypothetical extreme storm. As shown in Table 1, there are considerable discrepancies between the PMFs and the actual evidence of the maximum flood on each river within a period of several thousand years. This is especially true on the Salt and Verde Rivers, where the PMFs are almost five times larger than any flood in the extended record. In such cases, the geologic evidence provides an independent check on the estimated design floods and a rigorous test of the assumptions made in theoretical models (Baker et al., 1990).

The maximum paleoflood discharges from a series of rivers within the Colorado River basin approach an envelope curve encompassed of gaged peak discharges from drainages of various sizes throughout the basin (Fig. 3). This consistent relationship supports the concept of a physical or hydroclimatological limit on the magnitude of the maximum flood that can

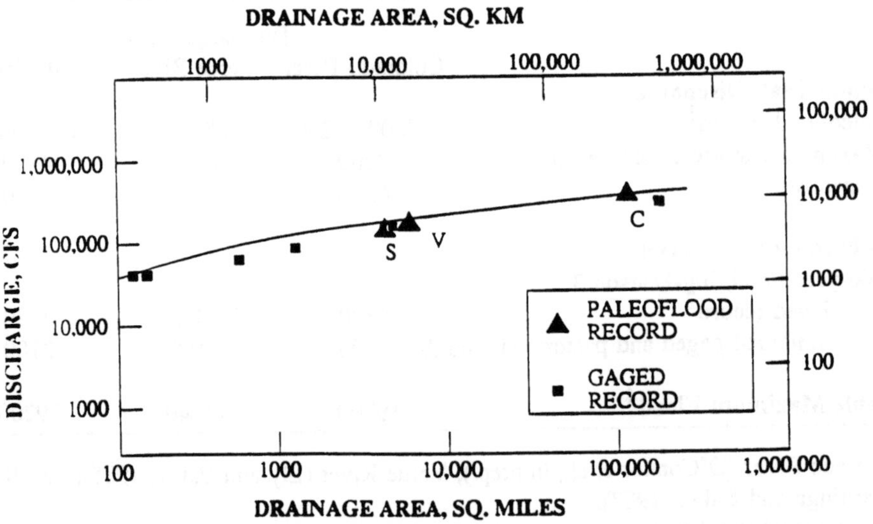

Figure 3. Envelope curve encompassing maximum gaged peak discharges on drainages throughout the Colorado River Basin. The selected gaged discharge values shown here represent only those that define the curve. The maximum paleoflood discharges from the Colorado, Verde, and Salt Rivers fall very close to the regional curve, but do not exceed it. C=Colorado River, V=Verde River, S=Salt River.

be expected in a given drainage (Costa, 1987). Our observation suggests a questioning of exceedance probability estimation by standard statistical procedures, which assume no upper bound on flood magnitudes.

CONCLUSIONS

Paleoflood records provide quantitative information on the magnitude and frequency of large, rare floods that can be effectively used in risk analyses associated with dams, bridges, and floodplain management. Integrating the paleoflood data with the gaged flow records results in a more accurate description of the frequency distribution by increasing the information on the occurrence of the largest events. In addition, identifying the magnitude of the largest flood within a period of hundreds to thousands of years provides substantial evidence about the maximum flood that can be reasonably expected to occur on a particular river.

ACKNOWLEDGMENTS

We thank E. Wohl, L. Stevens, T. Melis and V. Kale for help in field work and data analysis. The research was supported by the Engineering Directorate, Natural and Manmade Hazards Mitigation Program, National Science Foundation, Grant BCS-8901430. Additional support was provided by the Salt River Project, Tempe, Arizona, and the U.S. Bureau of Reclamation. This publication is Contribution No. 8 of the Arizona Laboratory for Paleohydrological and Hydroclimatological Analysis (ALPHA).

REFERENCES

Baker VR (1987) Paleoflood hydrology and extraordinary flood events. Jour. Hydrol. 96:79-99.

Baker VR, Kochel RC (1988) Flood sedimentation in bedrock fluvial systems. *In* Baker VR, Kochel RC, Patton PC (eds) Flood Geomorphology. Wiley New York, p. 123-137.

Baker VR, Pickup G, Polach HA (1983) Desert palaeofloods in central Australia. Nature 301:502-504.

Baker VR, Ely LL, O'Connor JE (1990) Paleoflood hydrology and design decision for high-risk projects. *In* Chang HH, Hill RH (eds) Hydraulic Engineering, Proceedings of the 1990 National Conference, San Diego CA. Am. Soc. Civil Eng., New York, p. 433-438.

Costa JE (1987) A comparison of the largest rainfall-runoff floods in the United States with those of the People's Republic of China and the world. Jour. Hydrol. 96:101-115.

Ely LL, Baker VR (1985) Reconstructing paleoflood hydrology with slackwater deposits, Verde River, Arizona. Phys. Geogr. 6:103-126.

Ely LL, Baker VR (1990) Large floods and climate change in the southwestern United States. *In* French RH (ed) Hydraulics/Hydrology of Arid Lands, Proceedings of the 1990 International Symposium, San Diego CA. Am. Soc. Civil Eng. New York, p. 361-366.

Ely LL, O'Connor JE, Wohl EE, Stevens LE, Melis TS, Kale VS, Baker VR (1991) 4000-year record of extreme floods on the Colorado River in the Grand Canyon. Geol. Soc. Amer. Abst. with Programs 23,4:20.

Enzel Y, Cayan CR, Anderson RY, Wells SG (1989) Atmospheric circulation during Holocene lake stands in the mojave desert: evidence of regional climate change. Nature 341:44-47.

Hua S-Q (1987) A general survey of flood-frequency analysis in China. Jour. Hydrol. 96:15-25.

Kochel RC, Baker VR (1982) Paleoflood hydrology. Science 215:353-361.

O'Connor JE, Ely LL, Wohl EE, Stevens LE, Melis TS, Kale VS, Baker VR (in prep.) 4000-year record of large floods in the Grand Canyon.

Partridge JB, Baker VR (1987) Palaeoflood hydrology of the Salt River, Arizona. Earth Surf. Process. Landforms 12:109-125.

Stedinger JR, Cohn TA (1986) Flood frequency analysis with historical and paleoflood information. Water Resour. Res. 22:785-793.

Stedinger JR, Surani R, Therivel, R (1988a) MAX users guide: a program for flood frequency analysis using systematic-record, historical, botanical, physical paleohydrologic and regional hydrologic information using maximum likelihood techniques. Dept. Environ. Eng. Cornell Univ., Ithaca, New York.

Stedinger JR, Therivel R, Baker, VR (1988b) The use and value of historical and paleoflood information in flood frequency analyses. Proceedings, 8th Annual U.S. Committee on Large Dams Lecture Series, Phoenix, Arizona, p. 4.1-4.31.

U.S. Bureau of Reclamation (1990) Colorado River basin probable maximum floods Hoover and Glen Canyon Dams. U.S. Dept. Interior, Washington D.C.

Webb RH, O'Connor JE, Baker VR (1988) Paleohydrologic reconstruction of flood frequency on the Escalante River, south-central Utah. *In* Baker VR, Kochel RC, Patton PC (eds) Flood Geomorphology. Wiley, New York, p. 403-418.

Wells LE (1990) Holocene history of the El Niño phenomenon as recorded in flood sediments of northern coastal Peru. Geology 18:1134-1137.

Wohl EE (1988) Northern Australian paleofloods as paleoclimatic indicators. PhD, Univ. Arizona, Tucson.

Part 2.

QUANTIFICATION OF RISKS IN WATER QUANTITY & QUALITY PROBLEMS

STOCHASTIC DYNAMIC SYSTEM THEORY: A CHALLENGE FOR NATURAL RESOURCES MANAGEMENT

Eric Parent
Department of Applied Mathematics and Computer Science
ENGREF, 19 av du Maine, 75015 Paris,France.

ABSTRACT

Most of stochastic modeling methods in natural resources are based on a set of analytical model equations, mainly transport and mass-balance equations including stochastic behavior of model variables. These techniques belong to system analysis and control theory and they have become a common practice in engineering studies although the decision making implications and limits of such models are not often explicitly realized. This paper gives a general overview of sequential optimization and system analysis applied to the field of water resource and environmental engineering. It underlines the advantages but also the limits and the weak points of such an approach especially with regards to risk management and multicriterion decision making.

1.- INTRODUCTION

Why do scientists elaborate probalistic models for engineering applications? Methodological advices to design such models can be found in Kree and Soize (1983). First, risk managers may think of using models to derive properties of interest such as asymptotic behavior (transiency, stationary processes and so on) or probability density functions that are useful for design purposes. For instance one would investigate hydraulic structural safety given by a probabilistic model of streamflows with parameter uncertainty as in Plate and Duckstein (1987).

Second, such models are of practical importance to mimic the physical characteristics of the phenomena that is modelized. To do so once a class of realistic models has been selected, various parameter estimation techniques can be performed to qualify a model using real world data. Monte Carlo methods can then generate synthetic series which respect the statistical properties of the model. This random generation is perhaps the best and the easiest way to visualize on numerous runs which risks the system manager can take.

Last but not least, probabilistic models should be the basis for proposing an optimal control law in environmental engineering studies. The fundamental role of engineers is to act onto a system and to modify its behavior so as to meet various criteria (among which the most enlightened one has often been cost reduction), to respect constraints and to absorb stochastic perturbations. The first part of this paper gives the classical framework of stochastic

NATO ASI Series, Vol. G 29
Water Resources Engineering Risk Assessment
Edited by J. Ganoulis

natural resources management problems. Basic concepts in the deterministic case can be found in BELTRAMI(1987). In this quest for optimal control laws, one should recognize that the engineer has often done worse than bad in the field of environmental issues. Although we have created sophisticated technical devices and processed every day an increasingly huge amount of data, we have displaced equilibrium points, transformed stable equilibria into unstable ones and greatly increased the potential risk of systems failure. Nowadays in most people's minds, human policies are always associated with an increase of risk and our will to rule over renewable natural resources often reveals more clumsiness and even foolishness than engineering rationality.
How is it that we can use optimal control theory to send space shuttles to the moon, to make planes commonly land and take off in the thickest fog, to optimize the maintenance policy and the replacement date of manufacturers' production machines but, at the same time, there still exists obvious waste in water management, non rational allocation of renewable resources and a bitter feeling of helplessness to face natural uncertainties? Maybe casting natural resource operation into the general framework of system analysis and control theory is a difficult task for two reasons :

1) in nature driven systems, there always remain uncertainties in the form of dynamic transition functions, mostly because the variables do not evolve according to any "good" type of a fixed and rational Newtonian law;

2) the aims that the system manager wants to attain through a system design and control are not always clear; anyhow they are in most cases conflicting and multidimensional, difficult to formulate and to assess. To underline the broad gap and the few bridges between system theory and practice in natural resources management, this paper is organized as follows :

- in the first part, we will give a general overview of stochastic control theory and its potential applications to natural resources management by means of dynamic programming. The advantages and disadvantages of such methods will be briefly discussed.

- in a second part, we will review some advanced topics from a dynamic programming viewpoint that can provide interesting decision support tools for system managers, especially when taking risk into account. We will focus on structural properties of dynamic programming that allows for other combinations of reward functions than the well known sum of rewards. Since risk management can be understood as a multicriterion decision making method as it will be explained in that part, we will give a presentation of multicriterion dynamic programming techniques and its potential extension to resources management problems will be stressed.

2 GENERAL FRAMEWORK FOR SYSTEM ANALYSIS AND CONTROL THEORY

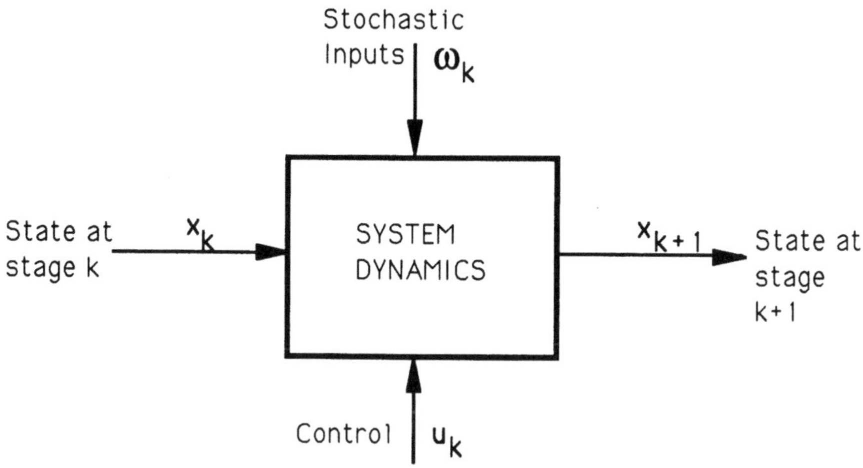

Figure 1: General framework for system analysis

This approach has been detailed in ROBIN and FAURRE (1987), FRIEDLAND(1986) , ALEXEIEV(1987) and, as it is shown on figure 1, the designer willing to use such state space model has to identify :

1) the decision vector u that contains the control variables . Reservoir releases, crop pattern or the imposition of adjusted water prices are examples of such control variables. They represent the means that the system manager can activate at every stage to master the system and to keep it within guidelines.

2) the stochastic inputs ω occurring at each step of the process: these can be rainfall or streamflows, uncertain water demand consumption, or any random variable interfering with the evolution of the system.

3) the state vector x that keeps at hand the dynamic variables of interest: reservoir levels, river levels or agricultural indices related to the present state of crop evolution and so on. Indeed, in our dynamic view of a system, a state is a summary of the prior history of the process that is sufficiently detailed to enable evaluation of current decisions. A transition function $f(x, u, \omega, t)$ is used to describe the dynamic behavior of the system from stage to stage.

In discrete time: $x(t+1) = f(x, u, \omega, t)$..(1)

In continuous time: $\frac{dx}{dt} = f(x, u, \omega, t)$..(2)

This approach is non specific; TAPIERO(1988) presents a comprehensive view of stochastic dynamic modeling in the management sciences. The approach includes at each stage: studies in hydrology and hydraulics and establishment of mass balance equations . Indeed what the decision maker looks for, is how to design a feedback rule by constructing a mapping from the state space to the set of possible decision. In other words, the system manager sequentially makes his own decision at time t according to what he knows about the system at time t, consisting of the coordinates of the state vector. How can he choose a policy that will give a satisfactory trajectory for the system? At this point we shall introduce a criterion to be optimized, which is a function J of the control law u(t) often expressed as :

in discrete time $J(u) = \sum_{t=0}^{T} L(x, u, t) + V_T(x_T)$..(3)

or, in continuous time, $J(u) = \int_{t=0}^{T} L(x, u, t)dt + V_T(x_T)$....................................(4)

$L(x, u, t)$ is the instantaneous loss or reward, $V_T(x_T)$ is the final cost or benefit. For example, in an hydro-power plant system, the maximization of energy generation would typically be $L(x, u, t) = h(x_t)u_t$, where the control variable $u(t)$ is the release at time t and $h(x_t)$ would be the waterfall height that depends of the storage volume x_t which belongs to the state space. Other reward functions can be built as we will see in the next section of this paper, but now suppose that we have at hand a dynamic system (by Equations (1) and (3)) and we are looking for a feedback policy so as to maximize a criterion consisting of the sum of reward functions.

3 THE PRINCIPLE OF BELLMAN

The principle of Bellman (1957) is commonly used to derive such a policy. Let us quote Bellman: "An optimal policy has the property that whatever the initial conditions are, the remaining decision must constitute an optimal policy with regards to the state resulting from the first decision..."

We illustrate this principle on the following schematic discrete stage and space example. Let suppose we want to go from town G to town A using the longest path in figure 2. The basis of Bellman principle is exposed in figures 3,4,5,6: we sequentially decide where to go from stage to stage. At each of the four steps , we have indicated the reward to go from the given present town to the ending town A. We can recursively compute the optimum reward by combining the immediate reward and the future one obtained once the next town is reached. Let V(x,k) be the maximum benefit from town x at stage k to town A and L(x,u,k) be the immediate reward for going from town x to town u. The sequential way of reasoning can be summarized under the form:

$x(k+1) = u(k)$

$V(x,k) = \operatorname{Opt}_u \{ V(u,k+1)+L(x,u,k)\}$..(5)

and the complete solution to this illustrative longest path problem obtained in this manner is given in figure 7. It is remarkable that this sequential calculus gives at each stage a feedback solution:

u = u(x,t) ....(6)

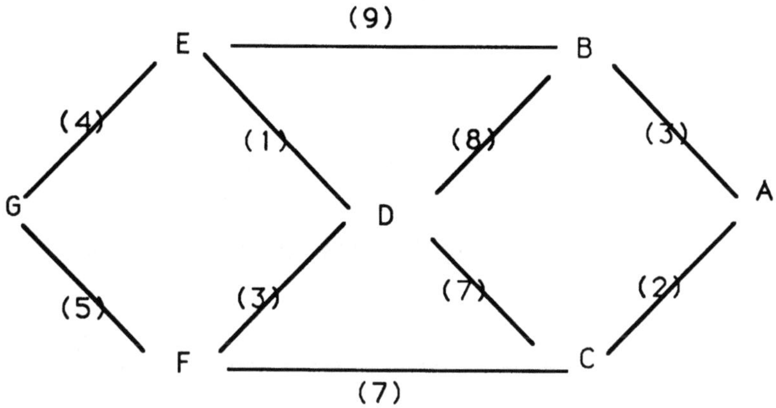

Figure 2: What is the longest path from G to A

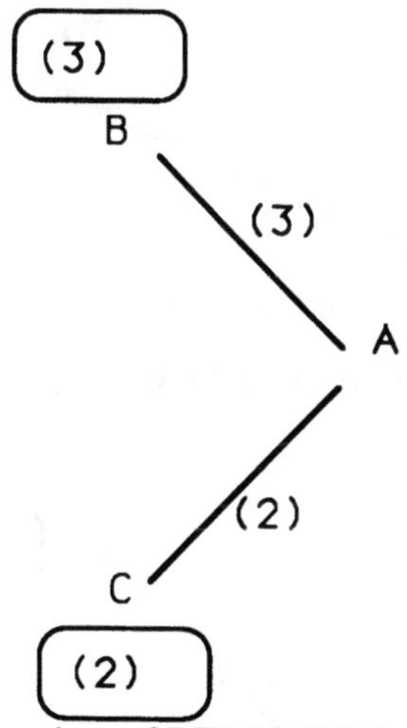

Figure 3: Illustration of BELLMAN 's principle; Step1

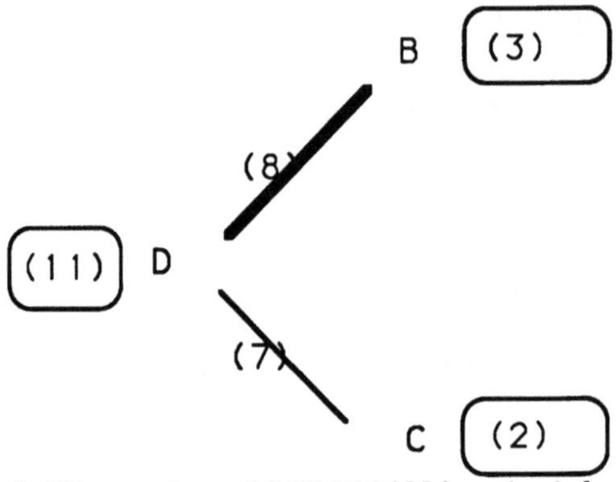

Figure 4: Illustration of BELLMAN 's principle; Step 2

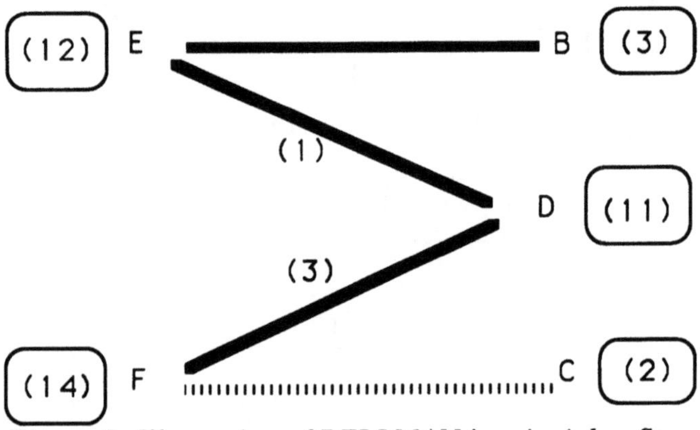

Figure 5: Illustration of BELLMAN 's principle; Step 3

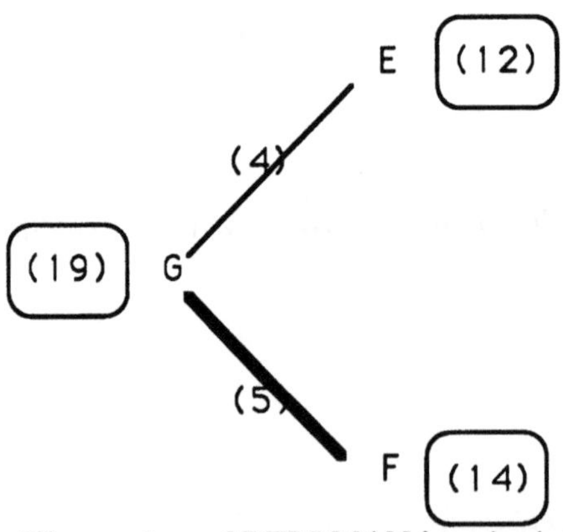

Figure 6: Illustration of BELLMAN 's principle; Step4

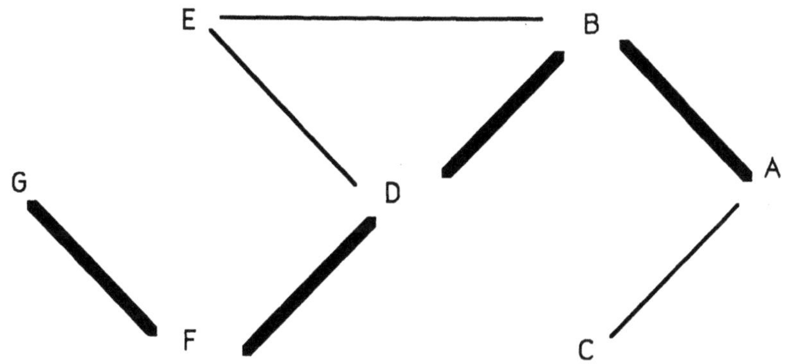

Figure 7: Feedback solutions to the longest path problem given by DP

4 ADVANTAGES AND LIMITS OF DYNAMIC PROGRAMMING

Illustrative case of a Linear Quadratic Gaussian reservoir

In the stochastic case , this sequential decomposition principle can be applied the same way provided that the criterion is now the expected value of the sum of rewards. When the optimum feedback strategy $u(t) = u(x(t),t)$ is applied the state process is then a Markov chain of order 1. To find this optimum strategy let:

$$V\big(x(t_0)\big) = \text{Max } J(u(.)) = \text{Max}\left\{E \sum_{t=t_0}^{T} L(u(t),x(t),t)\right\} \quad \text{.......................(6)}$$

The maximum is to be calculated for $u(t), t \in [t_0, T]$

The Bellman principle can be applied to reformulate this problem by separating time t_0 and the future periods for which V given in Equation (6) is supposed to have already been computed from T to t_0+1:

$$V\big(x(t_0)\big) = \text{Max}\,(E\{ L(u(t_0)x(t_0),t_0) + V\big(x(t_0+1)\big) \} \quad \text{.......................(7)}$$

$$V(x\,(t_0))=\text{Max}(E\,\{L(u(t_0)x(t_0),t_0) + V(f(t_0,u(t_0))\,,x(t_0),\,\omega)\,\}\,) \quad \text{.................(8)}$$

This time, the maximization operates only on $u(t_0)$ that is expressed as a function of $x(t_0)$. Equation (8) gives a backwards functionnal recurrence for V(.) which is known as the Bellman function. DENARDO(1982) gives numerous examples and theoretical demonstrations of this sequential decomposition principle.

Since the works of MASSE(1959), dynamic programming methods have been widely used in water resource operation and planning. YAKOWITZ (1982) and YEH (1985) have given extensive state-of-art reviews of these sequential

optimization techniques. We will solve a reservoir problem in continuous time and space under uncertainty for illustrative purposes.
Consider a single reservoir as in figure 8 (storage x(t)) with stochastic inflow and a release u(t) that is to be regulated.

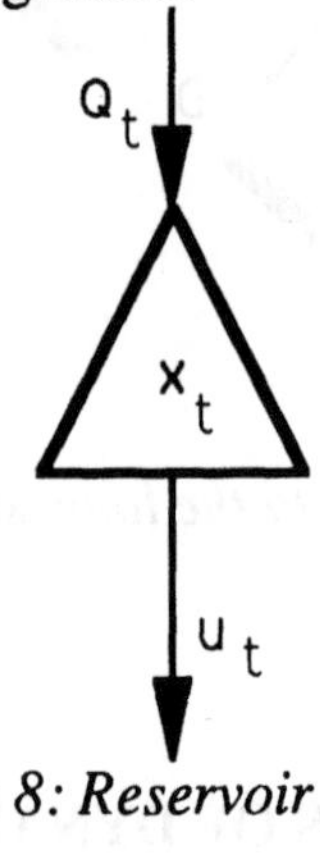

Figure 8: Reservoir problem

The mass balance equation takes the form of a stochastic differential equation:
$dx = (-u+Q)dt + \sigma\, dW$ ....(9)
$x(0) = x_0$ is the given initial level of the storage, Q is the mean value of the streamflow and σ is the variability of the flows that is assumed to be stationnary and independent so that dW is a Brownian motion. Theory and applications of such stochastic models are detailed in ARNOLD(1974).
We look for a policy u(t) such that a quadratic criterion is minimized:

$$\text{Min } J(u(.)) = \text{Min}\left\{E \int_{t=0}^{T} \left[(u-u_0)^2 + (x-x_0)^2\right] dt \right\} + m\, e^{rT} x_T \qquad (10)$$

This minimum $V(x_0)$ is to be calculated for $u(t)_{t \in [0,\, T]}$

Applying Bellman's principle and Ito's differentiation rule, we obtain he following partial differential equation for V(x(t),t):

$$\frac{\partial V}{\partial t} + \underset{u}{\text{Min}}\left\{\frac{\partial V}{\partial x}(Q-u)+(u-u_0)^2\right\} + (x-x_0)^2 - rV + \frac{1}{2}\frac{\partial^2 V}{\partial x^2}\sigma^2 = 0 \qquad (11)$$

A feedback solution u^* is given by:

$$u^* = u_0 + \frac{1}{2}\frac{\partial V}{\partial x} \qquad (12)$$

so that :

$$\frac{\partial V}{\partial t} + \frac{\partial V}{\partial x}(Q-u_0) - \frac{1}{4}\left[\frac{\partial V}{\partial x}\right]^2 + (x-x_0)^2 - rV + \frac{1}{2}\frac{\partial^2 V}{\partial x^2}\sigma^2 = 0 \qquad (13)$$

this will lead to a Riccatti equation whose solution can be found by setting:

$$V(x(t),t) = a(t)\, x(t)^2 + b(t)\, x(t) + c(t) \qquad (14)$$

For the case $r = 0 = u_0 = x_0$, we see that the functions a(t),b(t), c(t) must verify the following equations

$$\dot{a} + 1 - a^2 = 0 \;;\; a(T) = 0 \qquad (15)$$

$$\dot{b} - ab + 2aQ = 0 \;;\; b(T) = m \qquad (16)$$

$$\dot{c} - \frac{1}{4} b^2 + \sigma^2 a + Qb = 0 \;;\; c(T) = 0 \qquad (17)$$

The solutions can be expressed as:

$$a(t) = \tanh(t-T)) \qquad (18)$$

$$b(t) = 2Q + (m-2Q)\, \mathrm{coshh}(t-T) \qquad (19)$$

$$c(t) = -\int_t^T \{ \frac{1}{4} b^2(\tau) - \sigma^2 a(\tau) - Qb(\tau) \}\, d\tau \qquad (20)$$

Equation (12) can be written as

$$u^* = a(t)\, x(t) + \frac{b(t)}{2} \qquad (21)$$

It is easier to give an interpretation in term of a compromise between input and storage volume if the optimal release u^* is written under the form:

$$u^* = \tanh(t-T)\, x(t) + Q\,(1 - \cosh(t-T) + m \cosh(t-T) \qquad (22)$$

However one can also notice in this case that the stochastic nature of this inflow does not appear in the calculation of u^* which would be the same if σ were equal to 0 so that in such a model risk consideration is of no importance to determine the control law. Using the terminology of LI and HAIMES(1987) who classify systems according to their reaction to an increase of input variability, this system is risk neutral.

The curse of dimensionality

To summarize, even in greater dimensions it is a general property of Gaussian linear quadratic systems that the optimal policy is a linear feedback on the state which can be evaluated considering only the mean process of the stochastic Gaussian variable. Calculations are not so easy when one of the following conditions is not fulfilled:

1) The stochasticity of the system belongs to the Gauss Markov process family, for which under mild general conditions there is a stochastic differential equation such as Equation (9) to represent the phenomenon and its variability.

2) The transition equation is linear.

3) The instantaneous reward function is quadratic.

4) No constraint is binding.

In such cases, the stochastic characteristics influence the control policy, the Bellman function of the reward to go can be of any shape, the feedback is no longer a linear function of the state. Numerical resolution of Bellman Equation (4) has to be achieved by discretization. It does not simply consist in finding the solution of a Riccatti equation for which algorithms are commonly available in software packages . KLEMES (1977) has underlined the unsound consequences of inappropriate discretization for stochastic reservoir optimization. Although the software program consists mainly in three loops (one over time, one for discretized values of the state, one for each possible decision) it becomes rapidly intractable as the dimension of the state increases, which is known as the curse of dimensionality. Even with regards to present computer performances, such numerical resolution techniques cannot break the limit of three or four dimensional state. YAKOWITZ (1982) whose words are still up to date states that:"...present day computational capabilities allow the solutions of systems up to two or three state variables...". Researchers have worked to bypass this obstacle, and progress has been made especially in the field of water resources management. TURGEON (1981) has proposed aggregation-disaggregation methods to obtain a quasi optimum policy ; SAAD and TURGEON (1988) have tried to reduce the state dimension using principal component analysis. KITANIDIS (1987) has developed a first order approach by considering the decision in uncertain future as a Taylor series expansion of the optimum policy that one would obtain under average deterministic conditions. FOUFOULA-GEORGIOU and KITANIDIS (1988) have also proposed to approximate the Bellman function by cubic Hermite polynomials. Gal (1979) has applied iterative techniques using alternatively simulation and optimization to avoid the curse of dimensionality.

Why use Dynamic Programming?

These technical considerations explain why the use of optimal control theory has not been widespread in natural resources management. Nevertheless we will discuss in this section some of the main reasons why such methods should be of interest in practical applications.

An integrated approach

First of all, stochastic dynamic system theory can respect the various levels of decision and provide tools fit for each level. Consider, for illustrative purposes of this hierarchical approach , the very common case of a regulated irrigation area as indicated in figure 9. An automatic device works in real time operation: the control variable is the release from the reservoir (the level of which belongs to the state vector) and the system operates so as to maintain a level at a gauge downstream as close as predetermined to a prefixed target. This release policy is obtained as indicated in the previous example using Equation (10) to minimize the expected sum of squared deviations from the target, taking into account irrigation demand stochasticity and the propagation delay along the river. At this level of decision this system does not incorporate any risk considerations: schematically it simply opens the gate to increase the release when there is not enough water downstream and lowers the release in the opposite case.

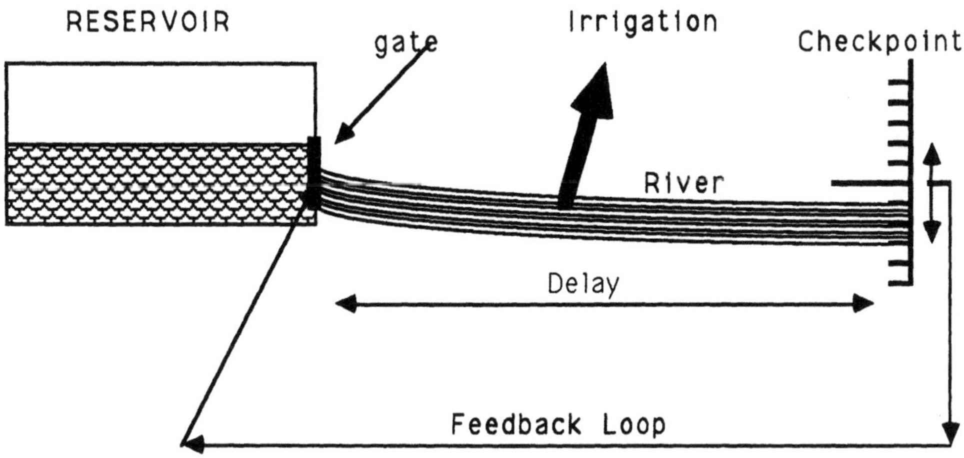

Figure 9: Real time regulation of a reservoir for an irrigated area

A second level of decision, on a weekly basis for instance, can work together with the first level of hourly operation. At this second level the previous real time operation target will become a control variable , the state will include more complete information about hydrological conditions in addition to the available resources in the storage reservoirs. The risk management of the system will consist in increasing the target at the downstream gauge in wet conditions (not too much so as to avoid the risk of emptying the reservoirs too early) and in lowering it in dry conditions (but with great care due to the ecological balance in the river and potential resources that may occur in the future). PARENT et al (1991) have developed for this case a solution based on system analysis and control theory.

Computing risk related indices

A second remarkable advantage of control theory is that it provides a general framework for risk analysis in the sense of the definitions given by DUCKSTEIN et al.(1987). Indeed, risk indices as defined for example in ASKEW (1974) or SIMONOVIC and MARINO (1984), happen to be a spillover effect of sequential optimization: an extensive and realistic display of the risks the system manager may encounter can be computed at practically no cost. Suppose that in the previous irrigation example we focus on the event that the reservoir dries out, defined as an alert. Let u be the chosen policy, x_k^u the state at stage k when the decision maker follows the policy u . We can define the following performance indices (PI):

<u>P.I. of risk :</u>

$W(x_k, k)$ is the probability that there occurs at least an alert after time k

$$W(x_k, k) = \text{Prob}\left\{x_k^u \le 0\right\} + \text{Prob}\left\{x_k^u \ge 0\right\} W(x_{k+1}, k+1) \quad (23)$$

P.I. of vulnerability

$D(x_k, k)$ is the average length of an alert if we suppose that when the reservoir dries out it remains empty until the horizon step T

$$D(x_k, k) = \text{Prob}\left\{x_k^u \le 0\right\} \{T - t\} + \text{Prob}\left\{x_k^u \ge 0\right\} \quad (24)$$

Because dynamic programming investigates all the possible trajectories of the system, risk performances indices given in Equations (23) and (24) can be sequentially evaluated at the same time as the optimal policy is computed. These equations present the same form as Bellman's equations and can be computed at each stage. No simulation is needed to evaluate these performance indices that can be given to the decision maker so as to provide a more complete view of the system evolution and the potential consequences encountered when following the optimal sequential decision track.

Building risk index objective

In addition, the criterion we want to optimize can belong to a broader class than the classical sum of rewards and can therefore be fit for risk analysis: in this way dynamic programming can provide a decision support tool to suggest new alternatives that integrate risk and reliability concepts. What kind of criteria can be considered? Let us look at a two step discrete deterministic system with state x_n and control u_n at stage n . We will try to explore other combinations of rewards than the classical sum, so we will note $L_1 \bullet L_2$ the resulting combination of instantaneous reward L_1 at stage 1 and instantaneous reward L_2 at stage 2. V_k is the benefit to go until the end of the operating horizon. . The sign "• " is be used for a combination operator for rewards that can also be understood considering the function ϕ so that ϕ (a, b) = a • b. What are the properties necessary for the combination operator "• "?

To apply valid dynamic programming techniques, taking into account the transition function $x_n = f(x_{n-1}, u_{n-1})$

we can write :

$$V_1(x_1) = \sup_{u_1 u_2} \left\{L_1(x_1, u_1) \bullet L_2(x_2, u_2)\right\} \quad (25)$$

$$V_1(x_1) = \sup_{u_1} \left\{\sup_{u_2} \left[L_1(x_1, u_1) \bullet L_2(x_2, u_2)\right]\right\} \quad (26)$$

We would like to write :

$$V_1(x_1) = \sup_{u_1} \left\{ L_1(x_1, u_1) \bullet \sup_{u_2} \left\{ L_2(x_2, u_2) \right\} \right\} \quad \text{.......(27)}$$

$$V_1(x_1) = \sup_{u_1} \left\{ L_1(x_1, u_1) \bullet V(x_2) \right\} \quad \text{.......(28)}$$

Let

$\phi\,(a, b) = a \bullet b$

$a = L_1(x_1, u_1)$

$b = L_2(x_2, u_2)$

$b^* = \sup_{u_2} L_2(x_2, u_2) = V_2(x_2)$ so that $b^* = \operatorname{Sup} b$

The necessary condition to go from Equations (26) to (27) can be expressed as:

$$\sup_{b} \phi\,(a, b) = \phi\,(a, b^*) \quad \text{.......(29)}$$

One can easily verify that a sufficient condition is

$$b < b^* \Rightarrow a \bullet b < a \bullet b^*, \ \forall\, a \quad \text{.......(30)}$$

This property is known as the temporal monotonicity condition.

For instance , this condition will hold if ϕ is continuous and differentiable such that $\frac{\partial \phi}{\partial b}(a,b) > 0, \ \forall\, a$

More specifically, let us examine different candidates for "$\bullet$ " the operation of combining rewards:

- if $a \bullet b = a + b$, "$\bullet$ " verifies (30) and this leads to a criterion

$J = \sum L_i$ which is the classical sum of rewards

- using $a \bullet b = a \,.\, b$, $a > 0$ also verifies (30) and this leads to a criterion

$J = \prod L_i$. This product of rewards is very important in reliability theory because the global reliability index for systems in series is expressed in such a multiplicative form.

- $a \bullet b = b$ is the right absorption for which we obtain the terminal condition criterion $J = L(x_{final})$.

- for $a \bullet b = \mathrm{Min}(a, b)$, the criterion is $J = \underset{u_i}{\mathrm{Min}} \left\{ L_i \right\}$: we can see that the sufficient condition for $\bullet$ is fulfilled. Considering the three possible cases for a quantity a compared to b it is shown below that one always gets :

	$\phi(a, b)$	$<$	$\phi(a, b^*)$
$a < b < b^* \rightarrow$	a	$<$	a
$b < a < b^* \rightarrow$	b	$<$	a
$b < b^* < a \rightarrow$	b	$<$	b^*

This criterion gives also a valid dynamic programming formulation and can provide an interesting objective to guarantee a minimum value in resources allocation: that is the case for instance in firm energy generation with hydropower plant.

Multicriterion dynamic programming

Another main advantage to search optimal feedback policy by dynamic programming is that it can naturally be extended to multicriterion sequential decision making . SNIEDOVITCH(1986) underlines the advantages of applyiung such an approach to management problems. Technical arguments and theoretical justifications can be found in MITTEN (1974).

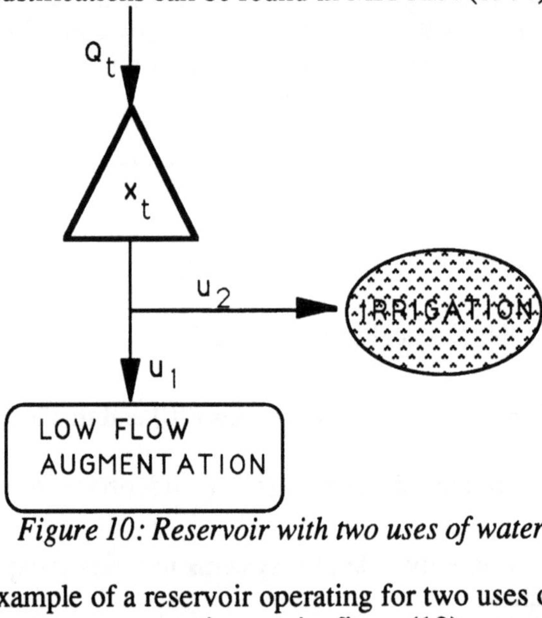

Figure 10: Reservoir with two uses of water

If we take the example of a reservoir operating for two uses of water , namely irrigation and water augmentation as in figure(10), we see that for each allocation vector $\begin{pmatrix} u_1 \\ u_2 \end{pmatrix}$ we can define a partial order or preference relationship:

$$\begin{pmatrix} u_1 \\ u_2 \end{pmatrix} \underline{\text{is preferred to}} \begin{pmatrix} u_1' \\ u_2' \end{pmatrix} \text{if}$$

$u_1 > u_1'$ (more water for quality) and

$u_2 > u_2'$ (more water for irrigation)..(31)

but situations of incomparability exist $\left(u_1 > u_1' \text{ and } u_2 < u_2'\right)$

In this case the notion of optimum is generalized in the concept of Pareto set that contains all the non dominated solutions at which a given criterion cannot be improved without compensation from at least one of the other criteria.

Let us now introduce the combination of rewards to examine how the dynamic aspect of the system interferes with this preference relationship. Suppose that the augmentation global index V_1 is the addition of instantaneous reward $l_1(u_1,x,t)$ and , due to agronomical considerations related to crop growth, the global yield V_2 takes a multiplicative form of immediate responses $l_2(u_2,x,t)$. The dynamic combination of rewards can be rewritten as:

$$\begin{pmatrix} l_1 \\ l_2 \end{pmatrix} \bullet \begin{pmatrix} V_1 \\ V_2 \end{pmatrix} = \begin{pmatrix} l_1 + V_1 \\ l_2 V_2 \end{pmatrix} \qquad (32)$$

We note that this operator • respects a temporal monotonicity condition (30) in the sense that for any pair (a, b>0) we can prove that :

$$\begin{pmatrix} V_1 \\ V_2 \end{pmatrix} >> \begin{pmatrix} W_1 \\ W_2 \end{pmatrix} \Rightarrow \begin{pmatrix} V_1 \\ V_2 \end{pmatrix} \bullet \begin{pmatrix} a \\ b \end{pmatrix} >> \begin{pmatrix} W_1 \\ W2 \end{pmatrix} \bullet \begin{pmatrix} a \\ b \end{pmatrix} \qquad (33)$$

because:

$$\begin{matrix} V_1 > W_1 \\ V_2 > W_2 \end{matrix} \quad \Rightarrow \quad \begin{matrix} V_1 + a > V_2 + a \\ W_1 \,.\, b > W_2 \,.\, b \ (b > 0) \end{matrix}$$

this will make it possible to give a valid dynamic formulation to the problem:

$$\begin{bmatrix} V_1\left(x_{(k)}, k\right) \\ V_2\left(x_{(k)}, k\right) \end{bmatrix} = \underset{\vec{u}_{(k)}}{\text{Opt}} \left\{ \begin{bmatrix} L_1\left(x_{(k)}, u_{(k)}\right) \\ L_2\left(x_{(k)}, u_{(k)}\right) \end{bmatrix} \bullet \begin{bmatrix} V_1\left(x_{(k+1)}, k+1\right) \\ V_2\left(x_{(k+1)}, k+1\right) \end{bmatrix} \right\} \qquad (34)$$

An illustrative example of multicriterion dynamic programming

We illustrate this sequential approach to a hypothetical case study given in figure 11 that works in the same spirit of our agronomical reservoir example. Each arc linking two nodes represents a possible decision to reach a new state from a given state at the next stage and the two numbers on each arc represent bicriteria rewards that are combined according to Equation (32).

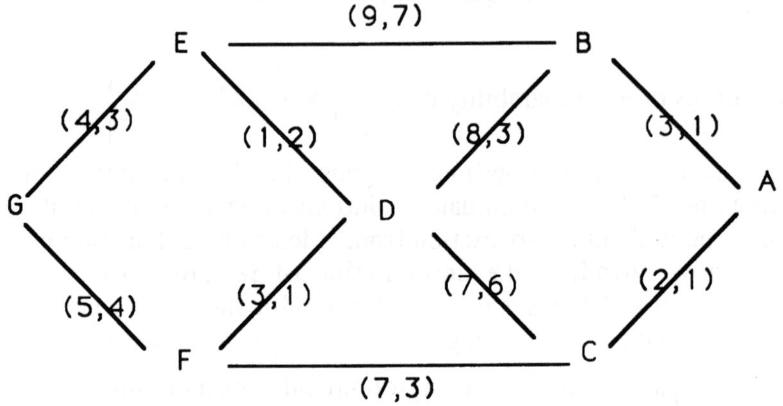

Figure 11: A dynamic bicriterion problem

Applying Bellman's principle at each stage , we keep recursively the non dominated solutions for the rewards to go as indicated on figures 12,13,14 and15.

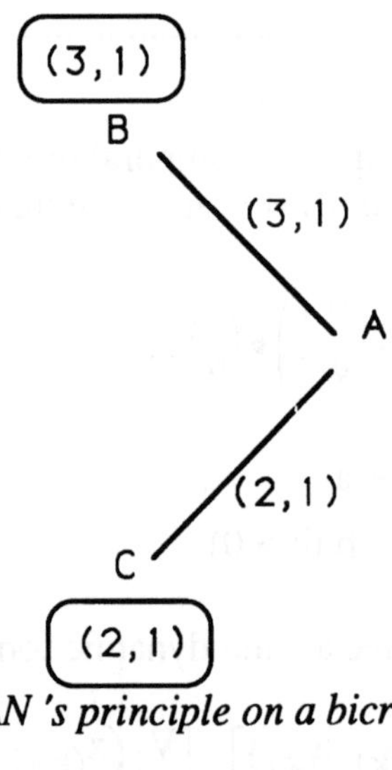

Figure 12: BELLMAN 's principle on a bicriterion problem; step 1

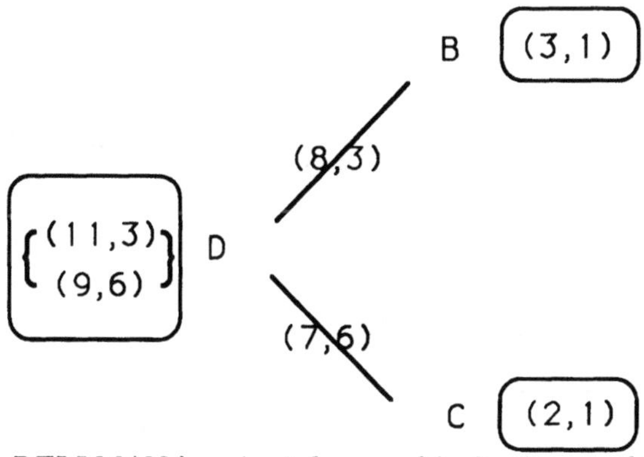

Figure 13: BELLMAN 's principle on a bicriterion problem; step 2

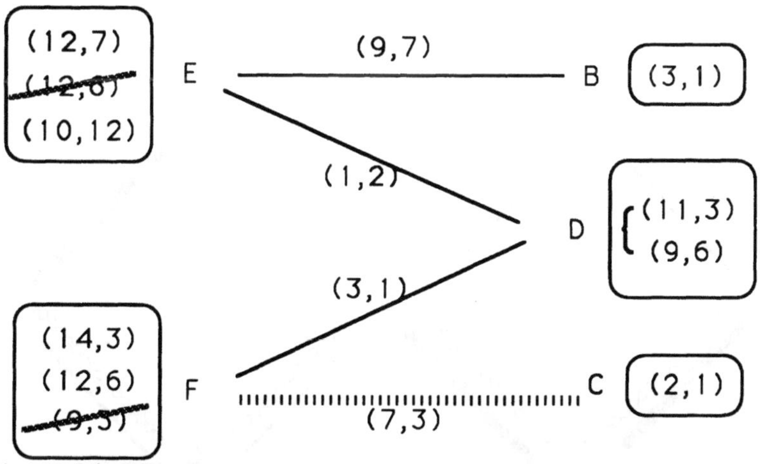

Figure 14: BELLMAN 's principle on a bicriterion problem; step 3

For instance in figure 14 , the decision linking state F to state C is not kept because it is dominated by one of the solutions obtained by choosing to go from F to D.

Figure 16 shows the three Pareto solutions given by this sequential construction. The property expressed in Equation (33) guarantees that these three are the only possible ones.

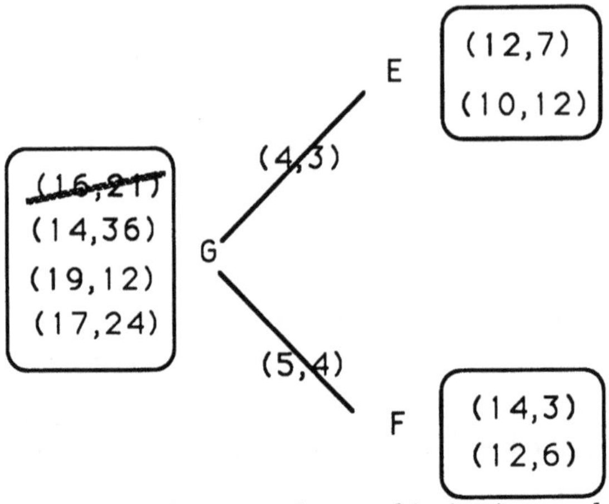

Figure 15: BELLMAN 's principle on a bicriterion problem; step 4

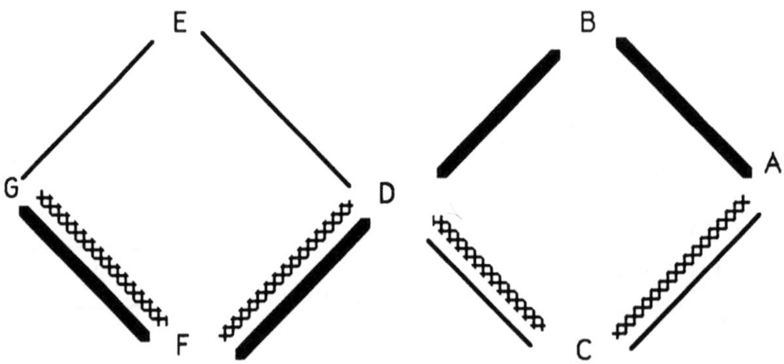

Figure 16: Pareto solutions for a bicriterion problem given by DP

More detailed considerations and algorithmic shortcuts can be found in HENIG(1985).

Risk management in a multicriterion framework

This natural outgrowth of dynamic programming to multicriterion sequential problems is very promising for risk management because one can ultimately say that reliability analysis is definitely a matter of multicriterion decision making. Indeed the methodology of risk management presented in DUCKSTEIN et al.(1987) may synthetically be schematized as follows:

-list all the incidents that can happen in a system by asking the decision maker what he wants to avoid by a satisfactory policy;

- for each potential incident, three indices can be defined:

* one for risk, that is the probability that a non desirable event occurs,

* another one for resilience, a concept connected to the system recovery time necessary once such an incident has occurred;

* the last index is for vulnerability, a notion related to the severity of the damages that can be caused.

Consequently, after starting a vague discussion with the system manager who had no clear definition of what he wanted at first, the operations research engineer ends up with a lot of criteria that have to be traded off considering the sequential aspect of the decision making process.

Unfortunately the previous multicriterion dynamic programming approach does not work if stochastic transitions are involved. New theory has to be developed in this case and this will provide a general framework to risk analysis for dynamic systems. Recent efforts in this new field of research have been proposed in KARLSSON and HAIMES (1988) . Such developments have to be compared to aggregation methods incorporating the attitude of the decision maker toward risk. For example multiattribute utility functions have been applied to agro-hydraulics engineering studies in KRZYSZTOFOWICZ(1982) to give a "satisficing" multiuse allocation of water.

5 CONCLUDING COMMENTS

A systems analysis approach can be performed to design a model for natural resources management. Its foundations lies on known techniques of control theory. Bellman's principle can be used to find an optimal policy respecting the sequential aspect of the decision process. Such an approach is very well fitted for simulation because it is composed of elementary blocks that are gathered in a transition equation as (1) or (2) to describe the system's dynamic evolution. It can help to derive efficiently optimal feedback rules of operation and can routinely deal with complex decisions. Its main drawback is due to the curse of dimensionality that allows only simplified systems or the very particular case of quadratic linear Gaussian models to be efficiently solved in practice.

At the cost of forgetting some necessary hypotheses for application of rigourous mathematical theorems, such defaults may not seem essentially relevant for application purposes. JACOBSON and MAYNE's techniques(1970) essentially "linearize" the transition function and "quadraticize" the cost function.

But we have to accept the cost of enforcing the assumptions made with regards to physics, hydraulics, hydrology and so on; in particular, the Markovian property of the controlled system is certainly an important point to be critically examined.This class of models relies on short memory assumptions and can of course represent neither long memory nor persistence effect nor hysteresys in the physical evolution of the phenomena under consideration.

On the other hand, it provides a means to get an optimal policy as long as the system manager accepts the necessity to formulate his objective function consistently with dynamic programming (i-e stage separable and "•_additive"). Model uncertainties may also influence or disguise the real interactions between system response and stochastic dynamics so that one should handle with care the concept of optimality in such conditions. In water resources management for instance, KLEMES (1978) has proved that model uncertainties can have the same order of magnitude as the computed "optimized" results themselves. More specifically to the field of risk management, this approach offers very interesting potential developments for the following reasons:

1) Risk performance indices can be sequentially computed indirectly if they can be expressed under the form of a Bellman-like backward relationship or they can be directly used as a criterion to be optimized, provided the dynamic structure of this criterion respects Equation (30) or Equation (33) in the multicriterion case.
2) Multicriterion sequential decision making opens a promising field of research for risk management. New techniques and theoretical advances should be made in stochastic multicriterion dynamic programming and compared to present techniques.
3) Such models are designed to serve only as decision making aids. Their most important contribution is a rational way of establishing constantly updated multidimensional risk estimations that can help the system manager to select a "satisficing" operation rule.

BIBLIOGRAPHY

Alexéev V., Galéev E., Tikhomirov V. (1987), Recueil de problemes d'optimisation, Editions Mir.
Arnold L. (1974), Stochastic differential equations: Theory and applications. Wiley.
Askew J.A. (1974), Optimum reservoir operating policies and the imposition of a reliability constraint. Water resources research 10, 1, p.51-56.
Bellman R.E. (1957), Dynamic programming. Princeton University press.
Beltrami E. (1987), Mathematics for dynamic modeling. Academic Press.
Denardo E.V. (1982), Dynamic programming. Prentice Hall.
Duckstein L., Plate E.J. and Benedini M. (1987), A System framework in "Engineering reliability and risk in water resources "(Duckstein & Plate), Dordrecht.
Faurre P, Robin M. (1984), Elements d'automatique. Dunod.
Foufoula-Georgiou E. and Kitanidis P. (1988), Gradient dynamic programming for stochastic optimum control of multidimensional water resources systems. Water resources research.
Friedland B. (1986), Control system design. McGraw-Hill, NY.
Gal S. (1979), Optimal management of a multireservoir water system, Water resources research 15, p.737-749.
Hashimoto T., Stedinger J.R. and Loucks D.P. (1982), Reliability, resiliency, and vulnerability criteria for water resource system performance evaluation. Water resources research 18, 1, p.14-20.
Henig M. (1985), The shortest path problems with two objective function, European J. of operation research 25, p.281-291.
Jacobson D. and Mayne D. (1970), Differential dynamic programming. American Elsevier Cie.
Karlsson P. and Haimes Y. (1988), Probability and their partitioning. Water resources research 24, 1, p.21-29.
Kitanidis P.K. (1987), A first order approximation to stochastic optimum control of reservoirs, stochastic hydro hydraul., p.169-182.
Klemes V. (1977), Discrete representation of storage for stochastic reservoir optimization. Water resources research 13, 1, p.149-158.
Klemes V. (1978), The unreliability of reliability estimates of storage reservoir performance based on short streamflow records. In "Water resource management, Water Resource Publications, Fort Collins, Colorado", p.193-205.
Kree P. and Soize C. (1983), Mecanique aleatoire. Bordas.
Krzysztofowicz R. (1982), Utility criterion for water supply: Quantifying value of water and risk attitude. In "Decision-making for hydrosystems-Forecasting and operation" (T. Unny), 43-62.
Li D. and Haimes Y. (1987), Risk management in a hierarchical multiobjective framework. In "Towards interactive and intelligent decision support system" (eds. Y Sawaragi, K. Inoue and H. Nakayama) Springer Verlag.
Masse P. (1959), Le choix des investissements. Dunod.
Mitten L. (1974), Preference order Dp., Management science 21, p.43-46.
Parent E., Lebdi F. and Hurand P. (1991), Gestion strategique du systeme Neste. to appear in Revue des sciences de l'eau.
Saad M. and Turgeon A. (1988), Application of principal component analysis to long term reservoir management, Water resources research 7, 1.
Simonovic S.P. and Orlob T.G. (1984), Risk-reliability programming for optimal water quality control. Water resources research 20, 6, p.639-646.
Sniedovith M. (1986), A new look at Bellman Principe of optimality, Journal of optimization theory and optimization.
Tapiero C.S. (1988), Applied stochastic models & control in management. In "Advanced series in management", V12. North-Holland.
Turgeon A. (1980), Optimal operation of multireservoir power systems with stochastic inflow. Water resources research 16, 2, p.275-283.

Turgeon A. (1981), Optimal short term hydro scheduling from the principle of progressive optimality. Water resources research 17, 3, p.481-486.

Yakowitz S. (1982), Dynamic programming applications in water resources. Water resources research 18, 4, p.673-696.

Yeh W.W.G. (1985), Reservoir management and operations models: A state-of-the-art review. Water resources research 21, 12, p.1797-1818.

PROBABILISTIC MODELLING OF WATER QUALITY IN RIVERS

Erich J.Plate, Institute for Hydrology
and Water Resources Planning
University of Karlsruhe
Federal Republic of Germany

Abstract

The problem of determining the concentration of pollutants in rivers from point and non-point sources is discussed formally as a problem of risk analysis. First, the case of the origin of pollutants from non-point sources is considered, and the analysis is concentrated on the case of a pollutant which is adhering to the surface soil and which is thus carried into a river. Basis of the analysis is the Universal Soil Loss Equation in modified form, or the MUSLE.

Secondly, a stochastic approach to transport of pollutants in a river is presented. Discharge and pollutant input are considered random variables, which are convected downstream and dispersed. The cases of continuous and of event input of pollutants are treated separately. For the event based case, a formulation, due to Beltaos, is selected, by means of which the dispersion can be quantified through simple parameters.

Input models and river models are combined to determine the probability of exceeding a safe concentration. As an example the critical safe distance is determined for the case of an accidental spill, and a formal presentation is given for the case of a continuous (normally distributed) random input of pollutant mass.

INTRODUCTION

Surface runoff from fertilized fields is considered an important source of pollution for surface waters ranging from rivers to lakes. For remedying potentially critical pollutant situations, it is necessary to quantify the amount of pollutant carried by the waters. A quantification of the pollutant load must be done by means of stochastic models, because both input of fertilizers etc and land use is continuously changing in space and time. These changes cannot be predicted with certainty. The most complete stochastic approach requires a long term simulation model, based on a deterministic process model with stochastically variable inputs and parameters, as described for example in Plate and Duckstein (1990), and which is feasible only if the physical situation is suitably simplified. The simplification is in part based on considering only special types of pollutants. Typical classes of pollutants that must be considered are those that adhere to the soil surface (like phosphates), those that act like simple tracers (such as salt NaCl), and those that interact physically and chemically with the soil and with other substances (like nitrates). Here, only the case of erosion generated pollutant is considered.

NATO ASI Series, Vol. G 29
Water Resources Engineering Risk Assessment
Edited by J. Ganoulis

Further simplifications depend on the time and space scales of the situation: a model for annual pollutant load can be based on less detailed models, with appropriate stochastic variability, than a model which is intended for the purpose of determining the exceedance of a critical concentration in a river. A compromise between models for long term averages and models for event based long term simulation is a class of models based on uncorrelated events, which permit the determination of probability density functions for pollution inputs during events. In this paper, such an event based stochastic model is outlined, which is intended to guide the analysis of field data, and which can be calibrated by suitably chosen data.

The structure of a stochastic river pollution model

We consider the situation depicted in Fig.1. A river basin exists whose area is partly covered with agricultural fields. Crops grown on each field are chosen by the farmers, and so are amounts of fertilizer, insecticides, pesticides and herbicides. These substances become pollutants if they reach the creek or river into which rainfall from fields is draining. The paths of pollutants to a point of interest denoted by P in Fig.1 - for example, a point of water extraction - depend not only on natural topographic and geological conditions of the area, but also on actions taken by farmers, and by the interaction of soil and crops and rainfall. All these factors combine to yield the time variable concentration in the river, which is described by concentration c(t) at P.

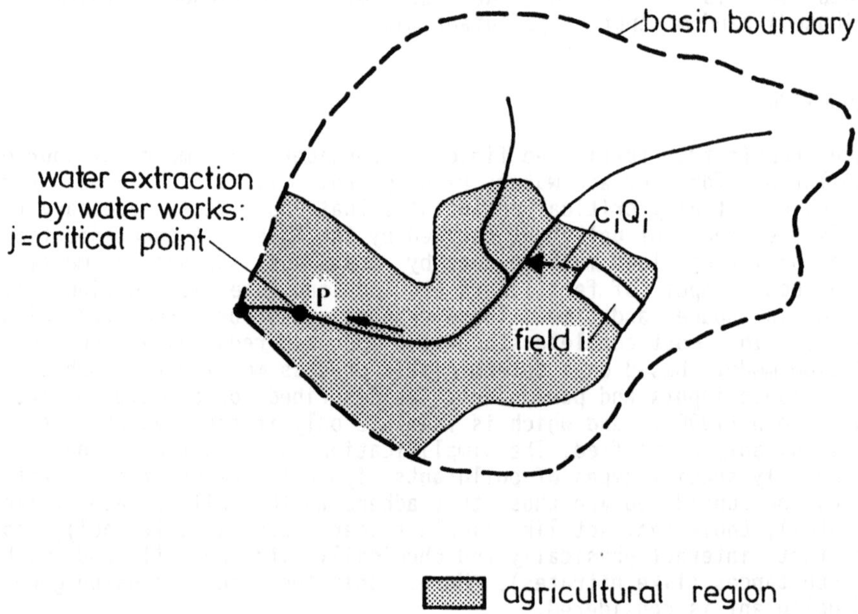

Fig.1: Schematic representation of pollutant runoff from fields

A stochastic non - point pollution model for this situation must consist of three parts, as is shown in Fig.2. The first is the input model, in which the inputs are specified. They consist of two components. The pollutant is generated by fertilizer inputs into each field, which is identified as M_i for the i-th field. The pollutant is transported to the river by runoff generated from space and time distributions of rainfall, because rainfall provides the transporting agent for the pollutant transport, The input model also includes process models for the fields, which convert rainfall into runoff, and fertilizer mass into fertilizer concentration c_i in the runoff from field i. The input model provides the loading of the river.

The second part is a transport model for the river, by means of which transport and mixing in the river is specified. This is also a process model, because it incorporates transport by convection and by diffusion between point of discharge from each field i into the river and point P. This model yields a time series of the concentration c(t) and of discharge Q(t) in the river at point P.

The third model expresses the objectives which should be satisfied by the model and is called the decision model. In its most elementary form, the overall model consists of mass balance relationships with statistical parameters, and decisions +X'X level will thus have to be made on a statistical basis. The problem associated with decision making on the basis of pollutant concentration in a river may then be taken as a design problem based on failure probability, to which standard techniques of reliability analysis (Plate and Duckstein, 1988) or of stochastic design (Ang and Tang, 1984, Duckstein and Plate, 1987) can be applied.

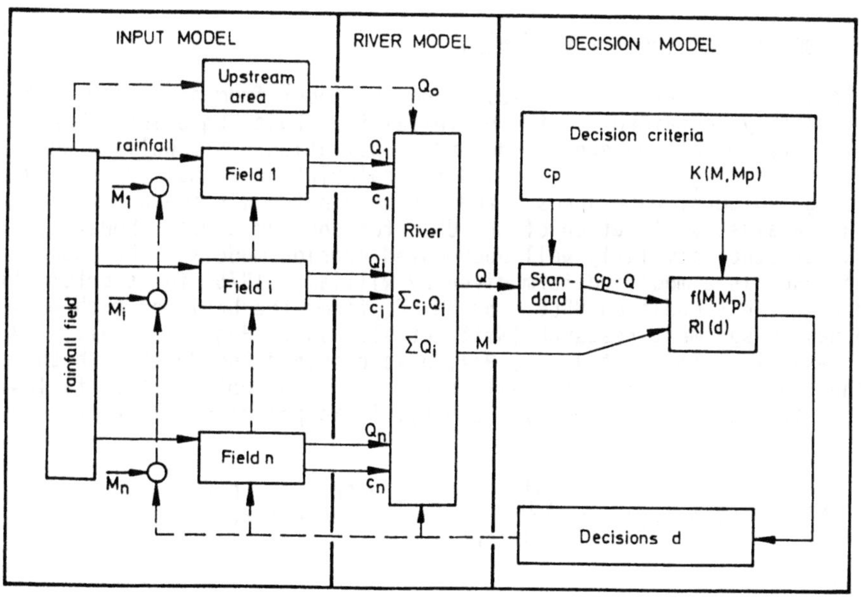

Fig.2: System of a river pollution model for non - point sources

We see from this discussion that each of the three models consists of an input x(t), a transfer function g(x,t), and an output y(t). Let us consider the three models as they are integrated into a stochastic water pollution model. It shall be the purpose of the next sections to specify x(t), g(x,t) and y(t) for each of the three submodels.

INPUT MODELS

The purpose of the input model is to provide time distributions of pollutants at the inflow points along the river, for a given rainfall and pollutant input. Therefore, rainfall and pollutant application represent the input vector, the conversion of these inputs into output concentration is caused by the transfer function, which depends on the properties of the land and of the path through which the pollutant reaches the river. There exist different methods for describing the transfer process, most of which are of more or less limited validity. Here, we shall be concerned with two types of inputs. The first one is an event based input, which is caused by a short duration point source, such as an accidental spill. For this input, we need to know the total mass M which enters the river, and to determine its downstream progress and development. The most important aspect of this type of model consists of an appropriate description of the transfer process in the river, which we shall consider in the next chapter.

The second type of input is that from agricultural fields, which is caused by pollutants adhering to the soil which erodes from the fields. This situation will concern us next.

Input models for input from fields

The best tested method for calculating the input from agricultural areas is probably obtained if the Universal Soil Loss Equation (USLE) is applied for modelling erosion dependent pollution. This well known equation in its original form is valid for calculating annual sediment yield from agricultural watersheds, and its originator Wischmeyer expressedly warns against application of the USLE for shorter times. However, for lack of other similarly well documented erosion models, the USLE has been used in a modified form given by Williams (1975) for shorter time periods. This modified USLE, or MUSLE, forms the basis of much work on stochastic sediment research (Smith et al., 1977, Bogardi et al., 1985). For example, Smith et al. (1977) have combined the MUSLE with the SCS method of calculating rainfall abstractions to obtain the following expression for the sediment yield M(r), in kg per rainfall event r:

$$M(r) = c \cdot w \left[\frac{a_0 y_r^4}{(y_r+f)^2 \; (a_2 \cdot T_r + T_c)} \right]^{0.56} = c \cdot w \cdot z_r^{0.56} \qquad (1)$$

In Eq.1, c is the concentration depending on the application of pollutants to the field, w and a_0 are conversion factors incorporating size of field, land use etc, y_r = effective rainfall depth per event r (in mm), for r =1,2,..N, where N is the number of events per season, f is the constant describing total infiltration (in mm), a_2 is a dimensionless constant, T_r = rainfall event duration (in hours), and T_c is the concentration time, i.e. the time which elapses between the beginning of the rainfall and the occurrence of the peak runoff. For details of the generation of this model and for numerical values of the model parameters, the original paper by Smith et al. (1977) or Bogardi et al. (1985) should be consulted.

The structure of Eq.1 makes it clear that the input for this model must be an event based rainfall. The actual rainfall event is approximated by a block rainfall with basis T_r and height y_r. In this way, the rainfall event is described by the two random variables **T** and **y** and thus, the random distribution of rainfall events can be approximated by the joint probability density $f_{yT}(y,T)$. Eq.1 has been solved numerically (by Smith et al., 1977) with rainfall event inputs which have a joint pdf for rainfall intensity and rainfall duration described by a bivariate exponential pdf $f_{yT}(y,T)$, with Poisson distributed number N of events per season. It has also been solved by simulation in Bogardi et al.(1985). Both groups of authors also considered the uncertainty of the parameters which are used both in the input as well as in the process models.

If the function $f_{yT}(y,T)$ is known either theoretically, or experimentally, then the method for finding the erosion rate is based on the random variable z, which is obtained as schematically illustrated in Fig.3 . The random variables **y** and **T** are the coordinates of a plane in which isolines of the joint pdf $f_{yT}(y,T)$ are indicated. In this plane, the condition z = constant calculated from Eq.1 leads, for different values of the constant, to a family of curves which is also shown schematically in Fig.3. The cumulative distribution $F_z(z)$ is calculated for each z = Z by integration of $f_{yT}(y,T)$ over the area in which $z < Z$, as is indicated by the cross hatched region of Fig.3. Finally the probability density function $f_z(z)$ is found from $F_z(z)$ by differentiation.

As an example, let us consider a simplified version of this problem for illustrative purposes, in which it is assumed that for each rainfall event r rainfall depth y_r and duration T_r are uncorrelated exponentially distributed variables.

Example 1: Erosion based on a simplified form of the MUSLE, and with uncorrelated inputs of rainfall events.

<u>1.Problem formulation and basic equations:</u> We assume that rainfall duration T and height y can be modelled by an exponential distribution. Experimental evidence for this has been provided by Julien and Frenette (1984), as is shown in Fig.4.

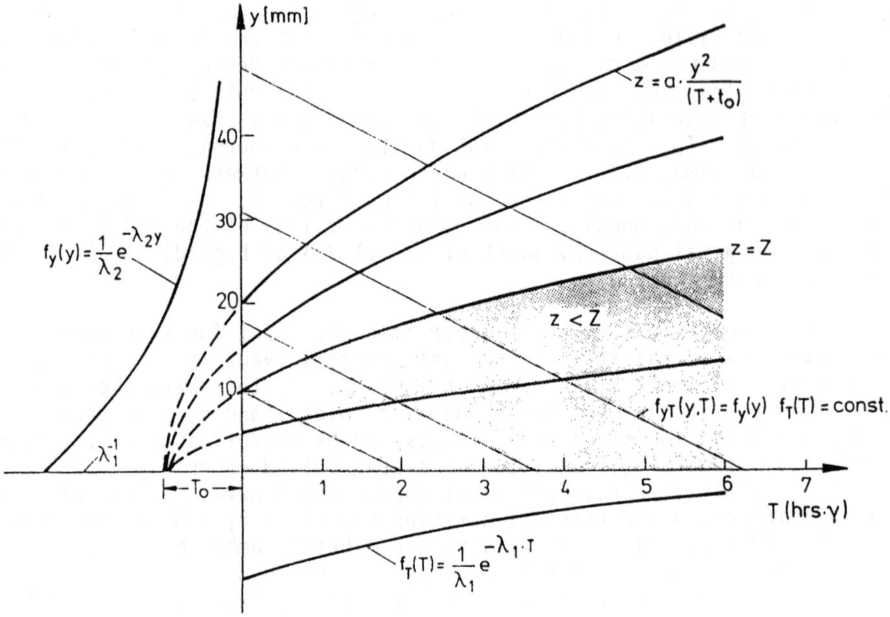

Fig. 3: Schematic representation of the function z in the plane of the random variables T and y. The parabolas are curves of constant z, the straight lines are lines of constant joint probability $f_{yT}(T,y) = f_T(T) \cdot f_y(y)$

$$f_T(T) = \frac{1}{\lambda_1} e^{-\lambda_1 \cdot T} \tag{2}$$

$$f_y(y) = \frac{1}{\lambda_2} e^{-\lambda_2 \cdot y} \tag{3}$$

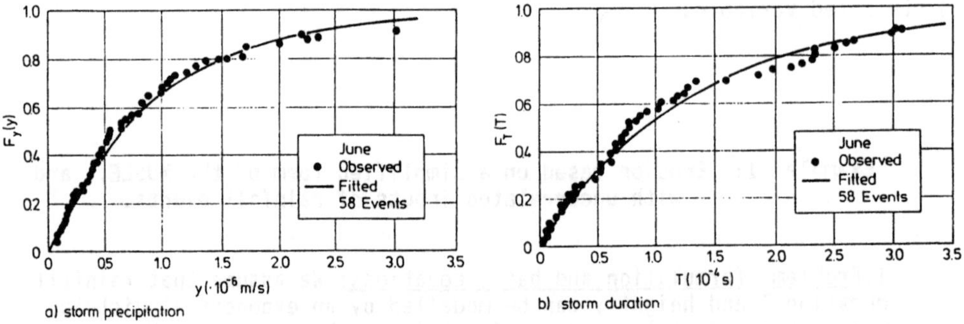

Fig.4: Rainfall depth and rainfall duration, from Julien and Frenette, (1984)

Modelling the erosion by the modified Universal Soil Loss Equation, according to Bogardi et al,(1978) and Smith et al.(1977) takes place by means of Eq.1. Here, we consider the case where Eq.1 is simplified by making the following assumptions:

$$\left.\begin{aligned} & y + f = a_1 \cdot y \\ & a = \frac{a_0}{a_1 \cdot a_2} \qquad t_0 = \frac{T_c}{a_2} \end{aligned}\right\} \tag{4}$$

where a can be considered a general scaling factor, into which all effects of area specific parameters are lumped, except for the event specific quantities T and y.

In this way we get:

$$z = a \cdot \frac{y^2}{T + t_0} \quad \rightarrow \quad y^2 = \frac{z}{a} \cdot (T + t_0) \tag{5}$$

Also, T and y are assumed to be uncorrelated, so that:

$$f_{Ty}(T,y) = f_T(T) \cdot f_y(y)$$

The situation is shown formally in Fig.3. We can see that the lines of constant z are parabolas, which in the region for positive T become almost straight lines: a result which is even better satisfied if one uses $z^{0,56}$ instead of z, which according to Eq.1 is almost linear. From Fig.3 we infer that the probability density $f_z(z)$ can be determined by integration over the cross hatched region, which yields the probability function $F_z(z)$ and from which $f_z(z)$ is obtained by differentiation. This leads to the function for the probability density $f_z(z)$ for z which is found from Equ.6 (Papoulis,1965):

$$f_z(z) = \int_{-\infty}^{\infty} f_{T,y}(y^* | T) \cdot f_T(T) \cdot \left| \frac{dy^*}{dz} \right| \cdot dT \tag{6}$$

where y* is the value of y resulting from z and x. For stochastically independent variables T and y Eq.6 becomes:

$$f_z(z) = \int_{-\infty}^{\infty} f_y[y^*(T,z)] \cdot f_T(T) \cdot \left| \frac{dy^*}{dz} \right| \cdot dT. \tag{7}$$

<u>2.Determination of $f_z(z)$:</u> For determining $f_z(z)$, the solution for Eq.7 is found as follows. First we consider the function y^*, for which we find:

$$y^* = \sqrt{z \cdot (T + t_0)/a} \tag{8}$$

and therefore:

$$\frac{dy^*}{dz} = \frac{\sqrt{(T + t_0)/a}}{2 \cdot \sqrt{z}} \tag{9}$$

If we insert Eq.8 and 9 into Eq.7 we obtain with:

$$f_y(y^*) = \frac{1}{\lambda_2} e^{-\lambda_2 \cdot \sqrt{z \cdot (T + t_0)/a}} \tag{10}$$

the result:

$$\begin{aligned} f_z(z) &= \int_0^\infty \frac{1}{\lambda_1 \cdot \lambda_2} e^{-\lambda_1 \cdot T} \cdot e^{-\lambda_2 \cdot \sqrt{z \cdot (T+t_0)/a}} \cdot \frac{\sqrt{a \cdot (T+t_0)/a}}{2 \cdot \sqrt{z}} \cdot dT \\ &= \frac{e^{+t_0}}{\lambda_1 \cdot \lambda_2} \cdot \int_{t_0}^\infty e^{-\lambda_1 \cdot T'} \cdot e^{-\lambda_2 \cdot \sqrt{z \cdot T'/a}} \cdot \frac{\sqrt{T'/a}}{2 \cdot \sqrt{z}} \cdot dT' \end{aligned} \tag{11}$$

where:

$$T + t_0 = T'.$$

Combining the exponents and completing the square yields:

$$-\lambda_1 \cdot \left[t' + \frac{\lambda_2}{\lambda_1} \cdot \left[\frac{z}{a} \cdot t'\right]^{1/2}\right] = -\lambda_1 \cdot \underbrace{\left[t' + \frac{\lambda_2}{\lambda_1}\left[\frac{z}{a} \cdot t'\right]^{1/2} + \frac{1}{4} \cdot \frac{\lambda_2^2}{\lambda_1^2} \cdot \frac{z}{a}\right]}_{\xi^2} + \frac{\lambda_2^2}{4\lambda_1} \cdot \frac{z}{a} \tag{12}$$

Therefore: $$\xi = \sqrt{\lambda_1} \cdot \left(\sqrt{t'} + \frac{1}{2} \cdot \frac{\lambda_2}{\lambda_1} \cdot \sqrt{\frac{z}{a}}\right) \tag{13}$$

$$\left.\begin{aligned} \text{and:}\qquad d\xi &= \frac{\sqrt{\lambda_1}}{2\sqrt{t'}}\cdot dt' \\ \sqrt{t'} &= \frac{\xi}{\sqrt{\lambda_1}} - \frac{1}{2}\cdot\frac{\lambda_2}{\lambda_1}\cdot\sqrt{z/a} \\ \xi &= \sqrt{\lambda_1}\cdot\left(\sqrt{t_0} + \frac{1}{2}\cdot\frac{\lambda_2}{\lambda_1}\cdot\sqrt{z/a}\right) \end{aligned}\right\} \tag{14}$$

Substituting these expressions into Eq. 12 yields:

$$f_z(z) = \frac{1}{z\cdot\lambda_2\sqrt{a\cdot\lambda_1^3}}\cdot e^{t_0}\cdot e^{4\cdot\lambda_1\cdot a}\int_{\xi_0/\sqrt{\lambda_1}}^{\infty} e^{-\xi^2}\cdot\left[\frac{\xi}{\sqrt{\lambda_1}} - \frac{1}{2}\cdot\frac{\lambda_2}{\lambda_1}\cdot\sqrt{z/a}\right]^2\cdot d\xi \tag{15}$$

The details of the additional calculations are omitted. It can be seen that the integral can be decomposed into integrals of the form of incomplete gamma functions or of error functions.

An advantage of using Eq.1 is that it can be used also with experimentally determined rainfall functions $f_{yT}(y,T)$ which have been studied extensively for other purposes, such as urban hydrology (Xanthoupolos, 1990). In Germany, it has been found that a few types of empirical functions $f_{yT}(y,T)$ are characteristic of large areas. An example is shown in Fig.5 from which it follows that the bivariate exponential is not a very satisfactory distribution. (The example of Fig.5 is for illustrative purposes only, and no attempt was made to fit this distribution to data from erosion experiments). In order to utilize such information more exactly, it will be necessary also to describe the other factors appearing in the USLE, such as a, as random variables.

However, for statistical purposes it has been found that assumptions as the ones used here are acceptable, because Bogardi et al (1985) have shown for the case of a number of reservoirs in Hungary, that the expected value of the annual sediment yield (which was obtained by multiplying the expected value of the sediment yield per event by the average number of events per year) was in reasonable agreement with the observed sediment depositions.

A number of authors have attempted to obtain better quantitative agreement of results from measurements with those calculated by means of a physically based and more accurate functional relation between rainfall and sediment runoff. Such models for the erosion process have been suggested, for example, by Julien (Julien and Frenette,1986, Julien and Dawod,1987), who used a more elaborate input hydrograph and an overland erosion model based on hill slope hydraulics. It is however, not easy to see how such models can really improve the quantification of pollutant

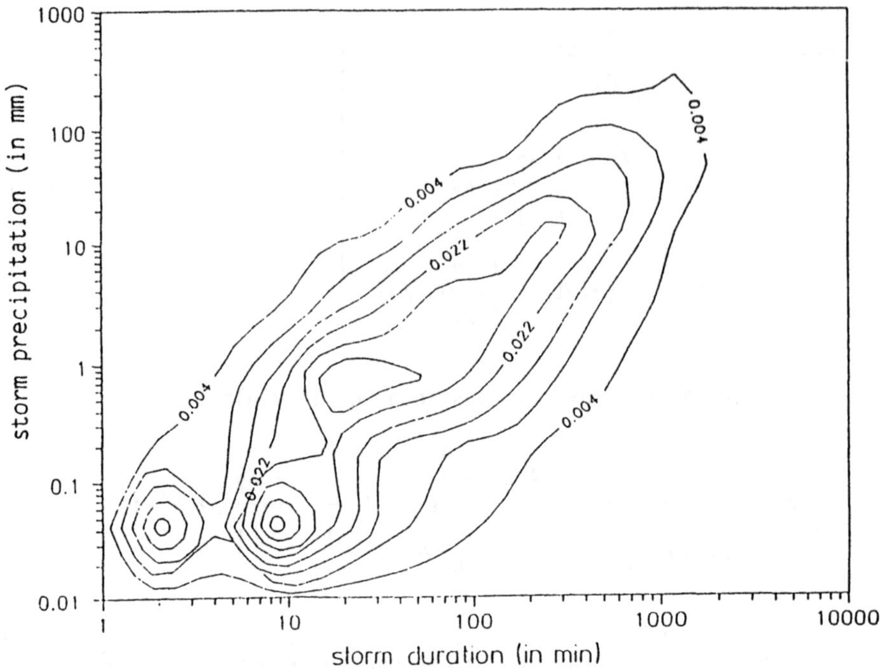

Fig.5: Typical plot of experimentally obtained joint probability density of rainfall duration T and rainfall height y (for station Essen - Steele, made available by A.Bardossy)

runoff from fields. The randomness of all parameters add so much uncertainty that the exercise of calculating exact probability distributions from physical models may not improve agreement of measurement and theory by very much.

The calculation of the total amount M(r) of pollutant from event r requires multiplication of the sediment rate with concentration c_r, as has been indicated in Eq.1. In this way water quality discharges which are directly proportional to the rate of surface erosion can be obtained. As an application of this approach, Bogardi et al.(1978) have calculated from Eq.1 the phosphorus load transported from a land surface. However, uncertainty is introduced by the fact that the amount of fertilizer or other pollutant that has been applied is not known with certainty, and that the distribution of this fertilizer over the top layer of the soil is not known. Both these quantities are likely to be random variables, so that the runoff of pollutant has an added variability as compared with the water and sediment runoff.

An alternate approach to pollutant runoff is to use a "pollutograph". This is a function constructed in the same manner as unit hydrographs h(t), i.e. the pollutant applied to the land serves as input into an impulse response function, $h_s(t)$ which is the pollutograph (Jury et al.,1986). The simplest shape of $h_s(t)$ is obtained if we assume proportionality between rainfall and pollutant, so that $Q_S(t) = a \cdot Q(t)$. However, other types of "pollutograph" are also useful: it is often noticed, for example, that due to the processes described above the peak of the pollutant wave arrives faster than that of the flood wave, which would imply a unit pollutograph with a shorter rise time than the flood unit hydrograph.

In general, unit hydrographs and "pollutographs" are used deterministically, i.e. a deterministic input function is used in conjunction with a deterministic h(t) function. However, it is not difficult to use a deterministic (and linear) h(t) with random event inputs, by employing simulation techniques. A special case is a random input consisting of a time series with normally distributed input magnitudes; in this case the output also is a normally distributed variable, with a variance which is determined through the spectrum function of the input process and the Fourier transform (or system function) of h(t). Some use of these relations have been made for groundwater models, see for example Geldner (1980).

Erosion models have been included into catchment models consisting of a network of channels and unit hydrographs of the type described by Plate et al. (1988) in order to calculate the rate of erosion from soil surfaces and deposited in reservoirs (Bogardi et al. 1985, Hrissanthou, 1986). Both authors used simulation methods to obtain the historical sequence of event based erosion yields, which were then summed over the year as the time unit. For the model the basin was subdivided into many small, approximately field size subbasins, to each of which the MUSLE has been applied. An annual precipitation index K was used, but event dependence was introduced by the use of a daily value of precipitation from which the sediment inflow into the reservoir was calculated for the daily effective rainfalls obtained from historical records. The differences between the two studies were in the way the MUSLE was used for the subbasins, and how they were linked for the total yield. For each partial area the assumptions made by Hrissanthou (1986) were about the same as for the model of Smith et al (1977), which has been described above. However, instead of using the SCS method for determination of the effective rainfall, he used a modification of the SCS method developed by Lutz (1982). Furthermore, the MUSLE included a channel sediment routing subroutine (Williams, 1975). In both cases the results were compared with measurements of annual reservoir deposition, and good agreement was found for average values, but considerable scatter for the individual values.

The MUSLE or analytical approaches like the one of Julien apply to a field and can in principle be used without new data, although no model exists at this time which can do without calibration. Of course, this holds also for pollutographs and distributed area conceptual models. Experiments are needed for this, and also to verify the concepts which have been expressed in this paper, and therefore, a large scale experimental program on pollutant runoff has been initiated at the University of Karlsruhe.

Input model for pollutant input along the river

For a river model we must determine the distribution of pollutant input along the river as function of distance and time. This is not directly obtained from the erosion model for a field. The result of the rainfall pollutant runoff model is a point load M_j of sediment which leaves the field during the rainfall event, and this has to be converted into a model for calculating the pollutant load for the river. We need to know how much of the sediment which is eroded from the field will reach the river, where it will reach the river, and how it is distributed over time. We need to have component models which combine many fields into inputs, and this requires that we know the delivery ratio in addition to the soil erosion from individual fields, i.e. we must have an estimate of the losses of eroded sediment in the course of the transport from field to river.

The erosion rate and the pollutant adhering to it from many fields enters the river at a point j, as is indicated schematically in Fig.1. The river originates in an unpolluted area, so that there is zero concentration in the river before it reaches point j. We assume that the rural area consists of many different fields, each of which is fertilized. However, application of fertilizer, land use, and fertilizer uptake by crops are not constant; for example, the rates which are applied to each field vary from field to field in a random manner.

A complete simulation procedure which should be based on such input models requires not only that the temporal variability of all input variables be considered appropriately, but also that physical processes which contribute to the conversion of input into pollutant loaded runoff be well known and quantifiable. In general, it will be impossible to quantify exactly all necessary processes, and the result may show a high degree of uncertainty. We therefore need a certain amount of stochasticity in all water quality models. Stochastic water quality models have been reviewed in Plate and Duckstein, (1990).

A feasible concept employing such a model is to describe the fertilizer runoff model for (daily) mean quantities, and to relegate all other dynamic effects to an uncertainty band around the mean. It is likely that such a strictly statistical model can provide results of similar quality as physically more developed deterministic models, provided it can be calibrated with reliable data.

It is furthermore assumed that only little runoff of fertilizer takes place during dry conditions: critical conditions for concentration c to exceed concentration c_p occur only during or after rainfall events. Let c_i be the concentration in the runoff Q_i from field i during a rainfall event, so that the total amount of pollutant from field i which reaches the river at point j is $c_{ij} \cdot Q_{ij}$. The total amount M of pollutant substance reaching point j on the river is also a random variable, which, with given discharge Q and average concentration c is given as:

$$M = \sum_{i=1}^{I} c_i \cdot Q_i \qquad i = 1, 2, \ldots, I \qquad (16)$$

where the subscript has been dropped, and the summation has to be extended over all I fields. Since the number of fields involved in producing M is assumed to be large, we may invoke the central limit theorem of statistics and assume that M is a normally distributed random variable, with probability density function:

$$f_M(M) = \frac{1}{\sigma_M \cdot \sqrt{2\pi}} \exp\{\frac{(M - \mu_M)^2}{2 \cdot \sigma_M^2}\} \qquad (17)$$

Experimental evidence, for example by Marani and Bendoricchio (1986), indicates that event based runoff from individual fields can be represented by a gamma distribution, or a generalized gamma distribution. Clarke (1990) has used the log-normal distribution for describing both Q_i and c_i, in which case $c_i \cdot Q_i$ is also log-normally distributed. All these distributions have the property based on the central limit theorem of yielding the normal distribution, if many variables of the same distribution are added. We shall assume that data are given on the basis of daily values, conditional on the occurrence of a rainfall event which is strong enough to cause erosion.

PROCESS MODELS FOR RIVER TRANSPORT

Continous processes of pollutant transport

During days of erosion causing rainfall the average concentration of pollutant in the river caused by runoff of sediment from fields is c, which is calculated as:

$$c = \frac{\sum_{i=1}^{I} c_i \cdot Q_i \cdot b_i}{Q} \qquad (18)$$

In this equation, Q is the discharge of the river. Let us assume that this discharge is unaffected by runoff from fields. However, since this runoff is generated by rainfall events which also increase discharge Q in the river, we must assume that a certain dependence exists between M and Q. The simplest type of a model which describes such a dependency consists of a one to one relationship between an input variable x and an output variable y of the form:

$$y = g(x) \qquad (19)$$

Such a relation is found for many water quality processes. They usually connect the variable y describing the transport of pollutant with the variable x describing the flow of transporting agent, which can be water directly, or indirectly through the flow of sediment runoff. The most important relations for process models of pollutants are of the form:

$$y = a.x^n \tag{20}$$

where a and n are (often empirical) parameters. For the case of pure dissolved pollutants, x is the discharge or the rainfall, n = 1, and y is the output concentration (Haith and Tubbs, 1981). For the case of solid phase pollutant , x is the eroded soil, and y is the outflow of solid pollutant, and n = 1. If we consider the case of pollutant adhering to the soil, and the soil is transported as suspended load by the river, then we must also use a relationship between river discharge (x = Q) and suspended sediment load (y = QS). This relationship is often found to agree with Eq.20, with n ranging from 1.5 to 3, (see for example Walling,1977). Therefore, for such cases we can describe pollutant outflow by Eq.20:

$$M = a \cdot Q^s, \tag{21}$$

where M is the mass of pollutant per time unit, and c is its average concentration during the time unit, which is defined as M/Q, and thus given to:

$$c = a \cdot Q^{s-1}. \tag{22}$$

For improving the fit of experimental data to this curve, it is useful to consider a also to be a random variable, so that c is the result of a product of two random variables. We shall consider an example.

Example 2: Pollution transport from log-normal populations

<u>1.Problem formulation and basic equations.</u> Let a and Q be log-normally distributed and uncorrelated random variables, with properties μ_a and σ_a^2, and μ_Q and σ_Q^2, where μ and σ^2 are mean values and variance of the random variable. Find the probability distribution of c, and its parameters μ_c and σ_c^2.

<u>Solution for $f_{lnc}(lnc)$:</u> We take the logarithms of both sides of Eq.22 and obtain:

$$\ln c = \ln a + (s-1) \cdot \ln Q \tag{23}$$

The right side is a linear combination of normally distributed random variables, and therefore ln c is also a normally distributed random variable with mean and variance:

$$\left.\begin{aligned} \mu_{lnc} &= \mu_{lna} + (s-1)\cdot\mu_{lnQ} \\ \sigma^2_{lnc} &= \sigma^2_{lna} + (s-1)^2\cdot\sigma^2_{lnQ}. \end{aligned}\right\} \qquad (24)$$

We therefore obtain the probability density for lnc:

$$f_{lnc}(lnc) = \frac{1}{\sigma_{lnc}\sqrt{2\pi}} \cdot e^{-\frac{(lnc - \mu_{lnc})^2}{2\sigma^2_{lnc}}} \qquad (25)$$

The simplest, albeit not the most unbiased procedure for estimating the parameters of the probability distribution for c is to work with measured data c_m and Q_m in lnc space, in which one obtains μ_{lna} and (s-1) as intercept and regression coefficient, respectively, of a linear regression between lnc and lnQ. The random variable lna is then obtained by subtracting $(s-1)\cdot lnQ_{si}$ from lnc_{mi}, where index i stands for the pair of simultaneuosly measured values of c and Q, respectively. (A more accurate method of parameter estimating is obtained by estimating parameters from the mean and variance of the original data, see for example Aitchison and Brown, 1957).

Note that we have assumed that c is proportional to the suspended sediment load Q_s, so that for estimating parameter (s-1) we could replace c by Q_s.

Short duration input processes

For short duration processes, concentrations calculated from Eq.21 are subject to additional variability due to mixing in the river. This is expressed through b_i in Eq.17 which is the function by means of which c_i is routed to the point of extraction. For a fluid mechanically accurate model it is necessary to write a dispersion equation for the flow between injection point and the point where concentrations are measured, for example at a station where drinking water is taken from the river. This only applies for processes in which the input is of short duration and where only a point source of pollutant is considered, as is shown in Fig.6. In such a case, the pollutant input M is conveyed through the river by means of the discharge Q. Due to convection and dispersion the input is moved downstream, thereby spreading in the longitudinal and lateral direction. We assume a uniform distribution of pollutant over the width of the river, so that the quantity M released at the time t = 0 of the accident is spread out in time and space from an initial concentration of $c_0 = M/\Delta x\cdot A$ where A is the area of the cross section and Δx is the distance along the river in which the mass is injected.

We find the concentration c(x,t) downstream from point P in Fig.6 by means of a model of the process of convection and dispersion for a given discharge Q and given mass M. This model is the process model to be used, whose output is the concentration c(x,t), and whose input is the (two dimensional) input vector M,Q.

There exists no completely successful process model for calculating the dispersion of pollutants in a river. The traditional diffusion model for emission from point sources, in which it is assumed that diffusion is a linear process is not likely to be applicable to natural rivers. Nevertheless, in order to illustrate the dispersion effect, it is useful to consider the dispersion from a point source, and to consider the possibility of superposition. When this has been done, one will be better prepared to determine whether or not dispersion must be considered.

Spreading of pollutants from a point source takes place in three zones, which are depicted in Fig.6. In each of these zones, diffusion is governed by different fluid mechanical processes. The first zone exists near the source. In this zone, diffusion is governed by the configuration of the source, and by the rate of lateral spreading. In particular when emission takes place from the shore, a considerable distance is needed before the pollutants are uniformly distributed across the river (Naudascher and Fink, 1983). This process is modelled by assuming a virtual origin for the source, which is located at a virtual distance x_0 upstream from the actual point of emission. We shall only state this effect.

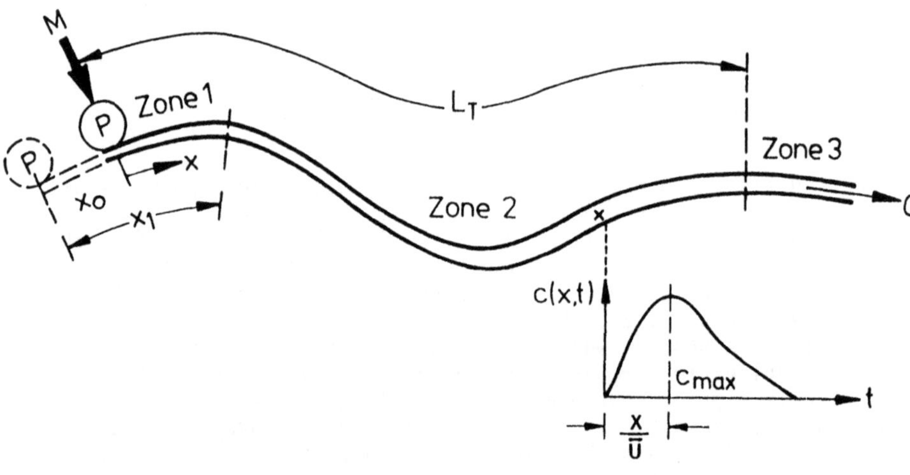

Fig.6: Zones of dispersion in a river downstream of a point source of pollutant input

Both the second and the third zone have the property that the concentration is constant across the river, so that only longitudinal dispersion needs to be considered. It results from two different effects: from diffusion caused by small scale turbulence, which is generated on banks and river bottom, and from dispersion, which is caused by velocity gradients in the vertical and horizontal directions. The effect of dispersion has first been formulated by Taylor (1954) for pipe flow. It was applied by Fischer (1973) to rivers, by including the transfer processes induced by horizontal velocity gradients. However, these models yield very crude approximations to observed dispersion in actual rivers, which might deviate by more than three orders of magnitude from theoretical predictions. A reason for this is to be found in large eddies which are generated on large scale bottom features (Grimm-Strehle, 1983), but also by exchange processes in dead water zones, which are generated by river training structures or by natural bays (Valentine and Wood, 1977), or by overbank flow over wooded flood plains (Pasche and Rouve',1986). Furthermore, the curvature of the rivers also has an effect (Friedrich and Plate, 1973).

For our calculations we use a recently proposed process model by Beltaos (1980). According to this model, the concentration c(x,t) can be calculated from the equation:

$$c(x,t) = \frac{M}{A \cdot x \cdot \beta \cdot \sqrt{2\pi}} \cdot \left[\frac{u \cdot t}{x} \cdot \exp\{1 - u \cdot t/x\}\right]^{1/\beta^2} \tag{26}$$

Since only the maximum concentration needs to be considered, which occurs at a time $t = x/u$ after release, we can write:

$$c_{max}(x) = \frac{M}{A \cdot x \cdot \beta \cdot \sqrt{2\pi}} \tag{27}$$

The factor β is the result of an analysis by Beltaos (1980) which is shown in Fig.7. From this figure we find, that we can express, to a first approximation, β by a linear relationship, which is also shown in Fig.7 , and which can be expressed analytically by:

$$\beta = 0.33 \cdot \frac{u_*}{u} + 0.03 \tag{28}$$

where for a wide river $u_* = \sqrt{g \cdot h \cdot J}$ is the shear velocity, and $u = Q/A = Q/B \cdot h$ is the mean cross sectional velocity, and discharge Q and depth h are related through Manning's equation:

$$Q = \frac{1}{n} \cdot h^{5/2} \cdot J^{1/2}. \tag{29}$$

An important aspect of the model of Beltaos is the large variability of the parameter β, as is evident from Fig.7. It is interesting to see that the data in Fig.7 are roughly bounded by parallel lines in a log-log presentation. This implies a constant coefficient of variation C_β of ß. If it is assumed (roughly) that the parallel lines have a distance of $\pm 2\sigma_\beta$ around ß, then we can assign a value of C_β = 0,35. Due to the variability of ß alone, the maximum concentration c_{max}(ß) therefore has a variation:

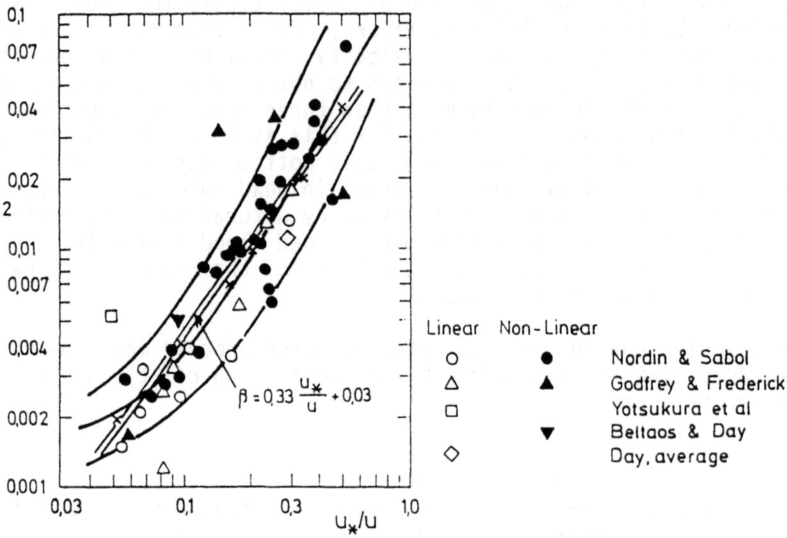

Fig. 7: Relationship of ß and u*/u for the dispersion model of Beltaos (1980). The names on the curves correspond to data sources used by Beltaos.

$$c_{max}(\beta) = c_{max}(\beta)\cdot\left(1 \pm \frac{d\beta}{\beta}\right) \tag{30}$$

where to a first approximation dß/ß has a normal distribution with zero mean and standard deviation of 0.35.

DETERMINATION OF FAILURE PROBABILITIES

From the previous section it follows that the process model by means of which transport and dispersion of the pollutant in the river is calculated is not exact and subject to uncertainty, and we shall have to consider the process model as a stochastic model which for the same input, if applied repeatedly, may lead to a probability distribution of concentration downward from the point of input. Formally, we have given the method by means of which these probability distributions can be obtained. These distributions are a measure of the uncertainty of the process model, caused both by the uncertainty of the model itself, as well as by the uncertainty in the data and in the parameters. This uncertainty combines with the natural variability of inputs, both of pollutant and of discharge, which have to be considered in applications.

We consider a concentration as harmless, if it is below a critical value c_p, and as dangerous if $c>c_p$. In this section we shall be concerned with calculating the probability:

$$P_F = P\{c>c_p\}, \tag{31}$$

that a dangerous concentration exists. We call this probability the "failure probability", in analogy to the usage in reliability theory, whose methods become applicable to solutions of river quality problems. We shall only consider the case of natural variability, for incorporating parameter uncertainty into the analysis, the methods of multidimensional reliability analysis (Ang and Tang, 1984) may be applied.

Calculating the probability of exceeding standards

The failure condition $c > c_p$ is a stochastic event, which in terms of the naturally random variables Q and $M = c \cdot Q$ can be rewritten by multiplying the failure condition with Q. Thus we obtain the inequality in terms comparing the injected mass M with the theoretically permissible mass $c_p \cdot Q = M_p$:

$$M > c_p \cdot Q \tag{32}$$

The left side of Eq.32 is a random variable due to concentration runoff from fields, the right side is a random variable due to variability of discharge Q. The model expressed by Eq.32 is equivalent to the classical case of calculating the probability of failure for random resistances and loads, $r = c_p \cdot Q$ and $s = M$, as illustrated in Fig.8. Shown is the two dimensional probability density distribution $f_{rs}(r,s)$ for resistance and load, where it is assumed that statistical dependence exists between both variables. Marginal distributions $f_r(r)$ and $f_s(s)$ are also plotted on the axes. It is to be noted that $f_r(r)$ is identical to the probability density of Q except for the scaling factor c_p, whereas $f_s(s)$ is the density of pollutant mass loading into the river.

In this diagram, condition $r = s$, which corresponds to the failure surface, is a straight line bisecting the s - r quadrant. The probability of failure is that of finding a combination of r and s in the shaded region of the r - s quadrant. It is calculated from the probability distribution of pollutant excess z defined as:

$$z = s - r \tag{33}$$

with mean value:

$$\mu_z = \mu_s - \mu_r. \tag{34}$$

The equation for $f_z(z)$ can be obtained by integration of $f_{rs}(r,s)$ over the region $-\infty < z < 0$ according to Fig.8, using the Eq.7 in the form:

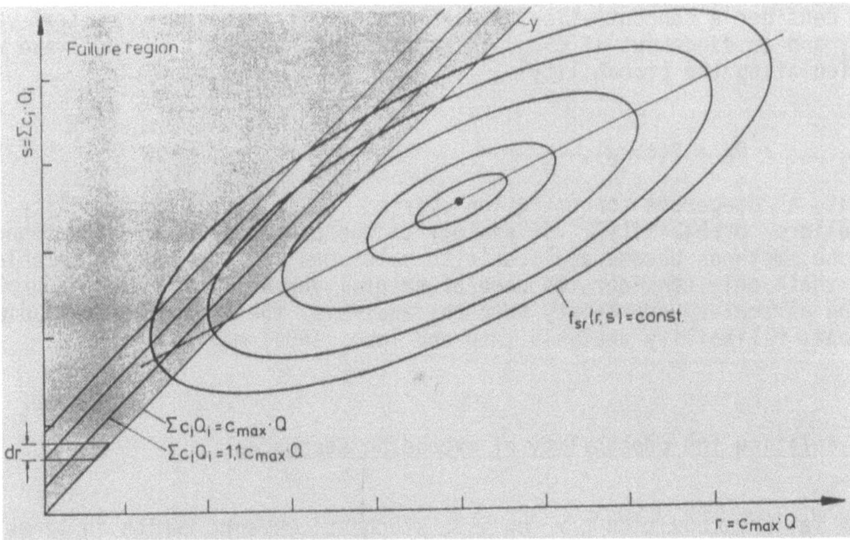

Fig.8: Probability diagram for resistance r and load s, with failure surface indicated by z = 0.

$$f_z(z) = \int_{-\infty}^{\infty} f_{rs}(s^*,r) \cdot \left|\frac{ds^*}{dz}\right| \cdot ds \qquad (35)$$

where s* is the transformation of s(z) for condition z = constant. In our case s* is simply:

$$s^* = z + r. \qquad (36)$$

Consequently we obtain:

$$f_z(z) = \int_{-\infty}^{\infty} f_{r,s}[(z+r)|r] \cdot f_r(r) \cdot dr \qquad (37)$$

where $f_{r,s}(r|s)$ is the conditional probability density for r given s. A straightforward solution of Eq.8 cannot be obtained in general. Let us consider, as a first example, a case where the solution has to be found by numerical integration, which has been investigated by Plate (1988).

Example 3: Failure due to accidental spill. Consider the case of a possible spill from a chemical plant into a river.

<u>1.Formulation of the problem and basic equations.</u> It is assumed that a toxic chemical is stored in a tank at the beginning of a production process and is slowly worked up until the tank is empty and is then refilled. At any time the tank may be destroyed by an

accident, spilling its content into the river. Consequently, the (conditional) probability distribution for the mass released when an accident should happen is a uniform distribution with density:

$$f_M(M) = \frac{1}{M_{max}} \qquad 0 < M < M_{max}. \tag{38}$$

This is a conditional distribution, since it depends on the occurrence of an accident. We want to determine the conditional probability of finding toxic concentrations c exceeding the toxidicity level c_p at a distance of 80 km downstream from the location of the chemical tank, conditional on the occurrence of an accident. For numerical values, we assume that M_{max} = 500 kg, and that the threshhold value of toxicity is a concentration of c_p = 0,1 mg/l.

Let us assume that the discharge Q of the river has a Weibull probability distribution:

$$F_Q(Q) = 1 - e^{-\alpha \cdot Q^{\gamma}}. \tag{39}$$

For numerical values we assume μ_Q = 540 m^3/s, and σ_Q^2 = (300 $m^3/s)^2$, and from these values we find for the coefficients of the Weibull distribution: $\alpha = 7.53 \cdot 10^{-6}$ and γ = 1,839. Furthermore, we assume:

a.the river has a width of 150 m, which we assume to be very wide, so that its hydraulic radius is equal to the depth h,

b.during the accident the river discharge remains constant (steady flow),

c.the value of Manning's n is 0,04 and the channel slope is 0,0006.

<u>2.Formulation of the limiting condition</u> The critical condition for the threshold of toxicity at point x is:

$$c_{max} = c_p \tag{40}$$

with dangerous conditions existing for $c_{max} > c_p$. Using the equations for Q, c_{max}, and ß from example 2, we obtain the limiting condition after some manipulations:

$$M \le (x \cdot c_p) \cdot \left[\frac{2.59 \cdot b^{1/2}}{(1/n)^{3/2} \cdot J^{1/4}} \cdot Q^{1/2} + \frac{0.075 \cdot b^{2/5}}{(1/n)^{3/5} \cdot J^{0.3}} \cdot Q^{3/5} \right] \tag{41}$$

or after inserting the numerical values of the example, we obtain:

$$M \le (x \cdot c_p) \cdot [1.427 \cdot Q^{1/2} + 0.643 \cdot Q^{3/5}] = (x \cdot c_p) \cdot R \tag{42}$$

As will become evident when we look at the probability distribution of the event $c_{max} < c_p$, we find it convenient to invert the relationship between Q and R. For this purpose, the relationship has been plotted on log-log paper (not shown), and a simpler empirical expression has been found:

$$Q = 0.286 \cdot R^{1.8}. \tag{43}$$

<u>3.Calculating the probability of safety.</u> The probability of a spill not causing concentrations exceeding the toxicity level c_p downstream at distance x is given by considering the joint probability distribution of $L = M/(c_p \cdot x)$ and R. In the two dimensional space of L and R, we find the limit condition L = R as a straight line bisecting the L-R plane under 45^0, as is shown in Fig.8. Due to the fact that R and L are uncorrelated random variables, the two-dimensional probability density $f_{LR}(L,R)$ in this plane is found as product of the two marginal densities, and we obtain:

$$f_{LR}(L,R) = f_L(L) \cdot f_R(R) \tag{44}$$

With this, we obtain the probability:

$$P_Z = P\{c_{max} < c_p\} = P\{L < R\} \tag{45}$$

as follows:

$$P_Z = \int_0^\infty \left[\int_0^R f_R(R) \cdot f_L(L) \cdot dL \right] \cdot dR = 1 - \int_0^\infty f_L(L) \cdot F_R(L) \cdot dL \tag{46}$$

which with our numerical values and formulas yields:

$$P_Z = 1 - \frac{c_p \cdot x}{M_{max}} \int_0^{L_{max}} (1 - e^{-7.53 \cdot 10^{-7} \cdot L^{3.32}}) \cdot dL. \tag{47}$$

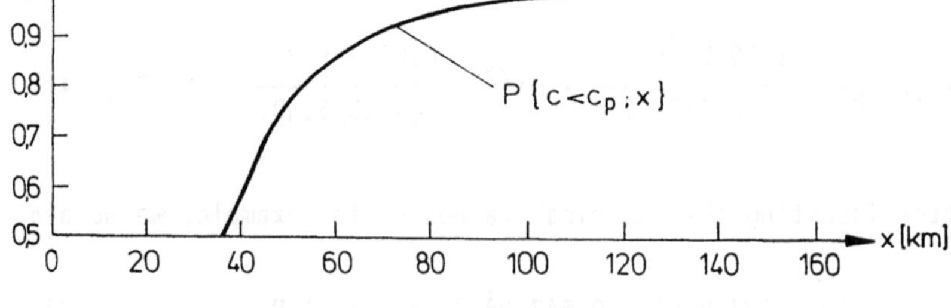

Fig.9: Probability $P\{c_{max} < c_p\}$ as function of distance x

The probability $P\{c_{max} < c_p\}$ has been calculated numerically and plotted in Fig.9 as function of x, which can be used to assess the probability of safety against toxicity at any point x. For our example with x = 80 km, we obtain the result $P\{c_{max} < c_p\} = 0.92$, as a solution to our problem.

Continuous input processes

Example 3 serves to show how we can use river quality modelling in a stochastic context. However, if we return to the problem of pollutant from agricultural fields, this approach cannot be used. Assessment of long term water quality is done by taking samples, at regular or irregular intervals from the river: therefore, it seems unnecessary to be concerned with the fluid mechanical details described in the previous example, and the whole fluid mechanics may be imbedded in the calibration factor b_i, which also may be a random variable, and thus contributes to the uncertainty of the model.

As an approximation, the pollution level can be estimated by assuming that the mass M along the river follows a normal distribution (central limit theorem). For simplicity, we also assume that Q is normally distributed. Since discharge and mass are caused by the same rainfall conditions, a certain amount of correlation must exist between the two variables, and variables r and s have a correlation coefficient ρ. The calculation of the probability of failure for this case can be done in closed form. This shall be illustrated by an example.

Example 4: Normally distributed discharges and pollutant loads. Assume both Q and M are normally distributed random variables. Then the joint probability distribution of $r=c_p.Q$ and $s=M$ is described by:

$$f_{rs}(r,s) = \frac{1}{2\pi\cdot\sigma_r\cdot\sigma_s\cdot\sqrt{1-\rho^2}}\exp\{-R\} \tag{48}$$

where:

$$R = \frac{1}{2(1-\rho^2)}\left[\frac{(r-\mu_r)^2}{\sigma_r^2} - \frac{2\rho}{\sigma_r\cdot\sigma_s}\cdot(r-\mu_r)\cdot(s-\mu_s) + \frac{(s-\mu_s)^2}{\sigma_s^2}\right] \tag{49}$$

and after inserting this expression into Eq. 37 and some algebraic manipulations we obtain for the density of z:

$$f_z(z) = \frac{1}{\sqrt{2\pi}\sigma_z}\exp\{-\frac{(z-\mu_z)^2}{2\sigma_z^2}\} \tag{50}$$

i.e. it is seen that z is also a normal distribution with mean as given by Eq.34, and variance:

$$\sigma_Z^2 = \sigma_S^2 + \sigma_r^2 - 2 \cdot \rho \cdot \sigma_S \cdot \sigma_r \tag{51}$$

The probability distribution $F_Z(z)$ can thus be obtained from standard tables of the error function $\phi(\eta)$ applied to the reduced variate:

$$\eta = \frac{z - \mu_Z}{\sigma_Z} \tag{52}$$

The condition of failure corresponds to condition z = 0, so that we obtain for the failure probability:

$$P_F = F_Z(0) = \phi(-\frac{\mu_Z}{\sigma_Z}) = \int_0^{\mu_Z/\sigma_Z} \frac{1}{\sqrt{2\pi}} \cdot e^{-\eta^2/2} \cdot d\eta \tag{53}$$

This is the standard expression used in the second moment approach to reliability analysis (Ang and Tang, 1984), therefore methods developed in the context of structural safety analysis can be applied, including uncertainty analysis. Also, if the assumption of normality of distributions or that of perfect (error free) data cannot be maintained, then other approaches towards refined second moment analysis can also be used.

DECISION MODELS FOR RIVER POLLUTION

Because the general problem illustrated in Fig.2 is of extreme complexity, simplifications are in order if design or management decisions must be made. Decision models have components whose outputs can be modified by means of decision variables, i.e. those variables which can be influenced directly or indirectly by human actions. The type of model to be used in water quality decision processes must depend on the decisions which are contemplated. For example, a model which predicts the peak concentrations of pesticides in a river at a particular point, can have a different structure from a model which predicts the amount of pesticides which penetrates into the ground water in a subarea of the basin.

The generalized risk, as defined for example in Duckstein et al,(1987) which quantifies such decisions, is therefore an important figure of merit for the value of pollutant transport models. We shall briefly review how the generalized risk is defined and how it can be calculated.

Risk as figure of merit

In the context of stochasticity of decision models we define, in the sense of decision theory (Berger, 1985) risk as follows. Let **y** = (y_1, y_2, ...y_I) be those variables from the output vector of a WQM which can be manipulated by decisions, so that their values are conditional on the decisions **d** = (d_1,d_2,...dJ) (**d** = vector of decision variables). Let these variables occur in combinations determined by the joint probability density function (joint pdf) given by:

$$f(\mathbf{y},\mathbf{d}) = f(y_1,y_2,\ldots y_I,\ldots d_1,d_2,\ldots\, d_J) \tag{54}$$

Furthermore, let K(**y**,**D**) be the function which describes the consequences of the occurence of the combination y_1,y_2... for a given decision vector **d** = **D**. Then the risk is defined in general as the expected value of K over the (conditional) pdf f(**y**|**D**), or:

$$RI(D) = \int_{-\infty}^{\infty} K(y,D)\cdot f(y|D)\cdot dy. \tag{55}$$

Example 5: Linear consequence function. To illustrate the concept of a consequence function, consider the case of a single output variable y with pdf f(y|**d**) = f(y) (i.e., we assume that y is stochastically independent of **d**), with a consequence function K(y,**d**) = K(y,d) (i.e., **d** is assumed to be a one-dimensional variable). Assume further that y_1 = z is the difference between actual concentration c and permissible concentration c_p, with probability density function $f_z(z)$ given by Eq.50 where z is the decision variable which can be manipulated by controlling the amount of fertilizer applied to the land. We further assume that a linear consequence function exists:

$$K(z|a) = a\cdot z \tag{56}$$

which quantifies the damage due to application of excessive fertilizer to the river. Risk then is defined as the expected value of the damage to the system for a given value of a. The expectation is a function of a:

$$RI(a) = E\{K(z|a)\} = \int_{-\infty}^{\infty} K(z|a)\cdot f_z(z)\cdot dz \tag{57}$$

$$= \int_{-\infty}^{\infty} a \cdot z \cdot f_z(z) \cdot dz$$

The situation is depicted in Fig.10. The following solution in terms of the error function is obtained:

$$RI(a) = \left[\frac{\sigma_Z}{\sqrt{2\pi}} \cdot e^{-\mu_Z^2/2\sigma_Z^2} + \mu_Z \cdot \Phi\left(\frac{\mu_Z}{\sigma_Z}\right) \right] \cdot a \tag{58}$$

This result depends on three parameters: a, μ_Z and σ_Z which are assumed to be known, but which might also be subject to uncertainty.

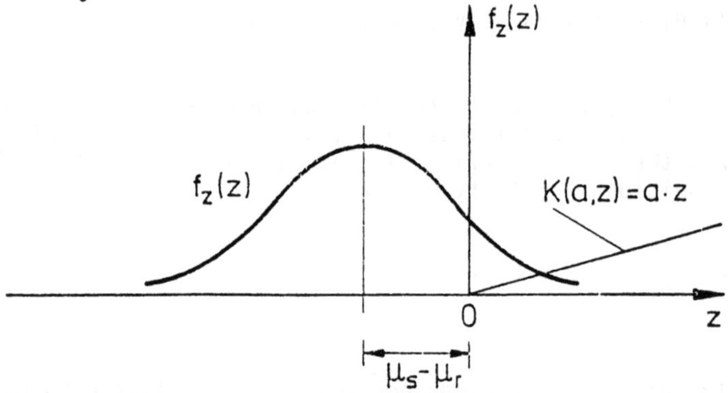

Fig.10: Linear consequence function with a normal probability density function $f_Z(z)$.

Note also, that Eq.58 only quantifies the consequences of the decision through the risk, which is an input into a decision model, not its result. The extension of this model to a decision situation, where an optimum decision has to be formulated, is beyond the scope of this paper (see Berger, 1985).

Generalized decision models

The probability of failure is a particular FM, based on a consequence function K(..) = 1. It may not be a useful criterion for assessing the merit for pollution control, (obtained for example, by changing the prevalent practice of fertilizer application), because it does not include the consequences of exceeding c_p. An analysis which includes the consequence may have two different uses. First it serves to evaluate alternatives and the necessary actions to optimize decisions - for example by deciding which region should be kept free of fertilizers, or in which areas the amount of fertilizer application should be reduced by how much. Second, a decision analysis can be used to select a model for a given purpose, and to determine how far one should go in improving simple models. Decision models have components whose outputs can be modified by means of decision variables, i.e. those variables which can be influenced directly or indirectly by human actions. The type of model to be used in water quality decision processes must depend on the objectives. For example, a model whose purpose is to predict the peak nitrate concentrations in a river at a particular point can have a different structure than a model which predicts the nitrate amount that percolates into the ground water in a subarea of the basin.

Uncertainty in decision models

Because various uncertainities could limit the quality of the predictions obtained from a model, it is useful to consider the trade-off of model uncertainity against natural variability. If the model variance is small compared to that of the data, then it is not worthwhile to improve the model.

More generally, the value of a model can best be assessed if a numerical or at least an ordinal preference value is assigned to the consequences of a decision based on the model. If a clearly advantageous decision can be made which is independent of the model quality, then there is no operational reason in further improving the model. If, on the other hand, costly investments would be necessary for preparing against uncertain consequences, then model improvements, or improvements in the data basis may be in order.

In general, stochastic models are subject to many uncertainties. Uncertainty has the effect that risk RI(**d**) becomes a random variable. It is a conditional function f(RI{**d**|**PAR**} = **g**(**d**|**PAR**), also called goal function, where the vector **PAR** consists of all uncertain parameters on which RI may depend. From this pdf we must estimate the best value of RI to be used for our decision, which could be for example the expected value.

Example 6: Mean value of uncertain decision consequence To illustrate the treatment of uncertainty consider only randomness of the factor a of the consequence function. Therefore, the probability density function of $RI\{\mu_Z, \sigma_Z^2, a\}$ is equal to $f_a(a)$, and has the expected value:

$$E\{RI(a)\} = \int_{-\infty}^{\infty} RI(a)\cdot f_a(a)\cdot da$$

$$= \left[\frac{\sigma_Z}{\sqrt{2\pi}}\cdot e^{-\mu_Z^2/2\sigma_Z^2} + \mu_Z\cdot\phi\left(\frac{\mu_Z}{\sigma_Z}\right) \right]\cdot\mu_a \qquad (59)$$

where μ_a is the mean value of the random variable a.

DISCUSSION AND CONCLUSION

A concept was presented by means of which the failure probability for a concentration of substances in a river can be determined from a stochastic model. The complete model consists of an input part, a river part, and a decision part. For some cases, the problem can be formulated as a problem of determining the failure probability, or the exceedance probability. The use of normal distributions simplifies this problem, but it is not a necessary condition for the applicability of the concept. An important aspect of this formulation, (and also of deterministic formulations), is that the transfer functions which describe the process of rainfall conversion to mass runoff M need to be quantified. It is shown how consideration of uncertainty can be used to trade uncertainty in the model formulation into stochastic uncertainty. For many problems the natural variability caused by the stochasticity of the river discharge Q, and by the randomness of the pollutant input is a most important factor. Statistical properties of Q have been extensively studied, whereas the input from pollutants requires investigation of the pollutant sources. For example, it is a problem to find the parameters determining the fertilizer mass which enters the river from a field i. They result from the amount of pollutant that is applied or generated in soil and groundwater, and from the processes which determine the chemical and biological changes of the pollutant during its path from source to river.

It appears that numerous field studies are required to determine the variability of the processes involved in the problem. At the University of Karlsruhe we have embarked on such a study whose ultimate purpose is to develop a simple model for describing the transport of pollutants caused by agriculture from fields to a river. Fifteen Institutes cooperate in this study, which is financed by the German Ministry of Science and Technology. Our procedure is to start from a complete description of the process of non-point water pollution, and then to simplify the model until only the essentials for the decision situation remains. In an exemplary manner, this model will be developed in a combination of theoretical and field studies, which is called the "Weiherbach Project". We have selected an area of some 6 square kilometers situated in the Kraichgau area North East of the city of Karlsruhe, in which we measure and theoretically model the transport processes from rainfall and fertilizer application on every field to runoff and concentrations of substances ranging from fertilizers to pesticides and heavy metals in the "Weiherbach" creek, from which the study derives its name.

Although we initially intend to describe the "Weiherbach" situation by a model, which is as deterministic as possible, it is our long term objective to only use this model as a test case for statistical models calibrated and tested against "ground truth", or against results from an accurately calibrated and accurately formulated deterministic model. It will give us the opportunity to find out which are the essential processes contributing to the mass transfer from land to water, and which processes can be considered secondary and thus relegated to statistical variability.

This paper is an exploratory study on how to formulate aspects for a stochastic river pollution model for use in the above study. The support of the German Federal Ministry of Research and Technology is gratefully acknowledged.

REFERENCES

Aitchison,J. and J.A.C.Brown, 1957: "The log - normal distribution" Cambridge University Press

Ang,A.H.S. and W.H.Tang, 1984: "Probability concepts in engineering planning and design: Vol.2" J.Wiley, New York

Beltaos,S. 1980: "Longitudinal disperion in rivers" Proc ASCE, Vol.106, J. Hydraulics Division, pp.151-172

Berger,J.B. 1985: "Statistical decision theory and Bayesian analysis" Springer, New York

Bogardi,I. and L.Duckstein 1978: "Input for a stochastic control model of P-loading" Ecological Modelling, No.4 pp.173 -195

Bogardi,I. A.Bardossy, and L.Duckstein, 1985: "Effect of parameter uncertainty on calculated sediment yield" Advances in Water Resources, Vol.8 pp.96-101

Clarke, R.T. 1990: "Statistical characteristics of some estimators of sediment and nutrient loadings", Water Reources Research, Vol. 26, pp. 2229 - 2234

Duckstein,L., E.J.Plate, and M.Benedini 1987: "Water engineering reliability and risk: a system framework" in L.Duckstein and E.J.Plate (eds): "Engineering reliability and risk in water resources" NATO Advanced Study Series, Martinus Nijhoff, pp.1-20

Geldner,P. 1981: "Deterministische und stochastische Methoden zur Bestimmung der Selbstdichtung von Gewässern" Mitteilungen No.49, Institute for Hydraulic Engineering, Stuttgart University, FRG.

Grimm-Strehle,J. 1983: "Diffusion u. Dispersion in Flüssen" Die Wasserwirtschaft, Vol.73, pp.380-385

Fischer,H.B. 1973: "Longitudinal dispersion and turbulent mixing in open channel flow" Annual Reviews of Fluid Mechanics, Vol.5, pp.59-77

Friedrich,R. and E.J.Plate 1973: "The effect of bends on longitudinal dispersion of floating particles" Proceedings, XV th Congress International Association for Hydraulic Research, Istanbul, Vol.2, pp.61-69

Haith,D.A. and L.J.Tubbs 1981: "Watershed loading functions for nonpoint sources". Proc. ASCE, Vol.107 J. Environ-mental Engineering Division pp.127-137

Hrissanthou,V. 1986: "Computation of sediment yield from a large watershed in middle Europe" Proceedings, Third Internatl. Symposium on River Sedimentation, Jackson, Miss.

Julien,P.Y. and A.M.Dawod 1987: "On predicting upland erosion losses from rainfall depth" Stochastic Hydrology and Hydraulics, Vol.1 pp.127-134

Julien,P.Y. and M.Frenette 1984: "Modeling of rainfall erosion" Proc. ASCE Vol.111, J. Hydraulics Division pp.1344-1359

Jury,W.A. G.Sposito and R.E.White 1986: "A transfer function model of solute transport through soil: 1.Fundamental concepts" Water Resources Research, Vol.22, pp.243-247

Marani,A. and G.Bendoricchio, 1986: "Models of statistical distributions for NPSP (non-point source pollution) concentrations" in A.Giorgini und F.Zingales (editors) "Agricultural non - point source pollution: model selection and application" Elsevier, Amsterdam pp.363 - 382

Naudascher,E. and L.Fink, 1983: "Ausbreitung bei Einleitung in Strömungen" Die Wasserwirtschaft Vol.73, pp.369 -379

Pasche,E. and G.Rouve, 1985: "Overbank flow with vegetatively roughened flood plains", Proc. ASCE, Vol.111, J. Hydraulics Division pp.1262-1278

Papoulis, A. 1965: "Probability, random variables and stochastic processes", McGraw Hill Co.,N.Y.

PLate,E.J. 1988: "Sicherheit gegen Schadstoffeinleitungen in Flüsse" Schweizer Ingenieur und Architekt, pp.153-155

Plate,E.J. and L.Duckstein 1988: "Reliability based design concepts in hydraulic engineering" Water Resources Bulletin, AWRA, Vol.24 pp 235-245

Plate,E.J. and L.Duckstein 1990: "Stochastic aspects of water quality modeling for non-point sources" in (DeCoursey, Ed.) Proceedings of International Symposium on Water Quality Modelling of Agricultural Non-point Sources, Pt. 2, US Agriculture Research Service, ARS pp.631-654

Plate, E.J., J.Ihringer and W.Lutz 1988: "Operational models for flood calculations" Journal of Hydrology, Vol.100 pp.489-506

Smith,J.H., D.R.Davis, and M.Fogel 1977: "Determination of sediment yield by transferring rainfall data." Water Resources Bulletin, Vol.13, pp.529-541

Taylor,G.I. 1954: "The dispersion of matter in turbulent flow through a pipe" Proc. Roy Soc. London, Serie A, Vol.233, pp.446-468

Valentine,E.M., and I.R.Wood 1977: "Longitudinal dispersion with dead zones" Proc. ASCE, Vol.103, J. Hydraulics Division, pp.975-970

Walling,D.E. 1977: "Assessing the accuracy of suspended sediment rating curves for a small basin" Water Resources Research, Vol.13, pp.531-538

Williams,J.R. 1975: "Sediment yield predictions with universal equation using runoff energy factor" Agricultural Research Service ARS-S-40, USDA, Wash. pp.244-252

Xanthopoulos,C. 1990: "Methode zur Entwicklung von Modellregenspektren für die Schmutzfrachtberechnung" Schriftenreihe des Instituts für Siedlungswasserwirtschaft, Vol.57, University of Karlsruhe

RISK IN COASTAL POLLUTION

Jacques Ganoulis
Division of Hydraulics & Environmental Engineering,
School of Engineering,
Aristotle University of Thessaloniki,
54606 Thessaloniki, Greece

ABSTRACT

Pollution in marine coastal areas is considered from point and non-point land-based sources, such as rivers, drainage ditches, submarine outfalls and coastal cities. Various uncertainties in physical, hydrodynamic, chemical and biological characteristics of the marine environment influence the engineering design of wastewater treatment plants and marine disposal strategies. This is formulated as a problem of assessing the engineering risk, to meet the environmental quality standards.

The probability of the fate of pollutants in coastal areas with randomly varying currents is studied using two methods (a) a direct simulation based on time series of the recorded currents and (b) a random walk modelling of the dispersion and decay processes.

The methodology is illustrated for the design of the optimum sewage treatment and disposal scheme in the bay of Thessaloniki, where currents and water quality characteristics have been monitored during the period 1984-90.

INTRODUCTION

In the design of wastewater treatment plants and the disposal of effluents into the sea the risk in marine pollution is subject to several types of uncertainties. These are related to the high variability in space and time of the involved hydrodynamic, chemical and biological marine processes.

The principal task of the engineering risk analysis is twofold
(a) to assess the probability or risk to comply with the water quality standards in the coastal areas to be protected, (b) to quantify the probability to exceed the maximum receiving capacity of the coastal area.

The environmental objectives are defined on the basis of frequency of occurence in relation with the water uses and the most critical pollutants.For example, in shellfish growing waters, bacterial contamination is the most critical type of pollution.

NATO ASI Series, Vol. G 29
Water Resources Engineering Risk Assessment
Edited by J. Ganoulis

EC standards require that 80% of the samples does not exceed a number of fecal coliforms equal to 70 in 100 milliliters (Council of European Communities, 1979).

The maximum receiving capacity of the marine environmental for a given pollutant, is the maximum quantity of this pollutant which can be eliminated by physical, chemical and biological mechanisms, without any perturbation of the coastal ecosystems (Ganoulis, 1988b; 1989). This part of the design requires a good knowledge of the biological coastal environment such as phytoplankton, zooplankton and benthic populations.

In the general case, land based treatment of wastewaters is to be considered together with the design of the location of the pipeline outlet. For example, a short outfall can be sufficient when tertiary treatment of sewage is provided, involving not only primary decantation and biological treatment but also nitrification and denitrification. When only primary treatment is used, a long outfall could be added in order to meet the environmental objectives. Various local constraints are usually limiting factors for the design of submarine outfalls. These are related to the regional development of the area, the land uses, the economic capabilities, the local situation etc. However, the use of advanced engineering tools related to the risk analysis and computerized mathematical modelling and techniques can reduce several uncertainties in the design.

In this paper, the uncertainties in physical, hydrodynamic and ecological coastal processes are described first. Then the risk to exceed the environmental standards is quantified by mathematical simulation. This is based on the probability that a pollutant particle travelling with random velocity occur at a given site. Numerical simulation techniques and examples of application in the bay of Thessaloniki show how time series of current recordings and random walk simulations can be used for the assessment of risk.

UNCERTAINTIES IN COASTAL HYDRODYNAMIC AND WATER QUALITY PROCESSES

The fate of pollutants in coastal areas is influenced by the combination of three mechanisms: (a) advection by currents (b) turbulent diffusion and (c) chemical, biological or other interactions. In relatively shallow coastal areas and with weak tides wind is the principal mechanism for generation of currents. The shear stresses applied at the sea surface create a three dimensional circulation, greatly influenced by the action of the Coriolis and inertia forces (Fischer et al., 1979; Baines and Knapp,1965).

The large extent in space of the areas of interest (bays, coastal areas) and the relatively large time scales of interest (days, seasons), generate high variabilities of the involved processes both in time and space.

Fig.1 shows a typical time history of the intensity and direction of currents recorded at one point on the sea.

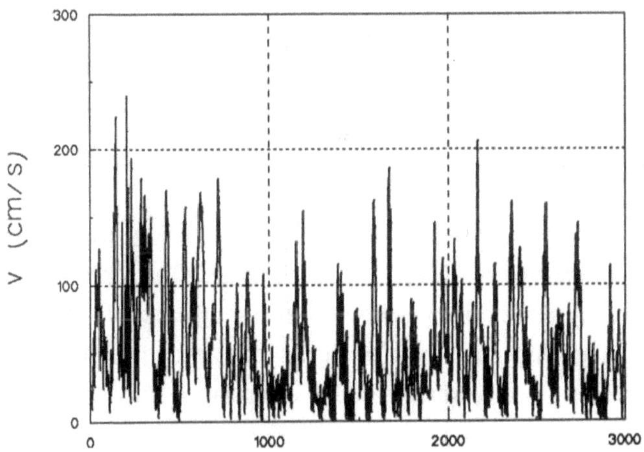

(a)

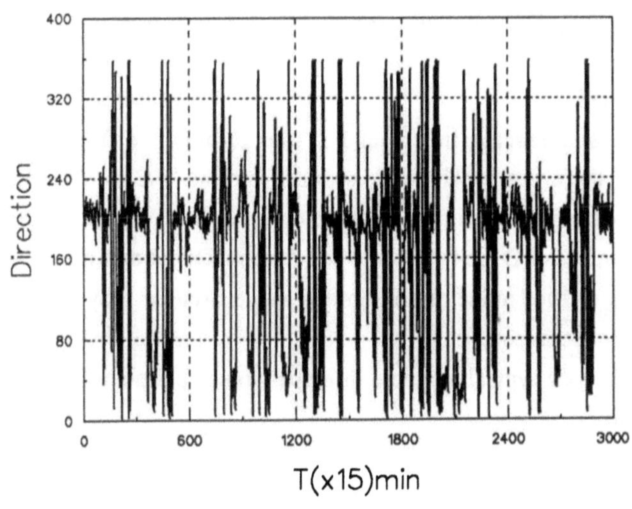

(b)

Fig. 1: Time series of current intensity (a) and direction(b)

Time and space scales of observation are very important for analysing the stochastic characteristics of the hydrodynamic and water quality phenomena (Ganoulis, 1987). Marine pollution occur on three scales: the macro-, meso- and microscale or engineering scale. Some characteristics of these scales are summarized in Table 1.

Table 1: Space and time scales in coastal dispersion

Scale	Characteristic horizontal length	Characteristic time	Example
macro-scale	10^3 km	season/year	major ocean currents
meso-scale	10^2 km	month/week	dispersion around an island in the Aegean sea
microscale or engineering scale	1-10 km	hour/day	dispersion in a bay or coastal area

Macro-scale dispersion refers usually to quasi-steady global ocean currents associated with synoptic eddies. Meso-scale currents and dispersion are characterized by density stratification and horizontal fronts, generating unsteady eddies. Micro-scale currents and pollutant dispersion are of particular interest for coastal engineering. This is the scale for the design of submarine outfalls, portuary works and coastal protection structures.

Dispersion characteristics of coastal areas are highly variable in space and time. Usually the dispersion coefficient D (Taylor,1921) is introduced in the convective dispersion equation. For seas with strong tides the following correlation has been proposed

$$D = (0.002 - 0.01)\, L^{4/3}$$

where L is the space scale in cm and D the turbulent diffusion coefficient in cm^2/s.However, there is a considerable scatter from the "4/3 rule" and its validity is questionable in shear dominated flows close to the shore line. When L varies within some kilometers, the values of D are between 1 and $10 m^2/s$. Direct measurements of D using floating drogues or rhodamine are of limited value because they reflect the specific conditions of the experimental site.

Figure 2 shows a typical example of dispersion of pollutants discharging from a submarine outfall.

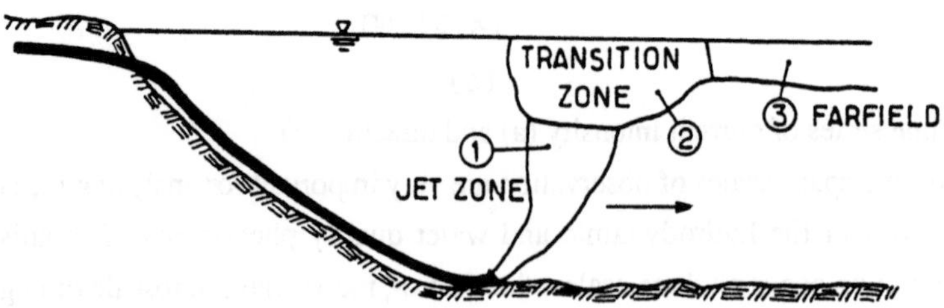

Fig. 2: Dispersion from a submarine outfall

The near-field or the jet zone (1) is distinguished from the far-field or the dispersion zone (3). In the jet zone fluid from the ambient region is entrained by turbulent flow and dilution by mixing of the wastewaters by a factor up to 10^3 occur. In the far-field the wastewaters are transported by currents and mixing is done by turbulent diffusion. The dilution is one or two orders of magnitude smaller than in the jet zone. For non-conservative pollutants additional dilution takes place by chemical and biological interactions (Fischer et al. 1979).

Although good experimental and mathematical information exists for the near-field problem, many questions still remain open concerning the simulation of the far-field dilution. This is important for the outfall design purpose, because pollutants can be convected by local currents and reach regions of particular ecological interest. The time scales in this process vary from a few hours to a day, the rate of dilution is slow compared to that in the near-field, but the bacterial decay can be important.

Randomly varying currents generate significant variations in physical and water quality parameters. As an example (Ganoulis, 1988a,1988b,1990), the time series of the water temperature and dissolved oxygen measured in three different depths (E: surface, M: mean depth, B: bottom) at the same point is shown in Fig.3. The frequency of sampling in this case is seasonal, with irregular time intervals between the samplings.

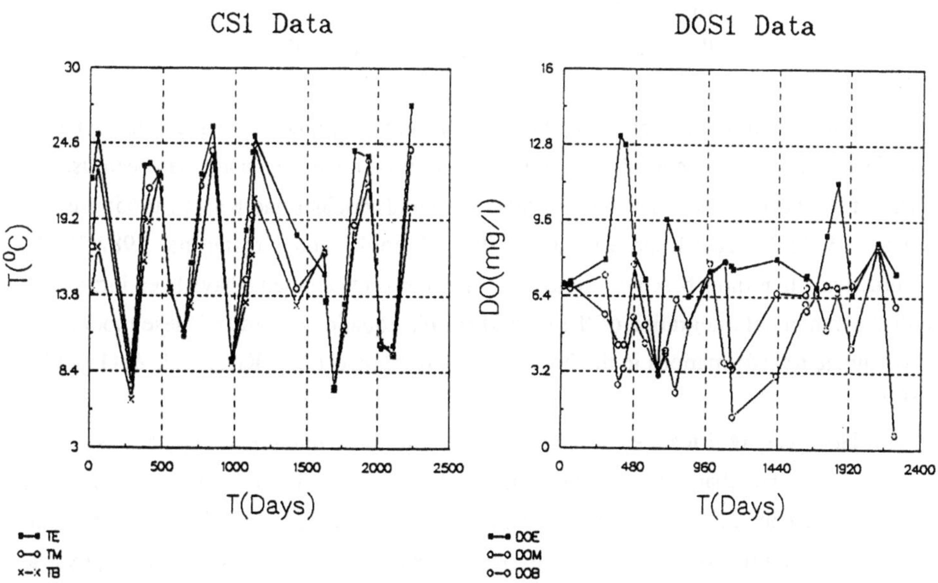

Fig. 3: Time series of the water temperature (T) and dissolved oxygen (DO) in a coastal area

COMPUTERIZED MATHEMATICAL MODELLING AND RISK ASSESSMENT

Deterministic versus stochastic approach

In the far-field, the convective-diffusion equation is used to represent the space and time variation of the pollutant concentration (Brooks,1960; Fischer et al., 1979). Depending on the composition of the pollutant (organic, radioactive, microbiological), its interaction with the sea water is expressed as a function of the pollutant concentration. For non-conservative pollutants, in a first order approximation, the rate of decay is assumed proportional to the pollutant concentration. This gives the following mass conservation equation

$$\frac{\partial C}{\partial t} + \frac{\partial}{\partial x}(uC) + \frac{\partial}{\partial y}(vC) = D\left(\frac{\partial^2 C}{\partial x^2} + \frac{\partial^2 C}{\partial y^2} + \frac{\partial^2 C}{\partial z^2}\right) \pm \lambda C \pm W \qquad (1)$$

where:
$C(x,y,z,t)$: is the pollutant concentration (M/L^3)
$u(x,y,z,t)$, $v(x,y,z,t)$: are the current velocity components (L/T)
$D(x,y,z,t)$: is the dispersion coefficient (L^2/T)
$\lambda(x,y,z,t)$: is the decay coefficient (T^{-1})
$W(x,y,z,t)$: is the pollutant flux from a source or sink ($M/L^3/T$)

In the deterministic approach, the velocity field is obtained by using hydrodynamic models. The development of such models both in 2-D and 3-D space has been used to predict the current fields generated by tides and winds (Baines and Knapp, 1965; Fischer et al., 1979; Churchill and Csanady, 1982; Ganoulis and Krestenitis, 1982;1984). Numerical algorithms based on finite differences or finite elements have been introduced for the numerical integration of the hydrodynamic equations. Some of these models use coordinate transformations in the 3-D space (Krestenitis, 1987; Krestenitis & Ganoulis, 1987).

However significant errors are induced in all these numerical simulations. These are due to the fact that only a limited number of terms in the Taylor series expansions are taken into account. Explicit algorithms suffer from the so-called numerical diffusion. This is an artificial diffusion related to the truncation errors. It is superimposed on the physical diffusion and leads to an excessive attenuation of the input signals. Implicit finite difference algorithms introduce trailing effects, because the initial signals are propagated with celerities different than the physical ones.

In fact, the equation (1) is a random partial differential equation. Randomness is due to three main reasons.

(i) The random forcing mechanism related to wind generated current velocity components (u,v)

(ii) The random character of the pollutant flux W

(iii) Parameters such as the dispersion coefficient D and the decay rate coefficient λ are random variables.

As an example the kinetics of fecal bacteria is considered. It is known (Quétin and De Rouville, 1986; WHO/UNEP, 1985) that the number of bacteries N per unit water volume decreases exponentially according to the law

$$N = N_o \exp(-\lambda t) \tag{2}$$

where N_o is the initial number of bacteries per unit water volume. Introducing the T_{90} i.e. the time required for eliminating 90% of the bacteria, λ is related to T_{90} by the following expression

$$\lambda = \ln 10 / T_{90} = 2.3 / T_{90} \tag{3}$$

Various measurements of T_{90} from several locations indicate that T_{90} is not constant, but follows a log-normal distribution (Quétin and De Rouville, 1986). This means that the probability density of the logarithm of T_{90} i.e. $f[\log(T_{90})]$ is expressed by the formula

$$f[\log(T_{90})] = \frac{1}{\sigma\sqrt{2\pi}} \exp\left\{-\frac{1}{2}\left(\frac{\log(T_{90})-\mu}{\sigma}\right)^2\right\} \tag{4}$$

where μ and σ are the mean value and standard deviation of $\log(T_{90})$. By combining the equations (2), (3) and (4) and noting $x = t / \overline{T}_{90}$, $\overline{T}_{90}$ being the geometric mean, the following expression is deduced

$$N = N_o \int_0^\infty \exp(-2.3t / T_{90})\, f[\log(T_{90})]\, d[\log(T_{90})]$$

or

$$N = (N_o/\sqrt{2\pi}) \int_0^\infty \exp[-(2.3t / \overline{T}_{90}) \exp(-\sigma x) - x^2/2]\, dx \tag{5}$$

Knowing $\overline{T}_{90}$ and σ the graph in Fig. 4 gives the mortality ratio N/N_o of the bacteries, according to the formula (5).

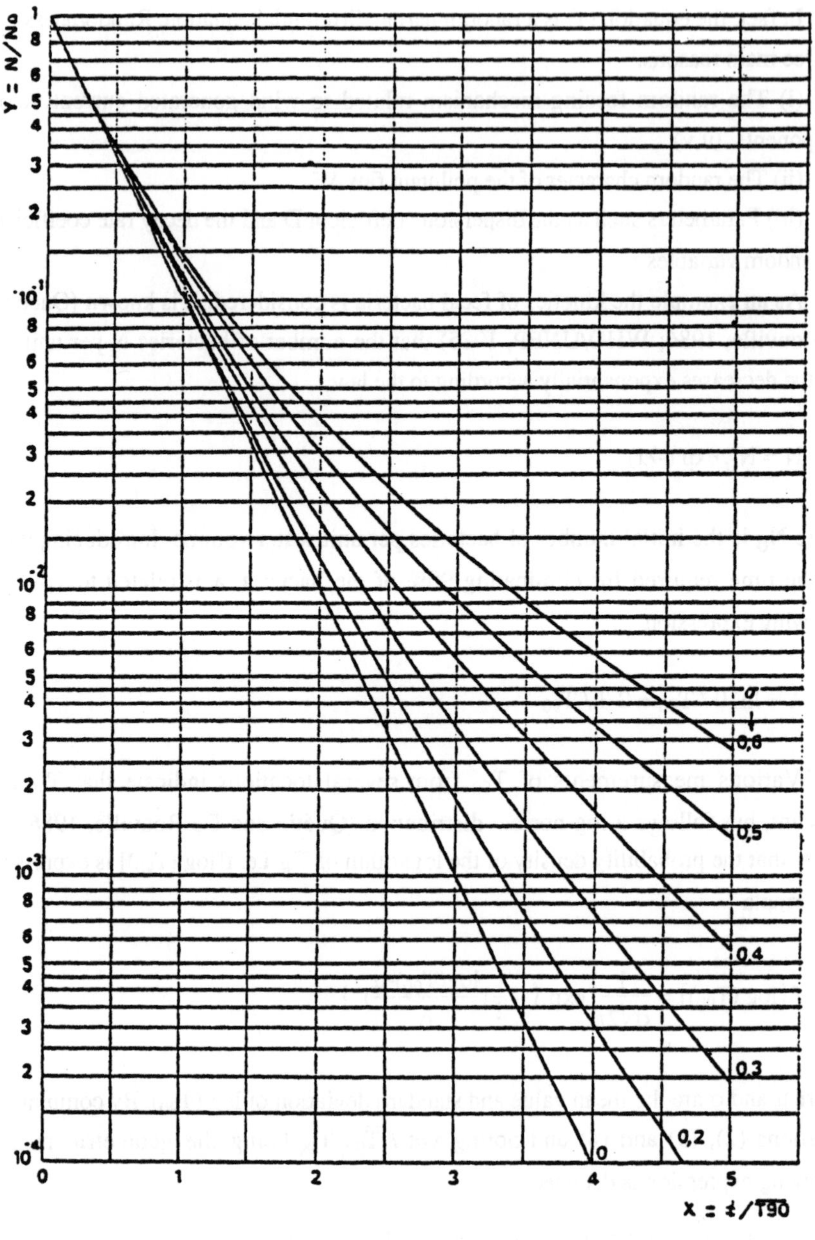

Fig. 4: Bacterial decay N/N_o versus $x=t / \overline{T}_{90}$ and σ. (Quétin et de Rouville, 1986)

Random walk simulation

Consider at time t=nΔt a large number of particles N located at the positions

$$r_p^n = (x_p^n, y_p^n) \qquad p = 1, 2, ..., N \tag{6}$$

According to the random walk principle the probability to find a particle at a given position after time Δt follows a Gaussian law of mean value 0 and variance $\sigma^2=2\Delta tD$, where D is the dispersion coefficient. Now the particles are moving from time t=nΔt to time t+tΔ=(n+1)Δt according to the relations

$$x_p^{n+1} = x_p^n + u\Delta t + \xi_1 \tag{7}$$

$$y_p^{n+1} = y_p^n + y\Delta t + \xi_2 \tag{8}$$

where u, v are the velocity components of the current and ξ_1, ξ_2 random variables following a normal distribution of mean value 0 and variance $\sigma^2=2\Delta tD$ (Larsen,1983).

The procedure is illustrated in the Fig. 5a for three particles initialy located at the same point A. Every particle is moving according to the relations (7) and (8). After 10 time steps the particles occupy three different positions A_1, A_2 and A_3.

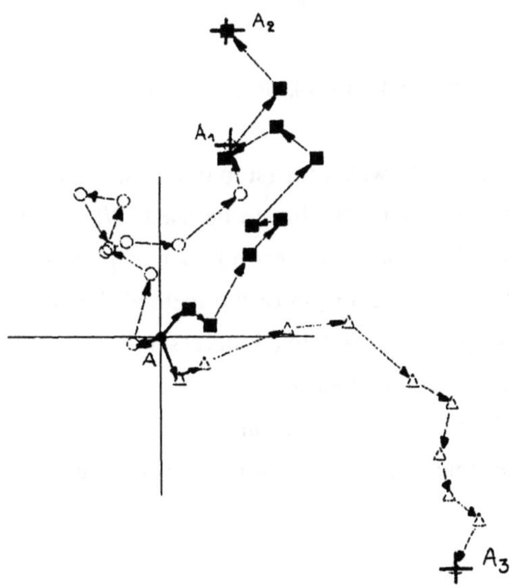

Fig. 5a Random walk of three particles after 10 time steps

To evaluate probabilities and concentrations of the particles, the area is covered by a regular grid (Fig. 5.b). Knowing the velocity components u, v at the grid points, particle velocities are computed by linear interpolation. The probability for reaching a given grid cell and consequently the particle concentrations are evaluated by counting the number of particles which fell within the grid square.

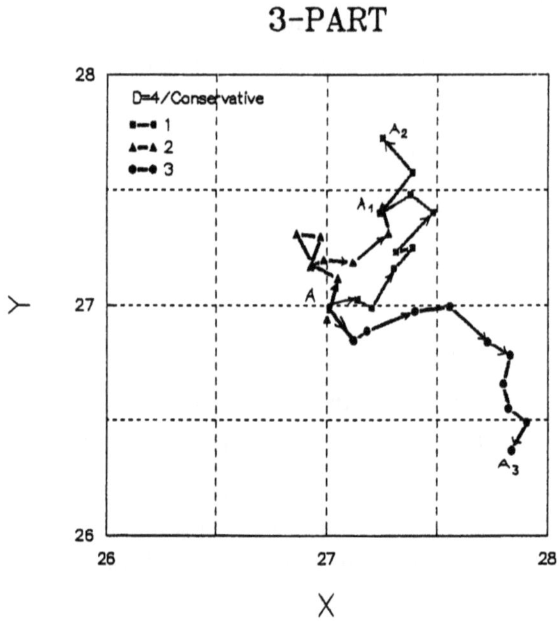

Fig. 5b: Grid overlay in the random walk of three particles

This method has some drawbacks: first to obtain statistically meagnigful results a large number of particles at least 10^3 has to be used. When continuous emissions of pollutant sources take place the necessary number of particles becomes very large. Second, because of the statistical origin of the method the concentration field shows oscillations and averaging in space and time is necessary.

When the deterministic current velocity field is used in the equations (7) and (8), the solution obtained tends to that of the deterministic convective-dispersion equation. The decay term $-\lambda C$ can be simulated by eliminating at every time step a number of particles proportional to λ.

Simulation by using random wind generated currents

Impact risks from wastewater discharges in the far field is more realistically assessed by using the time recordings of currents. Time series of wind generated currents taken over one season are usually *stationary*. This means that all the *statistical properties* of the *random variables , such as the current velocity and direction (see fig. 1 for example) are independent from the time origin.*

To illustrate this property the time series of current directions over 2000 time intervals Δt is considered. Δt comes from an AANDERAA type currentmeter fixed to 15 min. The original time series is divided into two subseries each of 1000 Δt time length, as shown in Fig. 6.

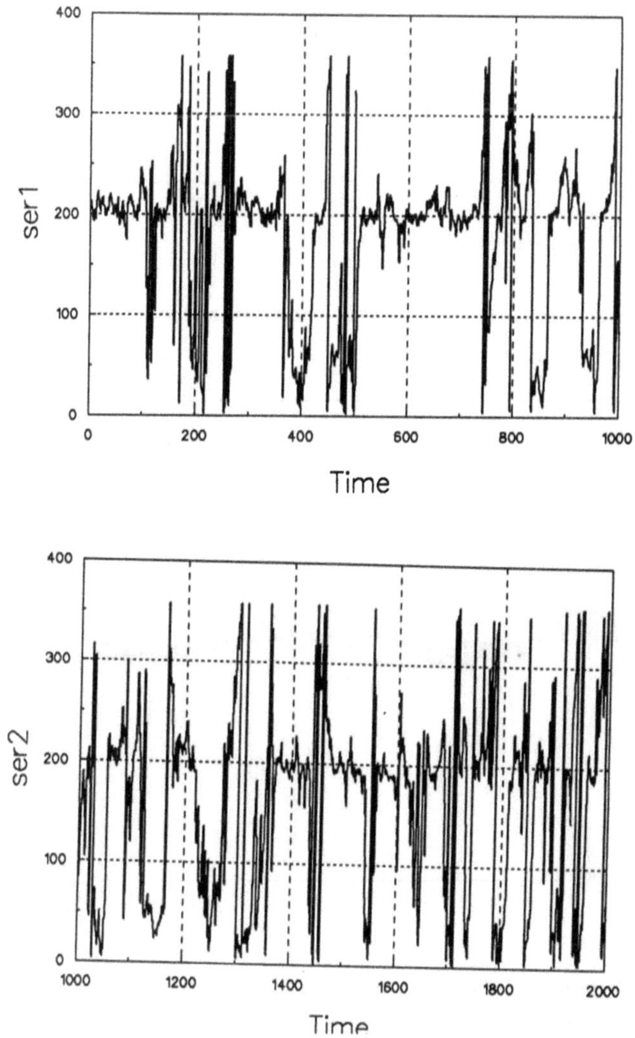

Fig. 6: Two time subseries of recorded directions of currents

The statistical properties of the two subseries are almost identical, as indicated in Figure 7 for the two frequency histograms. The independence from the time origin is an indication that the time series are stationary.

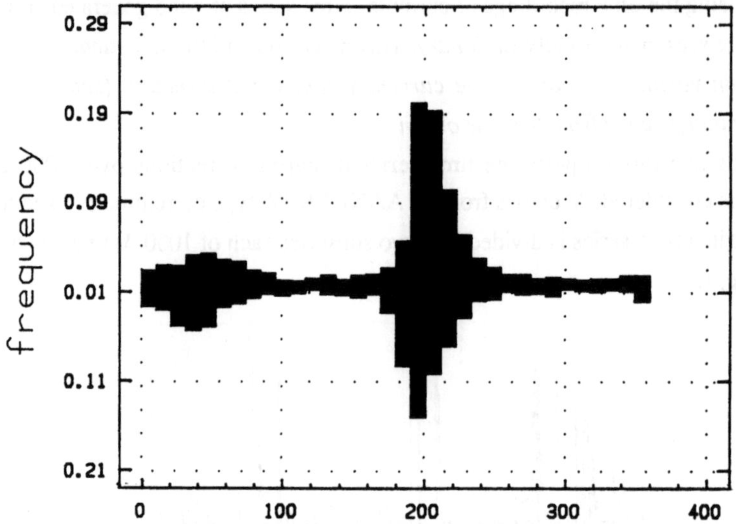

Fig. 7: Frequency histograms of the two subseries

The random character of the original series is reflected in the form of the autocorrelation function shown in Figure 8.

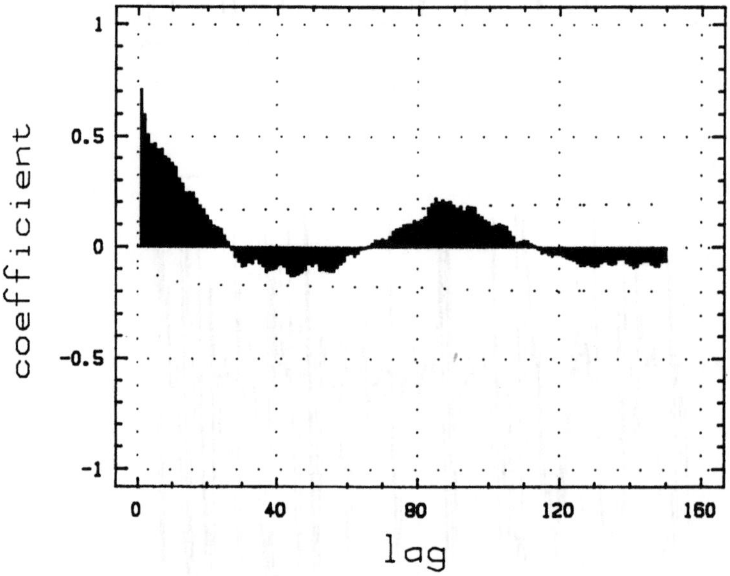

Fig. 8: Autocorrelation function

Now consider a large number of particles (Tayfun,1983; Csanady, 1983; Roberts, 1989) initially located at the same point (point source), but departing at different times $\tau=n\Delta t$. Each particle is moving during a given travel time $T>\tau$ where $T=\tau+m\Delta t=(n+m)\Delta t$. The final position $\mathbf{r}(\tau+T)$ of the particle after time T can be computed by using the randomly varying current velocity field $\mathbf{V}(t)$ as follows

$$\mathbf{r}(\tau,\tau+T) = \int_{\tau}^{\tau+T} \mathbf{V}(t)\, dt \tag{9}$$

It is obvious that the final position of every particle depends on the initial departing time τ. By allocating different values of τ to each particle, different final positions of the particles after time T are found. Because of the stationarity of the random process, the concentration field and the probability to reach a given location are independent of τ.

Having a relatively long record of time series of currents $\mathbf{V}_i$, the impact probabilities and consequently the risk assessment for a given location are evaluated numerically. After travel time $T=(n+m)\Delta t$, equation (9) takes the following form

$$\mathbf{r}(\tau,\tau+T) = \sum_{n}^{n+m} \mathbf{V}_i\, \Delta t \qquad i=n, \ldots n+m \tag{10}$$

The counting of particles at every location is done by using a grid overlay as in the case of random walk simulations. Pollutant concentrations are proportional to the number of particles located within every square of the grid.

The simulation procedure is illustrated for the case of the bay of Thessaloniki. The hydrodynamic characteristics of this area have been studied during the past six years by using floating drogues and currentmeters.Also several water quality data have been collected by sampling taken at twelve stations located in characteristic positions to cover the whole area of the bay. From these data, which have been collected with seasonal frequency, the distribution in space of different water quality environmental zones has been obtained. The pollution level is generally increasing from the open sea to the estuaries of the rivers and the city of Thessaloniki.

APPLICATION TO THE BAY THESSALONIKI

To demonstrate the capability of the simulation to handle real cases, an example of application is given for the Bay of Thessaloniki. This is a semi-enclosed coastal bassin located in the North-West corner of the Aegean sea (Fig. 9). Actually the bay is polluted mainly by large quantities of untreated sewage and some industrial effluents. Northerly blowing winds create a good mixing and dispersion of the pollutants, but the situation is very irregular. Problems such us oxygen deficiency, odours and eutrophication occur mainly in summer.

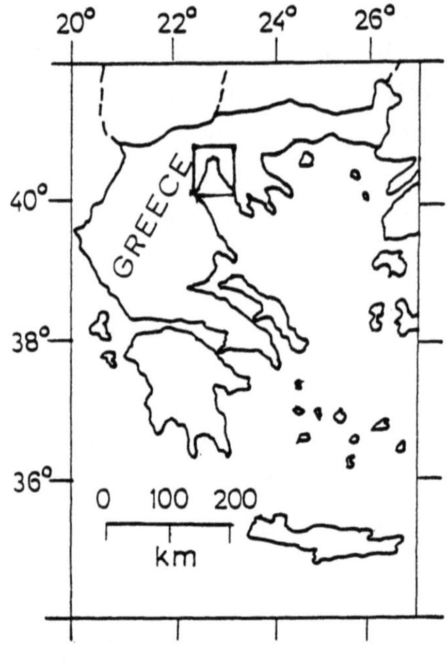

Fig. 9 The bay of Thessaloniki

An extended monitoring program of currents and water quality parameters has been developed during the period 1984-today (Ganoulis, 1990). Also numerical simulation of the currents and the impacts from wastewater discharges have been studied.

To obtain a sufficient precision in the simulations in specific sites within the bay, rectangular overlapping grids with variable grid size have been used (Fig. 10).

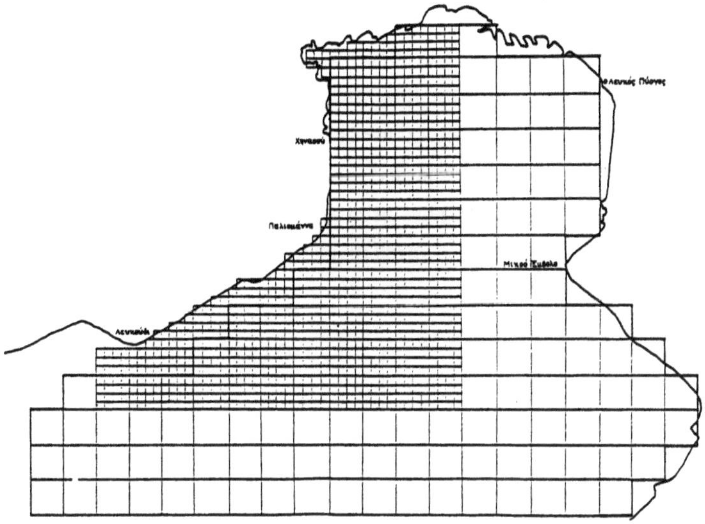

Fig. 10: Grid of variable size used in the bay of Thessaloniki

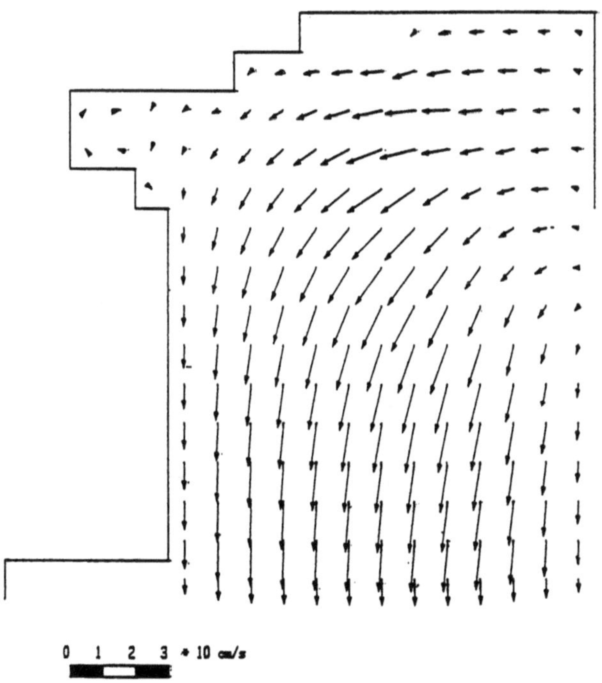

Fig. 11: Current velocity field generated by Northern wind

This type of current fields has been used for the application of the random walk simulation method.The algorithm has been implemented on an IBM/PS microcomputer. 10^4 particles have been used over the total time simulation period. Small oscillations due to the statistical character of this method are not very important for the applications.

For testing the simulation method, a point source discharging with a given concentration of E-coli (10^4/100ml) has been assumed at the point A (Fig. 12). As shown in Figs. 12, a,b,c for T_{90}=6h the coliforms reach a steady state at time t=8h.

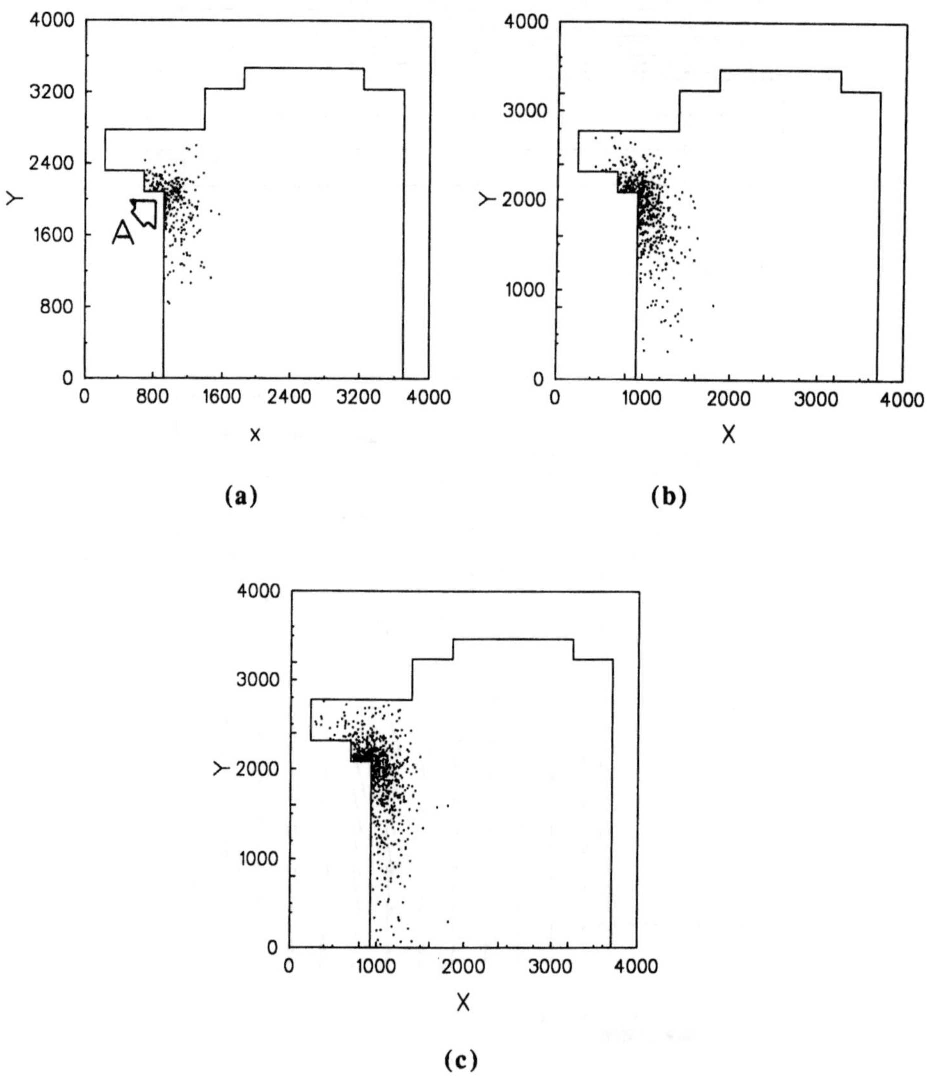

(a) (b)

(c)

Fig. 12: Time evolution of fecal coliforms discharging from point A after time 3h (a), 6h(b) and 8h (c)

By using a fixed grid and counting the number of particles located in a given grid cell, the lines of equal concentration are obtained (Fig. 13(a) and (b)).

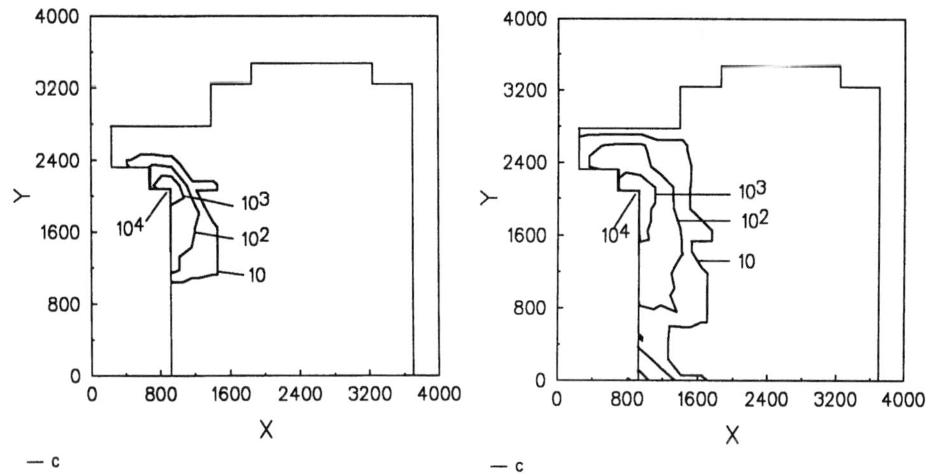

(a) **(b)**

Fig. 13: Concentration fields of fecal coliforms after time t=3h (a) and 8h (b)

To validate the random walk simulation algorithm, the actual situation near the site Paliomana has been simulated. Using the results from in-situ measurements and calibration of the model (Ganoulis, 1990), the actual situation of fecal coliform distribution has been computed (Fig. 14).

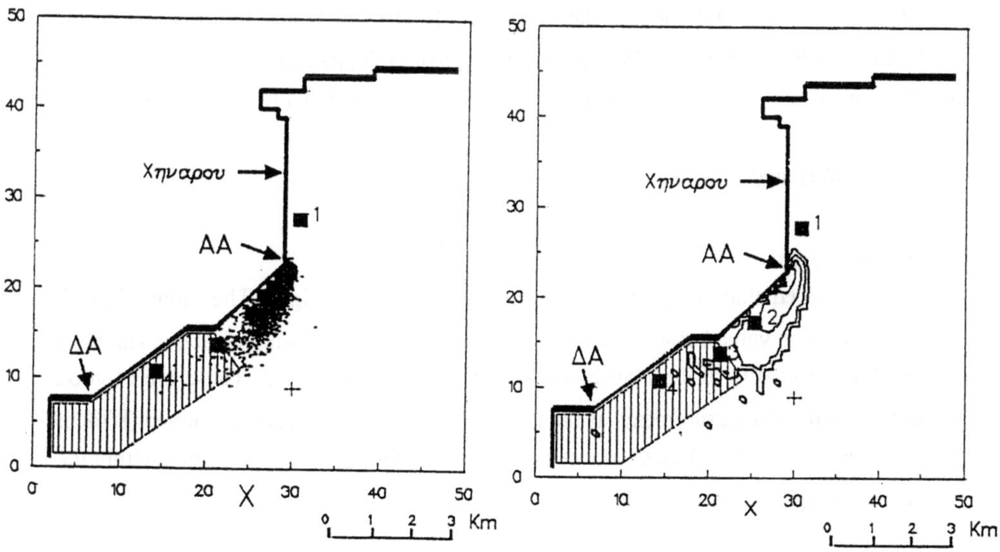

Fig. 14: Simulation of the actual situation

In this area of the bay, there is a risk of contamination of shelfishes by the E-coli. If chlorination has to be used very carefully for sewage desinfection (Ben Amor et al., 1990), a short submarine outfall gives satisfactory dilution.

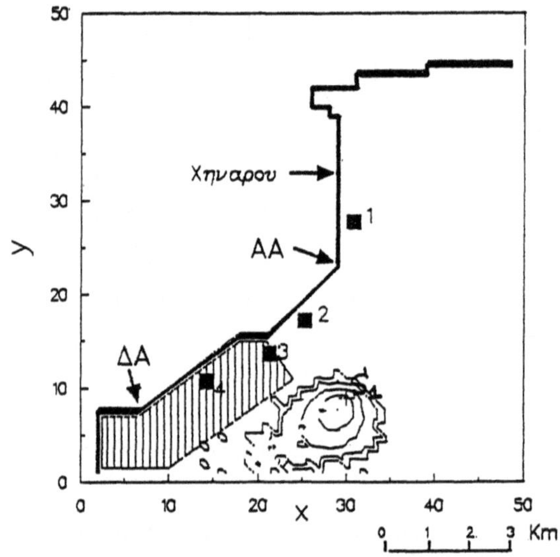

Fig. 15: E-coli concentrations from a submarine outfall

The assessment of risk using the random walk simulation is based on a choise of a "mean" value of two parameters: the dispersion coefficient D and the time T_{90} of the bacteria mortality. When data from sampling are available, then the best choise of these coefficients is made by calibration. This is the case shown in Fig.14, where the values

D=4 m2/s and T_{90}=5h

have been chosen.

It is noticed that sampling has made also during the night. The value T_{90}=5h represents a mean value between day and night time situations. Using the above values of D and T_{90} the results of simulation are shown in Figs. 19 and 20. These are in good agreement with the measurements as far as the C80 concentrations are concerned (80% of the samples have $C < C_{80}$).These concentrations must comply to EC standards within the oysters growing area.

When time series of currents are available, the direct simulation method based on the displacement of particles with the random velocities of currents gives more realistic results. The method has been used to evaluate the risk of pollution from two different discharge sites. Risks from discharge site A are shown in Fig.16 and from site B in Fig.17.

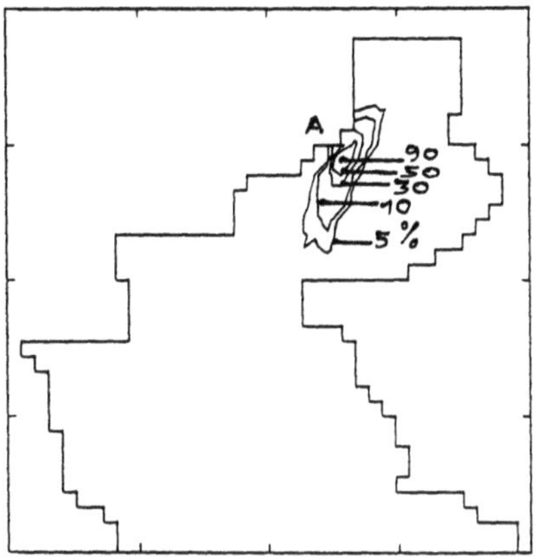

(a)

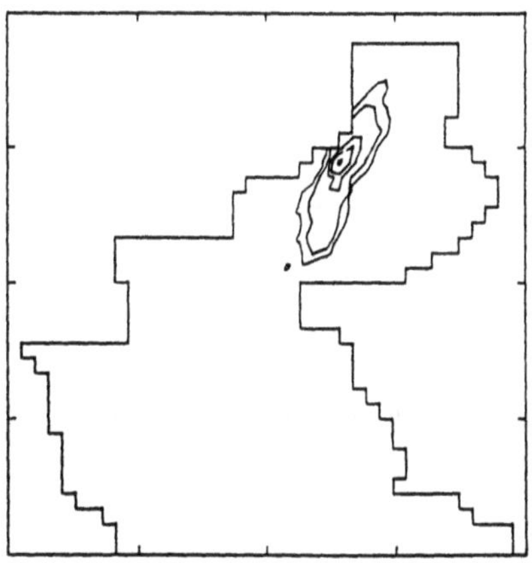

(b)

Fig. 16: Contours of probability of wastewater to contaminate the bay within T=3h (a) and T=6h (b) after release. Discharge site A.

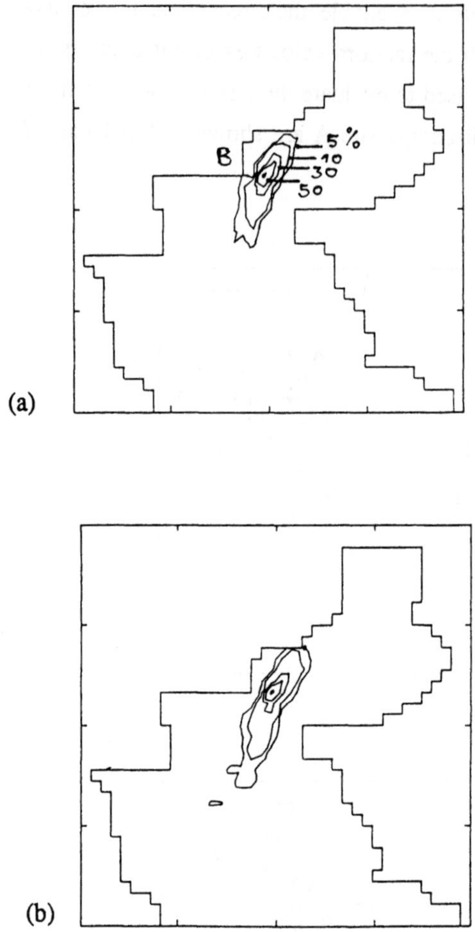

Fig. 17: Contours of probability of wastewater to contaminate the bay within T=3h (a) and T=6h (b) after release. Discharge site B.

CONCLUSIONS

For the assessment of risks of coastal pollution and the design of submarine outfalls, reliable numerical simulation methods are needed.

Lagrangian algorithms based on random walk simulation need a calibration for the dispersion and kinetic coefficients. The reliability and the precision of the obtained results have been tested in the bay of Thessaloniki, where experimental information is available on E-coli concentrations.

The probability to meet the water quality standards is more realistically quantified by using the randomly varying currents as recorded by currentmeters.This methodology has been applied for the design of the optimum sewage disposal scheme in the bay of Thessaloniki .

REFERENCES

Baines, W.D., Knapp, D.J.(1965), Wind driven water currents, *J. Hydr. Div.*, ASCE, HY2, pp.205-221

Ben Amor, H., De Laat, J.et Dore, M. (1990). Oxydation d' un acide humique aquatique par le bioxyde de chlore. *Revue des Sciences de l' Eau*, 3:83-105

Brooks, N.H. (1960), "Diffusion of sewage effluents in an ocean current". *Waste disposal in the marine environment*. Pergamon Press.

Churchill, J.H., Csanady, G.T. (1982), "Nearsurface measurements of quasi-lagrangian velocities in open water". *J. Physical Oceanography.*

Council of European Communities (1979), Directive 76/923/EEC on the quality required of shellfish waters. *Official Journal of the European Communities*, No. L281/4

Csanady, G.T. (1983), Dispersion by Randomly Varying Currents. Fluid Mech., 132: 375-394

Fischer, H.H., List J., Koh R., Imberger J. and Brooks, N. (1979). Mixing in Inland and Coastal Waters. *Academic Press*

Ganoulis, J. and J. Krestenitis (1982), Choix du site optimum d' évacuation des eaux usées dans une région côtiére. *XVII émes Journées de l' Hydraulique, SHF, Nantes*

Ganoulis, J. and J. Krestenitis (1984), Modélisation numérique du transport côtier des polluants en Méditerranée, *XVIII émes Journées de l' Hydraulique, SHF Marseille*

Ganoulis, J. (1987). Pollutant Dispersion in Oceans *In: Disorder and Mixing*, Guyon E. et al. (Eds.), *Kluwer Academic*

Ganoulis, J. (1988a), Integrated studies against pollution in the Mediterranean: the case of the Thessaloniki Bay. *2e National Symp. of Oceanography and Fisheries,* Athens and Technica 30, 34 (in Greek)

Ganoulis, J.(1988b), Modelling the sewage assimilative capacity of coastal waters in the Mediterranean. *Rapp. Comm. int. Mer Medit.*, 31 (2), 321

Ganoulis, J.(1989). Marine pollution assessment from wasterwater discharges. *Proc. IAHR Congress*, D.123-129, Ottawa

Ganoulis, J. (1990). Environmental Impacts from sewage works in the Bay of Thessaloniki. *Reports to the Ministry of Environment and Public Works* (in Greek)

Quetin, B. et De Rouville, M. (1986). Submarine Sewer Outfalls-A Design Manual. *Marine Pollution Bulletin*, 17, No. 4

Krestenitis, J. and J. Ganoulis (1987), Forecasting coastal currents using curvilinear coordinates. *Technica Chronica*, A 7(4), 33

Krestenitis, J. (1987). Numerical study of three-dimensional circulation using coordinate transformations. *Ph.D. Thesis, AUT* (in Greek)

Larsen, T. (1983). Sea Outfall Design Based on a Stochastic Transport/Dispersion Model. *Symp. on Modelling of Environmental Flows, ASME and ASCE, Boston*

Roberts, P.J.W. (1989), Dilution and Transport Predictions for Ocean Outfalls, *Wat. Sci. Tech.* 21, *Brighton*: 969-979

Tayfun, M.A.(1973), "Monte Carlo simulations of Oil slick movements". *J. of Waterways, Harbors and Coastal Eng. Div.*, ASCE, Vol. 99, No. WW3

Taylor, G.I. (1921). Diffusion by Continuous Movements. *Proc. London Math. Soc.* Ser. A 20: 196-211

WHO/UNEP (1985). Guidelines for the computations concerning marine outfall systems for liquid effluents, Athens

PROBABILISTIC SIMULATION FOR RELIABLE SOLUTE SOURCE IDENTIFICATION IN HETEROGENEOUS POROUS MEDIA

Amvrossios C. Bagtzoglou
Department of Civil Engineering
University of California, Irvine, CA 92717, USA

Andrew F. B. Tompson
Earth Sciences Department
Lawrence Livermore National Laboratory, Livermore, CA 94550, USA

David E. Dougherty
Department of Civil Engineering and Mechanical Engineering
University of Vermont, Burlington, VT 05405, USA

ABSTRACT A probabilistic framework to identify solute sources in heterogeneous porous media is the theme of the present paper. Monte Carlo analyses of reversed time solute transport are conducted with the help of stochastically generated hydraulic conductivity fields, kriging, and the random walk particle tracking method. The methodology is capable of pinpointing the most probable solute source and assessing, in a rational manner, the relative liability of each source.

INTRODUCTION

Backward or reversed time simulations have never been successfully applied to solute source identification for the following reasons:

1) Finite difference and finite element methods constitute numerical mass transport models of unstable behavior, when used for backtracking of present contaminant configurations. The stability criteria are, for all practical purposes, restrictive enough to make their use non-feasible.

2) Particle tracking methods, though effective in handling time reversal for purely advective contaminant transport, are incapable of treating reverse dispersive and sometimes reactive transport [Bagtzoglou et al., 1991].

Typically, forward-time simulations are used in conjunction with optimization algorithms to identify the most probable source of contamination [Gorelick et al. (1983), Lindstrom and

NATO ASI Series, Vol. G 29
Water Resources Engineering Risk Assessment
Edited by J. Ganoulis

Boersma (1989)]. This paper presents an alternative methodology that is capable of performing such an identification based on backward-time simulations directly.

STOCHASTIC FORMULATION FOR SOLUTE SOURCE IDENTIFICATION

Consider the case of a heterogeneous, random porous formation. A solute particle, which at time t=0 is located at **Xo**, moves at its new location **X** at time t. This new location **X** can be considered a random variable and similarly the evolution of the trajectory **X**=**X**(**Xo**,t), which the particle followed, is a random process. According to Dagan (1982), the random vector **X** is characterized by a probability density function fx(**X**,t;**Xo**,0) such that the probability for a particle, originally released from **Xo** at time t=0, to end up within the elementary volume dΩ surrounding **X** at time t is given by

$$P = \int_{d\Omega} f_x \, dx \tag{1}$$

Alternatively one can write

$$\int_{d\Omega} f_x \, dx = P(\mathbf{X} \in d\Omega) \tag{2}$$

and

$$\int_{d\Omega} f_{xo} \, dx = 1 \tag{3}$$

It follows from above that the solute concentration c(**X**,t) is also a random variable, and its probability density function (pdf) fc is entirely determined in terms of fx. Dagan (1982) proves that the expected value of concentration resulting from the release of mass M in an aquifer of porosity n and initially at concentration 0 can be given by

$$< c(\mathbf{X},t) > = \frac{M}{n} f_x(\mathbf{X},t;\mathbf{X_o},0) \tag{4}$$

where <c> is the expected value of the concentration field for a series of random hydraulic field realizations. Equation (4) establishes the direct proportionality between <c> (or <c>/M) and fx. Since our goal is to formulate a solute source identification methodology, the roles of source and target locations can be reversed. We, therefore, seek the pdf describing the probability that a particle located at **X** at time t (now, say) originated at position **Xo** at time t=0. We denote this probability by $f_{Xo}(\mathbf{X}_o,0;\mathbf{X},t)$. The pdf f_{Xo} is obtained by solving the governing partial differential equation of solute transport with the velocity field reversed. The net pdf identifying solute sources must be the convolution of the actual f_X, obtained from the observed contaminant plume at time t, and the reversed process, f_{Xo}. The methodology involves repeated (Monte Carlo) solute transport simulations, with an operator splitting based random walk method, and use of stochastically generated hydraulic conductivity field realizations [for details see Bagtzoglou (1990) and Bagtzoglou et al. (1991)].

THE RANDOM WALK PARTICLE TRACKING METHOD

The balance of a reactive solute in a porous medium of constant porosity may be expressed as

$$\frac{\partial c}{\partial t} + \nabla\cdot(\mathbf{v}c) - \nabla\cdot(\mathbf{D}\cdot\nabla c) = r(c,t) \tag{5}$$

where c is the solute concentration, **v** is the fluid velocity, **D** is a velocity dependent dispersion tensor, the porosity is assumed to be constant, and r(c,t) is a net species production rate.

Over the years the advection-diffusion-reaction equation (ADRE) has been extensively investigated and numerically approximated by numerous methods. Pinder and Cooper (1970) were of the first to apply a particle-oriented model in hydrological applications, and studied the transient position of the salt water front in coastal aquifers. A large collection of computer-generated particles, each of which is assigned a value of concentration, moves along streamlines. Then, these concentration values are mapped on a grid and a standard Eulerian diffusion problem is solved via finite difference or finite element techniques. Practice has shown that this method often leads to mass conservation errors, especially during the first time increments. The method is, also, extremely sensitive to the initial number of particles and the time step [Konikow and Bredehoeft (1978)].

Neuman (1981) and Ewing (1988), among many others, reported the successful application of operator splitting techniques to advective-diffusive transport problems in porous media. Sorek (1988) used an Eulerian-Lagrangian scheme to reformulate the advection-dispersion equation. Wheeler and Dawson (1988) proved the convergence of such an operator splitting algorithm. A similar approach is followed in the present work. Equation (5) is treated using a time-splitting procedure according to which the change in concentration at a particular point in space, over time period Δt, is defined as the sum of the changes in concentration due to advection, diffusion, and reaction. This can be expressed as

$$\Delta c = \Delta c_{AD} + \Delta c_{R} \tag{6}$$

where Δc is the total concentration change, and Δc_{AD}, Δc_R are the changes in concentration due to advective-diffusive and reactive processes, respectively. Expanding the individual terms of the RHS of (6), we can update the concentration c^n at time t^n to concentration c^{n+1} at time t^{n+1} with the help of an intermediate concentration $^*c^{n+1}$ at the end of the advective-diffusive part of the time-splitting procedure.

According to Hockney and Eastwood (1988), "particle models" is a generic term used for the class of simulation models in which the discrete representation of physical phenomena involves the use of interacting particles. Each particle may have a set of attributes such as mass, charge, position, momentum, heat, or vorticity. The basic concept behind this method is to represent the distribution of an extensive quantity (in our case, the mass of the solute species) as a large collection of particles [Tompson and Dougherty (1988)]. Particle tracking methods have been used to solve partial differential equations in various areas of application, such as fluid dynamics involving large distortions of the fluid, supersonic flow of gases past cylinders, plasma gun simulations, and studies of spiral structure in thin disk galaxies. Since Ahlstrom et al. (1977) first applied this technique to numerical investigations of groundwater contaminant hydrology, many other investigators have used particle methods with success in various fields of research such as transport in fractured rock, stratified formations and anisotropic heterogeneous aquifers; bed sediment dispersion in natural and/or artificial waterways; alluvial fan deposition; and roughness of drainage networks. Unlike the conventional Eulerian methods, particle tracking methods have no unacceptable grid Peclet numbers or grid orientation effects, and are traditionally considered as well suited for advection dominated simulations. The time step size is practically limited by the ability of the algorithm to accurately track the flow characteristics.

One of the more popular particle methods, which has been extensively applied in solute transport in hydrology, is based on a random walk approach. The theoretical basis of the random walk model has its origin in Einstein's explanation of Brownian motion, describing the motion of suspended particles under the influence of a fluctuating force. The motion of a Brownian particle experiencing random changes in acceleration is described by the Langevin equation, given by [Srinivasan and Vasudevan (1971)]

$$m \frac{dv}{dt} = - \varphi v + \mathbf{F}(t) \tag{7}$$

where v is the velocity of the particle and m is its mass. The RHS of (7) represents the influence of the surrounding medium, comprising of a fluctuating force $\mathbf{F}(t)$ (random part), and a dynamical friction $- \varphi v$ (deterministic part). Equation (7) can be written in terms of any state variable $\mathbf{X}(t)$ which evolves probabilistically with time, t. Therefore, it can be written in the form of a "position" instead of a "force balance" equation. Then, changes of the variable $\mathbf{X}$ in time (i.e. the motion of the particle) will be modeled by

$$\frac{d\mathbf{X}}{dt} = \mathbf{A}(\mathbf{X},t) + \mathbf{B}(\mathbf{X},t) \cdot \boldsymbol{\xi}(t) \tag{8}$$

Assuming that $\mathbf{B}(\mathbf{X},t)$ varies very rapidly compared to $\mathbf{X}$, the Fokker-Planck equation (FPE) can be obtained [Haken (1983); Gardiner (1985)]

$$\frac{\partial f}{\partial t} + \nabla\cdot(\mathbf{A}\, f) - \nabla\cdot[\nabla\cdot(\frac{1}{2}\mathbf{B}\cdot\mathbf{B}^T f)] = 0 \tag{9}$$

where f is some extensive quantity; $\mathbf{A}$ is a deterministic forcing vector; and $\mathbf{B}$ is a deterministic scaling (square) matrix. In order to make (9) analogous to (5), the following assignments have to be made:

$$\begin{aligned} f = f(\mathbf{X},t) &<=> c \\ \mathbf{A} = \mathbf{A}(\mathbf{X},t) &<=> \mathbf{v} + \nabla \cdot \mathbf{D} \\ \mathbf{B} \cdot \mathrm{B}^T = \mathbf{B}(\mathbf{X},t) \cdot \mathbf{B}^T(\mathbf{X},t) &<=> 2\mathbf{D} \end{aligned} \tag{10}$$

Imposing the second assignment is necessary to correct the unrealistic/unreliable results obtained with commercially used random walk models, in regions of strong velocity gradients [Kinzelbach (1989)]. The difference form of (8), over a time period Δt, takes the form [Sposito et al. (1986)]

$$\mathbf{X}(t+\Delta t) - \mathbf{X}(t) = \mathbf{A}[\mathbf{X}(t),t]\, \Delta t + \int_t^{t+\Delta t} \mathbf{B}[\, \mathbf{X}(s),s\,] \cdot \boldsymbol{\xi}(s)\, ds \tag{11}$$

With the understanding that the relaxation time Δt is very large compared to the time in which random force fluctuations occur, the stochastic process described by a conditional probability function satisfying the FPE is equivalent to the *Ito* stochastic differential equation [Bhattacharya and Gupta (1983)]

$$d\mathbf{X}(t) = \mathbf{A}[\mathbf{X}(t),t]\, dt + \mathbf{B}[\mathbf{X}(t),t] \cdot d\mathbf{W}(t) \tag{12}$$

where it was inherently assumed that

$$\int_t^{t+\Delta t} \mathbf{B}[\, \mathbf{X}(s),s\,] \cdot \boldsymbol{\xi}(s)\, ds = \mathbf{B}[\, \mathbf{X}(t),t\,] \int_t^{t+\Delta t} \boldsymbol{\xi}(s)\, ds = \mathbf{B}[\mathbf{X}(t),t] \cdot d\mathbf{W}(t) \tag{13}$$

In a discrete sense the above equation can be written in the form of the following step equation

$$\mathbf{X}^{n+1} = \mathbf{X}^{n} + \mathbf{A}(\mathbf{X}^{n},t^{n})\,\Delta t + \mathbf{B}(\mathbf{X}^{n},t^{n}) \cdot \mathbf{W}(t^{n}) \tag{14}$$

where the index ^{n} indicates time t^{n} and $\mathbf{W}(t^{n})$ is a random forcing vector given by

$$\mathbf{W}(t^{n}) = \mathbf{Z}^{n}\sqrt{\Delta t} \tag{15}$$

where $\mathbf{Z}^{n}$ is a normalized random number, at time t^{n}, with mean of zero and unit variance. Equation (14) can be used to move particles around with the help of deterministic and independent, random displacements. If the experiment were to be conducted N times, the same particle will end up at N different locations. Then, the spatial density of the points, evaluated at $t=t^{n}$, will be $f(\mathbf{X},t)$, the extensive quantity involved in the FPE. The same results would be obtained if, instead, N particles all located at the same initial location at time t=0 were to independently move according to (14) up to time $t=t^{n}$.

In spatially varying flows, the choice of the time step is of importance. Selection of time steps small enough so that the Courant number is always less than 0.1, ensures that the maximum overshoot error is always less than $0.1\Delta x$. Within the particle method framework initial and boundary conditions are treated by constraining either the number or the movement of particles. Instantaneous spills can, readily, be simulated by introducing a certain number of particles within the domain. Dirichlet boundary conditions are imposed by simply assigning a certain number of particles in each cell of a boundary layer, and letting them move under the velocity influence. After all particle displacements have taken place, a fresh batch of particles is re-imposed within this boundary layer. No-flux conditions are assigned by treating the boundary as a reflection plane, where the particles are forced to stay within the domain. Finally, infinite domain behavior can be simulated by treating the boundaries as absorbing, letting the particles to exit the computational domain.

HYDRAULIC CONDUCTIVITY FIELD GENERATION

Monte Carlo simulations of stochastic transport require that sample fields of logconductivity be developed. Numerous methods of random field generation exist and have been applied successfully. The method used in the present work is the turning bands method. A thorough description of the method is given by Mantoglou and Wilson (1982) and by Tompson et al. (1987, 1989). Assuming that the field to be simulated is second-order stationary, a series of one-dimensional simulations along numerous lines is performed. These simulations use a unidimensonal covariance function corresponding to the actual two- or three-dimensional behavior. Once the series of simulations is completed, a weighted sum of the corresponding line process values leads to the stochastic field at each point of interest [Tompson and Gelhar (1990)]. The line processes were generated using the method of Shinozuka and Jan (1972).

CONDITIONAL SIMULATIONS

Kriging, according to de Marsily (1986), is "a method for optimizing the estimation of a magnitude, which is distributed in space and is measured at a network of points". It can be applied to a variety of applications one of which is the interpolation of hydrogeological parameters such as hydraulic conductivity, hydraulic head, and aquifer thickness. The method requires a variogram (equivalently, a covariance function) and some observations be known. In our work, kriging is applied to condition stochastically generated fields so that the variability of the phenomenon is preserved and the simulated values are consistent with the observed values at observation points. The procedure is based on the following decomposition [Delhomme (1978)]:

true value = kriging estimate + kriging error

$$z(x) = z^*(x) + [z(x) - z^*(x)] \tag{16}$$

Only at the measurement points the true value of z(x) is known. Thus, the kriging error is unknown outside these points. To simulate this error, a random field, called the sample realization, s(x), with the same covariance structure as z(x) is generated. Then, the kriging estimate $s^*(x)$ may be computed with the same kriging weights as for $z^*(x)$ and the observed data $s(x_o)$. As above, we can write

$$s(x) = s^*(x) + [s(x) - s^*(x)] \tag{17}$$

A conditional simulation of the field z is obtained by substituting $s(x) - s^*(x)$ for the kriging error $z(x) - z^*(x)$. The resulting field, the conditioned $z_c(x)$ field is defined as

$$z_c(x) = z^*(x) + [s(x) - s^*(x)] \tag{18}$$

and is a random field possessing the given covariance structure and satisfying the observed values $z(x_o)$.

RESULTS

Flow and transport in a square domain 50 × 50 m in area, discretized into 2,500 square elements of $\Delta x=\Delta y=1$ m, is considered.

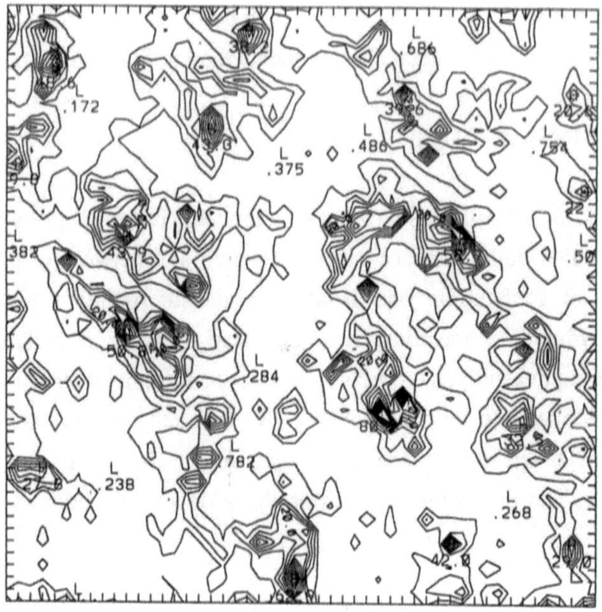

Figure 1 : Contour plot of the hydraulic conductivity field for σ_y=1.0

The hydraulic conductivity field realizations were obtained with the following parameters used: Effective hydraulic conductivity $K_{eff}=K_G$=5 m/d, standard deviation of the logconductivity field (σ_y) ranging from 0 to 4, the spatial correlation is assumed to be isotropic and exponentially decreasing, and the correlation length $\lambda_x=\lambda_y=2\Delta x$. Figure 1 depicts a contour plot of a typical hydraulic field realization with a K_G=5 m/d and σ_y=1.0.

The flow domain is subjected to two constant head boundary conditions (flow being from left to right) and two no-flow bondary conditions (top-bottom). The hydraulic head and velocity fields are determined simultaneously with the help of the mixed finite element method and a preconditioned conjugate gradient algorithm. Figure 2a shows a contour plot of a hydraulic head field for σ_y=2.0 and figure 2b presents a vector plot of the corresponding velocity field. It is worthwile noticing that as the heterogeneity of the porous medium increases, clusters of stagnation regions and preferential pathways develop. Figure 3 shows the aquifer system and three potential sources of contamination. In all simulations the spill incidents are assumed to be instantaneous, occurring simultaneously.

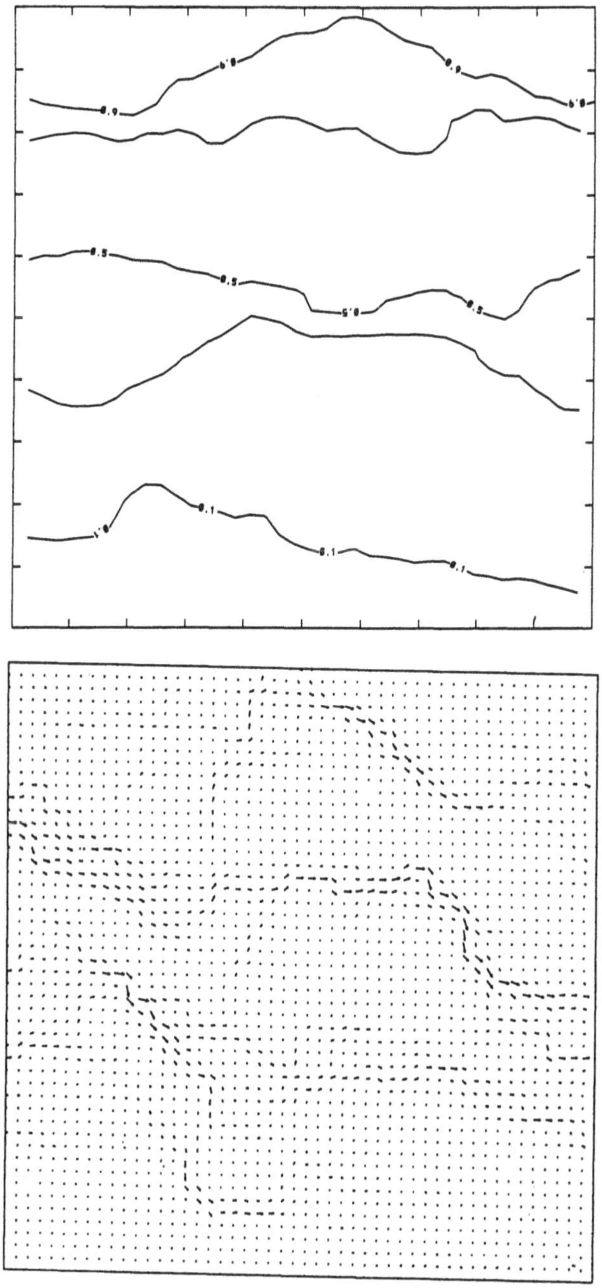

Figure 2 : a) Contour plot of the hydraulic head field for σ_y=2.0
b) Vector plot of the flow velocity field for σ_y=2.0

Numerical experiments based on heterogeneity level and type of contamination scenarios were conducted and the following parameters were used: time step Δt=0.25 d, a two-dimensional particle projection function was used with a support of $H=1.5\Delta x=1.5\Delta y=1.5$ m, 500 particles at each spill point, simulation time t_F=50 d, dispersivities α_L=0.5 m and α_T=0.05 m. A particle representation of the concentration field (at the present time) was evaluated. Assuming that the hydraulic conductivity field is known with certainty, the velocity field was reversed. Repeated reversed time solute transport realizations with evaluation of the first two moments of the concentration pdf were conducted. Figure 4 presents results from such a numerical experiment involving three active solute sources. The method identifies, within the resolution of one computational cell, all three sources as potentially responsible.

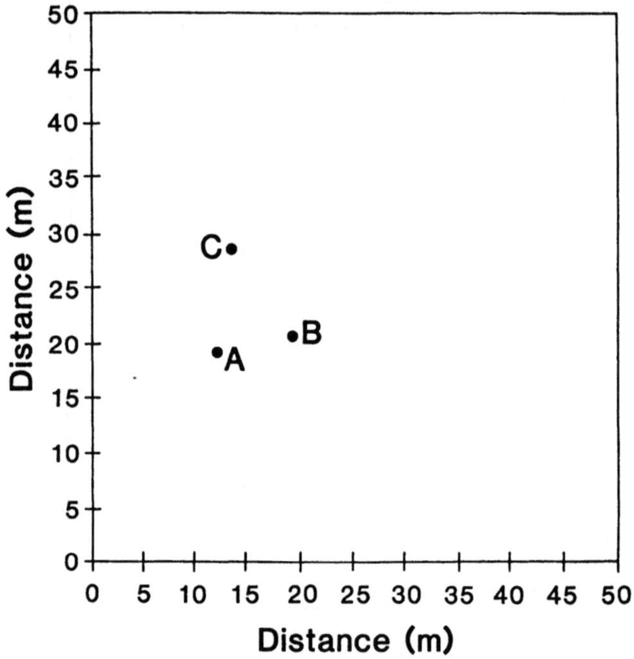

Figure 3 : Hypothetical aquifer system for solute source identification.

The same methodology was applied to flow fields where the only knowledge of the hydraulic conductivity field was the first two moments and the covariance structure of the logconductivity field. A dramatic decrease in identification power is introduced when unconditioned simulations are undertaken. Kriging was found to be an effective way of performing reliable source identification. Bagtzoglou (1990) and Bagtzoglou et al. (1991) have shown that the inter-measurement distance, relative to the correlation length, is the controlling parameter in such realistic cases. When information at an average inter-measurement distance of about 1 to 1.5 correlation lengths is available, the conditional simulation becomes effective enough to yield results comparable to those coming from use of the perfectly-known conductivity field.

A very rough estimate of the number of observation wells necessary to perform reliable solute source identification can be obtained using these results. Let Ap denote the area of the plume, determined by considering the map area with concentrations at or above a certain level

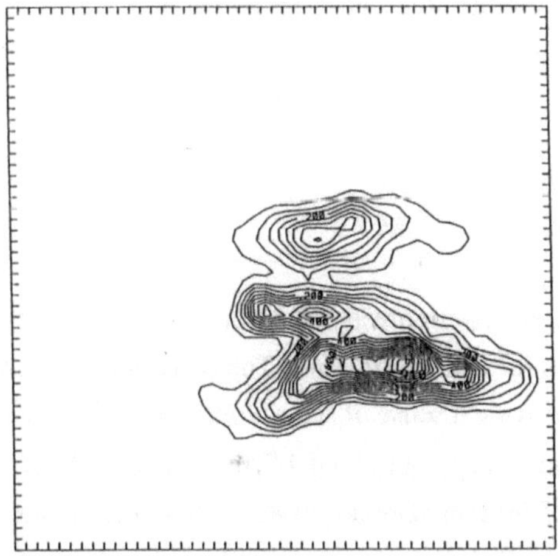

(a)

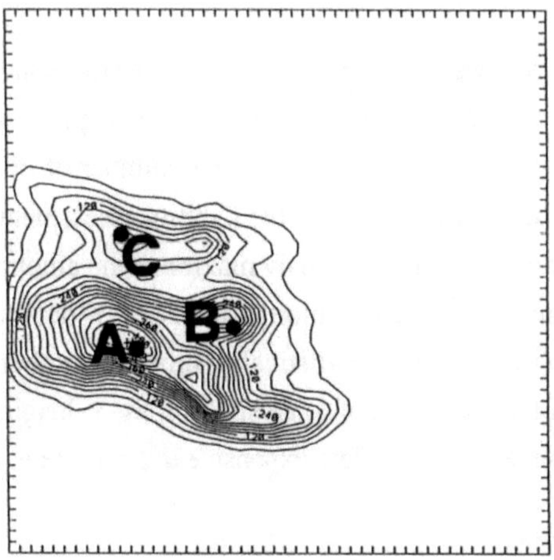

(b)

Figure 4 : Pdf values (%) for hypothetical solute source identification scenario with a σ_y=1 and all three sources being active. a) Forward-time simulation at time 50 days b) Backward-time simulation at reversed time 50 days.

of importance. Let As denote the area swept out by the plume area Ap as it is translated by the (negative of the) best estimate of the velocity field past the probable sources. Then, the number of wells (or observations), Nw, needed is bounded by

$$\frac{As}{2\lambda^2} \leq Nw \leq \frac{As}{\lambda^2} \tag{19}$$

For the hypothetical aquifer system presented before, and a contamination scenario involving one active source, Ap=190 m^2 and As=250 m^2 . The correlation length λ=2.0 m yields to $32 \leq Nw \leq 64$ wells. To provide a sense of scale, consider the following not-unrealistic case. A plume covers 1/8 mi^2 and sweeps out about 1/2 mi^2. A typical correlation length in a sandy aquifer is on the order of 50 feet (Gelhar, 1986). Therefore, a number of roughly 1,250 to 2,500 wells is necessary. This is excessive and cannot be even hoped for in real life applications.

CONCLUSIONS

A probabilistic methodology was developed that, based on stochastic methods, provides crucial information for the design of monitoring schemes in support of reliable solute source identification efforts. As it was shown, an excessive number of observation or measurement points is necessary in order to make any reliable liability assessment. Thus, methods to reduce the amount of *hard* data, and simultaneously maximize the amount of information extracted from each available point, are needed. Exploitation of *soft* data such as seismic or remote sensing data in conjunction with a realistically limited number of observation wells is imperative. Moreover, other methods of analysis such as co-kriging may provide additional flexibility to use measurements that are less expensive to obtain (e.g. piezometric head).

REFERENCES

Ahlstrom, S., Foote, H., Arnett, R., Cole, C., and R. Serne, "Multicomponent Mass Transport Model: Theory and Numerical Implementation", *Report BNWL 2127*, Batelle Pacific Northwest Laboratories, Richland, WA, 1977.

Bagtzoglou, A. C., *Particle-Grid Methods with Application to Reacting Flows and Reliable Solute Source Identification,* Ph.D. Dissertation, Department of Civil Engineering, University of California, Irvine, 1990.

Bagtzoglou, A. C., Dougherty, D. E., and A. F. B. Tompson, "Application of Particle Methods to Reliable Identification of Groundwater Pollution Sources", *Water Resources Management*, (in review), 1991.

Bhattacharya, R. N., and V. K. Gupta, "A Theoretical Explanation of Solute Dispersion in Saturated Porous Media at the Darcy Scale", *Water Resources Research*, 19(4): 938-944, 1983.

Dagan, G., "Stochastic Modeling of Groundwater Flow by Unconditional and Conditional Probabilities, 2. The Solute Transport", *Water Resources Research* , 18(4): 835-848, 1982.

Delhomme, J. P., "Kriging in the Hydrosciences", *Advances in Water Resources* , 1(5): 251-266, 1978.

de Marsily, G., *Quantitative Hydrogeology, Groundwater Hydrology for Engineers*, Academic Press, 1986.

Ewing, R. E., "Finite Element Techniques for Convective-Diffusive Transport in Porous Media", Advances in Water Resources, 11: 123-126, 1988.

Gardiner, C., *Handbook of Stochastic Methods for Physics, Chemistry and the Natural Sciences*, 2nd ed., Springer Verlag, 1985.

Gelhar, L. W., "Stochastic Subsurface Hydrology. From Theory to Applications",*Water Resources Research,* 22(9): 135S-145S, 1986.

Gorelick, S. M., Evans B., and I. Remson, "Identifying Sources of Groundwater Polution: An Optimization Approach", *Water Resources Research,* 19(3): 779-790, 1983.

Haken, H., *Advanced Synergetics: Instability Hierarchies of Self-Organizing Systems and Devices*, Springer-Verlag, New York, N.Y., 1983.

Hockney, R. W., and J. W. Eastwood, *Computer Simulation Using Particles*, Adam Hilger, 1988.

Kinzelbach, W., "Methods for the Simulation of Pollutant Transport in Ground Water - A Model Comparison", *Proceedings of NWWA Conference on Solving Groundwater Problems with Models*, Indianapolis, 1989.

Konikow, L. F., and J. D. Bredehoeft, "Computer Model of Two-Dimensional Solute Transport and Dispersion in Ground Water", in *Techniques of Water Resources Investigations of the United States Geological Survey*, Book 7, Chapter C2, United States Government Printing Office, Washington, 1978.

Lindstrom, F. T., and L. Boersma, "Two-Dimensional Transport and Fate of Chemicals Emitted by Arbitrarily Placed Sources in Confined Aquifers", *Water Resources Research,* 25(7): 1748-1756, 1989.

Mantoglou, A., and J. L. Wilson, "The Turning Bands Method for Simulation of Random Fields Using Line Generation by a Spectral Method", *Water Resources Research* , 18(5): 1379-1394, 1982.

Neuman, S. P., "A Eulerian Lagrangian Numerical Scheme for the Dispersion Convection Equation Using Conjugate Space-Time Grids", *Journal of Computational Physics*, 41: 270-294, 1981.

Shinozuka, M., and C. M. Jan, "Digital Simulation of Random Processes and its Applications" *Journal of Sound and Vibrations*, 25(1): 111-128, 1972.

Sorek, S., "Eulerian-Lagrangian Method for Solving Transport in Aquifers", *Advances in Water Resources*, 11: 67-73, 1988.

Sposito, G., Jury, A. W., and V. K. Gupta, "Fundamental Problems in the Stochastic Convection-Dispersion Model of Solute Transport in Aquifers and Field Soils", *Water Resources Research*, 22(1): 77-88, 1986.

BAYESIAN ANALYSIS OF ROBUSTNESS OF MODELS IN WATER AND ENVIRONMENTAL SCIENCES

J Bernier
Laboratoire National d'hydraulique
Direction des Etudes et Recherches d'Electricité de France
6 quai Watier
78400 Chatou
France

ABSTRACT

The effect of sampling uncertainties of applications of models is often overestimated. Far more important are the modeling uncertainties and the representativeness uncertainties. For dealing with such uncertainties methods are presented using Bayes techniques embedding the reference models class in larger ones. Examples of applications are given where design sea levels have to be estimated on one hand, river flow and temperature have to be simulated on the other hand.

1 - INTRODUCTION

The robustness of models in hydrological and environmental sciences concerns their behaviour with respect to various uncertainties in their estimation, validation and utilization processes. It is important to emphasize what the often vaguely defined word uncertainty means for us : it must not be confused with randomness. The latter is a natural aspect of the environment which cannot be reduced to zero whatever the information or knowledge about the physical process may be. Uncertainties (often called technological uncertainties) are generated by our present inability to understand the process (deterministic or random) or to collect enough information about it. However uncertainties can decrease and be reduced to zero, at least conceptually, by

NATO ASI Series, Vol. G 29
Water Resources Engineering Risk Assessment
Edited by J. Ganoulis

means of adequate experiments or observational processes as long as these have to be.

As far as these uncertainties are concerned we shall classify them in four categories :

- measurement uncertainties which appear in data used for estimation and validation. We do not deal with them in this report;

- sampling uncertainties inducing gaps between estimates based on limited samples and theoretical parameters or assumed models of the complete population. These are the classical uncertainties quantified in almost all studies up to now dealing with uncertainties;

- modeling uncertainties; whatever the justifications or validations of a model or class of models may be, errors in resulting choice are possible. The consequences of such errors can be important compared with classical sampling errors;

- representativeness uncertainties; there are various kinds of representativeness errors. For instance data can be subject to such errors which then can be included in measurement errors. Here we mean errors due to the gap between the model, as well chosen as can be using estimation and validation data sets representative of a particular "reality", and the"reality" of interest (eventually future "reality") on which the model will be applied for forecasting, simulation and more generally decision-making purposes. If, strictly speaking, such errors belong to the class of modelling uncertainties, it is useful to distinguish them.

In this report we insist on the importance of modeling errors compared with sampling ones and methods are presented for dealing with them in a quantification process using Bayes techniques.

2 - OUTLINES OF BAYES METHODS

The study in this section is an application of Bayes methods as presented at the previous NATO ASI on "Engineering Reliability and Risk in Water Resources" (Tucson - 1985). The steps of a Bayesian analysis are given in Table I (Bernier, 1987).

TABLE I
STEPS OF BAYES ANALYSIS

- 1 - Select a convenient Set Θ of states of nature θ

-2*- Define the Set A of decisions a

-3 - Compute the posterior probability density $P(\theta \mid x)$ given the data x and their probability density $f(x \mid \theta)$ via Bayes theorem (taking into account prior density $\pi(\theta)$) :

$$P(\theta \mid x) = \frac{f(x \mid \theta).\pi(\theta)}{\int_{\Theta} f(x \mid \theta).\pi(\theta).d\theta}$$

- 4 - Use posterior distribution P to calculate point or interval estimate for θ

- 5*- Use P and a loss function $L(a, \theta)$ adequately defined for choosing the best decision a* minimizing the posterior expected loss L*(a) with respect to the posterior distribution of θ:

$$L^*(a) = \int_{\Theta} L(a,\theta).P(\theta|x).d\theta$$

As shown in Table I there are two aspects in Bayesian methods : one incorporating decisions following the uncertainty analysis and explicit consequences of these decisions by means of assumed loss functions (steps marked with a star in Table I), the other aspect is non-decisionnal and limited to inference problems by means of posterior probability densities.

Together with the choice of decision set A and loss function L, the two parts where the analyst exercises his responsability are, on one hand the extent of the states of nature set Θ, and on the other hand the estimation of prior probabilities π.

The size of the set Θ depends on what kinds of uncertainties are asked to be incorporated in the analysis. Generally formal parametric models are used with a limited number of parameters respecting the parsimony principle. In this case the analysis of uncertainties is limited to the effects of the sampling errors of the parameters, each value of the vector of parameters defining a point in the set Θ. But if the form of the model itself is open to question, the analyst is free to enlarge the domain Θ introducing in it states of nature describing other models or (and) feared deviations around the previous model.

The principle of Bayes methods is to represent prior information, as vague as it may be, by means of prior probabilities of states of nature, the behaviour of which being the same as that of classical probabilities describing natural randomness. These principles are justified if a particular practical interpretation of probability is used, different from the usual one based on frequency concept. The foundations of this interpretation (see Savage, 1954; Raiffa, 1968; Berger, 1985) are based on a formalization of the behaviour of a decision maker in an uncertain environment. It is the reason why the use of Bayes methods is better understood in decisional contexts. This complete Bayesian analysis is sometimes reduced to the direct use of posterior probabilities for inference purposes.

In this paper the only information used is the data x explicitly introduced in the conditioning of posterior probabilities $P(\theta \mid x)$ via the Bayes formula. As prior probabilities we use non informative prior densities modeling the prior ignorance about states of nature; we adopted the so called locally uniform distributions introduced in Box and Tiao (1973) in the context of "knowing little a priori" relative to the information provided by the data.

3.- MODELLING UNCERTAINTIES IN GEOPHYSICAL EXTREMES FREQUENCY ANALYSIS

The study of sampling uncertainty in flood frequency estimation problems has a long history in Hydrology. For several decades the use of resulting classical confidence intervals has been known (with 95 percent of confidence size for instance). However this repeated use of such confidence intervals with such "high confidence levels" hides the fact that the consequence of sampling uncertainty is often overestimated in flood protection decision problems. In Bernier (1987) we presented the Figure 1 where, using the complete Bayesian analysis, the optimal protection level is shown, in terms of flood discharge, as a function of the available sample size that contains the information. More precisely the level overestimation Δ of level with respect to the level calculated in case of perfectly known parameters is given in this graph. The solution depends on hypotheses about the marginal flood damage i.e.the loss avoided by a unit of additional protection level. It can be seen that, whatever the sample size, Δ is small compared to the classical 95% confidence interval.when the marginal damage or loss is constant (s=1). It is only when the marginal loss is a strongly increasing function of the protection level (s=1.1) that the differences between the two curves decrease. Actual flood damages losses are not well known and it is sometimes argued that a high confidence interval is necessary to take into account so many uncertainties in the problem but this opinion goes against the necessary clarification of sharing responsibilities between the various uncertainties for improving decisions.

Furthermore the previous example deals only with sample uncertainty; in other words the flood stochastic model is assumed to be perfectly known. It is the classical "Peaks Over Thresholds" (POT) model with Poisson time distribution of flood events and exponential distribution of flood magnitudes over the threshold. In many cases the use of this particular model is questionable, especially concerning the distribution of magnitudes which can have important consequences when the sensitivity of the time process assumption is not so critical.Then the problem of model choice uncertainty is posed in this case.

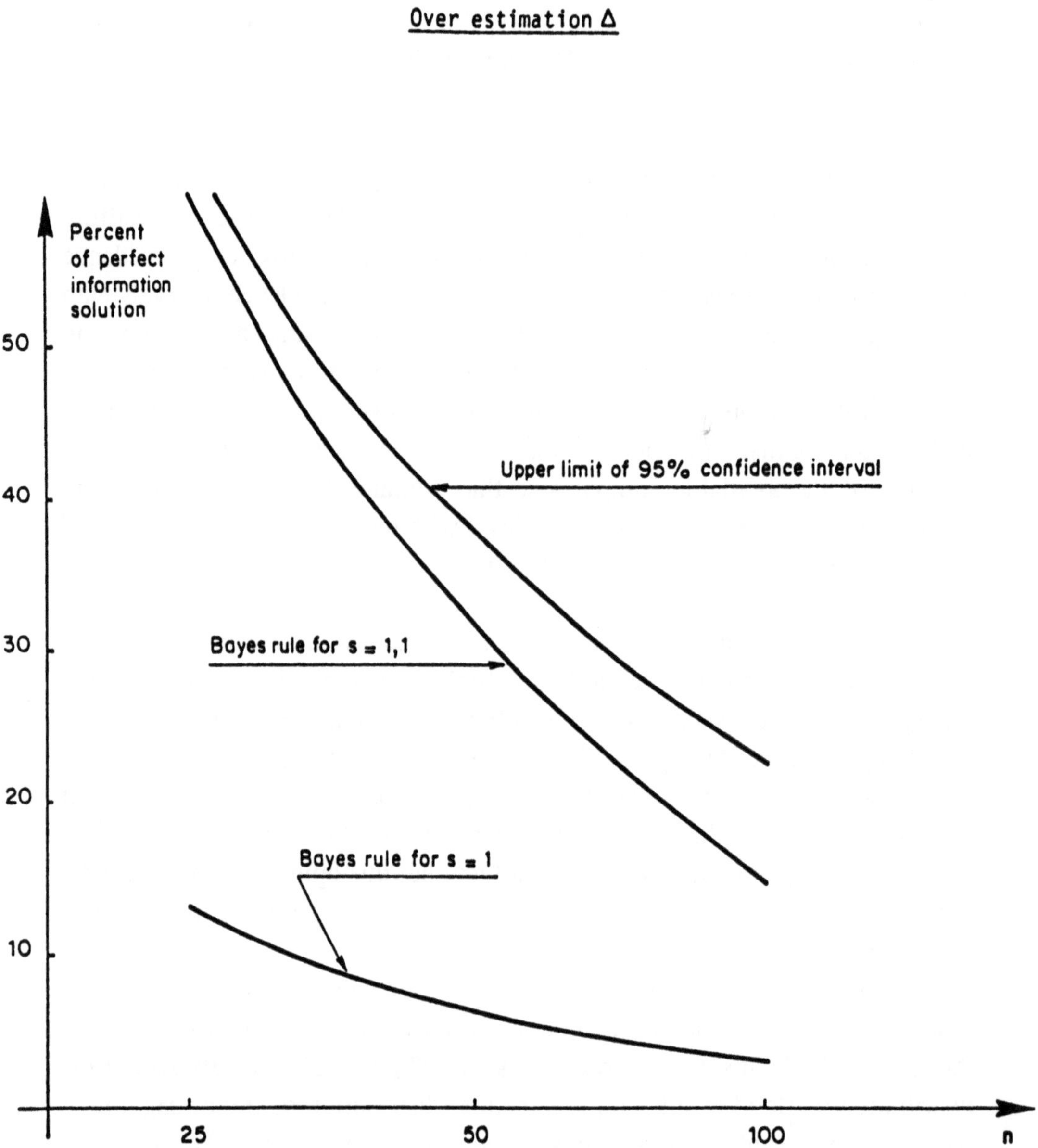

Figure I : Overestimate Δ of flood protection level as a function of sample size.

Extreme sea storm surges

Consider the problem of estimating sea rise extreme design values (Bernier and Manoha, 1991) which is closely linked to the corresponding flood problem : - comparable objective; design value of high return period T.for instance 1000 years storm surge setups (difference between sea level and astronomical tide level);

- almost comparable POT model characterized by :

• Poisson process of event arrivals with parameter μ interpreted as mean annual number of events over the threshold H_0,

• Weibull distribution of storm surge height difference $H - H_0$ with parameters ρ and p fitted to data which expresses the design value H_T (given return period T) as :

$$H_T = H_0 + \left[\left(\log\mu - \log \frac{T}{T-1} \right) / \rho \right]^{1/p} \qquad (1)$$

(this POT - Weibull model is also used for river floods in France),

- comparable lack of data even more drastic in sea rise problems where, for instance we have to estimate 1000 years storm surge at Paimboeuf site in Loire river estuary with 20 years of data.

A possible way to attempt to take in account modeling uncertainties with respect to the Weibull distribution choice is described in Table II.

Each distribution belonging to a particular parametric family (such as the Weibull one) can be characterized by a point θ, (such as the couple ρ, p) in the set Θ of parameters values of this family. It is useful to consider the Weibull family as a part of a larger family using more parameters. As far as model uncertainties are concerned, the larger the family, the more robust the estimation of design value.

Let us note that, in classical Statistics, very large families of distributions including all continuous densities are used to build robust non-parametric estimates; however these families are too large for distribution extrapolation purposes. It is necessary to limit ourselves to parametric classes.and the available data determines limits to the number of extra-parameters according to the principles of parsimony (a point which will be discussed later). For the time being let us only argue that the enlargement of the set Θ is useful only if the data bring significant information about the possible values of the new extra-parameters θ' and if one takes into account that θ' has significant consequences in decisional contexts.

TABLE II

UNCERTAINTIES ON SEA EXTREME DESIGN VALUES H_T

Sources of uncertainties :
- sampling errors
- choice of model (distribution formula)

Analysis of model uncertainties

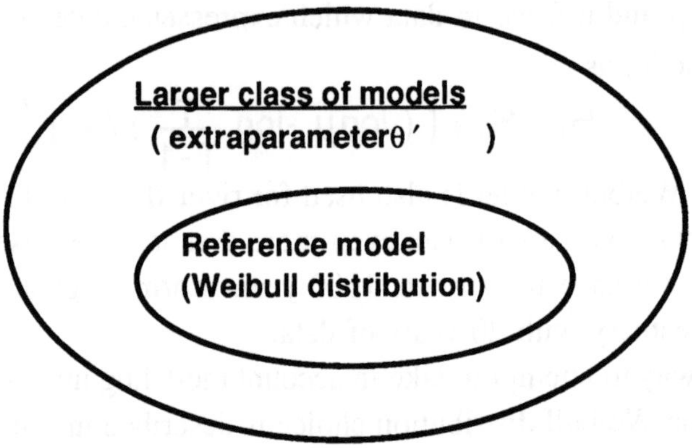

Example : log-generalized gamma distribution
(with density g(y)) :

$$g(y) = \frac{k^{k-.5}}{\sigma\,\Gamma(k)} e^{\sqrt{k}\frac{y-m}{\sigma} - e^{\left(\frac{y-m}{\sigma\sqrt{k}}\right)}}$$

$$\text{with} \qquad y = \frac{(H-H_0)^{\lambda} - 1}{\lambda} \qquad (\lambda \neq 0) \qquad (3)$$

$$y = \mathrm{Log}(H-H_0) \qquad (\lambda = 0)$$

- **sub-class : k=1 and λ=0 gives WEIBULL model**
- **fixing k=1 or λ=0 gives a larger class of models (with extra-parameter $\theta' = \lambda$ or k characterizing the distance from Weibull model)**

A way to measure the information in the data relative to θ' is by means of Bayes methods where parameters θ, θ' as well as H_T, function of them, are considered as random variables (cf Table I). Their posterior density $P(\bullet \mid x)$, given the data x and calculated according to the Bayes rule, takes into account their uncertainty; the basic relationship is :

$$P(H_T \mid x) = \int_{\Theta'} P(H_T \mid x, \theta') \cdot P(\theta' \mid x) \, d\theta' \qquad (2)$$

where Θ' is the domain of θ' values. In this formula the $P(\bullet \mid \bullet)$ are conditional probabilities that can be calculated according to Bayes rule and given the values of variables under the bar (|) symbol. This formula allows the computation of credible i.e. bayesian confidence intervals for design parameter H_T according TO Berger (1985), (see Figure 2). Another interest of formula (2) is the decomposition allowed for the posterior density of H_T which shows the effect of uncertainty about extra-parameters θ' measured by the posterior distribution $P(\theta' \mid x)$ separated from the posterior distribution of H_T given θ' and x.

These credible intervals are useful in non decisional contexts but it is interesting to note that they also have a decisional interpretation. As a matter of fact the upper limit of a credible interval with confidence probability equal to (f-1)/(f+1) represents the optimal decision of estimation when a constant marginal loss of underestimation is f times larger than the constant marginal loss of overestimation (a frequent situation in geophysical extremes problems where f can be large indeed). A further remark is that the corresponding estimated parameter H_T goes throughout the posterior domain of this parameter, as shown by the posterior density when the ratio f varies.

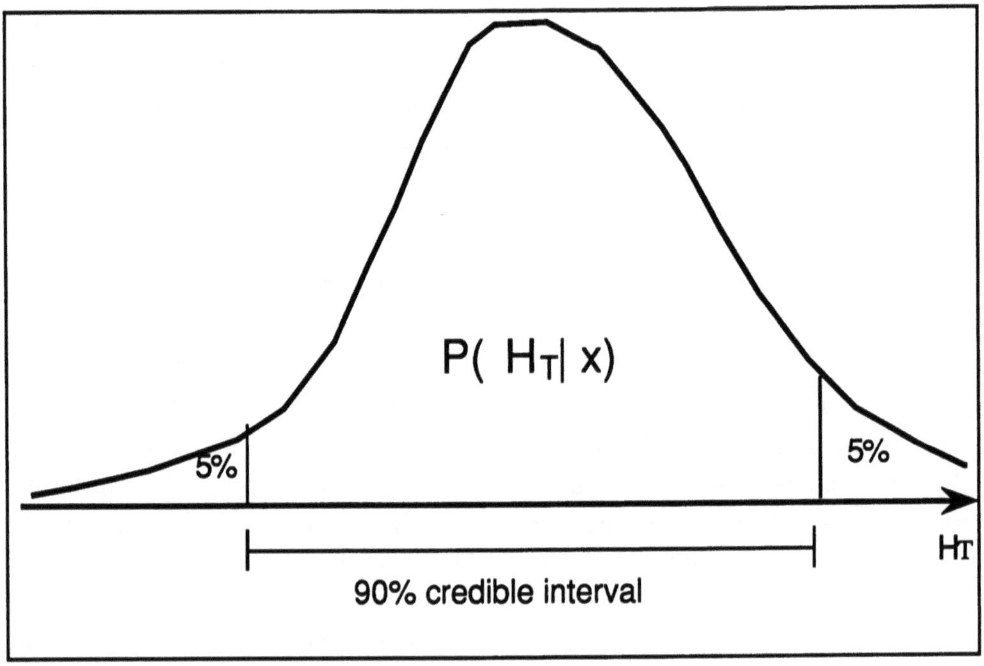

Figure 2 : Posterior density and credible interval

Proposed enlarged family generalizing the Weibull distribution

A suitable enlarged family generalizing the Weibull distribution and allowing extra-parameters λ and (or) k describing the distribution behaviour for large values of variable H is the so called log-generalized gamma distribution -LGG- (Bain and Engelhardt, 1981) slightly modifying the classical generalized gamma. The form of the density is given as formula (3) of Table II. Parameters λ and k are asymptotic behaviour parameters for large values of the return period. The reference Weibull model is obtained for $\lambda = 0$ and $k = 1$. Alternatively, fixing $\lambda = 0$ or $k = 1$, we generate two families of distributions where the extra parameter $k \neq 1$ (or $\lambda \neq 0$) measures the distance to the Weibull distribution. The parsimony principle is the reason why we use 3 parameters families instead of the complete 4 parameters family given by formula (3). It is also because the two parameters λ and k generate the same kind of asymptotic behaviour (at least in the presented applications).

Finally, using Bayes rule with locally uniform priors following Box and Tiao (1973) and conditional sample density (given the parameters) based on formula (3), we obtain the joint posterior density of m, σ, λ and (or) k. The integration of this density with respect to m and σ gives the posterior density of λ and k (Bernier and Manoha, 1991). More details can be found in Bernier (1988).

Applications

The examples presented elsewhere in Bernier and Manoha, (1991) concern storm surges in the Loire estuary (Paimboeuf site) and the problem of estimation of the 1000 years design height taking into account data over two thresholds 70 cm - sample I - and 50 cm - sample II - For comparison purposes a synthetic sample of 50 data points - sample III - was generated using Monte Carlo method according to Weibull distribution with parameters equal to those estimated with sample II.

Figure 3 shows the fitting of such distribution functions to sample II. It appears that convenient values of λ and k can represent nearly identical asymptotic behaviour. Figure 4 shows posterior densities of λ and k for the three samples. It can be seen that in samples I, II (real data), the values of $k = 1$ and $\lambda = 0$ (Weibull distribution) are far from being credible compared with the negative values of λ and values of k larger than 1. Sample I (corresponding to only 27 data points above the upper threshold) exhibits very large posterior variability of the asymptotic behaviour parameters. However, the variability is significantly reduced in case of sample II with more data (83 points). Case III (synthetic data with no model uncertainty) presents a parameter dispersion comparable to or lower than case II with less data.

Figure 5 shows 90% credible interval for the 1000 years storm surge H_T, conditioned upon given values of λ and k. The range of this interval is very sensitive to λ and k, especially in case of sample I. These later considerations and the corresponding large posterior variability of λ and k for this sample indicate the impossibility to obtain reliable estimation of a 1000 years storm surge with such a small sample.

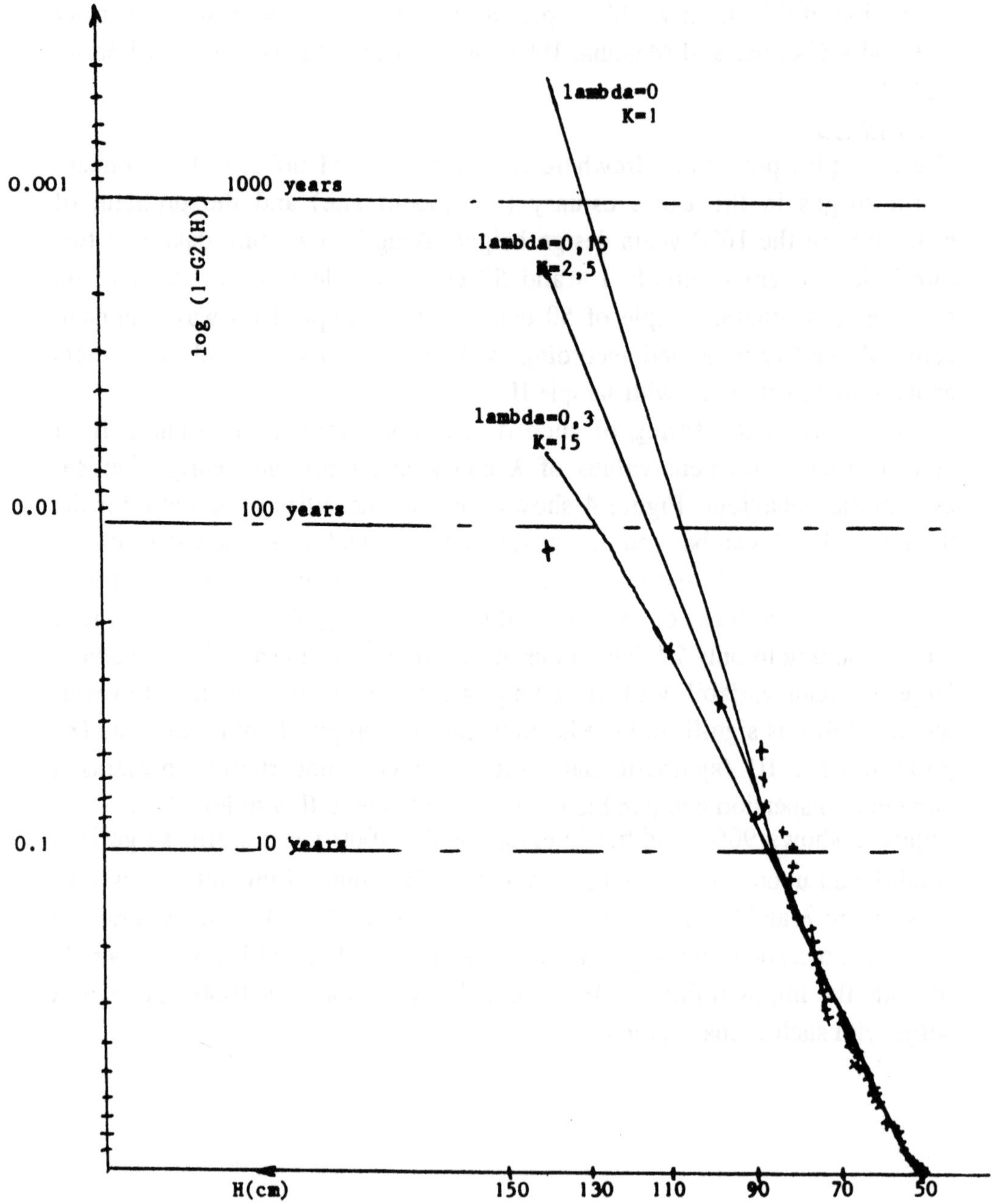

Figure 3 : Storm surges (Paimboeuf site 1959-1980; H_0=50 cm)
Fitting of Log-generalized gamma distributions.

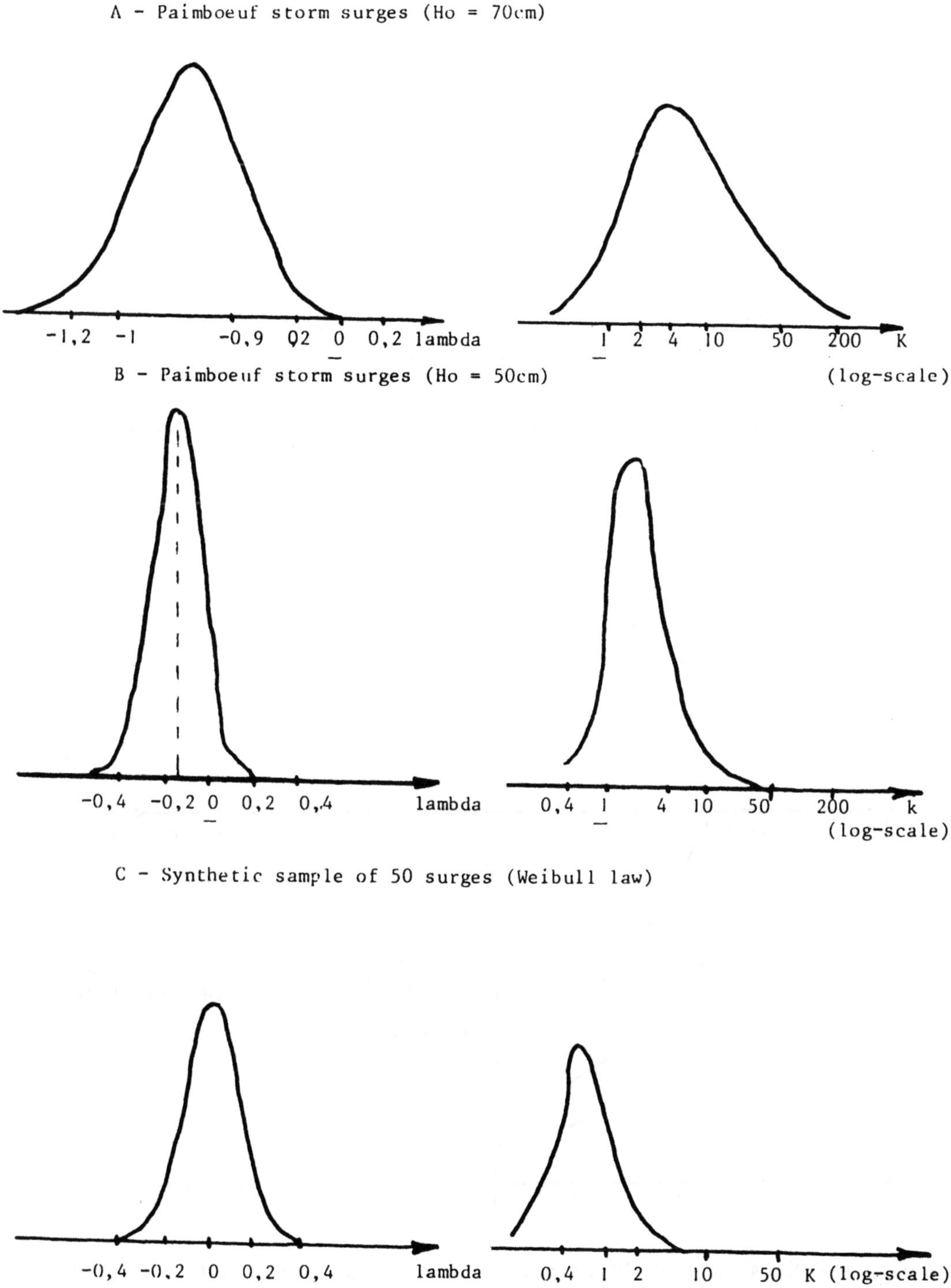

Figure 4 : Posterior densities of extra-parameters λ and k.

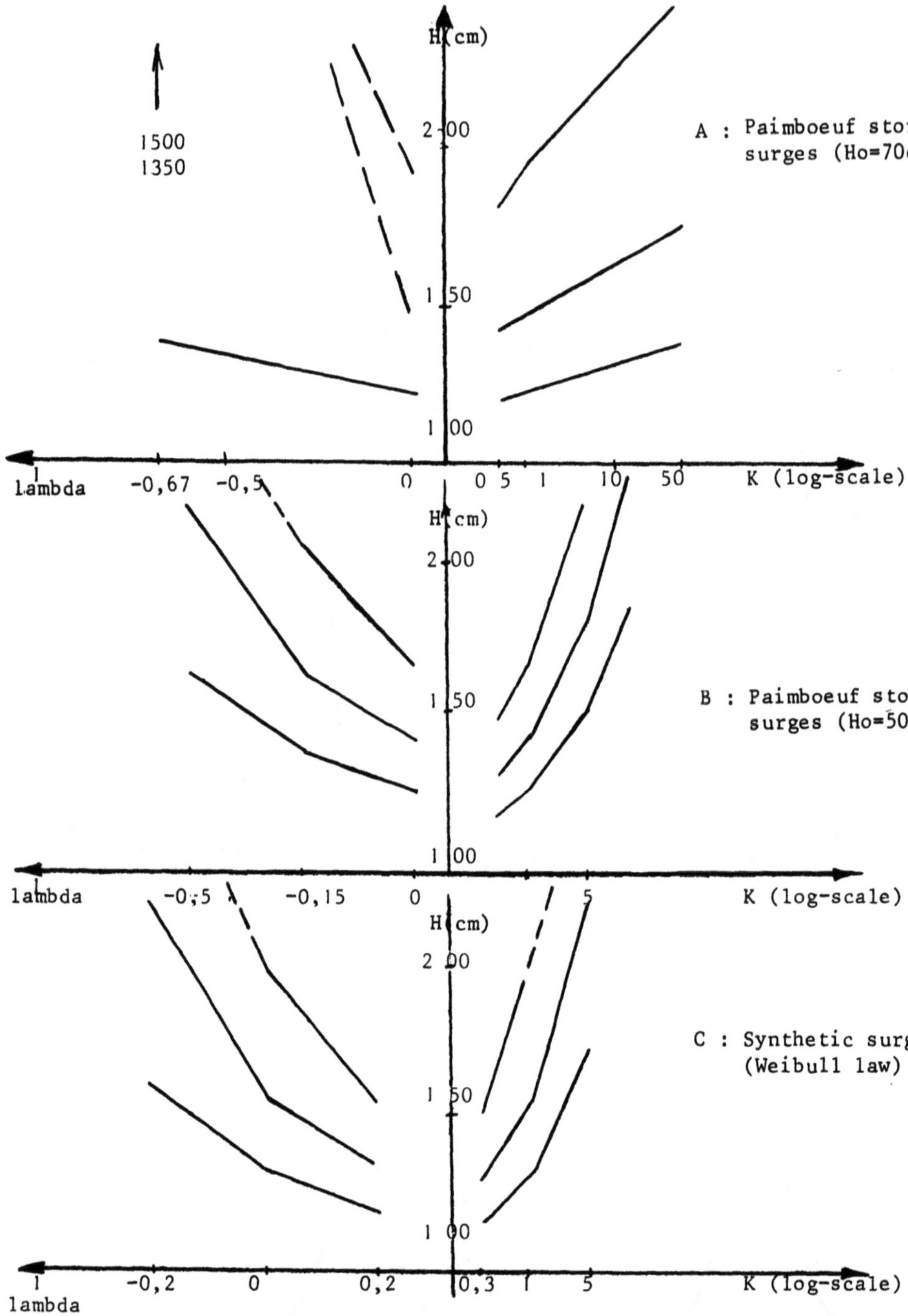

Figure 5 :Sensitivity of 1000 years design values and upper and lower limits of 90% credible intervals to variations of parameters λ and k.

Using formula 2 it is now possible to compute the overall 90% credible interval (i.e.Bayesian confidence interval) taking into account the uncertainty of λ and k; the results are given in Table III for samples II and III only. Note that the figures between parentheses are the ratio of interval widths to the corresponding widths of the classical confidence interval of the simple Weibull model, exhibiting the differences between the case with model uncertainty (sample II) and the case with no model uncertainty (sample III). So in this particular example, the uncertainty due to the choice of the distribution model can be very important (credible interval extending more than three times the usual confidence interval). The presented method allows more realistic interval ranges than the classical method with respect to the true uncertainty inherent to the problem.

TABLE III
1000-YEARS STORM SURGE SETUP AT PAIMBOEUF SITE
(90% CREDIBLE INTERVALS)

Model / Sample	fixed k=1 (cm)	fixed λ=0 (cm)
Sample II (with model uncertainty)	133-260 (3,17)	135-245 (2,75)
Sample III (without model uncertainty)	128-228 (1,51)	124-224 (1,51)

(The bracketed values are the ratios of the interval ranges to the corresponding ones relative to the simple Weibull model)

IV - ON THE USE OF PARSIMONY PRINCIPLE IN CHOOSING CLASS OF MODELS

Perhaps an important criticism against the proposed method is the following :

as a matter of fact, the modeling errors dealt with here are nothing but sampling errors with respect to a larger class of models where parameters may appear superabundant and violate the parsimony principle. This point is partly right and it must be discussed extensively.

Indeed some distribution models (used in flood studies for instance) are defined with too many parameters. Wakeby and TCEV distributions have 4 or 5 parameters and it can be certainly argued that these distributions violate the parsimony principle when, as it is the case usually, data samples are small. In case of POT models, applied to storm surges and to floods, the log-generalized gamma distribution adds 3 or 4 parameters to the Poisson parameter μ, the uncertainty of which has nevertheless little effect upon H_T estimation.

However we now want to point out that the parsimony concept is not an absolute principle; it is relative to the consequences of action following statistical analysis. Bayesian methods make this point clear. Consider an extra-parameter (such as λ in our example), may be superflous but can represent various possible asymptotic behaviours of the studied distributions. A parsimonious distribution model (such as Weibull one) may be chosen when it corresponds to the central part of the posterior density as long as the underestimation loss for λ is comparable to the overestimation loss for instance. When the underestimation loss is relatively too great, then a larger value of the parameter must be chosen even if its density value may be smaller than the density of more probable values.

The essential problem is the choice of a convenient larger family of distributions the behaviour of which <u>represents the uncertainty to be feared in an extrapolation decision</u> This choice corresponds to a "smaller degree of parsimony" than the classical one but there is a matter of judgment in it and it is always possible to refuse any model. For instance see sample I where there are too few data to use a log-generalized gamma or even a Weibull distribution.

V - MODELLING UNCERTAINTIES IN CONCEPTUAL MODELS

The previous principles and methods for studying errors in model choice are also applicable to conceptual models of hydrological or environmental time processes used for forecasting or simulation purposes. These models are often called deterministic because they are built with the help of physical, chemical or biological analogies. However these analogies are rather schematic compared with the complexities of natural phenomena so that large residual errors may be present. Usually these errors are statistically characterized by parameters like mean squarred errors but the present approach offers a deeper insight to these residual uncertainties.

Figure 6 gives a classical schema describing the structure of such conceptual models.

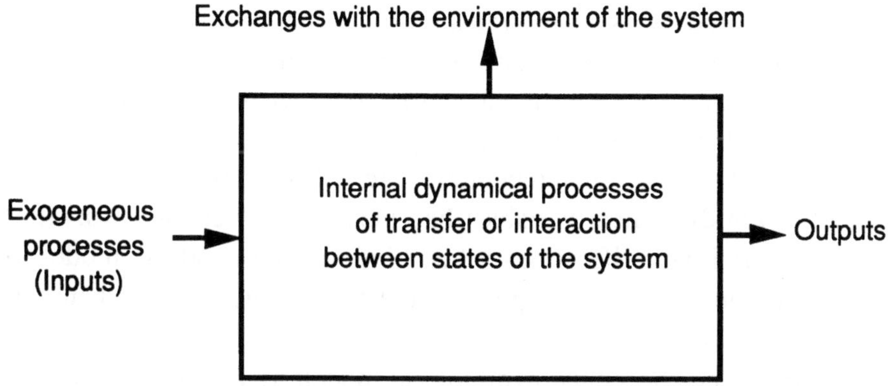

Figure 6 :Schematic modelled system

In terms of a time varying process, one can write the following equation :

$$Y_t = \mathbf{F}(Y^t, X_t, \theta) \qquad (4)$$

where Y_t is the modelled output process, Y^t is the part of the previous (prior to t) history of Y representing the dynamics of the system, X_t is a set of exogeneous variables and θ is a set of parameters.

Applying the principles presented above, it is useful to enlarge the class of models (4) in a more general family able to describe the possible deviations from the reference class of models. Such a generalized class including random variables is (among other possibilities) :

$$Y_t = b_0 + b_1 \mathbf{F}(Y^t, X_t, \theta) + \mathbf{G}(Z_t, \varphi) + \varepsilon_t(\rho) \qquad (5)$$

where b_0 and b_1 are respectively constant and proportional bias; Z_t is a possible concurrent input variable or factor incorporated in terms of a function **G** with parameters φ. **G** might be a statistical regression function The most important element in equation (5) is the radom deviation process ε_t.with a time structure described by parameter ρ. As shown in our examples based on so called deterministic models this time structure is very important to evaluate performance model. In fact the general model (5) uses mixed deterministic and probabilistic parts. It enables us to measure the information brought by the reference model compared with the stochastic part.

Two examples

These examples concern :

Example A : a hydrological runoff model (model CREC) with daily river runoff Y coming from a drainage basin where X is the precipitation; the exchange process is evapotranspiration. The internal tranfer process is described by means of 3 conceptual reservoirs and 7 parameters θ.

Example B : a river water temperature model where the forcing process X is the solar radiation and the exchange process depends on other meteorological variables like wind, air temperature, etc... The internal tranfer process use simple hydrodynamical variables like mean water velocity. Notwithstanding the involved simplifications, this kind of temperature model is often considered to include more physics and to be more reliable than conceptual hydrological model.

Concerning the conceptual hydrological modelling in general, Kuczera (1983) applied a Bayesian analysis, however not complete, to study sampling errors of parameters θ and the sensitivity of simulation results to those errors in θ. Without taking into account proportional bias he used ARMA models for the errors ε_t, only in view of adequately computing the variations of parameters.

In our examples we are interested in the performances of the reference models, as they stand with their estimated parameters, in terms of a particular loss.function. Therefore we consider these models as black boxes and focus on observed values YO_t and calculated values YC_t.

For illustration purpose we used simple and parsimonious models such as :

- Example A : $YO_t = b_0 + b_1 YC_t + \varepsilon_t(\rho)$ (6)

- Example B : $YO_t = b_0 + YC_t + \varepsilon_t(\rho)$ (7)

ε_t is a simple first order autoregressive process :

$$\varepsilon_t(\rho) = \rho . \varepsilon_{t-1} + v_t \qquad (8)$$

where ν_t is a normal random variable, independent from the previous ε_{t-1} with mathematical expectation 0 and standard deviation σ.

Some results of the study

Figure 7 gives the rough results of the two models applied on a day by day basis during a period of almost two years : runoff of Voulzie river (tributary of the Seine river, 1982-1983 - personal communication of a colleague J.Duplex), water temperature of Garonne river at Golfech station (1986-1988 - personal communication of a colleague J.Miquel).

Figure 8 gives the posterior probability densities of parameter ρ given the data. The graphs exhibit large enough significant values of autocorrelation of errors ε_t with most probable values of 0.78 for runoff of Voulzie river and 0.915 for temperature of Garonne river. The model errors seem to be relatively smaller for temperature than for runoff; the ratio of standard deviation of error ε to the mean value of variable YO is equal to 0.248 for Voulzie river and 0.085 for Garonne river. However the high values of probable autocorrelation coefficient make the final picture a little more precise as we will now show.

Adopting a loss function is useful to describe the consequences of the choice of a given model. Keeping in mind the purpose of simulation along a period of T days, we chose :

$$J = \sqrt{\frac{1}{T}\sum_{t=1}^{T} (YO_t - YC_t)^2} \qquad (9)$$

where YC_t can be calculated according to the reference model and to themore complete one (including simulation of errors).

Two comments are in order here. First, as the comparison is based on the posterior expectation of J or of any other function of J, it can be made with the data which were used for estimating the error model parameters. In our examples we do so. A second point is that J is not the only performance index that can be used. The very nature of each problem (a water ressources problem in example A or a power plant operation in example B) should be taken into account so as to build more realistic quantitative indices. In these cases the computation of posterior loss expectation can be made by means of Monte Carlo methods.

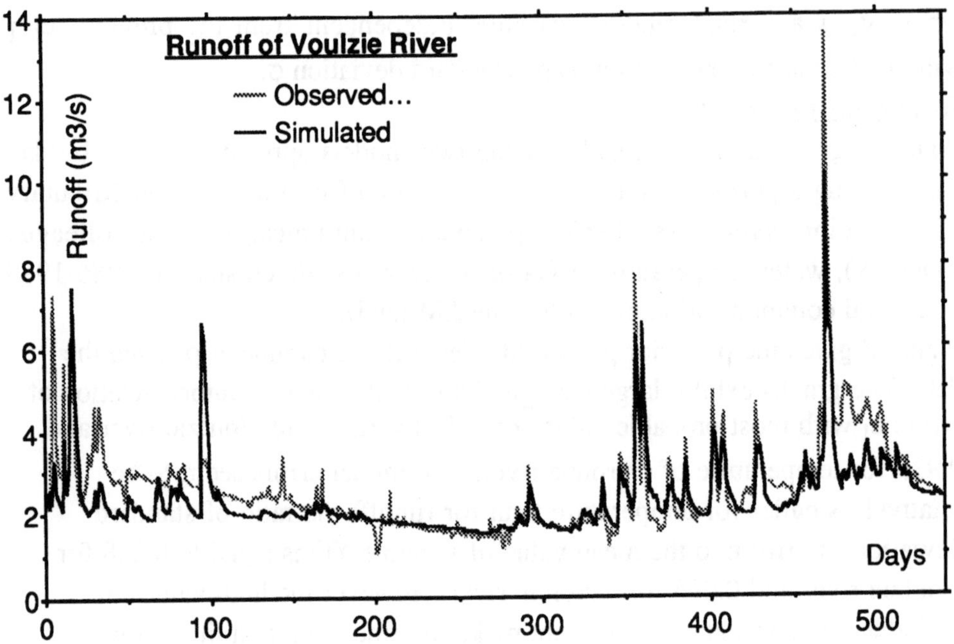

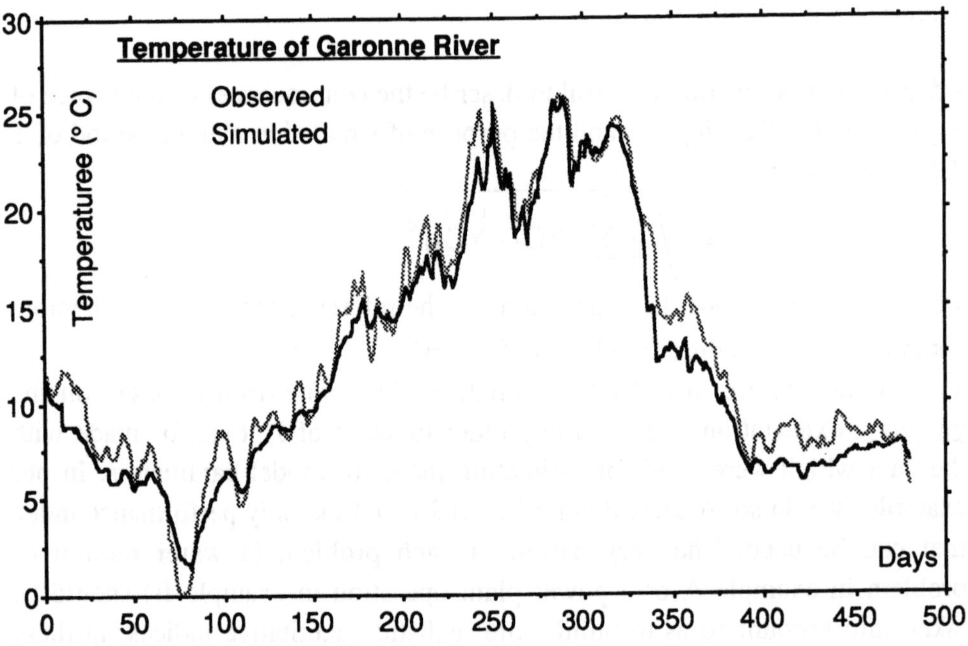

Figure 7 : Observed and simulated daily values for Voulzie river (between 1982-83) and Garonne river.(between 1986-88).

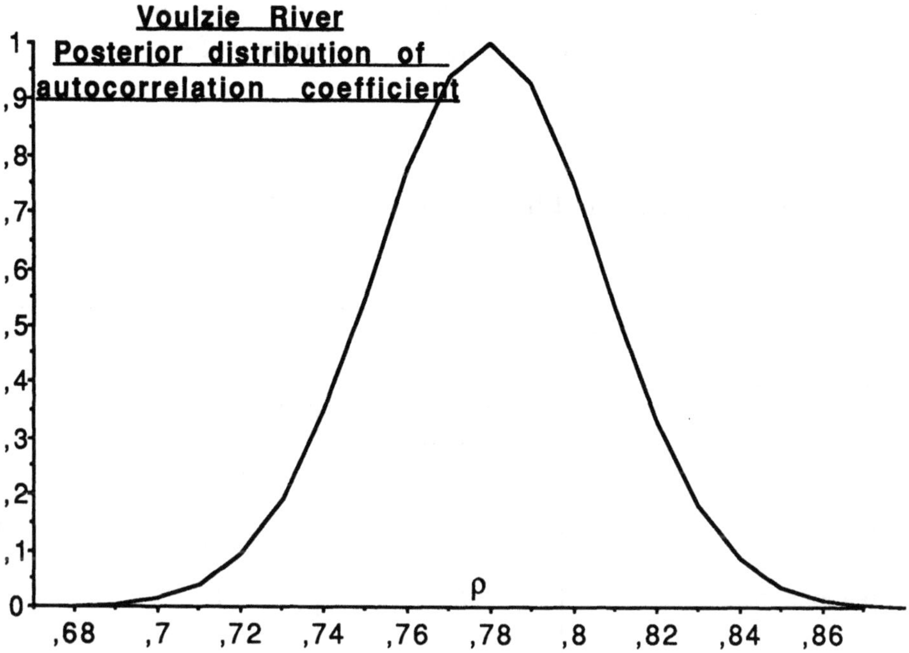

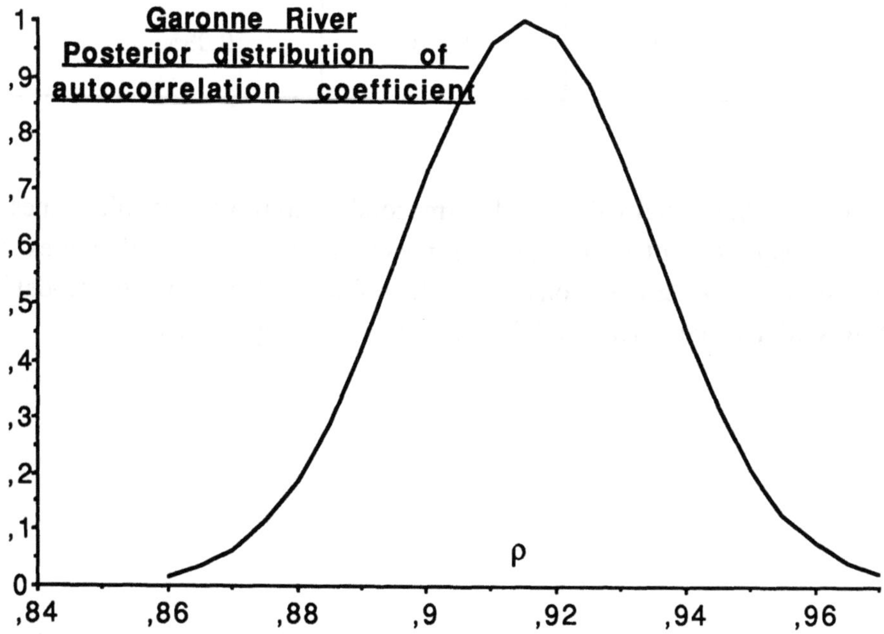

Figure 8 : Posterior distribution of autocorrelation coefficient ρ of model error

Table IV shows the most probable values of ρ, corresponding conditional expected values of b_0 and b_1 and the standard deviation v of ε.

TABLE IV
MOST PROBABLE VALUES OF ERRORS PARAMETERS

Parameters \ Example	A (Voulzie runoff)	B (Garonne temperature)
Autocorrelation ρ	0.78	0.915
Constant bias	-0.051	0.905
Proportional bias	0.127	0
Standard deviation of ε (v)	0.646	1.099
Ratio : v/mean value of Y	0.248	0.085

Table V gives the actual values of J corresponding to most probable values of ρ and the square root of the posterior expectation of J^2. All values are computed under two assumptions : first, the reference "deterministic model" is used, second, the "complete model" with error modeling is used.

TABLE V
COMPARISON OF QUADRATIC LOSS VALUES

Losses / Examples		A (Voulzie runoff)	B (Garonne temperature)
With reference model	Most prob. ρ	0.698	1.429
	Expected	3.524	0.705
With complete model	Most prob. ρ	0.404	0.444
	Expected	0.428	0.430

From Table IV it can be seen that the hydrological model exhibits a slight proportional bias and the temperature model has a constant bias (of course with respect to the period of study).
From table V, it is possible to calculate the ratios of J values of the "complete models" over the results of the reference models. The less is this ratio is better the complete model performs. In case of actual values of ρ the ratios are respectively 0.58 and 0.31 for the two examples. It seems that the temperature model is better improved with an error treatment like the one presented here. However in terms of expectation the ratios are better in the case of example A and not as good as example B. But the results concerning the expected losses must be considered with caution especially in the case of A because the effects of assumption of error normality which is not quite realistic in this example.

The errors models used are rather simple; furthermore relatively short periods of data have been chosen only for illustrative purposes but the Bayes method can be applied in more general contexts. It is possible to introduce the parameters θ (as presented by Kuczera -1983- in a non decisional context) in a complete decisional Bayes analysis together with errors parameters ρ and so on...
Such an approach could be applied to any conceptual, hydrological one as well as environmental model.

VI - REPRESENTATIVENESS UNCERTAINTIES

In presenting a more complete picture of uncertainty analysis, it is interesting to say a few words about representativeness uncertainties. The problem is important at this time when the questions about Glogal Change at the Earth scale and its consequences are asked. We think that Bayesian approach and the use of the so called hierarchical prior models (Berger - 1985) are also useful tools in these domains.

Without trying to deal with such general problems and to present a complete method, we only give a few ideas illustrated by a particular problem but an important one concerning the flood risk of Loire river in France. Figure 9 presents the time series of annual maximum floods discharges from 1833 to 1971 that were observed at Gien station.

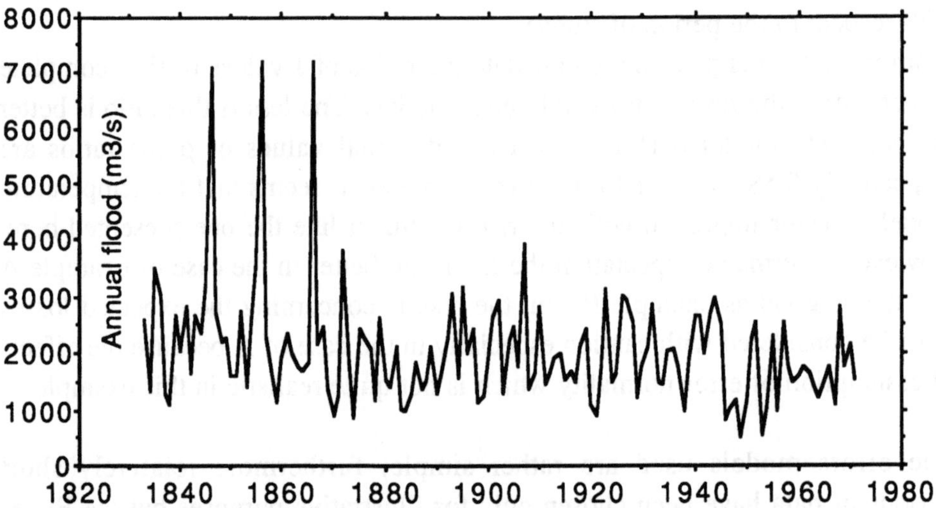

Figure 9 : Time series of annual floods of Loire River at Gien.

Time heterogeneity appears in this series and is statistically significant regardless of the probability distribution. In particular three huge floods were

observed in 1846, 1856, 1866 and since this last date no flood has occured larger than half the value of these events. No hydrological, meteorological factors, land or geomorphological variations seem to be able to explain this fact. What design flood must be used in future project along Loire river?
We already mentionned that no probability distribution can fit the Loire series. Among the many trials made, there is the log-normal distribution :

$$\text{Prob}(Q \leq q) = \frac{1}{\sqrt{2\pi}.\sigma} \int_{-\infty}^{\text{Log}\, q} e^{-\frac{(u-m)^2}{2\sigma^2}} du \qquad (10)$$

For generating a larger family of distribution the Cox-Box transformation can be used as given in the generalized gamma (formula 3) :

$$u = \frac{q^{\lambda} - 1}{\lambda} \quad \text{if}\ \lambda \neq 0 \quad (\ u = \text{Log}\ q \quad \text{if}\ \lambda = 0\ )$$

Then let us consider two sub-sample of Loire floods : sample I from 1833 to 1879, sample II from 1880 to 1971. Figure 10 shows the respective posterior densities of λ given each sample. The two posterior distributions have remarkable differences in expectation μ (positive in case of sample II and negative in case of sample I) and comparable standard deviation V notwithstanding the lenghts of the series (sample II is twice as long as sample I). These differences are too large for using single values of prior parameters. Following Berger (1985) it is possible to take into account the uncertainty about future flood distribution, represented here with uncertainty about future λ by means of hierarchical prior models :
for instance if a normal prior distribution is used for λ, the two parameters μ and V describing this prior, are second level parameters which can be estimated with the data using all possible prior information with the help of a second level prior. There are a lot of methods for doing this, such as empirical Bayes methods for instance. As shown in this paper, these methods apply to a variety of situations and of distributions which can be different from Log-normal distribution that we used in this section only for illustration purpose.

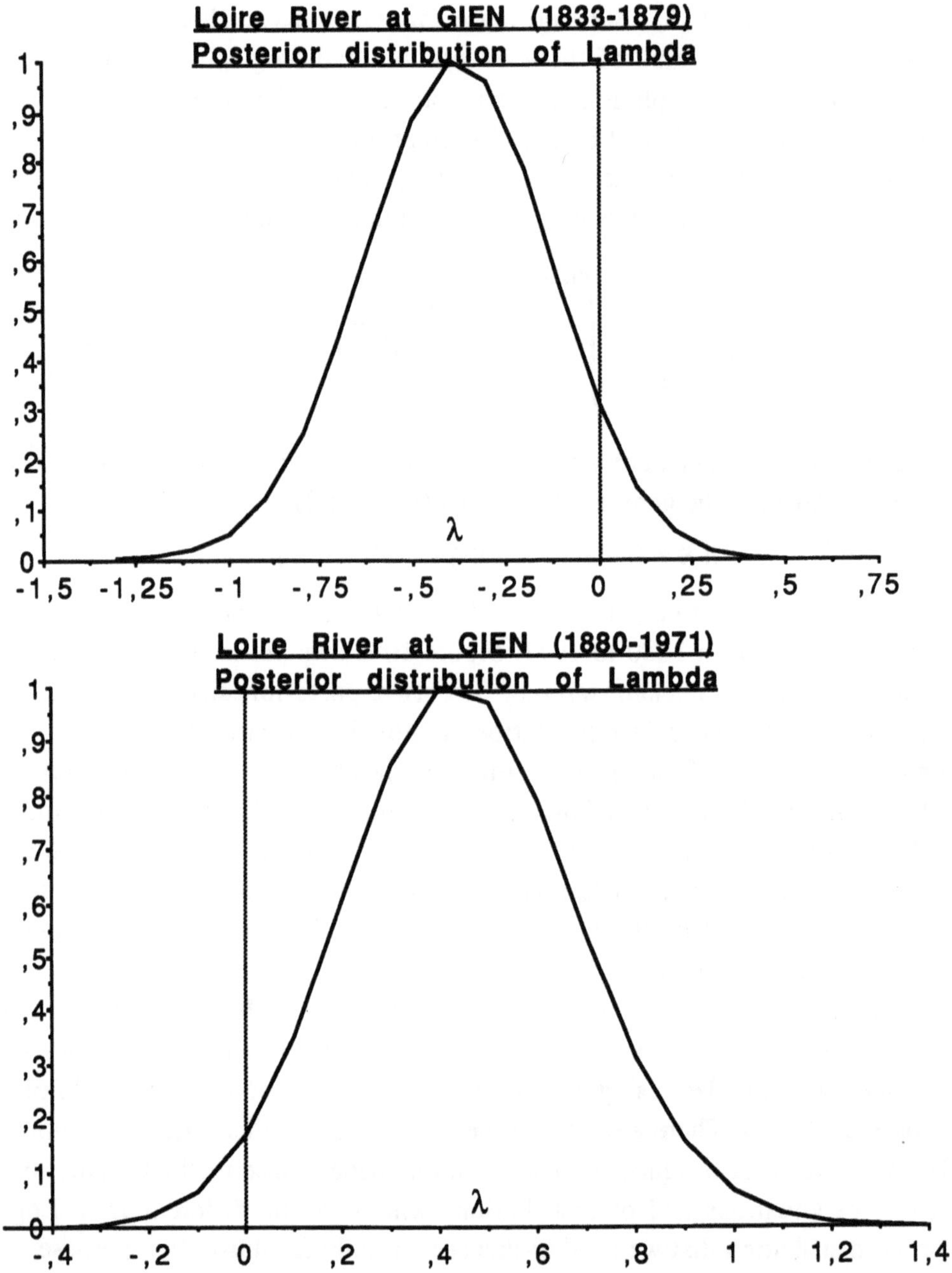

Figure 10 : Posterior distribution of Cox-Box parameter λ as a function of periods of observation of floods of Loire river.

VII - REFERENCES

Bain L.and Engelhardt M.(1981) : Simple Approximate Distributional Results for Confidence and Tolerance Limits for the Weibull Distribution Based on Maximum Likelihood Estimations; Technometrics Vol.23 - n°1.

Berger J.O. (1985) : Statistical Decision Theory and Bayesian Analysis, Springer - Verlag, Berlin.

Bernier J. (1988) : Intervalles de confiance pour les valeurs de projet. Rapport LNH : HE-40/88-15.

Bernier J. (1987) : Elements of Bayesian Analysis of Uncertainty in Reliability and Risk Models; in Engineering Reliability and Risk in Water Resources (L.Duckstein, E. Plate editors), NATO ASI Series; M Nijhoff-Dordrecht.

Bernieer J.and Manoha B.(1991) : "The Estimation of Uncertainties for Design Extreme Values of Wave and Sea Level", to appear in Reliability and Uncertainties in Hydraulic Design, American Soc. Coastal Eng.

Box G.E.P.and Tiao G.C.(1973) : Bayesian Inference in Statistical Analysis, Reading, Mass; Addison Wesley.

Raiffa H. (1968) : Decision Analysis : Introductory Lectures on Choices under Uncertainties. Addison Wesley, Reading.

Savage L.J.(1954) : The Foundations of Statistics; Wiley ,New York.

RELIABILITY WITH FUZZY ELEMENTS IN WATER QUANTITY AND QUALITY PROBLEMS

by

Lucien Duckstein* and Istvan Bogardi**

ABSTRACT

Uncertainty and imprecision in hydrologic reliability analysis procedures should be explicitly taken into account in both water quantity and water quality problems. A mix probabilistic/fuzzy logic based approach is presented, using the respective examples of flood protection reliability and control of pollutant loading reliability. In the flood protection example, the fuzzy risk-cost is calculated within the framework of a decision model with two states of nature S_1 and S_2, such that S_1: $X \geq X_o$ and S_2: $X<X_o$, X being the flood stage and X_o, the protection level. The probability of exceedance is calculated as a fuzzy number (FN) to account for model and sample uncertainty, as well as imprecision in the resistance X_o. The protection costs C and the flood losses K are also taken as FN's. The expected losses corresponding to each action can then be calculated by fuzzy arithmetic, also yielding FN's. The action that results in the smallest FN is selected. Techniques for ranking FN are briefly discussed, and the fuzzy mean is used as a ranking criterion.

The water quality example considers the probability of non-exceedance of total pollutant load M into a water body, such as an aquifer or a stream. This mass M is calculated as a linear combination of random loadings with fuzzy (number) coefficients. It is assumed, without loss of generality, that M is a normal variate, whose mean and standard deviation are FN's because the coefficients are FN's. The probability of non-exceedance is thus calculated by the extension principle, yielding the safety or reliability of control as a fuzzy number. As it was done in the case of flood protection, the comparison of actions may be made by means of ranking the fuzzy numbers corresponding to each action. Overall, a mixed fuzzy-probabilistic approach to risk and reliability analysis appears to provide another useful viewpoint for analyzing the effect of uncertainties in both water quantity and quality problems.

---

* L. Duckstein, Systems and Industrial Engineering Department, University of Arizona, Tucson, AZ 85721

** I. Bogardi, Civil Engineering Department, W348 Nebraska Hall, University of Nebraska-Lincoln, Lincoln, NE 68588-0531

NATO ASI Series, Vol. G 29
Water Resources Engineering Risk Assessment
Edited by J. Ganoulis

1. INTRODUCTION

The purpose of this development is to present a framework for reliability under fuzziness or imprecision (see Appendix). Reliability or safety occurs when the load Q does not exceed the resistance R. In general, load Q is a random variable but it also be fuzzy. The resistance may be fixed, random, fuzzy, or both random and fuzzy. Thus, the flood levee erosion case considered in Shrestha et al. (1990) possesses a fuzzy load (which may also be random) and a fuzzy resistance. Specifically, the load is the wind-generated wave erosion; the wave height above a datum is estimated by an empirical formula that uses fetch or wind direction, a very imprecise input. Randomness in the load is introduced when the datum is varied randomly according to the historical peak flood distribution. The resistance of the level to wave erosion is also fuzzy, because of the non-homogenity and irregularity of the structure. In such problems, as well as in the flood protection problem and the water quality example considered in the present study, it is convenient to define the difference M between load and resistance, and consider safety as the non-exceedance of M:

$$M = Q - R < m_o \tag{1}$$

In some applications, m_o may be fuzzy, in others it may be considered as a safety margin (a negative number in this case).

As pointed out in the system framework developed in Duckstein et al (1987) an exceedance event (complement of Eq. (1)) is called an incident, and a large incident becomes a failure. The concepts pertaining to the system model, which include performance indices such as resistance, vulnerability, and figures of merit such as risk, may be considered from a fuzzy logic or a mixed fuzzy-proabability viewpoint.

Fuzzy set and fuzzy number concepts are presented both in the Appendix and throughout the text. The next section deals with a flood protection example framed as a decision model with two states of nature (exceedance or no exceedance) and two decisions (protect or do not protect). An extension of the model to include the ecological effect of protection is also given. Section 3 consists of a water quality example, in which the non-exceedance probability of pollutant loading in a water body is estimated. In both flood and water quality cases, the consequence of a decision is a fuzzy number: ranking decisions thus consists of ranking fuzzy numbers.

2. FUZZY RISK ANALYSIS OF FLOOD PROTECTION

As described for example in Bogardi et al (1988), a risk analysis procedure generally involves the following steps:

(i) identification of a hazard- here the flood exceedance

(ii) definition of an incident and hence, of a failure event such as the breach of a flood protection structure

(iii) estimation of the probability of an incident or a failure

(iv) definition of the consequences of an incident (economic, human, environmental)

(v) evaluation of the perception of the consequences.

Uncertainty which includes non-controllable (natural) randomness and partially controllable informational uncertainty and/or imprecision, may be present in any of the above steps. The example given next illustrates this statement.

Consider the case of a binary decision to be made along a reach of the Santa Cruz River in southern Arizona, USA, to protect or not to protect against the maximum peak floods occurring during a 20-year horizon. This 2X2 decision problem is analogous to the one developed in Duckstein et al (1978) where a Bayesian approach was used: here a mixed probability/fuzzy set analysis will be used. Within the framework of this example, uncertainty and/or imprecision may be found in the following elements of the above risk analysis procedure

(1) randomness or natural uncertainty of peak floods

(2) informational uncertainty on flow model and loss functions

(3) levee resistance to floods due for example to non-homogneneity of soils

(4) the failure phenomenon itself, which may be "small" or "large"

It is possible to describe uncertainty of type (2), (3) and (4) above by either probabilistic, Bayes or fuzzy set analysis, or else, by a mix of those approaches. In particular, examples of problem types (3) and (4) analyzed by fuzzy sets (or mixed approaches) may be found in Kaufmann and Gupta (1988), Duckstein et al (1988) and Shrestha et al (1990). For background material on fuzzy sets and fuzzy numbers, the reader is referred to standard texts such as Dubois and Prade (1980), Kaufmann and Gupta (1985), and papers such as Zadeh (1965, 1973), Bardossy et al (1990).

Consider first a decision-theoretic formulation of the problem, with the folowing elements:

(a) two states of nature

$S_1 : X \geq X_o$ and $S_2 : X < X_o$

where X is the peak flood over the time horizon and X_o is the protection level

(b) A Bernoulli model of occurence of state S_i, $i = 1, 2$:

$P(S_1) = P(X \geq X_o | p) = p$

$P(S_2) = P(X < X_o | p) = 1 - p$

(c) Two actions: a_1 : protect and a_2 : do not protect

(d) Annualized flood losses K and protection costs C

The decision problem can be displayed in the loss matrix given in Table 1

Table 1 : Economic loss matrix

i	S_i	a_1 protect	a_2 do not protect	p (S_i)
1	S_1: $X \geq X_o$	C	K	p
2	S_2: $X < X_o$	C	O	1-p

Next, let the following uncertainties be taken into account and represented in the form of fuzzy numbers:

(i) the probability $\hat{p}$ of exceedance, reflecting sample and model/technique uncertainties as well as fuzziness in Xo

(ii) economic losses $\hat{C}$ and $\hat{K}$, because of the difficulty in estimating annualized construction, operation and maintenance costs $\hat{C}$, and flood losses $\hat{K}$.

Let the following triangular fuzzy numbers (TFN) represent, respectively, the fuzzy elements of the decision problem:

$\hat{p} = (7,8,10) \times 10^{-4}$ corresponding to a protection level of about 14.000 cfs in the Santa Cruz river basin, in Southern Arizona

$\hat{K} = (4,5,6) \times 10^3$

$\hat{C} = (3,4,5)$ corresponding to a cost of " about 4 "

The expected losses corresponding to actions a_1 and a_2 can now be calculated as

$$\hat{L}(a_1) = \hat{C}(\cdot)\, \hat{p}\, (+)\, \hat{C}(\cdot)(\hat{1}(-)\hat{p}) \tag{2}$$

$$\hat{L}(a_2) = \hat{K}(\cdot)\hat{p}$$

where $\hat{1} = (1,1,1)$ and (+), (-), ($\cdot$), indicate the fuzzy arithmetic operations of addition, subtraction and multiplication, respectively (Kaufmann and Gupta, 1985 and Appendix). One has:

$$\hat{L}(a_1) = \hat{C}(\cdot)\, \hat{p}\, (+)\hat{C}(\cdot)1(-)\hat{C}(\cdot)\hat{p} \quad = \hat{C}$$

$$\hat{L}(a_2) = (4,5,6)(\cdot)(7,8,10) \times 10^{-4+3} = (2.8,4,6)$$

To determine the optimum action a_1, the two fuzzy numbers $\hat{L}(a_1)$ and $\hat{L}(a_2)$ must be compared. This may be done by the method of Chen (1985) or else simply by using the fuzzy mean $\bar{A}$ of fuzzy number $\hat{A}$ defined on the domain of real numbers D(x)

$$\bar{A} = \frac{\int_{D(x)} x \; m\hat{A}(x)dx}{\int_{D(x)} m\hat{A}(x)dx} \tag{3}$$

If $\hat{A}$ is a TFN, (a,b,c), then algebraic manipulations yield:

$$\bar{A} = \frac{a+b+c}{3}$$

Accordingly, the fuzzy means of $\hat{L}(a_1) = (3,4,5)$ and $\hat{L}(a_2) = (2,8,4,6)$ are such that

$\bar{L}(a_1) = 4 < \bar{L}(a_2) = 4.27$.

Thus, on a purely economic basis action $a = a_1$, protect, is preferred to action a_2.

Next, consider the environmental or ecological effect of flood protection, and let this effect be measured as the number of species lost if a structure is built. Let the initial number of species be N_1 and the number after construction, which is not predictable precisely, be a fuzzy number $\hat{N}_2$. The ecological loss matrix shown in Table 2 can then be constructed.

Table 2: Ecological loss matrix

i	S_i	a_1 protect	a_2 do not protect	$p(S_i)$
1	S_1: $X>X_o$	$\hat{N}_2$	N_1	$\hat{p}$
2	S_2:$X<X_o$	$\hat{N}_2$	N_1	$1-\hat{p}$

In that Table, the simplifying assumption that the number of species depends only on the action, not the state of nature, had been made but this does not invalidate the methodology. Numerically, let: N_1=60 initial species, $\hat{N}_2=\hat{N}(a_2)$=(10,25,40), corresponding to about 25 species left, but not less than 10 and not more than 40.

To evaluate the two actions a_1 and a_2, a fuzzy trade-off relationship is constructed (Fig.1). Action a_1 (protect) is represented by a two-dimensional fuzzy number reflecting uncertain economic and ecological quantities. Action a_2 (do not protect) does not show ecological uncertainty, and is thus represented by a one-dimensional fuzzy number projected as the horizontal segment (2.8,6). The action closest to the ideal point is selected: the distance between

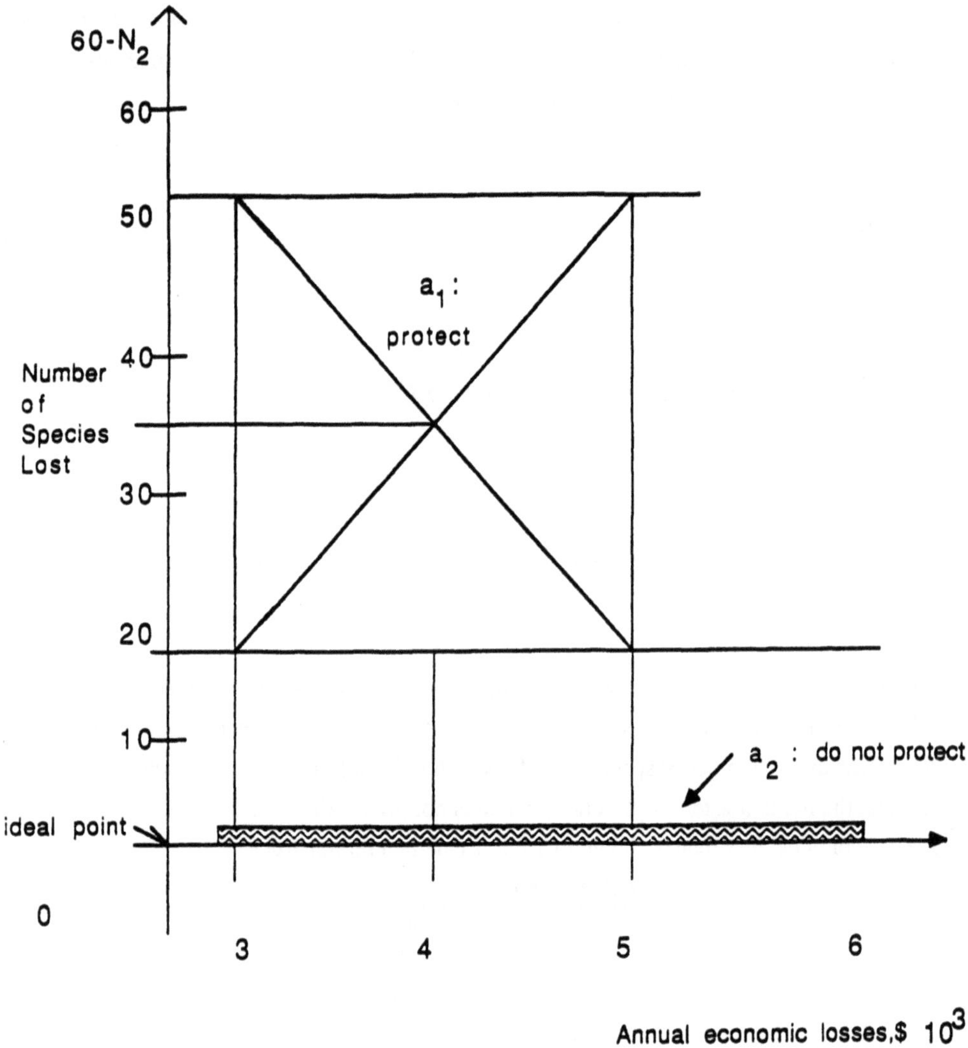

Figure 1

Trade off under uncertainty between ecological and economic effects of flood control.

two fuzzy numbers (here one point and one fuzzy number) must thus be defined. Taking again the mean value representation for a fuzzy number (Eq. (3)), it is found that action a_2 is preferred to action a_1. Thus, taking ecological effect into account leads to a reversal of the decision that would be taken on purely economic ground.

This approach can readily be extended to other examples of investigation of the effect of hydrologic uncertainties on the environment, where randomness (X), sample uncertainty (p), fuzziness (C and N_2) are jointly considered. These problems may be concerned with erosion, sediment yield, pollutant loading into surface water bodies and ground water. Some of the cited references deal with such case studies; so does the next section.

3. FUZZY RELIABILITY ANALYSIS OF POLLUTION LOADING

Consider, as in Plate and Duckstein (1991) a river subject to polluting substance loading from a set of subwatersheds indexed i=1, ..., I. Let the contribution of field i be the product of pollutant concentration c_i by discharge Q_i; then the total loading may be written as

$$M = \sum_i c_i Q_i \tag{4}$$

The uncertainty of M in Eq. (4) may be characaterized by natural uncertainty in Q_i together with c_i being either deterministic or random or fuzzy. Sample uncertainty in the pdf of Q_i or, if applicable, c_i, could also be considered. Examples in three case are given below, assuming that the Q_i have a common pdf.

(1) c_i deterministic:

The pdf of M has a mean and variance given by:

$$\mu_M = E(M) = \sum_{i=1}^{I} c_i E(Q_i) \tag{5}$$

$$\sigma_M^2 = VAR(M) = \sum_{i=1}^{I} c_i^2 VAR(Qi) + 2 \sum_i \sum_{\substack{j \\ i<j}} c_i c_j COV(Q_i Q_j) \tag{6}$$

Thus any two parameters of M is completely determined; in particular, if the number I of watersheds is large, then a normal approximation may be used:

$$f_M(m) = N(\mu_M, \sigma_M^2; m) \tag{7}$$

(2) c_i random:

In this case Eq. (4) is the sum of products $c_i \cdot Q_i$ of random variables. The determination of the complete pdf of M is in most realistic cases, theoretically but not computationally feasible ; thus approximate solutions have to be sought, such as linearization or second order analysis. For example, let each variable be defined as the sum of its expected value ($\bar{c}$, $\bar{Q}$) and α random fluctuation (c', Q') with mean zero:

$$c_i = \bar{c}_i + c_i, \quad Q = \bar{Q}_i + Q_i, \quad \text{with } E(c_i) = E(Q_i) = 0$$

Then,

$$E\,(c_i \cdot Q_i) = E[(\bar{c}_i + c_i)\,(\bar{Q}_i + Q_i)]$$

$$E\,(c_i \,.\, Q_i) = \bar{c}_i \,.\, \bar{Q}_i + E(c_i\, Q_i) \tag{8}$$

It is also possible to estimate the variance VAR($\Sigma c_i \,.\, Q_i$), $\sigma^2_{c_i}$ and $\sigma^2_{Q_i}$, if the higher order terms are neglected.

(3) c_i fuzzy:

Let the c_i's be given as fuzzy numbers $\hat{C}_i$ (Kaufmann & Gupta 1985; Bardossy et al, 1990. For example, the statement: "c_1 is about 20 kg/ha but not more than 35" may be represented by one of the membership m(c) of fuzzy number $\hat{C}$ sketched in Fig. 2. Membership function A is triangular, assigning degree of confidence proportional to the distance from the maximizing value 20. In contrast, function B means that values of c less than 20 corresponds to a low degree of confidence whereas values greater than 20 (and less than 35) correspond to a high degree of confidence.

Given that each c_i = 1, ..., I is represented as a fuzzy number $\hat{C}_i = (c_i, m_i(c_i))$ the means and variances given in Eqs. (5) & (6) are also fuzzy numbers which can be calculated by fuzzy arithmetic, as given in the Appendix. If triangular fuzzy numbers (TFN) are used, the arithmetic rules are very simple, as summarized below. Another technique for combining (not adding) fuzzy numbers is the so-called fuzzy extension technique (Bardossy, et al 1991) which is an application of the fundamental extension principle of fuzzy set theory (Zadeh 1965, 1973).

A TFN $\hat{A}$ may be represented as a triplet (a_1, a_2, a_3); for example, TFN $\hat{A}$ in Fig.2 is represented as $(a_1, a_2, a_3) = (0,20,35)$. The sum and product of TFN $\hat{A} = (a_1, a_2, a_3)$ and C = (c_1, c_2, c_3) are then calculated as

$$\hat{A}\,(+)\,\hat{C} = (a_1 + c_1, a_2 + c_2, a_3 + c_3) \tag{9}$$

$$\hat{A}(\cdot)\hat{C} = (a_1\, c_1, a_2\, c_2, a_3\, c_3) \quad \text{for } \hat{A} \geq 0 \text{ and } \hat{C} \geq 0$$

The difference ($\hat{A}$ - $\hat{C}$) and quotient ($\hat{A}$ / $\hat{C}$) of fuzzy numbers are calculated respectively by adding a negative TFN C^- or multiplying by an inverse TFN C^{-1} defined, as follows:

If $C = (c_1, c_2, c_3)$ then $C^- = (-c_3, -c_2, -c_1)$

$$\text{if } C = (c_1, c_2, c_3) \text{ then } C^{-1} = (c_3^{-1}, c_2^{-1}, c_1^{-1}) \tag{10}$$

Note that a crisp (non-fuzzy) number is a particular case of a fuzzy number with $a_1=a_2=a_3$; linear approximations to non-linear membership functions such as B in Fig. 2, are often in order.

The convex fuzzy extension technique would be valid for finding the mean contribution of a set of subwatersheds which is defined on the same domain as the individual contributions. If a normal approximation holds as in Eq. 8, then it is possible to express the probability of exceedance of nitrate loading as a fuzzy number.

$$\mu_M = \sum_{i=1}^{I} \hat{C}_i \mu_{Q_i}, \qquad \sigma_M^2 = \text{VAR} \left(\sum_{i=1}^{I} \hat{C}_i Q_i \right) \tag{11}$$

These quantities define the probability non-exceedance as a fuzzy number

$$\hat{P}(k) = \Pr \left[\hat{M} = \sum_{i=1}^{I} \hat{C}_i Q_i < k \right] \tag{12}$$

$$\hat{P}(k) = \Phi \left(\frac{k - \hat{\mu}_M}{\hat{\sigma}_M} \right)$$

where Φ is the standard normal cumulative distribution function (DF).

The membership function of $\hat{P}(k)$ is calculated by application of the extension principle (Appendix, Eq. (A-10)):

$$m_{P(k)}(p(k)) = \begin{cases} \max \{ \min (\mu_M(\mu), \sigma_M(\sigma)) \} & \text{if } p(k) = \Phi \left(\frac{k - \mu}{\sigma} \right) \\ 0 & \text{otherwise} \end{cases} \tag{13}$$

Using a cumulative standard normal table, the membership function of $\hat{P}(k)$ in Eq. (13) can be calculated as illustrated next. Let μ and σ be TFN's, $\hat{\mu}$=(0, 30, 50) and $\hat{\sigma}$=(10, 30, 50), and k=20. Fig. 3 exhibits these two TFN"s in the (μ, σ) plane. The so-called 0.5 level set of the joint membership function of two-dimensional fuzzy number (μ, σ) is the set of points in the (μ, σ) plane for which $m_{P(k)}$ ≥0.5. This set corresponds to the area inside the rectangle (H_1,H_2,H_3,H_4). Table 3a shows how the 0.5 level set of the safety probability $\hat{P}(45)$ is calculated: each element in that table is calculated as:

$$\frac{k - \mu}{\sigma} = \frac{45 - \mu}{\sigma}$$

and corresponds to a point H_i, i=1,..,4 in Fig.3. Thus, element (15, 40) corresponding to H_2 is equal to $\frac{45\text{-}15}{40}$ = 0.75. Then the maximum of the row minimum is found to be 0.75 and the maximized value of the column minimum, indicated by an arrow, is 0.25, yielding the limits of the 0.5 level set (0.25, 075). Table 3(b) shows the 0-level set calculations. The maximum of the membership function corresponds to point H(30), 30), for which $\frac{45\text{-}\mu}{\sigma}$ = 0.5. The membership function of $\hat{P}(45)$ given in Table 4 and sketched in Fig. 4 is thus obtained.

At a given safety level k, every decision d_j yields a fuzzy number. $\hat{P}(j,k)$, which may thus be considered as a figure of merit of the jth alternative j=1, .., J. To rank the alternatives, it is necessary to order the J fuzzy numbers resulting from the analysis, which may be done as in Dubois and Prade (1980), Chen (1985),or Kaufmann and Gupta (1988). As a first approximation, the ranking may be done on the basis of the fuzzy mean, defined in Eq. (3).

At this point, it may by noted that a formal analogy exists between a joint probabilistic approach and the fuzzy approach represented in Fig. 3 but the calculations are completely different: in fact they are much simpler in the latter than in the former case.

Table 3 Values of $\frac{k\text{-}\mu}{\sigma}$ for the calculation of h level set of safety probability P(k=45).

	3(a):		h = 0.5 level set
	σ = 20	σ = 40	row minimum
μ=15	1.5	0.75	0.75 ←
μ=40	0.25	0.125	0.125
column minimum	0.25 ←	0.125	
	3(b):		h = 0 level set
	σ = 10	σ = 50	row minimum
μ=0	4.5	0.90	0.90 ←
μ=50	-0.55	-0.10	-0.50
column minimum	-0.25	-0.10 ←	

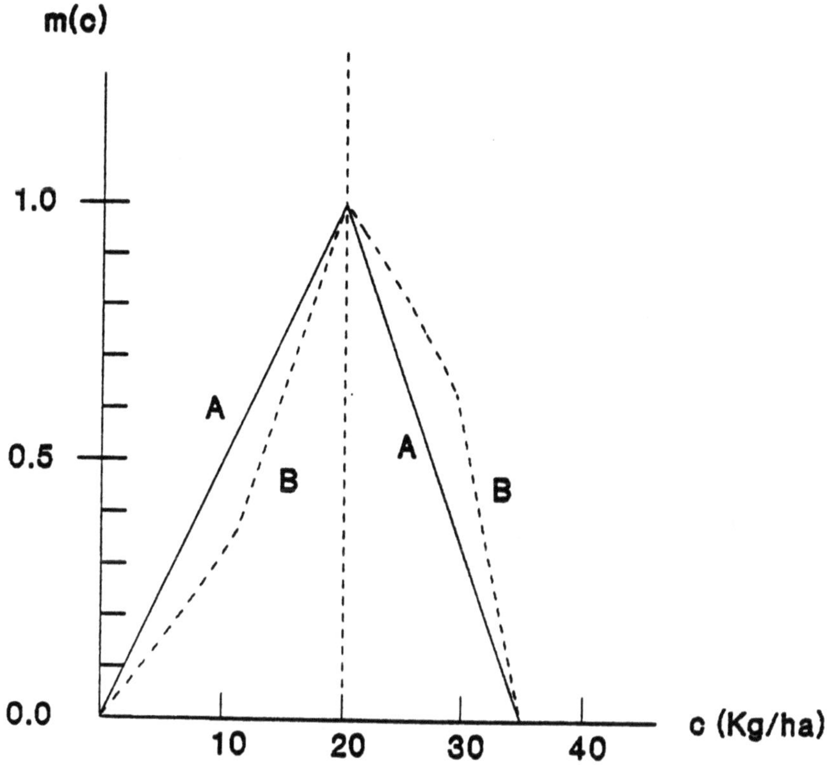

Figure 2: Two possible membership functions of the fuzzy number c " About 20 but no more than 35 "

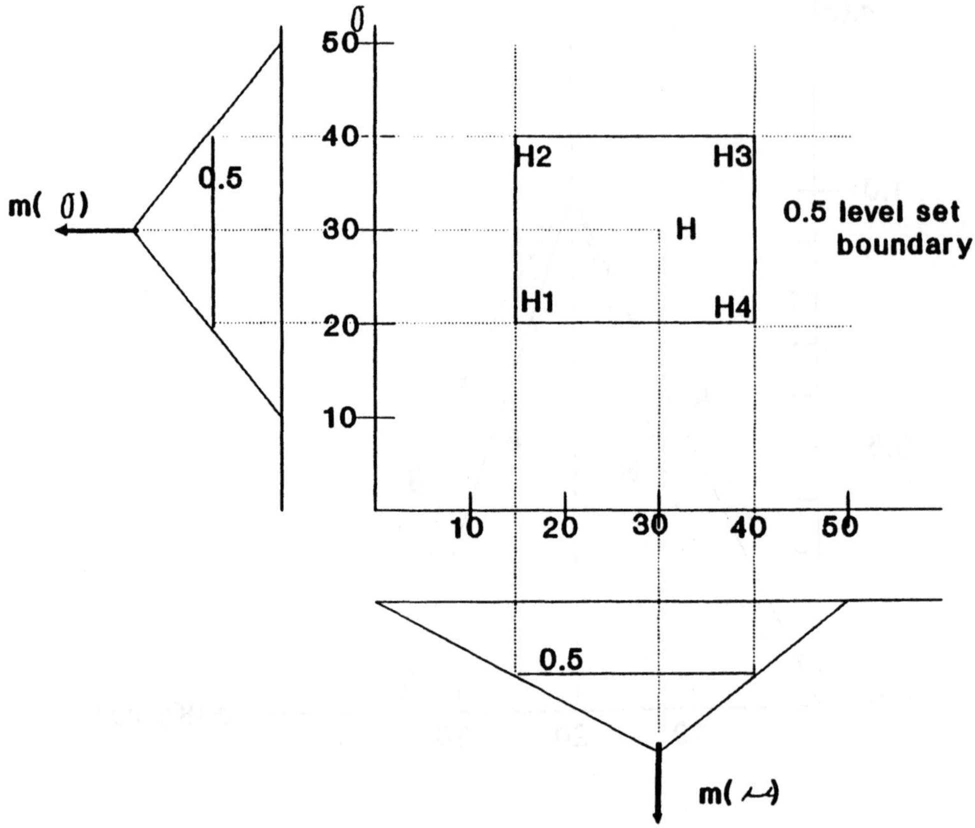

Figure 3: Application of the extension principle to calculate the membership function m(p(45))

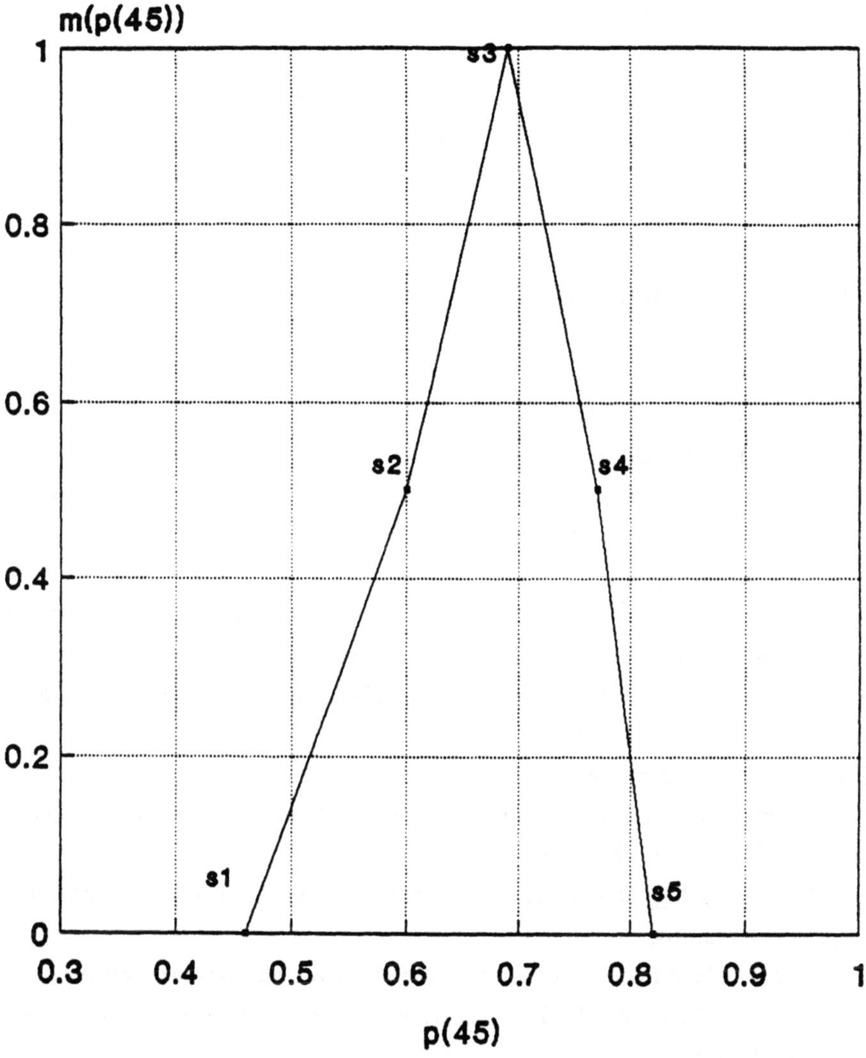

Figure 4: Membership function of safety probability p(45)

Table 4 Membership function of P(45) of the function of Fig. 4

$(45-\mu)/\sigma$	-0.10	0.25	0.50	0.75	0.90
$P(45)=\Phi((45-\mu)/\sigma)$	0.46	0.60	0.69	0.77	0.82
$m(\hat{P}(45))$	0.00	0.50	1.00	0.50	0.00
Point in Fig.3	s_1	s_2	s_3	s_4	s_5

4. DISCUSSION AND CONCLUSIONS

In this study, only simple cases have been considered in order to provide some hopefully clear introductory material on this relatively new topic of fuzzy reliability. In the flood protection example, the state of nature space can be extended to include floods of various magnitudes; the protection actions and probability can be extended in a similar way. In the water quality example, failure-related concepts other than reliability can be introduced; for example, fuzzy vulnerability may be defined as a fuzzy number representing the degree of an incident and fuzzy resilience, as the imprecise time, a fuzzy number, that the system needs to return to its initial equilibrium state after it has been disturbed, say, by a pollutant spill.

Concluding points may be made as follows:

1. The methodology can deal with any elements of reliability being uncertain (load, resistance, incident/failure phenomenon).

2. The methodology accounts for uncertainity or imprecision that may be present in the steps of risk analysis.

3. An example of flood protection by mean of a level in cludes fuzzy costs, losses and exceedance probability; it leads to a decision made on the basis of comparing two fuzzy expected losses.

4. Extension of this flood protection problem to include ecological damage (reduction in species) may be done by constructing a fuzzy trade-off.

5. The non-exceedance of a pollutant concentration in a water body may be modelled as a fuzzy reliability problem.

6. The fuzziness is encoded in a fuzzy mean and a fuzzy standard deviation.

7. The fuzzy reliability of water quality is calculated by the extension principle. Assuming a normal distribution makes it possible to use standard normal tables to construct the fuzzy number representing this reliability.

5. ACKNOWLEDGMENTS

The research leading to this investigation has been supported in part by grants from the US NSF # (ECS 8802350, ECS 8802920 and BCS 9016462. The help of K. BARATHAN, N. SHAFIKE and B. SHRESTHA in preparing this paper is also gratefully acknowledged.

6. REFERENCES

Ashkar F, B. Bobee and L. Duckstein (1990) "Alternative estimation methods of extreme phenomena for risk analysis, TIMS 12th Triennial Conference, Athens

Bardossy A, Bogardi I and Duckstein L (1991) "Fuzzy non-linear regression analysis of dose-response relationships", to appear, Eur. J. Op. Res.

Bardossy, A., Bogardi and L. Duckstein (1990) Fuzzy regression in hydrology. Water Resources Research, 26(7): 1497-1508, July.

Bogardi, I., L. Duckstein and A. Bardossy (1988) "Uncertainty in environmental risk analysis", in Risk Analysis and Management of Man-made and Natural Hazards, Y.Y. Haimes and E. Stakhiv (eds.), ASCE, New York, pp. 156-174, 1988.

Chen, S.H. (1985) "Ranking fuzzy numbers with maximizing set and minimizing set". Fuzzy Sets and Systems, 17, 113-129.

Dubois, D. and H. Prade (1980) Fuzzy Sets and Systems: Theory and Applications Academic Press, New York, 393 p.

Duckstein, L. and D. Davis (1976) "Applications of statistical decision theory", Chapter 11 in Stochastic Approaches to Water Resources, H.W. Shen (ed.), Water Resources Publications, Fort Collins, Colorado, pp. 11.1-11.52.

Duckstein, L., I. Bogardi, and A. Bardossy (1988) "A fuzzy reliability model of water supply during droughts", Fall AGU Meeting, San Francisco, December 1988, Available as paper 88-25, Systems & Industrial Engineering Department, University of Arizona, Tucson, Arizona 85721.

Duckstein, L., E.J. Plate and M. Benedini (1987) "Water engineering reliability and risk: A system framework", in Engineering Reliability and Risk Water Resources, L. Duckstein and E. Plate, (eds.), NATO ASI Series, Martinus Niijhoff, Dordrecht, The Netherlands, pp. 1-20.

Kaufmann, A. and M.M. Gupta (1985) Introduction of Fuzzy Arithmetic: Theory and Applications, Van Nostrand Reinhold, New York, 351 p.

Kaufmann, A. and M.M. Gupta (1988) Fuzzy Mathematical Models in Engineering and Management Science, North Holland, Amsterdam, 338 p.

Plate, E., and L. Duckstein (1991) Risk analysis for water supply from a river polluted by nitrate runoff, in press, Proceedings NATO Workshop "Nitrate Contamination; Exposure, Consequence and Control", North Holland, Amsterdam

Shrestha, B.P., K.R. Reddy and L. Duckstein (1990) Fuzzy reliability in hydraulics, Proceedings, The First International Symposium on Uncertainty Modeling & Analysis, University of Maryland, College Park, December.

Zadeh, L.A. (1965) Fuzzy Sets. Information and Control, (8): 338-353.

Zadeh, L.A. (1973) Outline of a new approach to the analysis of complex systems and decision processes. IEEE Trans SMC3: 28-44.

Zimmermann, H.S. (1985) Fuzzy Set Theory and its Applications, Kluwer Nijhoff Publishing, Dordrecht, The Netherlands.

APPENDIX*

Fuzzy Sets, Fuzzy Numbers and the Extension Principle

The purpose of this Appendix is to provide a summary statement on fuzzy sets and fuzzy numbers. We begin with some of the fundamental definitions underlying the concept.

Fuzzy subset

Set membership in an ordinary set is binary: an element either belongs or does not belong to a set. In a fuzzy subset, by contrast, there is a degree of membership associated with any element. The membership value of an element can range from 0 to 1; the higher the membership value, the more the element belongs to the set. A fuzzy subset then consists of a set of ordered pairs consisting of the element and its membership value.

Stated formally, a subset A is called a fuzzy subset of a universe X if the subset A comprises ordered pairs such that

$$A = \{(x, mA(x) : x.E.X; mA(x).E.[0,1]\} \qquad [A1]$$

mA(x) is the degree of membership of x in the set A.

If x were an ordinary or "crisp" number, it would have a membership function in the set {0,1}. The extent of membership would be equal to 1 if x were in A, and 0 if it were not. With fuzzy subsets, therefore, it is possible to deal with set membership that can be decided only in a relataively "vague" manner. For example the numbers 90, 108 and 120 all belong to the fuzzy subset: "About an order of magnitude greater than 10".

Since with each x in A, there is associated a level of membership, we can think of membership at different levels. Thus, the h-level of the fuzzy subset A will be defined as the set of all elements whose extent of membership is h or higher, that is,

$$A_h = \{(x, mA(x) \geq h: h.E[0,1]\} \qquad [A2]$$

Normal Fuzzy Subsets

A normal fuzzy subset is one where there exists at least one value of x.E.A such that mA(x) = 1. In other words, the maximum value of the membership function is unity.

*The contribution of K. BARATHAN in organizing and preparing this Appendix is gratefully acknowledged.

Convex Fuzzy Subset

A convex fuzzy subset is one where for all q between 0 and 1 and x.E.X y.E.X.

$$mA\,(qx + (1\text{-}q)y) \geq \min\,(mA(x), mA(y)) \qquad [A3]$$

Note the difference with ordinary convexity which is defined by comparing the left-hand side of Eq. (3) with the convex combination q.mA(x) + (1-q) mA(y). The interpretation of Eq. (3) is that the membership function possesses no local extrema.

These concepts of normality and convexity are used in the definition of a fuzzy number.

Fuzzy Number

A fuzzy number M is a normal, convex fuzzy subset of the set of real numbers, Figs.2 and 4 represent membership functions of fuzzy numbers. The property of convexity limits the shape that a fuzzy number can take: it is always non-decreasing to the left of the peak, and non-increasing to the right of the peak. As can be seen in Fig. 2 and 4, there are two values of x where the membership function reaches zero, and (at least) one where it reaches a value of 1. A fuzzy number can be characterized by these three points, and the shape of the curve defined by a pair of functions, one to the left and one to the right of the peak, since left-right symmetry is not a necessary condition. A real number is a fuzzy number whose elements comprise only one number with a non-zero membership value, which will be equal to 1. Such a number is described as a crisp number, as seen above.

Triangular Fuzzy Number

The simplest type of fuzzy number is triangular, that is, linear on either side of the peak. Fig. 2 gives an example of a triangular fuzzy number (TFN). The fuzzy number displayed in Fig. 2 can be described by the values of x at the points a_1=0, a_2=20, a_3=35. Thus, A = (a_1, a_2, a_3) completely characterizes the number.

It follows that any real or "crisp" number can be defined as a triangular fuzzy number, with a_1=a_2=a_3, as illustrated for example in Eq. (2), where 1=(1,1,1)

L-R Fuzzy Number

As stated above,whenever the fuzzy number is not triangular, i.e. is non-linear in shape, the characterization has to include the functional form of the functions to the left and the right of the peak. Thus:

A fuzzy set A on the set of real numbers is called an L-R fuzzy number if the membership of x can be calculated as:

$$m\hat{A}(x) = \begin{cases} L(\frac{A-x}{b}) & \text{for } x \leq A,\ b > 0 \\ R(\frac{x-A}{c}) & \text{for } x > A,\ c > o \end{cases} \qquad [A4]$$

where L and r are continuous strictly decreasing functions on [0,1] and:

$$L(z) = R(z) = 1 \text{ if } z \leq 0$$

$$L(z) = R(z) = 0 \text{ if } z \geq 1 \qquad [A5]$$

The L-R fuzzy number $\hat{A}$ can be written symbolically as:

$$\hat{A} = (A, b, c)\ (L\text{-}R) \qquad [A6]$$

A more general definition and further details may be found in Dubois and Prade, (1980) and Zimmermann (1985).

In general, fuzzy numbers are characterized by membership values measuring the level of confidence of the occurrence of that number. The level sets reflect different sets of numbers with a given minimum level of confidence of occurrence. We thus have the membership function at a level, say h.

Cartesian Product

If X(1),...,X(n) are sets, and A(1),...,A(I) are fuzzy subsets of X(1),...,X(I), then the Cartesian product A=A(1)X,...,XA(I) is the fuzzy subset
$\hat{X}(1) > <\hat{X}(2)> <...> <\hat{X}(I)$, with membership fucntion:

$$m\hat{A}(x(1),...,x(I)) = \min\ \{m\hat{A}(i)\ (x(i)) : i=1, ...,I\} \qquad [A7]$$

Extension Principle

The extension principle is a method of extending point to point operations to fuzzy sets, and is the basic tool of fuzzy arithmetic.

Let f be a point to point mapping from a set X to another set Y,

$$F = \{(x,y): x.E.X;\ y.E.Y;\ y = f(x)\} \qquad [A8]$$

Then f can be extended to the fuzzy set situation as follows:

Let $\hat{A}$ be a fuzzy subset of X with membership function mA. The image of $\hat{A}$ in Y is the fuzzy subset $\hat{B}$ with the membership function:

$$m\hat{B}(y) = \sup\{m\hat{A}(x);\ y = f(x):\ x.E.X\}$$

and mB(y) = 0 if there is no x.E.X such that f(x) = y [A9]

If in turn X is defined as a Cartesian product (Eq. [A7] then $m\hat{A}(x)$ in Eq. [A9] must be replaced by its expression in Eq. [A7]:

$$mB(y) = \sup\{\min\ \{mA(i)\ (x(i)):\ i=1, ...,I; y = f(x):\ x.E.X\}$$

and mB(y) = 0 if there is no x.E.X. such that f(x) = y [A10]

<u>Operations on Fuzzy Numbers</u>

Consider now the application of Eq. [A10] to the definition of operations on fuzzy numbers, Let represent a binary arithmetic operation, that is, may represent any of the operations +, -, ·, : Let $\hat{A}$ and $\hat{B}$ represent two fuzzy numbers on which the operation · is to be performed, and $\hat{C}$ be the result of that operation.

Then $\hat{C} = \hat{A} \cdot \hat{B}$

is the fuzzy number with membership function:

$$m\hat{C}(y) = \{\sup(\min(m\hat{A}(x1),\ m\hat{B}(x2)))\ \text{such that}\ \ y = x1 \cdot x2\},$$

and 0 otherwise [A11]

This formula can be readily expanded to include more than two fuzzy numbers. However, it may be noted that Eq. [A11] assumes that $\hat{A}$ and $\hat{B}$ have the same L-R functions. Where this is not the case, $\hat{C}$ will have new L and R functions, as described below, in the case when the operation *=+ (addition)

Let the two L-R functions of $\hat{A}$ and $\hat{B}$ be represented by L(1), R(1) and L(2), R(2) respectively. Then, the L-R representation of $\hat{C}$ will be:

$$L(\hat{C}) = \sum_{i=1}^{2} \left(\left(\sum_{j=1}^{2} \alpha(j) \right) L(i)^{-1} \right)^{-1}$$

and

$$R(\hat{C}) = \sum_{i=1}^{2} \left(\left(\sum_{j=1}^{2} \beta(j) \right) L(i)^{-1} \right)^{-1} \qquad [A12]$$

where $\hat{C}$ is the resultant fuzzy number. These formulae for $L(\hat{C})$ and $R(\hat{C})$ can be readily extended to include the case of operation on more than two sets.

Another notion useful for comparing fuzzy numbers is the fuzzy mean, defined as:

$$\bar{A} = \frac{\int_{-\infty}^{+\infty} tm \cdot A(t)\, dt}{\int_{-\infty}^{+\infty} mA(t)\, dt} \qquad [A13]$$

LIST OF SYMBOLS USED IN THE APPENDIX

A, B..:	sets
$\hat{A}$,$\hat{B}$..:	fuzzy sets, fuzzy numbers
h:	level of confidence of a fuzzy number
mA(x)	degree of membership of x in A
TFN:	triangular fuzzy number
sup (	supremum
min (	minimun
.E.	"element of"
> <	Cartesian product

Part 3.

RISK MANAGEMENT IN WATER RESOURCES & ENVIRONMENTAL ENGINEERING

CONSIDERATION OF RELIABILITY IN SYSTEM DESIGN FOR GROUND WATER REMEDIATION

W. Woldt, I. Bogardi[1], L. Duckstein[2]
Department of Biological Systems Engineering
University of Nebraska - Lincoln
253 L.W. Chase Hall
Lincoln, Nebraska 68583-0726

ABSTRACT

A remedial action design system is described that may be used to evaluate candidate remediation systems and select the preferred alternative under conditions of uncertainty. The remedial action design method is based on the ability to evaluate system alternatives within the framework of a multiple-objective decision making scheme. The methodology considers uncertainty in system performance as an element of the decision making process by using geostatistical conditional simulation to generate model input. A two-dimensional computer modeling of candidate remediation schemes is used to model the transport of contaminants, and predict the performance of each candidate system.

Composite programming is used for the multiple-objective decision making portion of the design system. Measures of system performance include average mass of contamination pumped and treated, average mass remaining in the aquifer after terminating operations, and average mass exiting the study area via boundaries. Uncertainty in system performance, or the risk of a system not performing according to predicted standards, is measured by computing the variance and coefficient of variation for each of the previous indexes. In addition, the operation horizon and modeled contamination levels at a hypothetical compliance surface serve as additional performance measures. Total costs for each option were estimated and incorporated into the remediation planning.

A case study is examined in which a preferred remedial action plan is identified from a set of alternatives for removing a plume of contaminated ground water emanating from an uncontrolled hazardous waste site near a radioactive waste reprocessing facility.

[1] Department of Civil Engineering, University of Nebraska, Lincoln, Nebraska, 68588
[2] Department of Systems Engineering, University of Arizona, Tucson, Arizona, 87521

NATO ASI Series, Vol. G 29
Water Resources Engineering Risk Assessment
Edited by J. Ganoulis

INTRODUCTION

Recent public awareness of the extent of ground water contamination at uncontrolled hazardous waste sites has prompted the creation of various governmental regulatory mechanisms to begin the remediation process. These regulatory mechanisms often take the form of water quality standards and the allowable time to achieve these standards. In the long run, the goal of government regulation is to protect public health. However, the limited resources available require that they be distributed as efficiently as possible. A significant element of making decisions with respect to funding allocation is the relationship between costs of remediation and system performance relative to regulatory requirements. Understanding these relationships provides insights into the most cost effective methods for risk reduction at a given site. A fundamental difficulty is the complexity of the economic and physical relationships combined with numerous goals of the different parties involved in a remediation. In addition, reliability of system performance prediction is hampered by uncertainty in parameters necessary for modeling remedial action alternatives. In the long term, a solution to this problem has the potential to provide a means of comparing different hazardous waste sites and guiding decisions related to a preferred sequence of site remediation.

The objective of this paper is to develop a methodology for allocating remediation resources at a given hazardous waste site within the framework of a multiple criteria decision making (MCDM) system. It is felt that an approach using MCDM provides the tools necessary to analyze the decision making process at a level that is able to consider the complex inter-relationships that occur during remedial action planning. The methodology is described in a general format, and should be applicable to any remedial action planning activities. A case study is provided to illustrate how the planning method might be used at an uncontrolled hazardous waste site.

PREVIOUS WORK

Early efforts to control and remediate ground water contamination in an optimal sense were directed toward linking deterministic ground water simulation models and optimization algorithms. Steady state management has been examined in Willis (1976), Futagami, et al. (1976), and Gorelick and Remson (1982a).

Management of contamination in a dynamic framework has been investigated by numerous researchers, including Willis (1979), Gorelick, et al. (1979), and Gorelick and Remson (1982b). The embedding and unit-response methods were used extensively with various attempts at reducing some of the difficulties associated with each of the methods. The importance of time in remedial action planning has been acknowledged in Lefkoff and Gorelick (1986) in which the objective was rapid removal of contamination.

Aquifer Parameter Uncertainty

An important element that is missing from early studies is the influence of parameter uncertainty on decisions related to management of ground water contamination. Early research on contamination management under conditions of parameter uncertainty can be found in Mercer, et al., (1983).

Wagner and Gorelick (1987) explored the effects of aquifer parameter uncertainty on ground water quality management. Important results indicate that remediation requirements can increase significantly due to parameter uncertainty. In a related study, Wagner and Gorelick (1989) extended the work presented in their earlier paper (Wagner and Gorelick, 1987). Results indicate that aquifer remedation using as few as 30 realizations of conductivity can yield a reliable management plan, and that there appears to be a relationship between volume pumped and reliability of cleanup.

In a recent investigation that departs from the traditional simulation-optimization methods, Andricevic and Kitanidis (1990) use a dual control methodology in which parameter estimation and plume management are considered jointly. The methodology considers time as an important element of plume management. The authors recommend a multi-criteria approach due to the difficulty in placing a monetary value on exceeding regulatory compliance levels.

Summary of Previous Work

Some general observations regarding the existing state of remedial action planning and management follow: Initial efforts were directed toward deterministic planning with a natural progression toward incorporation of stochastic solutions into the process. Classical linear and non-linear optimization methods have been used to solve the management problem with very limited objective functions, typically restricted to cost minimization. There has also been limited consideration of time in the planning process. In the few instances where time is recognized as an important element, it was usually incorporated into the problem as a cost consideration related to equipment operation and maintenance. In fact, time may enter into the decision process due to the actual, or perceived, level of hazard existing in the plume and possible regulatory requirements.

Another weakness of the existing approaches is the inability to consider risk within the decision making process. Each of the stochastic approaches previously described incorporate risk into the model as an integral element, in which the results are presented as a trade-off between cost and reliability. However, none of the methods provide a technique to make a decision on the final remedial action using the trade-off information obtained from the model. The methods are also very computer intensive, often solving numerous Monte Carlo simulations and non-linear optimization problems for each alternative option. This disadvantage should diminish as computing capabilities advance.

Finally, some of the methods allocate a large portion of their resources toward optimal parameter estimation. While the authors agree that accurate parameters are very important to the success of a remedial action, an interesting article by Marin (1986) questions the use of sophisticated parameter estimation algorithms. In his analysis, Marin has shown using two general cases that incorporation of optimal parameter estimators will not necessarily lead to better decisions compared to traditional parameter estimators. In view of these results, it seems logical that the acceptable level of uncertainty in results (due to parameter uncertainty) should be a decision variable incorporated into the planning process.

REMEDIAL ACTION SYSTEM DESIGN

The methodology described in this section implements some of the techniques previously described, while departing at the same time from the preceding base of work by introducing a different technique for management of ground water contamination. Similarities include the incorporation of aquifer parameter uncertainty within a Monte Carlo framework, prediction of solute transport using ground water modeling with advection and dispersion, and use of discrete system options and time frames. The methodology diverges from previous work by integrating an additional source of uncertainty into the analysis. In addition, it overcomes limitations imposed by restricted objective functions through application of a multi-criteria decision making procedure for selection of the "preferred" option. Description of the design system is based on a two-dimensional representation of the ground water system. However, the methodology is presented in a fairly general context and extension to three-dimensional representation simply involves additional computational effort.

Uncertainty in aquifer parameters, in this case transmissivity, is modeled using geostatistical simulation. Specifically, two-dimensional conditional simulation by turning bands (Brooker, 1985; Mantoglou and Wilson, 1982) is employed to generate possible realizations of transmissivity that exhibit statistics (first and second moments, spatial covariance) in concordance with data at sampled locations. Each realization of a transmissivity field is used as input for a ground water model to predict the performance of several discrete management options within a Monte Carlo framework. Management options include different system configurations and operating horizons.

An additional source of uncertainty is also included in the numerical analysis. It has been shown, by case study, that uncertainty in initial conditions may have a greater impact on subsequent modeling results than uncertainty in aquifer parameters (Woldt, 1990a). Consequently, contamination plume initial conditions are considered as an imprecise parameter in the ground water model. A unique initial condition can be generated for each ground water model run using conditional simulation by turning bands.

The ground water model is then executed numerous times with a different realization of transmissivity and initial conditions each time, for a given management option, in order to compute various indices of performance. The process is repeated for different management options. A schematic of the design system is presented in Figure 1. Resulting performance indices for each option may then be used in a MCDM framework to determine the final preferred option. In addition, the technique may be used to identify and select one or multiple "robust" solutions that are relatively insensitive to decision makers preferences. These robust solutions may then provide a foundation for further management option development.

Multi-Criteria Remedial Action Planning

Multi-criteria decision making techniques can be used to select the option that best satisfies numerous, often conflicting objectives. While there are many candidate methods of MCDM analysis available (Cohon, 1978; Haimes, et al.,1975), the methodology that has been selected for remediation system design is termed Composite Programming (Bogardi, et al., 1984).

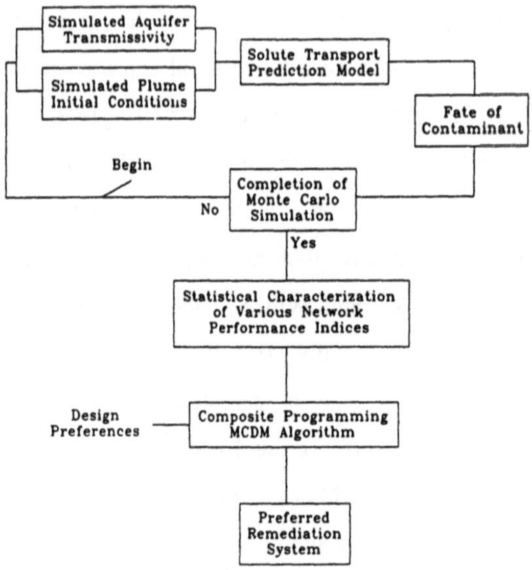

Figure 1. Schematic of Remedial Action Design System

Composite programming (CP) was selected because of its ability to consider and integrate completely different elements of remedial action planning into a single analysis. It also represents a compromise between mathematical sophistication and ease of use. The following paragraphs will describe CP from both an operational and theoretical point of view.

Composite programming is a distance based MCDM technique according to the definition of Szidarovszky, et al., (1986). Distance based methods require that the system designer/decision maker visualize the decision making process within the framework of a spatial coordinate system. In this reference system it is assumed that preferences for one element, or a group of elements, for a particular design can be expressed as a metric between two points. One of the two points represents an ideal situation coordinate, while the other represents a worst case coordinate. This process of constructive visualization within a spatial coordinate system is really a rather natural method for "weighing" alternatives.

Composite programming can be described as employing a hierarchial, normalized distance type of methodology as compared to compromise programming (Zeleny, 1973) which uses a single level analysis and non-normalized distance. The CP algorithm used in the current research is capable of handling up to 30 discrete alternatives in *n* dimensional space,

R^n. The hierarchial structure of CP provides a method of breaking down the n dimension space into a series of lower dimensions that can be visualized to gain a deeper understanding of the inter-relationships between the numerous remediation design elements.

Consider the family of metrics described by the following Minkowski class of distance measures (Klir and Folger, 1988).

$$d = \left(\sum_{i=1}^{n} |f_i(x) - f_{wi}(x)|^p \right)^{1/p} \tag{1}$$

where

d = distance measure,
$f_{wi}(x)$ = worst value of the ith system state,
$f_i(x)$ = i^{th} of n total system states, and
p = positive number, usually an integer.

When p is equal to 1 the distance measure is as special type of Minkowskian metric termed a Hamming distance, and when $p=2$ the distance is a Euclidean metric (Klir and Folger, 1988).

Now consider two modifications to the basic Minkowski metric (1). First, incorporate a weighting scheme for each of the n different system states (i.e., remedial design elements) and second, normalize the total range between ideal and worst cases to a zero to one interval. These alterations will change (1) to a family of l_p metrics used in compromise programming (Zeleny, 1973; Duckstein and Opricovic, 1980). For convenience variable definitions may change from (1) to reflect terminology used in CP.

$$l_p = \left[\sum_{i=1}^{n} a_i \left| \frac{f_{bi}(x) - f_i(x)}{f_{bi}(x) - f_{wi}(x)} \right|^p \right]^{1/p} \tag{2}$$

where

l_p = metric for objective of decision,
a_i = weight of the ith system element,
$f_{bi}(x)$ = ideal value of the ith system element,
$f_{wi}(x)$ = worst value of the ith system element,
$f_i(x)$ = state of the ith system element,
p = balancing factor.

Variables in (2) will be described in more detail in a later paragraph. At this point the hierarchial nature of CP will be described.

Composite programming is capable of analyzing a multi-dimensional, R^n, design problem. However, it would be very difficult to visualize a solution beyond R^3 dimension, with a preferred presentation of results in R^2 dimension. The methodology used in CP to overcome this problem is to break down a complex design process into its elementary components termed basic indicators. Selection of basic indicators may be based on both designers and managers prior knowledge about the site to be remediated, as well as previous experience with site remediations. They should characterize the most elementary aspects of the aquifer remediation design process. It should be noted that the type of basic indicator selected will have a tangible effect on the scale of analysis and characteristics of both the design system developed and selected remedial action alternative.

After the basic indicators have been identified, the process of building the design system from its elementary components by defining the structure of the CP model begins. This process can be characterized as grouping the basic indicators into clusters based on either similar characteristics or the desire to contrast different features through trade-off analysis. The initial set of indicators that result from grouping basic indicators are called first level indicators. After the first level indicators have been specified, the grouping process continues until the highest level indicators have been specified. In the case of remediation design, this final level may describe the trade-off between performance and cost. A result of breaking down a complex problem into its elementary components, and then building the problem back up by successive grouping of the components, is that a multi-dimensional problem can be characterized and analyzed within the preferred R^2 dimension.

In (2) the weight assigned to a particular remediation design element reflects the relative preference of that element compared to other elements within the same group. CP uses a double weighting scheme to reflect the importance of the maximal deviation between indicators. Therefore, the system designer or decision maker must specify a value for p, the balancing factor. A larger value of p will tend to restrict the allowable substitution of inferior options that may occur. As an example, consider a suspected groundwater plume of highly toxic material. A single weighting scheme may point to a poor management option if the cost of cleaning the plume is extremely high. Whereas a double weighted scheme will tend to minimize this type of error because it will recognize the importance of a reliable cleanup, even at a high cost.

The ideal and worst values for each of the basic indicators in (2) are final components needed to describe the CP model structure. Some indicators may have ideal values that are represented by numbers on the small end of a scale, while other indicators may have ideal values located at the large end of a selected scale. In either case the value for each indicator should be selected based on the ideal and worst conditions encountered in the region being considered with proper consideration for the scale of analysis. At times it may be advantageous to assign ideal and worst values after all remediation designs have been evaluated and the total range of basic indicators assessed.

After the structure of the design system has been formulated, an impact relationship, $R(x)$, for each of the basic indicators, R, and for each management alternative, x, can be calculated. The impact relationship is a measure of the basic indicator value for each of the design options and serves as input to the design system model. Values for basic indicators are obtained from the simulation modeling and estimation of costs for different alternatives.

Evaluation of the various management options is accomplished by calculating the composite distance within each group of indicators, beginning at the basic level and then progressing to successively higher levels. The preferred design option is then identified from the candidates by locating the system nearest the ideal point in terms of the composite distance. If design alternatives are represented by a continuous function, then the gradient method (Bogardi, et al., 1985) can be used to determine the option with minimum composite distance. If a discrete number of cleanup systems are examined, then total enumeration can be used to evaluate the options. As mentioned previously, this work will focus on the discrete design case.

Calculation of the composite distance within a discrete framework begins with normalization of the impact relationship $R(x)$. This is necessary since the units of basic indicators can be quite different, yet must be considered within an integrated system. Using the ideal (R_l) and the worst (R_s), the normalized value of R, defined as S, can be computed as:

$$S_i = \frac{R - R_s}{R_l - R_s} \qquad\qquad S_i = \frac{R_l - R}{R_l - R_s} \tag{3}$$

where the formula selected will cause S_i to be positive. It is important to note that the normalization process will always force the ideal coordinate to be (1,1) whether the objective is to be maximized or minimized. Substitution of (3) into (2) yields the following composite distance for the jth group of indicators:

$$L_j = \left[\sum_{i=1}^{n_j} a_{ij} S_{ij}^{p_j}\right]^{1/p_j} \tag{4}$$

where

L_j = the composite distance for first level group j basic indicators,
n_j = number of basic indicators in group j,
a_{ij} = weights expressing the relative importance of basic index i in first level group j such that their sum equals one,
S_{ij} = the normalized value of basic index i in the first level group j of basic indicators,
p_j = balancing factor among indicators for group j.

An illustration of the composite distance, L_j, is presented in Figure 2. In this case the composite distance is the distance between the point corresponding to the system state and the ideal point.

Following the calculation of composite distances for every first level indicator group, the second level composite distances are calculated using the first level distances. The formula is similar to the first level metric with details of the compositing procedure given in Woldt (1990b).

Combination of Simulation and CP

The only remaining step in the design system development is the combination of ground water simulation, other relevant input data such as costs, and CP into a single entity so that alternative designs can be evaluated. For the purposes of this discussion it will be assumed that the set of discrete remediation design options are synthesized by designers and policy makers. The task is to select the "best" of the candidate remediation systems.

The impact relationship $R(x)$, is a key link between CP and the problem to be considered. Impact relationships are a measure of basic indicator values for each design option. Post-processed output from the simulation model will function as impact relationships with respect to system performance, whereas system equipment and O&M costs will serve as impact relationships for expenditures. Once a $R(x)$ is obtained, it can be input to the CP algorithm as a mathematical expression in the continuous case, or expressed as a data table in the discrete case. In the current system, the user is expected to facilitate the link between

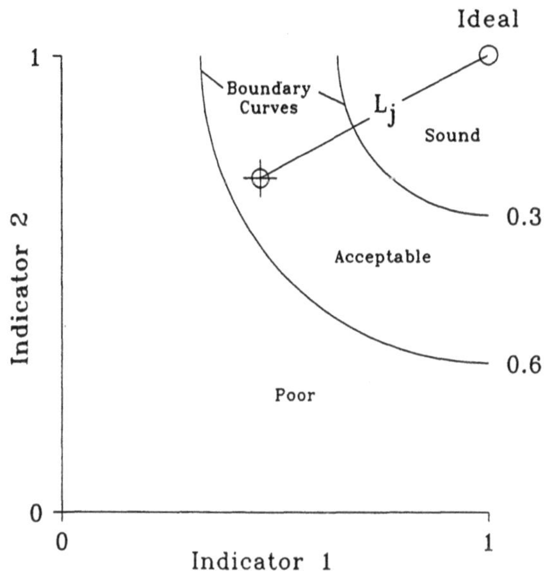

Figure 2. Composite Distance

the numerous impact relationships and the CP algorithm by entering the appropriate $R(x)$ data into the CP module.

Analysis of the results may proceed after all composite distances have been computed. Clearly, if L is small, the state of the system is close to the ideal state. Consequently it is possible to characterize the areas surrounding the ideal state based on the radial distance from the ideal point. As the state of the design for a given management option migrates away from the ideal point, it will pass through a series of zones that might be classified as sound, acceptable and poor designs. These designs can be defined, based on the limits of the composite distances that envelope each category. Although the limits are somewhat subjective, the values of B=0.3 and B=0.6 seem to be appropriate for most situations, however they may be modified (see Figure 2 for illustration of boundary curves).

Sensitivity analysis is an important aspect of the remediation design and should be considered as a necessary ingredient. The hierarchial nature of CP and the ability to examine design elements at the basic level as well as higher levels provides designers with the potential to investigate hidden relationships and preferences that may not be apparent.

CASE STUDY APPLICATION OF THE DESIGN SYSTEM

This case study deals with stochastic ground water modeling, combined with multi-criteria remedial action planning, at a well documented contamination site. The study area is located at a low-level radioactive waste processing facility near Wood River Junction, Rhode Island (Figure 3). A site investigation has been completed by the U.S. Geological Survey and documented in Ryan et al. (1984).

A plan view of the study area is presented in Figure 3 with the pollution source being the industrial processing site. The following site history has been condensed from Ryan and Kipp (1984). Contamination at the site is the result of 14 years of waste disposal from the industrial facility during the period 1966 to 1980. Liquid wastes containing both radio-nuclear materials and other chemical solutes have percolated from uncovered ponds and trenches located at the facility into an unconfined, highly permeable, sand and gravel aquifer. The resultant plume extends about 2300 feet from the source to the Pawcatuck River. Processing at the plant has been terminated. Material from the bottom of the ponds and trenches has been removed and shipped off site for proper disposal.

Available Data

Transmissivity estimates for the site have been obtained indirectly from well logs contained in Ryan, et al. (1984). Hydraulic conductivity values were assigned to separate geologic strata based on material composition. Conductivity values were taken from Table B in Peckenpaugh and Dugan (1983) and used for 37 selected wells distributed throughout the study area. An equivalent horizontal (i.e., 2-D spatial) hydraulic conductivity was then computed for each well using weighted averages, where weights correspond to thickness of each lithologic unit. The site was assumed to have a constant saturated thickness of 80 feet which was used to calculate transmissivity at each borehole. A frequency distribution of the transmissivity data is included in Figure 4.

Field data pertaining to ground water contamination have been collected from 70 wells distributed in three dimensions throughout the study area. Well clusters of 2 to 6 wells are located at eight of the observation wells, thereby providing an indication of vertical contamination distribution. Specific conductance is used as an indicator of the more hazardous Strontium-90 contamination. A relationship between elevated electrical conductivity and Strontium-90 levels has been confirmed by Woldt (1990b). The frequency

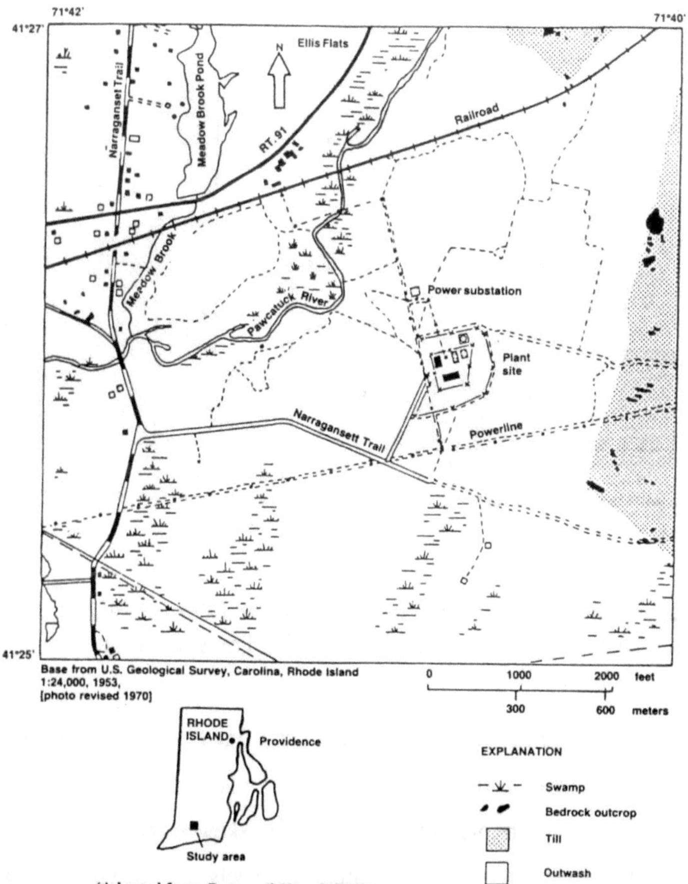

(Adapted from *Ryan and Kipp* [1984].)

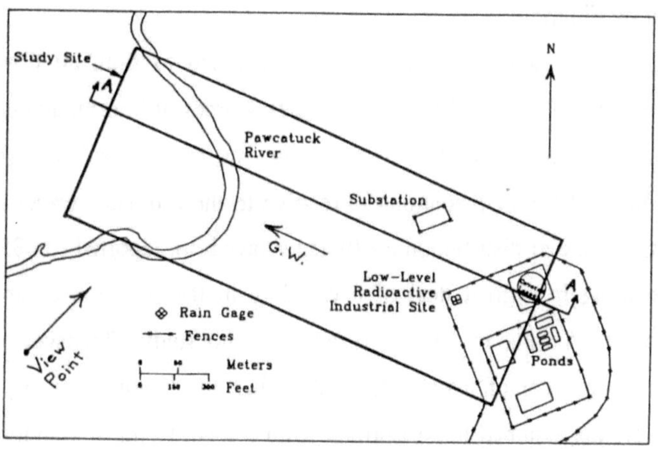

Figure 3. Study Site

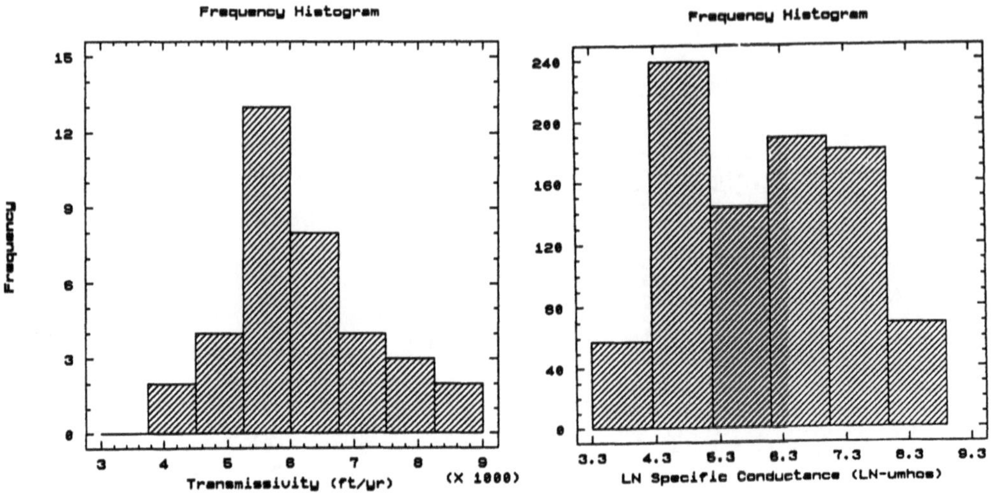

Figure 4. Transmissivity Data Distribution (feet/yr) and Ground Water Conductivity Distribution after Logarithmic (LN) Transformation

distribution of available electrical conductivity data for a fixed time frame equal to 450 days after initiation of data collection is included in Figure 4. A logarithmically (LN) transformed frequency distribution has been presented due to the skewed nature of the field data. Examination of the LN space frequency distribution reveals a somewhat symmetric shape closer to a normal distribution.

Additional aquifer parameters that must be specified to simulate flow and solute transport are presented in Table 1. The table indicates sources of information used to obtain estimates for the parameters. A first order decay factor of zero was selected with consideration for the half-life of Strontium-90 relative to the estimated travel time through the aquifer. A large range of distribution coefficients have been reported (0.085 to 10.0) for the study area resulting in a great deal of uncertainty in the selection of an appropriate retardation coefficient. A value of 1.0 was selected for this study. Transverse dispersivity was assumed to be one order of magnitude smaller than the longitudinal dispersivity. In addition, the aquifer was assumed to be isotropic with respect to transmissivity.

Table 1. Selected Aquifer Parameters

Parameter	Value	Source
Effective Porosity	0.38	Ryan and Kipp (1984)
Retardation Coefficient	1.0	Assumed
First Order Decay Factor	0.0	Assumed
Longitudinal Dispersivity	120 feet	Kipp, et al. (1986)
Transverse Dispersivity	12 feet	Assumed
Specific Yield	0.25	Ryan and Kipp (1984)
Transmissivity Anisotropy	1.0	Assumed

Transmissivity Simulation

Possible realizations of transmissivity were generated using conditional simulation by turning bands in two-dimensions (Brooker, 1985). The technique is similar to simulation by turning bands in three-dimensions (Journel and Huijbregts, 1978) in which several independent one-dimensional simulations are performed on lines radiating from a central point in space. Each one-dimensional simulation is computed from a convolution product (for the spherical covariance case) that has been derived from an intermediate one-dimensional analytical covariance function. Then any point in the three-dimensional grid to be simulated is projected onto each of these lines, with the simulated value being a function of the values of each line at the point of intersection.

In the two-dimensional case the intermediate one-dimensional covariance is a more complex analytical function than the 3-D case. This complexity precludes the use of a simplified analytical equation as part of a convolution product. Consequently, Brooker (1985) presents a derivation of the appropriate intermediate one-dimensional covariance function (for the spherical case), and suggests the use of Fourier transforms for solution of the equations thus obtained. This method is used for the case study with details of the derivation found in the referenced article. Simulated data are conditioned to observed data using kriging as described in Journel and Huijbregts (1978).

A two-dimensional, omni-directional experimental variogram has been computed using lag intervals of 90 feet. A plot of the experimental variogram is sketched in Figure 5, where a spherical variogram with an apparent nugget effect is also plotted. This theoretical variogram, together with the observed transmissivity data, form the basis for simulation using turning bands. Values were simulated over a uniform grid with spacing equal to 100' by 100'. A typical realization of a transmissivity field generated by conditional simulation is presented in Figure 6. Verification of the simulation algorithm has been completed by computing an experimental variogram from the simulated results illustrated in Figure 6. The experimental and theoretical variograms have been plotted in Figure 5. It can be seen that the algorithm is performing satisfactorily since the agreement between field and simulated variograms is fairly close.

Initial Condition Simulation

A slightly different approach was used to simulate possible realizations of initial conditions that describe the existing contamination plume location. Since log-transformed electrical conductivity is used as an indicator of contamination levels, it is those data that will be simulated. Since the observation wells are at discrete depths, three dimensional simulation by turning bands is used to simulate values over a 100' by 100' grid in the horizontal dimension with 10' depth increments down to a depth of 80 feet. Details of the simulation procedure and variogram parameter estimation can be found in Woldt, et al. (1990a). After values are simulated over the complete grid, they are transformed back to field data units of

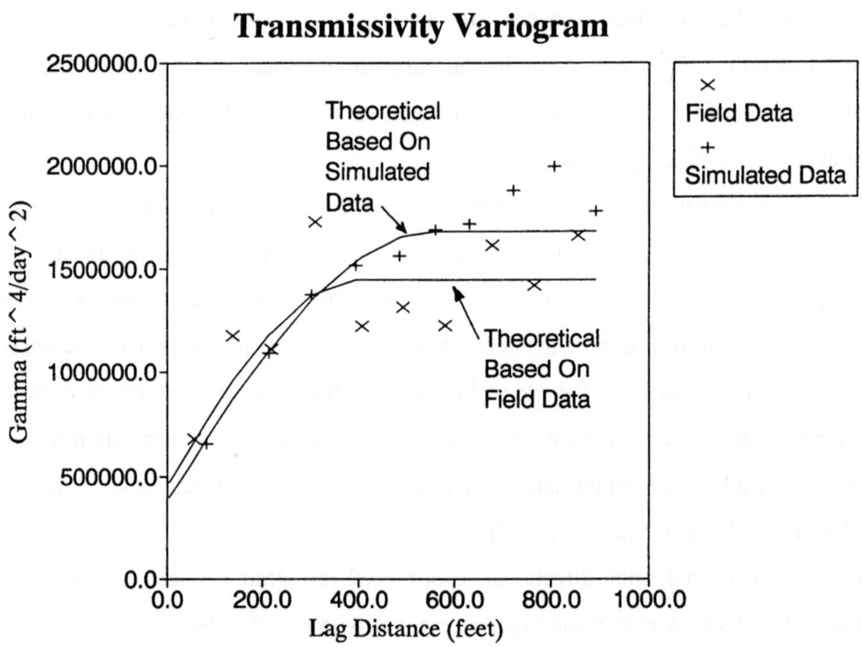

Figure 5. Variogram of Transmissivity

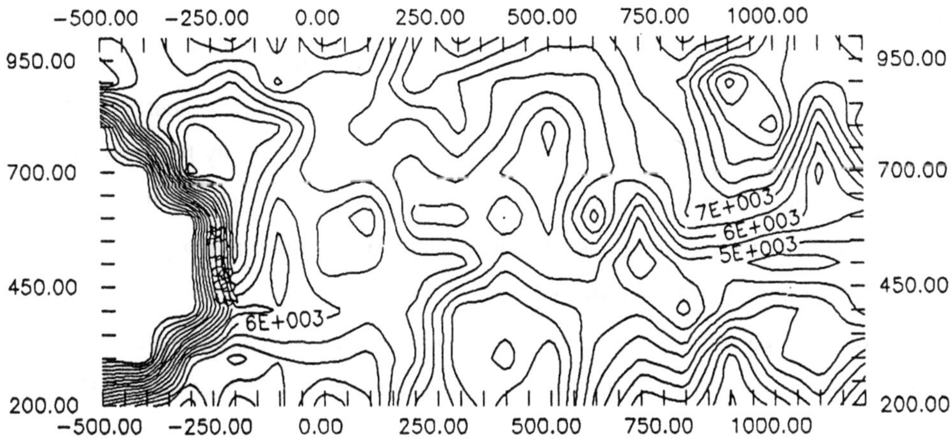

Figure 6. Typical Transmissivity Field (feet^2/day)

μmhos and averaged in the vertical direction to obtain conductivity in two-dimensional space. Finally, the electrical conductivity values are transformed to mg/L total dissolved solids using the relationship 1 mg/L = 1.56 μmhos (Todd, 1980). Simulated values of contamination in mg/L are then used as initial conditions in the ground water model with a typical simulated realization illustrated in Figure 7.

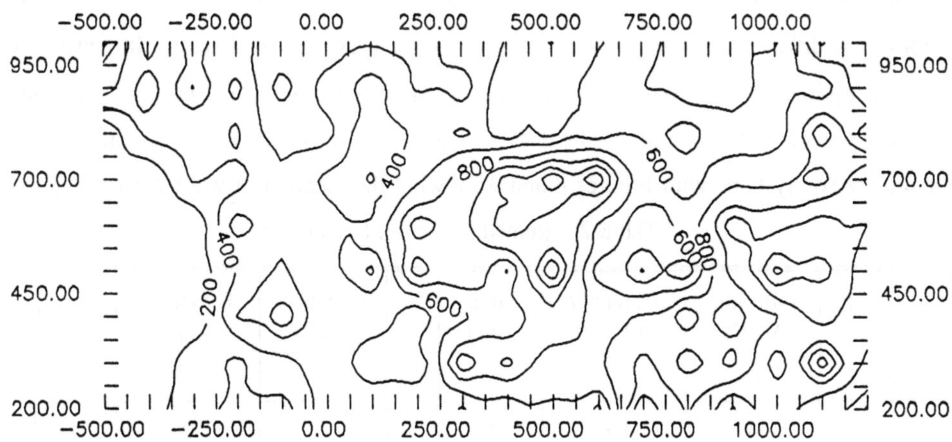

Figure 7. Typical Initial Conditions (total dissolved solids mg/L)

Ground Water Model

Prediction of solute fate in the aquifer was completed using the U.S.G.S. two-dimensional mass transport code (MOC), originally developed in Konikow and Bredehoeft (1978). The code was modified so that each run generated a simulated transmissivity field and simulated set of initial conditions, with a total of 100 runs being executed for each management option. Each transient run was set up to model contaminant transport over a four year period with output being produced at the end of each year of simulation.

Boundary conditions include constant head boundaries along the river nodes, and constant head conditions along the east-southeast boundary of the study area. Hydraulic gradient between the constant head boundaries is set to a value of 0.0053 as reported in Ryan and Kipp (1984). Hydraulic head boundary conditions at the beginning of each model run were determined by operating the model under steady state conditions without any stresses induced by clean-up alternatives.

Additional input data includes four time steps in a pumping period, one pumping period per run and an initial allocation of nine particles per node. Finite difference grid cells are square with dimensions of 100' by 100', and a maximum distance for each particle move equal to 50'.

Remedial Action Management Alternatives

A discrete set of four different pump, treat, and injection scenarios, plus the "do nothing option", were modeled as possible cleanup alternatives. Each pump and treat system employed a different number of wells, pumping capacities, and well sizes. In general, the range of options varied from a small number of large wells with high capacity, to a large number of small wells with reduced capacity. Table 2 identifies important characteristics of the alternative systems with Figures 8 and 9 showing the location of wells for each system.

Table 2. Remedial Action Systems

System	Pumping Wells	Capacity (gpm)	Size (in.)		Injection Wells	Capacity (gpm)	Size (in.)
0	0				0		
1	3	100	6		3	100	6
2	3	50	4		3	50	4
3	6	10	2		3	20	4
4	6	20	2		3	40	4

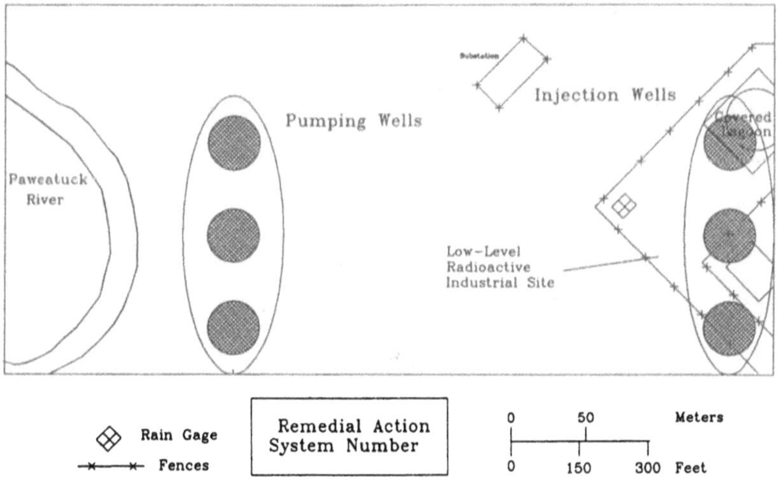

Figure 8. Remedial Action System Alternatives 1 and 2

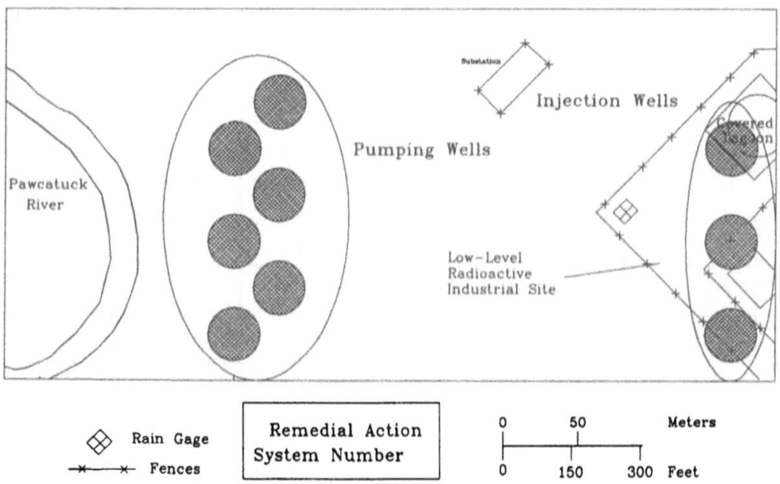

Figure 9. Remedial Action System Alternatives 3 and 4

Operational expense for each of the system alternatives is assumed to be a function of the number of wells, well pumping capacity, number of samples taken, laboratory analysis of field samples, energy to operate wells, and capacity of water treatment system. Unit costs for these expenses have been taken from Rishel, et al., (1984) and reflect best approximations for selected system equipment. Sampling expenses are based on an assumed sample rate per well of 25/year consuming a total of 12 labor days per year per well. Treatment costs are divided into two components: 1) A fixed cost per unit treatment capacity representing equipment acquisition costs. 2) A unit cost defined as a function of both treatment capacity and time in service, representing operation and maintenance costs. Assumed unit expenses are as follows:

Item	Cost
6" well	$ 11,452
4" well	$ 8,589
2" well	$ 5,726
Sampling	$ 1,894 / well-year
Energy Pump	$ 90 / well-year
Energy Inject	$ 20 / well-year
Treatment Equip.	$ 6,295 / gpm
Treatment O&M	$ 2,130 / gpm-year

Different combinations of number of wells in service, pumping rate per well (and therefore size), treatment plant capacity, and operating horizon generated a total of twelve different cleanup alternatives. Table 3 lists important attributes of each alternative.

Table 3. Management Options

Option	System	Capacity (gpm)	Operate Horizon (years)	Equipment Cost ($)	O&M Expenses ($)	Total Cost ($)
0	None	0	4	0	0	0
1	1	300	1	3,845,712	1,289,694	5,135,406
2	1	300	2	3,845,712	2,579,388	6,425,100
3	2	150	1	995,784	331,194	1,326,978
4	2	150	2	995,784	662,388	1,658,172
5	2	150	3	995,784	993,582	1,989,366
6	2	150	4	995,784	1,324,776	2,320,560
7	3	60	3	437,823	436,338	874,161
8	3	60	4	437,823	581,784	1,019,607
9	4	120	2	815,523	546,492	1,362,015
10	4	120	3	815,523	819,738	1,635,261
11	4	120	4	815,523	1,092,984	1,908,507

Selected Performance Indices and Composite Programming Structure

The performance indices designated for use in this case study were selected to provide values for basic indicators in the composite programming algorithm. Basic indicators were selected to provide representative measures of performance with respect to goals and objectives that are common to most remedial action planning efforts. The CP structure is formulated so that successively higher level indicators lead to a final trade-off between cost and performance. Formulation of the case study basic indicators and CP structure is presented for illustrative purposes only. It should be noted that composite programming is a general purpose MCDM algorithm and would lend itself to virtually any type of system performance measure. The case study employed 29 basic indicators that were aggregated through two higher levels to obtain the final trade-off between cost and performance. An illustration of the CP tree structure, including indicators, is presented in Figure 10 for reference in the discussion that follows.

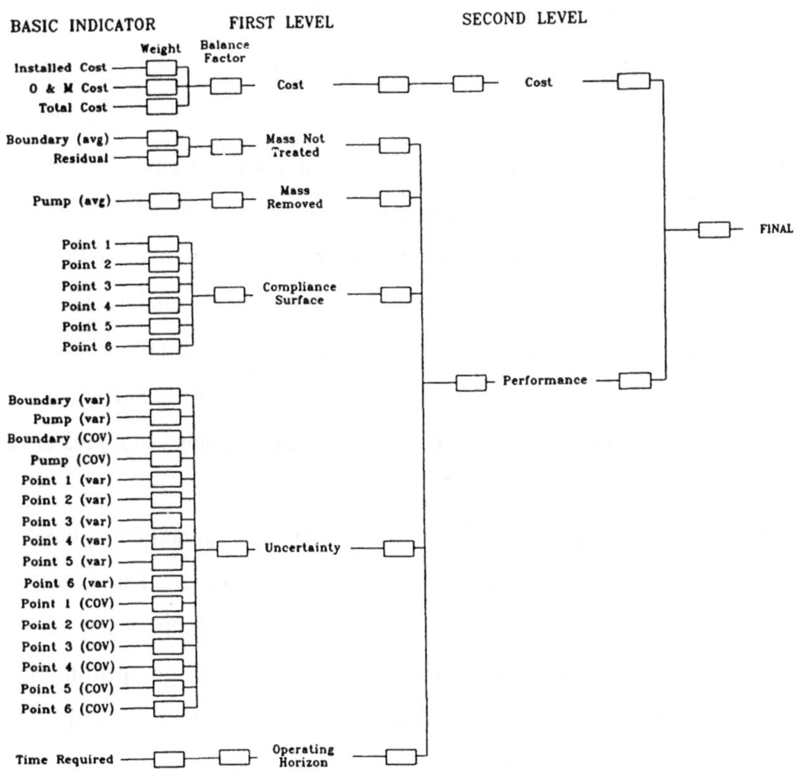

Figure 10. Composite Programming Structure

Cost indicators are expressed in monetary units and represent the aggregation of equipment costs, operation and maintenance expenses, and total cost. Expenses were divided into these categories so that designers can express a preference for equipment acquisition, or O&M expenses. A designer can also give a preference to total cost, without regard to how the total cost is allocated.

Two indices related to the performance of a particular system are the average mass of contaminant that leaves the study area via the boundary, and the mass remaining in the aquifer after remedial action has ceased. Another performance index is the mass of contaminant removed from the aquifer, treated and injected back into the aquifer. The contamination level at a hypothetical compliance surface (in this case the Pawcatuck River) is another index of performance, as well as the time frame during which the cleanup is operational. The model uses 6 individual grid nodes that correspond to the river nodes to represent the compliance surface as basic indicators. It also uses operational time measured in years as an index of cleanup time.

An important element of the design process is the uncertainty related to the effectiveness of the selected remediation option. Uncertainty is incorporated directly into the MCDM algorithm by computing the variance and coefficient of variation for each of the performance measures, except time, described in the previous paragraph. These statistical measures are derived from the stochastic modeling using conditional simulation as previously described. For example, model results may indicate that a given option is able to remove and treat 95% of the mass, yet there may be a high level of uncertainty in this estimate. Whereas, modeling results for a different alternative may yield a 80% mass removal and treatment with a greater level of certainty in the results. The decision must be made whether to use the efficient, but uncertain system, or the less efficient system with greater confidence in the results.

The CP structure displayed in Figure 10 reveals the aggregation of basic indicators into more general categories representing higher level indicators. Three basic indicators related to system cost are grouped in a single cost indicator. Contaminant mass (average) exiting the aquifer via boundaries, and residual contamination are grouped into a first level indicator measuring the mass not treated. Average mass pumped from the aquifer is assumed to be treated completely and is termed "mass removed". Contamination levels at the six nodes representing the Pawcatuck River are grouped into a first level indicator termed "river contamination". The variance and coefficient of variation for each of the statistically computed basic indicators are grouped into a single higher level indicator representing uncertainty in system performance. Operational time required to complete the cleanup remains the same at the first level.

A further aggregation of first level indicators into second level indicators brings the MCDM analysis to the final trade-off between cost and performance. In the present case, the first level indicators; mass not treated, mass removed, river contamination, uncertainty, and operational horizon are all used as measures of system performance at the second level. System cost is simply transferred to the second level without any further aggregation.

Analysis of Management Options

Selection of a preferred cleanup option is completed by determining which system is robust with respect to changes in preferences toward different criteria that measure system performance. These preferences may be expressed as various system scenarios. The following scenarios were investigated individually within the context of the CP package: 1) Emphasize minimum expenses with some concern for system performance. 2) Express a strong concern for limiting the amount of contamination exiting via the boundary and remaining within the aquifer, with little concern for costs. 3) Stress the importance of pumping and treating with some consideration for costs. 4) Introduce a concern for the amount of time required for the cleanup. 5) Emphasize the importance of uncertainty in the decision regarding final design selection. Preference weights that emphasize each of the overall objectives for a given scenario are given in Woldt (1990b).

Balancing factors were also modified for each of the 5 scenarios by sequentially modifying the values to 1, 2, and 3 for each of the scenarios. The combination of 5 different scenarios and 3 different balancing factors yields a total of 15 different composite programming outcomes, that can be used to determine a robust remediation system. Details of the results of the final trade-off between cost and performance are presented in Woldt (1990b) in the form of cost and performance coordinates. The results also provide a final distance from the ideal point and a sorted listing of each alternative.

Results

Three different techniques were used to assess robustness. First, all alternatives that rank number one were plotted on a pie chart (Figure 11a). It can be seen that option 3 is clearly the preferred option with option 6 a close contender using the number one ranking criteria.

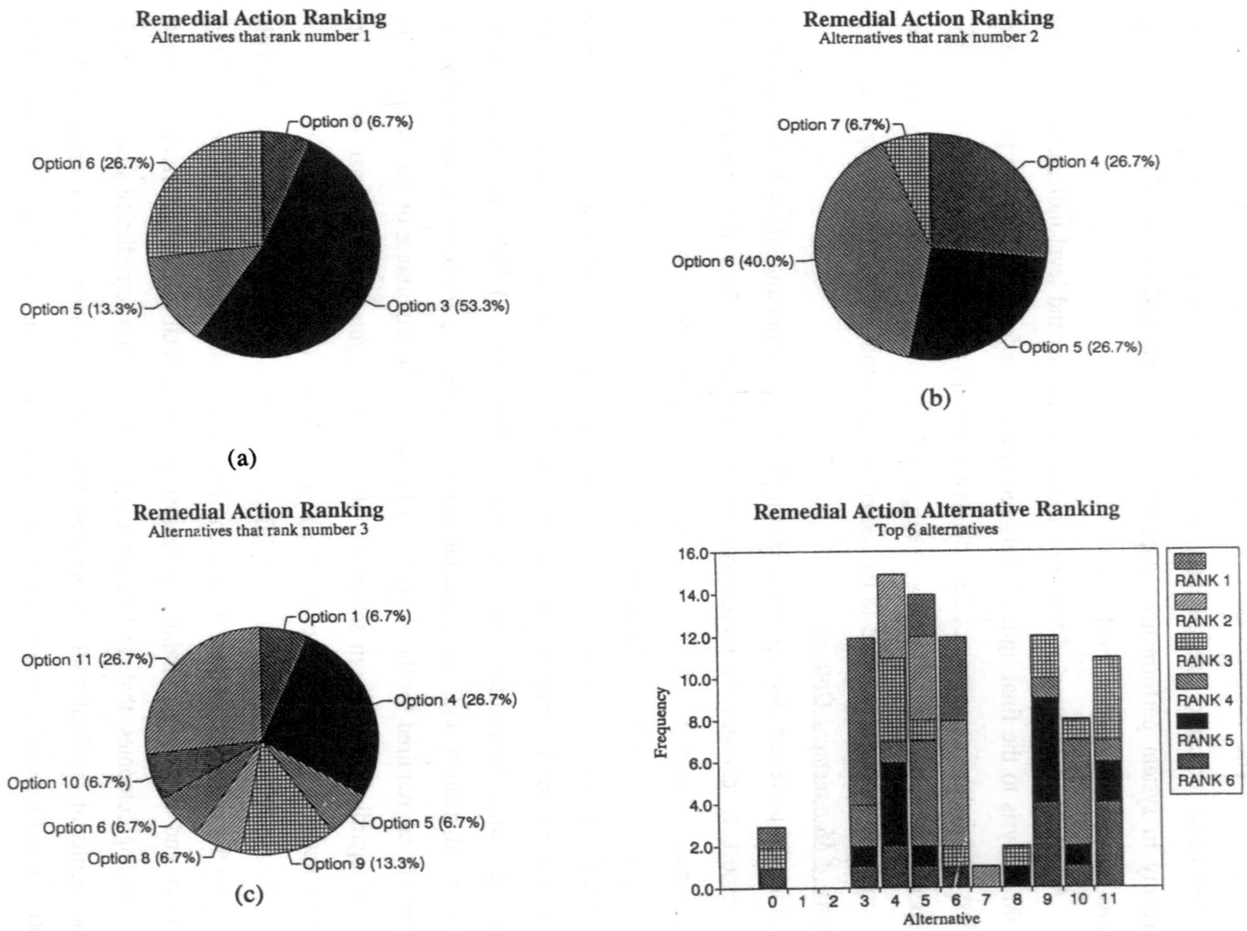

Figure 11. Alternative Remediation System Ranking

A second technique to assess robustness is to consider the alternative that ranks first and second most often. Alternatives that rank number 2 are presented in Figure 11b. Using the second method, option 6 appears to be preferred since it is a strong contender in ranking first and a leader in ranking second. The pie chart illustrating third rank (Figure 11c) becomes highly segmented without a clearly dominant alternative. A third method for assessing robustness is to find the alternative that has the greatest frequency of occurrence in the top 6 ranking. Figure 11d presents a stacked bar chart illustrating the frequency with which a given alternative ranks in the top six. Using this method it appears as though alternative 4 is a candidate system.

Based on these three methods to assess robustness, options 3, 6, and 4 would appear to be the most robust with respect to different preferences and therefore offer the performance desired while at the same time meeting cost considerations. It is interesting to note that alternatives 3, 6, and 4 are actually the same physical system (number 2) with different operating horizons.

These results lead to the conclusion that system design number 2 is the preferred design. As described previously, analysis of CP robustness tends to indicate solutions that have attributes that are attractive from all perspectives, with respect to different objectives. This leads to the observation that attributes of system 2 might be studied, and perhaps additional designs that incorporate some of the attributes could be investigated as part of a further refinement of the final design.

The decision related to operating horizon can be aided by reference to Table 4. Alternative 3 represents running system 2 for 1 year, whereas alternative 6 represents running system 2 for 4 years. Since it was assumed that an ideal value for operating horizon is equal to 1 year, alternative number 3 is the preferred option from a robust selection perspective.

SUMMARY AND CONCLUSIONS

Previous research related to remedial action planning and design has focused on limited objectives. Typical approaches have minimized pumping requirements as a surrogate measure of remediation costs. These strategies are a direct result of the numerical approaches (ie., linear and nonlinear programming) used to solve the management problem. In addition, there has been a gradual progression from deterministic to stochastic analysis in an effort to consider aquifer parameter uncertainties. Unfortunately, these efforts do not present a decision making method that incorporates the computed uncertainties.

The present work demonstrates that remedial action planning is, in reality, a multiple criteria decision making problem that may need to meet numerous objectives. Two primary objectives that often conflict are available financial resources and desired system performance. The final decision often involves a trade-off between these two objectives. It has been shown that measures of both primary objectives can be obtained using system performance indices that represent system operation at a more basic level. Indices include type of expenditures, migration of contamination through boundaries, residual contamination, mass of contamination treated, time required, and uncertainty in modeled results.

A case study has been presented to demonstrate the use of stochastic ground water modeling combined with composite programming for solving the system selection problem within a MCDM framework. It should be noted that the methodology is general in that it can be used with virtually any contamination site, ground water simulation model, and system performance indicators. The approach used to select a final design was to determine a system that is robust with respect to changing system objectives expressed as scenarios. The selected system is then placed in operation, or attributes of the most robust systems are examined and additional system alternatives are generated, with the design process repeated and subsequent system designs refined.

In summary, the present research extends previous efforts at remedial action design into the realm of multiple criteria decision making. It is felt that this extension provides a more realistic perspective of the design procedure. At the same time it allows the designer to manage uncertainties and time constraints within the decision making process. These features combined with the success of this research warrant further investigation into the MCDM approach.

ACKNOWLEDGEMENTS

The contents of this paper were developed under a grant from the Department of the Interior, U.S. Geological Survey. However, those contents do not necessarily represent the policy of that agency, and one should not assume endorsement by the Federal Government. The research has been supported under award number 14-08-0001-G1503 administered under the section 105 program. Additional support has been provided under the U.S. Geological

Survey section 104 program administered by the Nebraska Water Resources Center, University of Nebraska - Lincoln. Partial support by the National Science Foundation under grant number NSF-8919242 is also acknowledged.

REFERENCES

Andricevic, R., and Kitanidis, P.K., 1990, Optimization of the Pumping Schedule in Aquifer Remediation Under Uncertainty, Water Resources Research, Vol.26, No.5, pp. 875-885.

Bogardi, I., Duckstein, L., and Bardossy, A., 1984, Trade-Off Between Cost and Efficiency of Pollution Control, in Proceedings Sixth International Conference on Multiple Criteria Decision Making, Cleveland, Ohio.

Bogardi, I., Bardossy, A., and Duckstein, L., 1985, Multicriterion Network Design Using Geostatistics, Water Resources Research, Vol.21, No.2, pp. 199-208.

Brooker, P.I., 1985, Two-Dimensional Simulation by Turning Bands, Mathematical Geology, Vol.17, No.1, pp. 81-90.

Cohon, J.L., 1978, Multiobjective Programming and Planning, Academic Press, New York.

Duckstein, L., and Opricovic, S., 1980, Multiobjective Optimization in River Basin Developments, Water Resources Research, Vol.16, No.1, pp. 14-20.

Futagami, T., Tamai, N., and Yatsuzuka, M., 1976, FEM Coupled With LP for Water Pollution Control, Journal of Hydraulics Division, American Society of Civil Engineers, Vol.102, pp. 881-897.

Gorelick, S.M., and Remson, I., 1982a, Optimal Location and Management of Waste Disposal Facilities Affecting Groundwater Quality, Water Resources Bulletin, Vol.18, No.1, pp. 43-51.

Gorelick, S.M., and Remson, I., 1982b, Optimal Dynamic Management of Groundwater Pollution Sources, Water Resources Research, Vol.18, No.1, pp. 71-76.

Gorelick, S.M., Remson, I., and Cottle, R.W., 1979, Management Model of a Groundwater System with a Transient Pollutant Source, Water Resources Research, Vol.15, No.5, pp. 1243-1249.

Haimes, Y.Y., Hall, W.A., and Freedman, H.T., 1975, Multiobjective Optimization in Water Resources Systems: The Surrogate Worth Trade-Off Method, Elsevier Publishing Co., New York.

Journel, A.G., and Huijbregts, 1978, Gh.J. Mining Geostatistics, Academic Press, Orlando, Florida.

Klir, G.J., and Folger, T.A., 1988, Fuzzy Sets, Uncertainty, and Information, Prentice Hall, New Jersey.

Konikow, L.F., and Bredehoeft, J.D., 1978, Computer Model of Two-Dimensional Solute Transport and Dispersion in Ground Water, U.S. Geological Survey, Technical Water Resources Investigation, Book 7, Chapter C2.

Lefkoff, L.J., and Gorelick, S.M., 1986, Design and Cost Analysis of Rapid Aquifer Restoration Systems Using Flow Simulation and Quadratic Programming, Ground Water, Vol.24, No.6, pp. 777-790.

Mantoglou A. and Wilson, J.L., 1982, The Turning Bands Method for Simulation of Random Fields Using Line Generation by a Spectral Method, Water Resources Research, Vol.18, No.5, pp. 1379-1394.

Marin, C.M., 1986, Parameter Estimation in Water Resources Planning and Management: Optimal Actions or Optimal Parameters?, Water Resources Research, Vol.22, No.3, pp. 353-360.

Mercer, J.W., Silka, L.R., and Faust, C.R., 1983, Modeling Ground-Water Flow at Love Canal, New York, Journal of Environmental Engineering, Vol.109, No.4, pp. 924-942.

Peckenpaugh, J.M., and Dugan, J.T., 1983, Hydrogeology of Parts of the Central Platte and Lower Loop Natural Resources Districts, Nebraska, U.S. Geological Survey, Water Resources Investigations Report 83-4219.

Rishel, H.L., Boston, T.M., and Schmidt, C.J., 1984, Costs of Remedial Response Actions at Uncontrolled Hazardous Waste Sites, Noyes Publications, Park Ridge, New Jersey.

Ryan, B.J., DeSaulniers, R.M., Bristol, D.A., and Barlow, P.M., 1984, Geohydraulic Data for a Low-Level Radioactive Contamination Site, Wood River Junction, Rhode Island, U.S.G.S. Open File Report, 84-725.

Ryan, B.J. and Kipp, K.L., 1984, Low-level radioactive contamination from a cold-scrap recovery operation, Wood River Junction, Rhode Island, U.S.G.S. Open File Report, 84-066.

Szidarovszky, F., Gershan, M.E., and Duckstein, L., 1986, Techniques for Multiobjective Decision Making in System Management, Elsevier Scientific Publishing Co., New York.

Todd, D.K., 1980, Groundwater Hydrology, John Wiley and Sons, New York.

Wagner, B.J. and Gorelick, S.M., 1989, Reliable Aquifer Remediation in the Presence of Spatially Variable Hydraulic Conductivity: From Data to Design, Water Resources Research, Vol.25, No.10, pp. 2211-2225.

Wagner, B.J. and Gorelick, S.M., 1987, Optimal Groundwater Quality Management Under Parameter Uncertainty, Water Resources Research, Vol.23, No.7, pp. 1162-1174.

Willis, R., 1976, Optimal Groundwater Quality Management: Well Injection of Waste Water, Water Resources Research, Vol.12, No.1, pp. 47-53.

Willis, R., 1979, A Planning Model for the Management of Groundwater Quality, Water Resources Research, Vol.15, No.6, pp.1305-1312.

Woldt, W., 1990a, Simulation and Analytical Tools for Modeling Ground Water Quality, in Gambolati, G. et al., (Eds.), Computational Methods in Subsurface Hydrology, Proceedings of the Eighth International Conference on Computational Methods in Water Resources, Computational Mechanics Publications, Boston, pp. 253-262.

Woldt, W., 1990b, Ground Water Contamination Control: Detection and Remedial Planning, Ph.D. Dissertation, Department of Civil Engineering, University of Nebraska - Lincoln.

Zeleny, M., 1973, Compromise Programming, in Starr, M.K., and Zeleny, M., (Eds.), Multiple Criteria Decision Making, University of South Carolina Press, Columbia, pp. 101-145.

RISK-COST ANALYSIS UNDER UNCERTAINTY OF THE DISPOSAL OF CONTAMINATED DREDGED MATERIAL

John Stansbury, Istvan Bogardi, William E. Kelly
Department of Civil Engineering
University of Nebraska-Lincoln
Lincoln, Nebraska, USA 68507

Abstract

Risk-cost analysis can be used as a management tool to help determine which water resource management alternatives best satisfy both ecological and economic concerns. However, risk and cost assessments are often associated with very large uncertainties stemming from such factors as contaminant transport modelling and loss function assessment. If risk-cost analysis is conducted without considering these uncertainties, inappropriate management policies may result. On the other hand, considering these uncertainties may cause difficulties in decision making if they make management alternatives effectively indistinguishable. Disposal of contaminated dredged material is a water resource problem where both risks and costs are important considerations. Risks from disposal of contaminated dredged material may be reduced by incorporating measures to confine the contaminated dredged material; however, these confinement measures may increase disposal costs significantly.

Composite programming, a multicriterion decision making (MCDM) method, is used to trade off the risks to humans and ecological species and costs for a range of disposal alternatives. The composite programming algorithm makes it possible to encode uncertainties by means of fuzzy set membership functions. The final risk-cost trade-off distance for each management option is computed as a fuzzy number allowing the management options with their associated uncertainties to be compared and ranked.

1 Introduction and Purpose

The purpose of this paper is to develop a risk-cost trade-off methodology which directly incorporates uncertainties into the analysis, so that water resources management

NATO ASI Series, Vol. G 29
Water Resources Engineering Risk Assessment
Edited by J. Ganoulis

problems can be more effectively addressed. A complete description of the methodology can be found in Stansbury et al. (1991).

Multicriterion decision making tools such as Composite Programming, ELECTRE, and Multi-Attribute Utility Theory have been used successfully to assist decision makers deal with problems involving multiple and conflicting objectives (Goichoechea et al., 1982; Keeney and Wood, 1977). However, uncertainties inherent in these problems are often treated separately from the decision process itself. That is, the decision analysis is often carried out using point estimates (no uncertainty) of decision variables, and uncertainty is then characterized by "plausible intervals" (Cothern et al., 1987). The decision maker is then faced with either point estimates, which provide no characterization of uncertainty, or intervals showing bands of uncertainty which are often large and difficult to use in a decision framework.

Fig. 1 illustrates the importance and some of the difficulties of considering uncertainty in decision-making. For example, if either management option A or B must be selected on the basis of risk, then using point estimates of the risks (a method used today), management option B is superior because it clearly involves less risk. However, when uncertainties associated with risk assessment which often span several orders of magnitude (Dourson et al., 1983) are considered, it becomes difficult to determine which

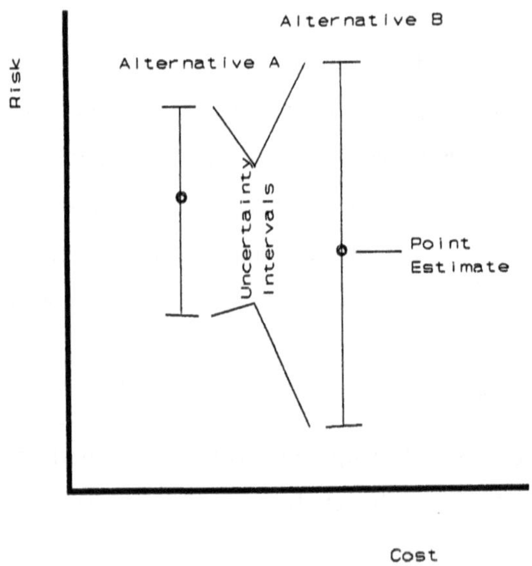

Fig. 1. Effects of uncertainty on decisions involving risks and costs

option is preferred. Further, if option B involves other negative aspects (e.g., more costly, loss of jobs, etc.) society may accept an economically inferior alternative (Option B) that may not provide greater protection from risk. Here, a method is developed using fuzzy set theory to rank uncertain numbers (i.e. risk and cost estimates) in a formal way so that decisions can be made while considering associated uncertainties.

Developing management policy for disposal of dredged sediments provides a good example of the interaction of risks, costs and uncertainty in water resources management decisions. The nation's major rivers and harbors need periodic dredging to facilitate waterborne commerce (U.S. Army Corps of Engineers (U.S.ACE), 1983). However, sediments from these harbors are often polluted with chemical contaminants from industry, agriculture, sewage, and urban runoff (NOAA, 1988). Disposal of these contaminated sediments can cause risks to ecological species (toxicological and physical) and risks to human health from consumption of contaminated seafood or water (Saucier, 1978; Tetra Tech Inc., 1987).

A case study site consisting of a bay and harbor in a heavily developed metropolitan area is selected (Fig. 2) where sediments are to be dredged from three dredging sites and disposed in any of five disposal sites (Stansbury et al., 1989). The sediments to be dredged are polluted with numerous carcinogenic and noncarcinogenic contaminants. The objective is to determine the best disposal site for the project on the basis of a risk-cost trade-off analy-sis, and to use uncertainties in the risk and cost estimates as an integral part of the analysis.

The disposal alternatives are: unconfined aquatic disposal (UAD) where the sediments are simply discharged to the bottom of the bay, capped aquatic disposal (CAD) which is similar to UAD except that the sediments are capped with several feet of clean sediments to contain any contaminants, near shore confined disposal facility (CDF) where sediments are pumped into a diked area near the shore while effluent water is allowed to escape, upland disposal (UPL) which is the same as CDF except that it is totally on land, and upland secure disposal (UPS) which is similar to upland disposal except that liners and other contaminant confinement measures are employed. Each of these disposal technologies is associated with a specific cost which generally increases with greater degrees of contaminant confinement (U.S.ACE, 1988).

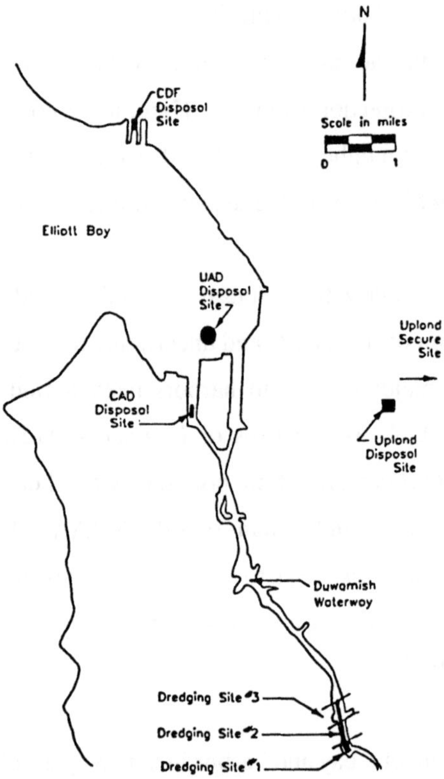

Fig. 2. Map of case study showing three dredging sites and five disposal sites

2 Sources of Uncertainty

Uncertainties are associated with all phases of risk-cost assessment. In the present case, a partial list of these sources of uncertainty might include:

1. Inaccurate modeling of fate and transport of contaminants in disposed material
2. Errors in estimating ecological population densities
3. Errors in estimating migratory and feeding habits of ecological species
4. Errors in estimating concentrations of contaminants in seafood and drinking water
5. Errors in estimating consumption rates of seafood and drinking water
6. Dose-response assessment

3 Computational Model

The computational model provides the framework for the risk-cost analysis of possible dredged material management alternatives. Indices of health risks are calculated individually for humans and ecological indicator species for each potential disposal site and associated disposal technology, and an overall composite risk index (human risk with ecological risk) is determined for the sites as well. This overall risk index is then traded off against the cost of each disposal alternative in order to determine the most acceptable management scheme.

Uncertainties inherent in all phases of the risk and cost assessments are incorporated into the analysis using fuzzy set methods (Zimmerman, 1985). That is, uncertain components of the risk and cost assessments are stated as fuzzy numbers. These uncertainties are propagated throughout the analysis using the vertex method (Dong and Shaw, 1987).

3.1 Overview of Fuzzy Set Theory

Fuzzy set theory (Zadeh, 1965; Zimmerman, 1985) is a mathematical method that may be used to characterize and propagate uncertainty and imprecision in data and functional relationships. Fuzzy sets are especially useful when insufficient data exist to characterize uncertainty using standard statistical measures (e.g., mean, standard deviation and distribution type). Numerous examples of use of fuzzy sets in engineering and risk applications can be found (Dong et al., 1988; Ayyub and Haldar, 1984; Yasunobu et al., 1987).

Fuzzy sets or fuzzy numbers are described by membership functions that characterize their maximum range of uncertainty or fuzziness (μ = 0.0-level membership) and their most likely range of values (μ = 1.0-level membership). Fig. 3 shows the fuzzy number that represents the unit cost of disposal at the UAD. The μ = 0.0-level membership is the interval between \$4.34/c.y. and \$12.22/c.y., while the μ = 1.0-level membership (most likely interval) ranges from \$7.14/c.y. to \$7.93/c.y.

A major element of using fuzzy set theory to characterize uncertainty is the assessment of membership functions for the uncertain parameters. Many different membership functions (intervals) could be suggested to define the uncertainty in the fuzzy unit-cost example. Several procedures have been developed to assess membership functions of

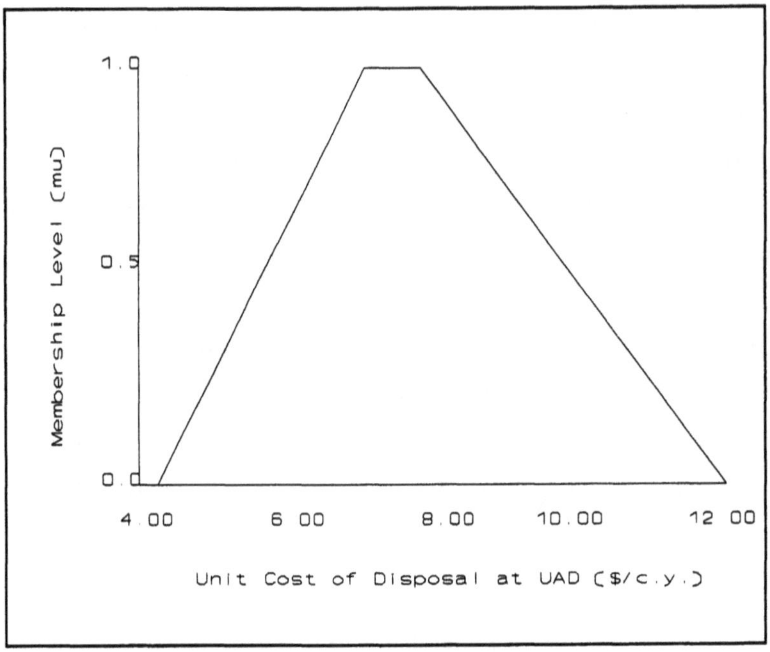

Fig. 3. Fuzzy number showing unit cost of disposal at the UAD

fuzzy numbers (Chameau and Santamarina, 1987). For the dredge material disposal problem, where fuzzy numbers are used to characterize uncertainty in physical parameters (e.g., population density, water consumption, etc.), "expert judgment" based on experience and observed measurement variability is used to assess membership functions (greatest and most likely intervals) for the uncertain parameters.

3.2 Economic Analysis

The purpose of the economic analysis is: (1) to determine the sources and quantities of dredged material and the available disposal sites, (2) to identify possible technological options for excavation, transportation, and disposal, and (3) to determine costs of operating the system.

The economic analysis will result in defining the unknown amount of material dredged at each excavation site, transported to and deposited in each disposal site. The present approach considers the quantities to be constant for the entire modeling period, therefore, a steady state analysis is used. First, two material balance constraints are defined: (1) the total volume of sediment must be excavated at all areas and transported to disposal sites, and (2) the total volume of material disposed at area *n* cannot exceed the capacity of the site.

The cost of operation involves the unit costs of excavation, transportation and disposal of dredge material. Unit costs are determined for each possible combination of transporting and depositing material from each dredging site in each disposal site (U.S.ACE, 1988; PTI Inc. 1989; Parametrix Inc., 1989).

Uncertainties in the costs are taken as the range of estimates given by the three sources, that is, the cost is described as a fuzzy number where the largest interval (μ = 0.0) is defined by the maximum range of the three estimates. The most likely cost range (μ = 1.0) is defined as the central estimate +/- 10% of the range from the central estimate to either extreme estimate. Table 1 shows the estimates for units cost of disposal at the UAD resulting as a fuzzy number.

Table 1. Unit costs (as fuzzy numbers) for disposal alternatives[a]

Disposal Sites	Lowest Estimate (μ = 0.0)	Central Estimates (μ = 1.0)	Highest Estimate (μ = 0.0)
UAD	4.34	7.14 -- 7.93	12.22
CAD	11.78	13.60 -- 15.09	26.76
CDF	25.10	26.40 -- 26.68	27.98
UPL	21.49	26.41 -- 27.01	27.43
UPS	266.67	273.40 -- 274.90	281.64

[a] Values in $/cubic yard.

Though in reality, these costs may not be linear functions of the amount of material moved, the linear assumption is generally made and unit costs are assumed constant through-out the entire operation (Marks and Liebman, 1970; Brill and Nakamura, 1978). The total cost of operation, C results from the unit costs times the quantities of dredged material removed from each source site and deposited in each disposal site according to the equation:

$$C = \sum_{l=1}^{L}\sum_{m=1}^{M}\sum_{n=1}^{N} g(l,m,n) * X(l,m,n) \tag{1}$$

where:

g(l,m,n) = unit cost of dredging at site l, depositing at site n, using technologies m,

x(l,m,n) = volume of dredged material excavated from site 1, and deposited at site n, using technologies m.

For simplicity only one set of technologies, m (excavation, transportation, and disposal technologies), will be considered for the remainder of this paper.

3.3 Risk Assessment

According to EPA (1984), risk assessment should follow the following four steps: (1) Hazard Identification, (2) Dose-Response Assessment, (3) Exposure Assessment, and (4) Risk Characterization. The method developed by Tetra Tech Inc. (1986) for risk assessment of dredged material management has been adapted for use in this study.

3.4 Hazard Identification

Hazard identification is divided into two parts: (1) toxicological hazard identification, and (2) physical hazard identification. Toxicological hazard identification consists of identifying chemicals (contaminants in dredged sediments) of concern according to procedures described in the Superfund Public Health Assessment Manual (ICF, 1985). Physical hazards such as turbidity effects, burial of benthic species, safety during transport of dredged material, and habitat destruction are also identified.

3.5 Dose-Response Assessment

According to the Federal Register (1980; 1984), the present method of dose-response assessment involves fitting a mathematical model (usually a linearized multi-stage model) to a set of dose-response data for which animal experimental doses have been transformed to equivalent human doses. The fit of the model to the data is optimized by maximizing the likelihood function. Since environmental exposures are likely to be in the low-dose region, the slope of the dose-response function in this region, represented by q_1 (the linear component) of the multistage model, is the quantity of interest. To account for uncertainties, q_1^* is defined as the upper 95% confidence limit of q_1. q_1^* is used as the carcinogenic potency factor for the chemical in question.

A problem with using q_1^* as the potency factor is that it conceals uncertainties that are present in the assessment. Since only the single value, q_1^* is reported as the carcinogenic potency factor, the uncertainty or the range of plausible values of the actual potency of a particular carcinogen is not evident. Therefore any uncertainties in dose-response assess-ment are not available for subsequent uncertainty analysis of the general risk assessment.

3.6 Dose-Response Assessment Using Fuzzy Regression

An alternative procedure is developed for dose-response assessment using fuzzy regression (Bardossy et al., 1991a; 1991b). This method produces dose-response estimates that are comparable to the standard method, but it keeps uncertainties evident so that they can be used in subsequent uncertainty analyses of the risk assessment. Further, the fuzzy regression method is generally simpler to use and more straight forward than the present statistically-based method. Finally, uncertainty analysis in the statistically-based framework can be difficult because of possible dependencies between parameters (Whitmore, 1984). These difficulties are avoided in the fuzzy regression framework.

3.7 Exposure Assessment

Exposure assessment is the determination or estimation (qualitative or quantitative) of the magnitude, frequency, duration, and route of exposure. For dredged material management, components of exposure assessment include describing the following four elements:

1. Environmental pathways and uptake routes for dredged material or chemical contaminants
2. Concentrations of chemical contaminants or the presence of contaminated sediment in various media (air, water, sediments) in space and time
3. Exposed populations and resources

4. Exposure dose or level (i.e., average rate of intake of a chemical contaminant by an individual of the exposed population)

The concentration of chemical contaminants (used for humans) or the presence of contaminated sediments (used for nonhumans) is accomplished using a fate/transport model. The fate/transport model must first consider the original concentrations of contaminants. Next, mechanisms controlling the fate of the contaminants and the sediments such as biodegradation or hydrolysis and factors that affect the movement of the contaminants or sediments are considered. These include caps and liners at confined disposal sites and various dredging techniques and physical factors such as tides and currents are also considered.

Exposure dose for humans is calculated using the equation:

$$ExposureDose = \frac{C*I*A}{W} \tag{2}$$

where:

C = concentration of contaminant in media (e.g., mg PCB/Kg fish),

I = intake rate (e.g., Kg fish/day),

A = absorption coefficient (e.g., contaminant absorbed/contaminant ingested),

W = weight of target organism (70 Kg for humans).

Exposure level for nonhumans is calculated as a probability of exposure to contaminated sediment. The probability of exposure to contaminated sediment for nonmigratory species is the fraction of the areal range of the population that coincides with the area affected by the disposal. On the other hand, the probability of exposure for migratory species would be a function of this fraction, as well as factors related to behavior patterns such as the time spent in the vicinity of the disposal site, feeding activity, and spawning activity.

Exposure assessment for humans from aquatic and near-shore disposal sites is accomplished using the equation:

$$Constant1(n,i,k) = \frac{T*MC*FC*MP*C}{Sc*W} \tag{3}$$

where:

Constant1(n,i,k) = exposure coefficient (Kg media/Kg human · MCY · day),

T = transport factor reflecting the efficacy of any confining measures such as caps.

MC = migration coefficient (0 - 1) representing the time a migratory species spends in the vicinity of the disposal site,

FC = food coefficient (0 - 1) representing the amount of feeding done by a migratory species while in the vicinity of the disposal site,

MP = percentage of media (fish) eaten (by various populations) which originates from the disposal site,

Sc = disposal site capacity (MCY),

C = media consumption rate (Kg media/day), and

W = weight of target organism (70 Kg for humans).

The transport factor is calculated using the equation:

$$T = CT\,(1-CE) + (1-CT) \tag{4}$$

where:

CT = fraction of time that the confinement mechanism (cap) is in place, and

CE = efficiency of the confinement mechanism (0 - 1).

Exposure assessment for humans for upland disposal sites is identical to that for aquatic sites except for an additional factor that defines effectiveness of the leachate confinement. When leachate tests are performed, the ratio of contaminant concentration in the leachate to concentration in the sediment is used. When leachate tests are not performed, estimates must be made considering the properties of the site liner and other factors affecting the leaching potential of the site.

The concentrations of contaminants in food organisms (caused by the dredging program) are estimated from bioaccumulation tests conducted on indicator species. Similar concentrations are assumed for similar nontest species (e.g., clams and oysters). When bioaccumulation test data are not available, simple bioaccumulation models are used to estimate the bioaccumulation stemming from the dredging program. The bioaccumulation model used for organic chemicals (and methylated mercury) in this study, developed in Lake et al. (1984) is:

$$C^{i}_{org} = \left[\frac{C^{i}_{sed}}{OC}\right] L \tag{5}$$

where:

C^{i}_{org} = Concentration of contaminant i in the organism,
C^{i}_{sed} = Concentration of contaminant i in sediment,
OC = Fraction of organic carbon in sediment,
L = g lipid in organism/g organism (dry wt).

The bioaccumulation model used for metals is based on estimated contaminant concentrations in interstitial water of the disposed sediments in which benthic species would reside. The interstitial concentration estimates, based on work by Pavlou (1984) are used with bioconcentration factors for various types of seafood species. The model is then stated as:

$$C^{i}_{org} = \left[\frac{C_{sed}}{K_{OC} * OC}\right] * BCF \tag{6}$$

where:

C_{sed} = Contaminant concentration in sediment,
K_{oc} = Sediment/water partition coefficient normalized for organic carbon content,
OC = Fraction of organic carbon in sediment, and
BCF = Bioconcentration factor for seafood group.

Contaminant concentrations in groundwater are estimated by determining dilution factors for leachate from a landfill entering a groundwater aquifer. The volume of leachate leaving the landfill is calculated using the equation:

$$Q = \frac{K * A * H}{L} \tag{7}$$

where:

Q = Volume of leachate leaving the landfill,

K = Hydraulic conductivity of bottom of landfill (m/d),

A = Bottom area of landfill,

H = Hydraulic head in landfill, and

L = Distance between bottom of landfill and top of aquifer.

The volume of groundwater in the aquifer passing beneath the landfill is calculated using:

$$Q = K * i * b * W \tag{8}$$

where:

i = Hydraulic gradient of the aquifer beneath the landfill,

b = Effective mixing thickness of the aquifer,

W = Width of landfill perpendicular to groundwater flow direction.

Contaminant concentrations in the groundwater are then calculated using:

$$C_{gw} = C_{iw} * \left[\frac{Q_l}{Q_a}\right] \tag{9}$$

where:

C_{gw} = Concentration of contaminant in the groundwater,

C_{iw} = Concentration of contaminant in interstitial water,

Q_l = Flow of leachate from the landfill, and

Q_a = Flow of groundwater beneath the landfill.

Contaminant concentrations in interstitial water are calculated using methods developed by Pavlou (1984):

$$C_{iw} = \frac{C_{sed}}{K_{oc} * OC} \tag{10}$$

where:

C_{iw} = Contaminant concentration in interstitial water,

C_{sed} = Contaminant concentration in sediment,

K_{oc} = Sediment/water partition coefficient normalized for organic carbon content, and

OC = Fraction of organic carbon in sediment.

Exposure assessment for nonhumans is calculated using the equation:

$$Constant2(n,r) = \frac{OM*PD*T*MC*FC}{Sc} \qquad (11)$$

where:

Constant2(n,r) = exposure coefficient (# organisms/MCY),

PD = population density (#/m^2), and

OM = area of population range that overlaps with a full disposal site (m^2).

3.8 Risk Characterization

In this step, results of exposure assessment and dose-response assessment are combined to estimate the extent of adverse effects associated with dredged material disposal. The risks are expressed as probabilities for carcinogenic risks, risk indices for noncarcinogenic risks, and expected consequence (e.g., mortality) for ecological risks.

Human carcinogenic risk for contaminant i and media k is calculated as a probability using the general equation:

$$CarcinogenicRisk_{ik} = f_i * ExposureDose \qquad (12)$$

where:

f_i = carcinogenic potency factor for contaminant i.

Specifically, human carcinogenic risks are determined for each disposal site using the equation:

$$R(n) = \sum_{i=1}^{I}\sum_{k=1}^{K} f_i \left[Constant1(n,i,k) \sum_{l=1}^{L} Cm(l)*X(l,n)\right] \qquad (13)$$

where:

R(n) = total carcinogenic risk at disposal site n from all contaminants i and media k,

f_i = carcinogenic potency factor for contaminant i (computed by fuzzy regression method),

X(l,n) = volume of sediment deposited at disposal site n from excavation site l,

Constant1 was defined previously in the Exposure Assessment section, and

Cm(l) = measured concentration of contaminant i in dredged material (for drinking water exposure) or seafood (from bioaccumulation tests for food exposure) from excavation site l.

Human noncarcinogenic risk for contaminant i and media k, is calculated as a hazard index using the general equation:

$$Hazard\ Index_{ik} = \frac{Exposure\ Dose_{ik}}{Rfd_i} \tag{14}$$

where a Hazard Index greater than unity indicates some risk while HI< =1 indicates a safe condition.

Specifically, noncarcinogenic risk at a disposal site is calculated using the equation:

$$HI(n) = \sum_{i=1}^{I}\sum_{k=1}^{K}\left[\frac{Constant1(n,i,k) * \sum_{l=1}^{L} Cm(l) * X(l,n)}{RfD_i}\right] \tag{15}$$

where:

HI(n) = total noncarcinogenic risk at site n for all contaminants i and all media k,

RfD_i = Reference Dose for contaminant i (as a fuzzy number including uncertainty factors),

X(l,n) = volume of dredged material deposited at disposal site n, from dredging site l.

Risks to nonhuman species are estimated for toxicity, burial, and habitat destruction. Risk to species r, from exposure to contaminated sediment from dredging site l, is calculated as an expected consequence (e.g., mortality) using the general equation:

$$Risk_r = Toxicity_r(l) * ProbabilityofExposure_r(l) \tag{16}$$

where:

$Toxicity_r(l)$ = toxicity to species r, caused by exposure to sediment from dredging site l, as determined by bioassay.

Specifically, the risk to nonhumans at each disposal site n is calculated using the equation:

$$R_r(n) = Constant2(n,r) \sum_{l=1}^{L} TF_r(l) * X(l,n) \tag{17}$$

where:

$R_r(n)$ = risk to species *r* at disposal site *n*,

$TF_r(l)$ = toxicity to species *r*, caused by exposure to sediment from dredging site *l*, as determined by bioassay.

$X(l,n)$ = volume of sediment from dredging site *l*, deposited at disposal site *n*,

Burial risk to species r is calculated using the general equation:

$$Risk_r = Burial\ Factor * Probability\ of\ Exposure_r(l) \tag{18}$$

where Burial Factor is the probability of mortality due to burial for species r. Specifically, the burial risk to species r due to burial is given by the equation:

$$R_r(n) = Constant2(n,r) \sum_{l=1}^{L} BF_r * X(l,n) \tag{19}$$

where:

$R_r(n)$ = risk to species r at disposal site n, and

BF_r = burial factor to species r.

Habitat destruction at disposal sites is computed using the following rationale and model. From an ecological perspective, habitat quality must be measured at a community level rather than a species level. That is, many habitat quality measures such as U.S. Fish and Wildlife Habitat Suitability Index (U.S. Fish and Wildlife Service, 1980) are designed to measure the habitat quality for a certain species. This runs counter to ecological philosophy as it assumes that the designated species (usually one important to man in some sense) is more important than the other species in the community. Therefore the model will measure habitat quality not in terms of any designated species, but in terms of the condition of the community. Specifically, habitat quality is measured relative to a "climax succession stage" community. Here a climax community is defined as that community that would inhabit the site with no anthropogenic disturbance.

The Habitat Loss Index model is:

$$HLI = \frac{I * \Delta C * RT * A}{Sc} * \sum X(l,n) \tag{20}$$

where:

I = importance index of the habitat at the disposal site,

dC = change in the condition of the habitat measured relative to "climax community,"

RT = recovery time index, time needed to recover to pre-disposal condition,

A = area of disposal site (m^2),

Sc = capacity of disposal site (m^3), and

X(l,n) = volume of material disposed at disposal site (m^3).

The habitat importance index is a qualitative measure based on the rating scale shown in Table 2. The maximum score for the habitat in an area is 1.0 indicating an exceptionally important area in terms of habitat.

The extent of disruption to the habitat is measured by the change in condition (dC) term. Here the original condition of the habitat is measured relative to the "climax community for the area in question. An area exhibiting a "climax community" is given a rating of 1.0 while a barren area is given a rating of 0.0. Areas with communities between climax and barren are given intermediate ratings as shown in Table 3. As shown, several methods of community succession can be used depending on availability of data.

Table 2. Habitat importance index

Attributes of Habitat	Score
A. Presence of sport or commercial species	0.2
B. Feeding, nursery, or support area for above	0.2
C. Unique or scarce habitat	0.2
D. Habitat for threatened or endangered species	0.2
E. Other factors making habitat important	0.2

Table 3. Habitat condition rating scale

Habitat Succession Stage	Benthic Succession[a]	Organism/ Sediment Index[b]	Condition Rating
Climax	III	8 to 11	1.0
	II/III	4 to 7	0.8
Disturbed	II	-1 to +3	0.6
	I/II	-5 to -2	0.4
Pioneering	I	-8 to -6	0.2
Barren		-10 to -9	0.0

[a] Benthic Succession Stages (I, II, III) are community structure forms that are more or less universal for benthic communities. Stage I consists of shallow-dwelling pioneering species while Stage III consists of deeper dwelling species (Rhoads and Germano, 1982; Rhoads and Boyer, 1982).

[b] Organism Sediment Index is a photographic method using the benthic succession stages paradigm and other benthic conditions in a rating scale (Rhoads and Germano, 1982).

Upon disposal, the habitat will be put into a barren successional stage. The change in habitat condition is then calculated as the difference between the original condition and barren stage. This value is modified by the fraction of the site that is significantly affected by the disposal event which is roughly proportional to the ratio of volume of sediments disposed to volume of site capacity. This ratio is supported by photographic surveys at aquatic disposal sites (SAIC, 1990).

The recovery time index is used to moderate habitat disruption estimated to be of short duration. The recovery time index is a measure of the time required for the habitat to recover to the pre-disposal condition. The recovery time index is assigned according to the scale Table 4.

3.9 Risk Management (Risk-Cost Trade-Off)

3.9.1 Principles

The purpose of risk management is to provide the decision maker with a range of alternative dredged material management measures which facilitates the choices or trade-

Table 4. Habitat recovery time scale

Recovery Time (yr)	Recovery Time Index
> 10	1.0
7 - 10	0.9
5 - 7	0.7
3 - 7	0.5
1 - 3	0.3
< 1	0.1

offs among desired objectives, i.e. elicits a compromise between the cost of operation and environmental risk. Composite Programming (Bogardi and Bardossy, 1983), an extension of compromise programming is used for the risk-cost trade-off assessment.

Composite Programming (CP) organizes a problem into the following format: (1) define management alternatives, (2) define basic indicators, (3) group basic indicators into progressively fewer, more general, groups, (4) define weights, balancing factors, and worst and best values for the indicators, and (5) evaluate and rank the alternatives.

The dredging alternatives are defined as depositing various quantities of sediments from each excavation site into each disposal site. Each alternative is associated with environmental impacts (human risks and ecological risks) and cost. Human risks, ecological risks, and cost are used as the basic indicators for this system.

The basic indicators are grouped into fewer, more general groups called first-level indicators. This grouping continues until only two indicators (risk and cost) remain. The indicators are grouped according to the structure shown in Fig. 4.

After the indicators have been grouped into this structure, weights for the indicators are assessed. These weights represent the relative importance of each indicator as viewed by the decision maker. For example, the risk to one species may be considered more important than the risk to another species at an upland disposal site.

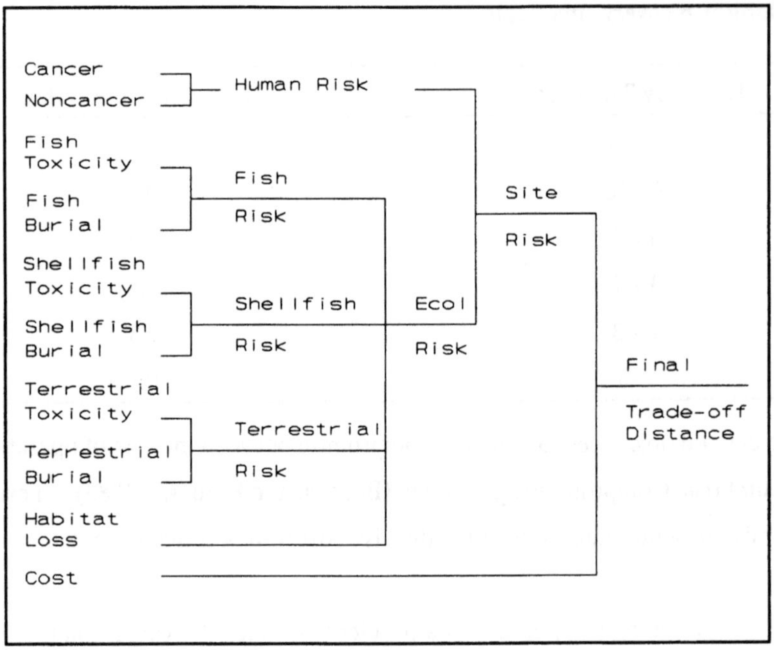

Fig. 4. Trade-off structure of criteria used in risk-cost analysis

Next the balancing factors are assigned for each group of indicators. Balancing factors reflect the relative importance that is assigned to the maximal deviations of the indicators and limit the ability of one indicator to substitute for another. In other words, with an appropriately high balancing factor, an indicator that must not be compromised, such as risk to an endangered species, will not be substituted for by one such as risk to an abundant, nonendangered species. In effect it reflects the strength of preference for a particular objective, defining its relative importance. Generally the balancing factors and weights are derived from public institutional inputs.

The "best" and "worst" possible values for the basic indicators for the particular study are then assessed. For instance, the "best" value for cost of operation would be the cost associated with the most inexpensive alternative, while the "worst" cost would be that associated with the most expensive alternative. Likewise, for human carcinogenic risk at site *n*, the "best" value would be zero risk (no material deposited at site n), and the "worst" value would be the risk level caused if all of the sediment from all excavation sites were deposited at site *n*.

3.9.2 Composite Programming Algorithm

Evaluation of the various management alternatives proceeds by computing composite distances using CP. The first step is normalization (placing into the 0 - 1 interval) of the basic indicator values (Z_i). This is necessary since the units of the various basic indicators can be quite different (e.g., probability, mortality, dollars) and difficult to compare directly. The ability to compare and evaluate variables having different units (eliminating the need to transform all units into a common unit) is one of the primary benefits of using CP. Using the maximum (Z_{i+}) and the minimum (Z_{i-}), the normalized value (S_i) of Z_i can be computed as:

$$S_i = \frac{Z_i - Z_{i-}}{Z_{i+} - Z_{i-}} \qquad (else) \qquad S_i = \frac{Z_{i+} - Z_i}{Z_{i+} - Z_{i-}} \tag{21}$$

Where the formula selected depends on whether the maximum (Z_{i+}) is the "best" or "worst" value. In other words, the formula should place the normalized value (S_i) the appropriate distance between the "best" and "worst" values, as well as in the 0 - 1 interval.

Next, the first-level composite distances are computed for each first-level group of basic indicators using the equation:

$$L_j = \left[\sum_{i=1}^{n_j} a_{ij} * S_{ij}^{P_j} \right]^{1/P_j} \tag{22}$$

where:

- S_{ij} = the normalized value of basic indicator *i* in the first-level group *j* of basic indicators (Fig. 4),
- L_j = the composite distance for first-level group *j* of basic indicators,
- n_j = the number of basic indicators in group *j*,
- a_{ij} = the weights expressing the relative importance of basic indicators in group *j* such that their sum equals one, and
- p_j = the balancing factor among indicators for group *j*.

Second-level composite distances are calculated using the first-level distances. The process of computing successively higher levels of composite distances is repeated until the final composition between the two highest level indicators is reached.

4 Results

4.1 Overview

Risks to human health and to ecological species are estimated for the five potential dredged material disposal alternatives. The costs stemming from these alternatives are also estimated. The parameters leading to each risk and cost estimate are characterized as fuzzy numbers. The component risks (e.g., cancer and noncancer for humans, or toxicity and burial for shellfish) from each disposal alternative are traded off using composite programming (CP) until a final composite risk index is computed. Finally this composite risk index is traded off with the cost of disposal. In this way, a risk-cost trade-off distance is computed for each disposal alternative. These trade-off distances can then be used by decision-makers to evaluate the effects of risk and cost in determining the preferred dredged material disposal plan.

Since the uncertainties are incorporated as fuzzy numbers into the risk and cost estimates, the alternatives appear as pyramids on the trade-off graphs (Fig. 5). If uncertainty were not considered, the alternatives would appear as points. A thorough description of the study can be found in Stansbury et al. (1991).

4.2 Trade-Offs

Trade-offs are made at all levels of the CP structure (Fig. 4). For example, Fig. 5 shows the trade-off of cancer risk to humans versus cost for the five disposal alternatives. It can be seen that, at the upper range of estimated risks, several of the alternatives exceed 1 X 10^{-4} lifetime cancer risk. However these upper range estimates use very conservative assumptions such as high bioaccumulation, or high consumption rates, and should be viewed as a maximum plausible risk estimate for a high-risk sub-population. From Fig. 5, it can be seen that the UPS (hazardous waste facility) provides the greatest protection from cancer, but it does so at a very high price. On the other hand, disposal at the UAD is the least expensive, but there is a potential for a relatively higher risk. Fig. 5 also shows how the uncertainties in the cancer risk estimates overlap. For example while disposal at the CAD may provide relatively low cancer risk given a certain set of exposure assumptions, it may also provide a relatively high risk given a different, yet plausible set of assumptions.

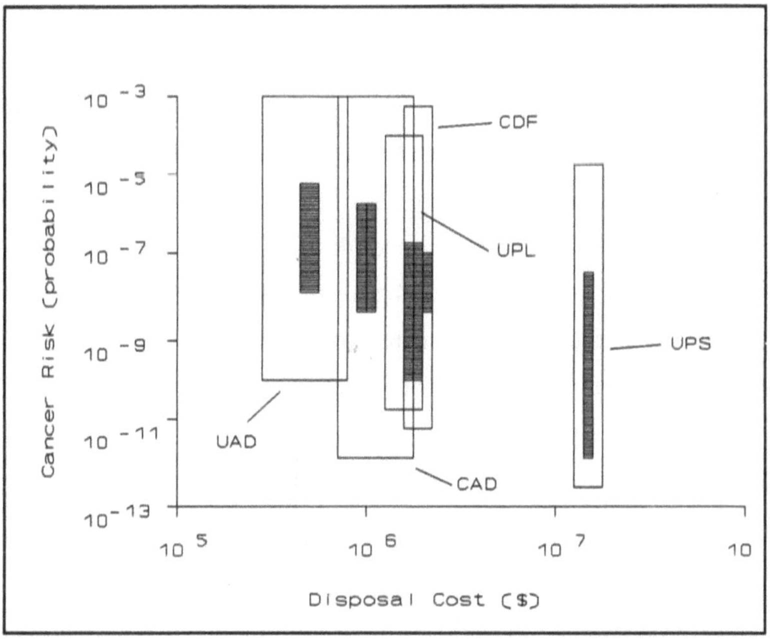

Fig. 5. Lifetime cancer risks traded off with disposal costs for the five disposal alternatives

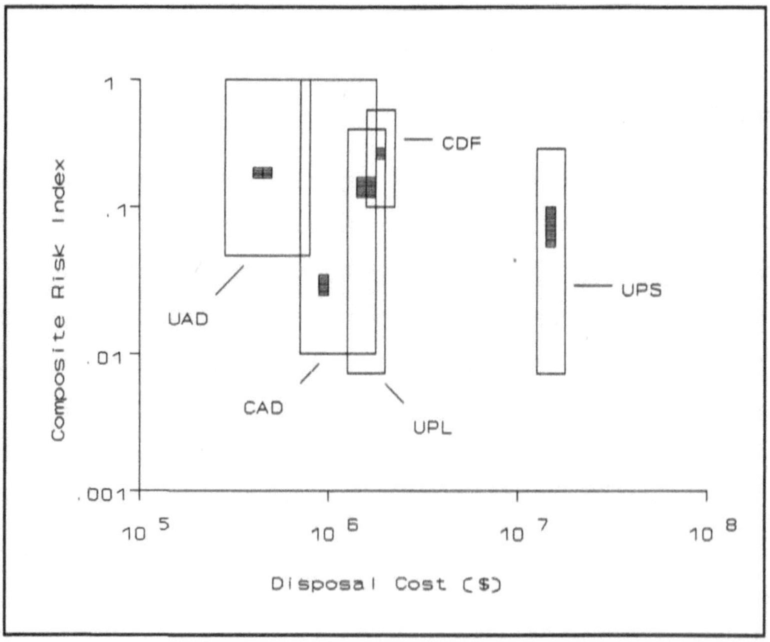

Fig. 6 Final trade-off of composite risk index with cost for the disposal alternatives.

The final composite risk-cost trade-off is shown in Fig. 6. It can be seen that the UPS site provides good risk protection, but it is very expensive. The UAD site is the least expensive alternative, but it allows the greatest risk. The CAD and the UPL sites appear to yield about equal composite indices. The UPL has generally lower risk and less uncertainty in the risk estimates (note the log scale) than does the CAD site. However the CAD site is less expensive. Also note the "distributions" of the risks: for the CAD site the most likely (μ = 1.0) risk index ranges from 0.025 to 0.026, while for the UPL site the μ = 1.0 level for the risk index ranges from 0.088 to 0.145. So while the CAD site has a potential for higher risk, the most likely estimate is lower than for the UPL site.

4.3 Ranking of the Alternatives

Since the alternatives appear on the final trade-off graph as fuzzy pyramids, it is difficult to see which alternative is preferred (i.e., has the minimum risk-cost trade-off distance from the origin of the graph). Since the distance from the origin for each alternative is a fuzzy value (Fig. 7). Therefore a method developed in Chen (1985) is used to rank these fuzzy distances and determine which alternative is preferred. The results of the ranking given in Table 5 show that the UPL site has the smallest fuzzy distance from the origin and is thus the preferred alternative given the weights used by the investigators.

4.4 Sensitivity Analysis

Since the choice of the preferred alternative may depend on the choice of weights in the CP trade-off program, a sensitivity analysis of those weights is conducted. Here a wide range of weighting schemes is used to represent decision makers with widely differing priorities (i.e., those more concerned about risks, those more concerned about costs, etc.). The results of the sensitivity analysis are given in Table 5, which shows that the ranking of the CAD and the UPL sites are nearly equal. In other words, depending on the weighting preferences of the decision makers, either the CAD site or the UPL site is preferred. As shown in Fig. 7, the UPL has less uncertainty than the CAD site, but the "most likely" trade-off distance for the CAD is less. The other three alternatives maintain their respective ranking for all weight sets.

5 Discussion and Conclusions

A risk-cost trade-off analysis has been conducted to evaluate dredged material disposal alternatives. The analysis incorporates uncertainties in the risk and cost assessments directly into the trade-off so that decision makers can readily evaluate the effects of those uncertainties on management decisions. Fuzzy set methods are used to characterize these uncertainties which typically are difficult to characterize using statistical methods.

The advantage of incorporating uncertainties directly into the trade-off analysis can be seen in Fig. 6. For example, if only the most likely risk and cost estimates were available to a decision maker, the CAD site would appear to be clearly superior to any of the others, because it would have a lower composite risk index than any other alternative, and is a relatively low-cost alternative. However with the use of the fuzzy trade-off methodology, it can be seen that the CAD site might involve relatively high risks. In addition, the UPL site with its lower overall uncertainty, becomes essentially equivalent to the CAD, a fact that would not be noticed in a non-fuzzy trade-off format.
Several conclusions may be drawn from this study.

1. Risk-cost analysis is a viable tool for evaluating dredged material management alternatives.
2. Multicriterion decision making tools such as composite programming are useful in aiding the evaluation of management issues with conflicting objectives.
3. Uncertainties in risk and cost assessments can have significant impacts on management decisions.
4. Uncertainties can be readily incorporated into the analysis with fuzzy set methods where statistical methods are not appropriate.
5. Incorporating uncertainties into the trade-off analysis instead of using post-analysis assessment of uncertainty, facilitates understanding of the impacts of the uncertainties on management decisions.

Acknowledgements

The authors gratefully acknowledge the U.S. Army Corps of Engineers Institute for Water Resources and the University of Nebraska Center for Infrastructure Research for supporting this project through the Water Center of the University of Nebraska-Lincoln.

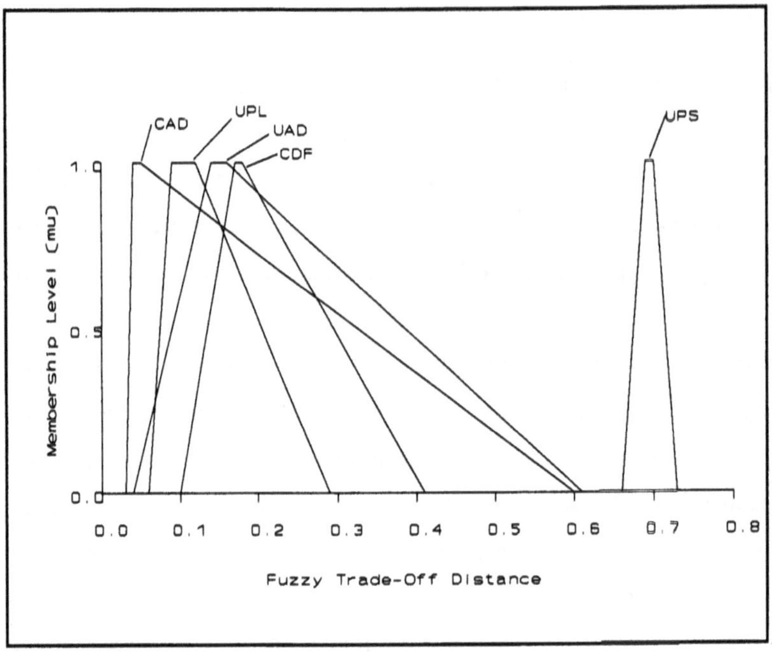

Fig. 7. Fuzzy trade-off distances for the five disposal alternatives using investigators' preferred weighting scheme

Table 5. Sensitivity analysis of weighting schemes and ranking of alternatives

Weighting Scheme		Disposal Alternatives				
		UAD	CAD	CDF	UPL	UPS
I	Distance	.63	.46	.59	.39	1.86
	Ranking	4	2	3	1	5
II	Distance	.74	.51	.67	.52	1.47
	Ranking	4	1	3	2	5
III	Distance	.65	.45	.57	.46	1.83
	Ranking	4	1	3	2	5
IV	Distance	.64	.51	.52	.51	1.45
	Ranking	4	1	3	1	5
V	Distance	.48	.37	.39	.36	1.90
	Ranking	4	2	3	1	5

References

Ayyub BM, Haldar A (1984) Project scheduling using fuzzy set concepts. Journal of Construction Engineering and Management, ASCE 110(21)
Bardossy A, Bogardi I, Duckstein L (1991a) Fuzzy non-linear regression to define dose-response relationships. European Journal of Operations Research, in press
Bardossy A, Hagaman R, Duckstein L, Bogardi I (1991b) Fuzzy least squares regression. Working paper 91-8, Systems and Industrial Engineering Department, The University of Arizona, Tucson
Bogardi I, Bardossy A (1983) Application of MCDM to geological exploration. In: Hansen P (ed) Essays and Surveys on Multiple Criterion Decision Making, Springer-Verlag, Berlin Heidelberg New York
Brill ED, Nakamura M (1978) A branch and bound method for use in planning regional wastewater treatment systems. Water Resources Research 14:109-118
Chameau J, Santamarina JC (1987) Membership functions: comparing methods of measurement. International Journal of Approximate Reasoning 1:287-301
Chen SH (1985) Ranking fuzzy numbers with maximizing set and minimizing set. Fuzzy Sets and Systems 17:113-129
Cothern CR, Coniglio WA, Marcus WL (1987) Uncertainty in population risk estimates for environmental contaminants. In: Proceedings of the Society for Risk Analysis International Workshop on Uncertainty in Risk Assessment, Risk Management, and Decision Making, Knoxville
Dong W, Shaw HC (1987) Vertex method for computing functions of fuzzy variables. Fuzzy Sets and Systems 24:65-78
Dong W, Shaw HC, Wong FS (1988) Condensation of the knowledge base in expert systems with applications to seismic risk evaluation. In: Adeli (ed) Expert Systems in Construction and Structural Engineering. Chapman and Hall
Dourson ML, Stara FF (1983) Regulatory history and experimental support of uncertainty (safety) factors. Regulatory Toxicology and Pharmacology 3:224-238
Goichoechea A, Hansen D, Duckstein L (1982) Introduction to Multiobjective Analysis with Engineering and Business Applications. John Wiley & Sons, New York
ICF Inc (1985) Draft superfund health assessment manual. prepared for US EPA Office of Emergency and Remedial Response Washington DC by ICF Inc. Washington DC
Keeney RL, Wood EF (1977) An illustrative example of the use of multiattribute utility theory for water resources planning. Water Resources Research 13:4
Lake JL, Rubenstein N, Pavignano S (1984) Predicting bioaccumulation: development of a simple partitioning model for use as a screening tool for regulating ocean disposal of wastes. In: Dickson K, Maki A, Brungs W (eds) Fate and Effects of Sediment-Bound Chemicals in Aquatic Systems, Proceedings of the Sixth Pellston Workshop, Pergamon Press, NY
Pavlou SP (1984) The use of the equilibrium partitioning approach in determining safe levels of contaminants in marine sediments. In: Dickson K, Maki A, Brungs W (eds) Fate and Effects of Sediment-Bound Chemicals in Aquatic Systems, Proceedings of the Sixth Pellston Workshop, Pergamon Press, NY
Parametrix Inc (1989) Confined disposal of contaminated sediments documentation of standards development (draft report). Prepared for Washington Department of Ecology by Parametrix Inc in association with Ogden Beeman and Associates Inc, Hart-Crowser Inc, Science Applications International Inc, Pacific Groundwater

Group, Janet N Knox Inc
PTI Inc (1989) Multiuser confined disposal sites program study Draft Report. Prepared for Washington Department of Ecology Sediment Management Unit, Olympia WA
Rhoads DC, Boyer LF (1982) The effects of marine benthos on physical properties of sediments: a successional perspective. In: McCall PL Tevesz MJS (eds) Animal-Sediment Relations, Plenum Geobiology Series Volume 2. Plenum, NY
Rhoads DC, Germano JD (1982) Characterization of organism-sediment relations using sediment profile imaging: an efficient method of Remote Ecological Monitoring of the Seafloor (REMOTS System). Mar Ecol Prog Ser 8:115-128
SAIC Inc (1990) PSDDA monitoring: post-disposal sediment profile surveys of the Elliott Bay and Port Gardner disposal sites draft report. Submitted to Washington State Department of Natural Resources Division of Aquatic Lands, Olympia WA
Saucier RT (1978) Executive overview and detailed summary dredged material research program. Technical Report DS-78-22. Environmental Lab, Vicksburg MS
Stansbury J, Bogardi I, Kelly WE (1989) Risk-cost analysis for management of dredging material. In: Haimes YY, Stakhiv EZ (eds) Risk-Based Decision Making in Water Resources. ASCE, NY
Stansbury J, Bogardi I, Kelly W, Bardossy A (1991) Risk analysis of dredged material management options part 3: decision-making under uncertainty. Report to U.S. Army Corps of Engineers, Institute for Water Resources, Washington DC
Tetra Tech Inc (1986) A framework for comparative risk analysis of dredged material disposal options Task 5b draft report. Prepared for Resources Planning Associates for US Army Corps of Engineers Seattle District by Tetra Tech Inc, Bellevue WA
US Army Corps of Engineers (1983) Dredging and dredged material disposal. EM 1110-2-5025, US COE Office of the Chief of Engineers
US Army Corps of Engineers (1988) Disposal site selection: technical appendix -phase I. Puget Sound Dredged Disposal Analysis Report
US EPA (1984) Proposed guidelines for carcinogenic risk assessment: request for comments. Federal Register. Vol 49 No 227 Part VIII 46304-46312
US Fish and Wildlife Service (1980) Ecological services manual, Implementation of habitat evaluation procedures. Division of Ecological Services US Fish and Wildlife Service Department of the Interior Washington DC
Whitmore RW (1984) Methodology for characterization of uncertainty in exposure assessment. Report to Exposure Assessment Group Office of Health and Environmental Assessment US EPA, Research Triangle Institute, Research Triangle Park, North Carolina
Yasunobu S, Sekino S, Toshitsugu H (1987) Automatic train operation and automatic crane operation systems based on predictive fuzzy control. In: Proceedings Second Congress of the International Federation of Fuzzy Sets and Applications, Tokyo
Zadeh LA (1965) Fuzzy Sets. Information and Control 8:338-353
Zimmerman HJ (1985) Fuzzy Set Theory and its Applications. Kluwer-Nijhoff, Hingham, MA

RISK-COST ASPECTS OF SEA LEVEL RISE AND CLIMATE CHANGE IN THE EVALUATION OF SHORE PROTECTION PROJECTS

E. Z. Stakhiv[1], S. J. Ratick[2] and Wei Du[2]

ABSTRACT

U.S. Federal coastal protection project planning can incorporate forecasts of sea level rise and storm frequency changes due to climate change through the application of risk and uncertainty analysis techniques. Incorporating future sea level rise into current projects implies building projects that are too large for existing conditions. Also, changed future conditions such as increased storm frequency and intensity, which increase recurring project maintenance costs tend to favor structural-type projects. Based on applications of a suggested risk-cost evaluation approach adapted from a simplified model for river flood damage assessment, results suggested that uncertainties in predicting future economic damages dominated the uncertainties in the physical regime, i.e. sea level rise. In terms of planning current projects, preliminary indications are that adverse impacts of sea level rise and climate change occur too slowly and too far into the future to have much influence on the choice of the type and scale of coastal protection projects. This is especially true given the higher interest rate used in present value calculations. Therefore, in the foreseeable future, the U.S. is likely to rely on nonstructural and land use management solutions that are administered by state and local agencies.

INTRODUCTION

The impending threat of climate change and sea level rise has brought calls from various sectors for governmental institutions among which is the U.S. Army Corps of Engineers (Corps), to prepare for this slowly evolving and uncertain natural hazard. In its responsibility for managing various aspects of a diverse water resources and shoreline protection program, the Corps recognizes that its activities are likely to be affected by the hydrologic, meteorologic, and oceanographic consequences of global warming and expected climate changes (Figure 1). One response has been the formal introduction of risk analysis to aid the evaluation and selection of alternative plans and project components to deal with natural hazard extremes and the mitigation of their social and economic consequences. Risk

---

[1]U.S. Army Corps of Engineers, Institute for Water Resources, Casey Building, Fort Belvoir, VA 22060.

[2]Clark University, Center for Technology, Environment and Development. Worcester, MA 01610

NATO ASI Series, Vol. G 29
Water Resources Engineering Risk Assessment
Edited by J. Ganoulis

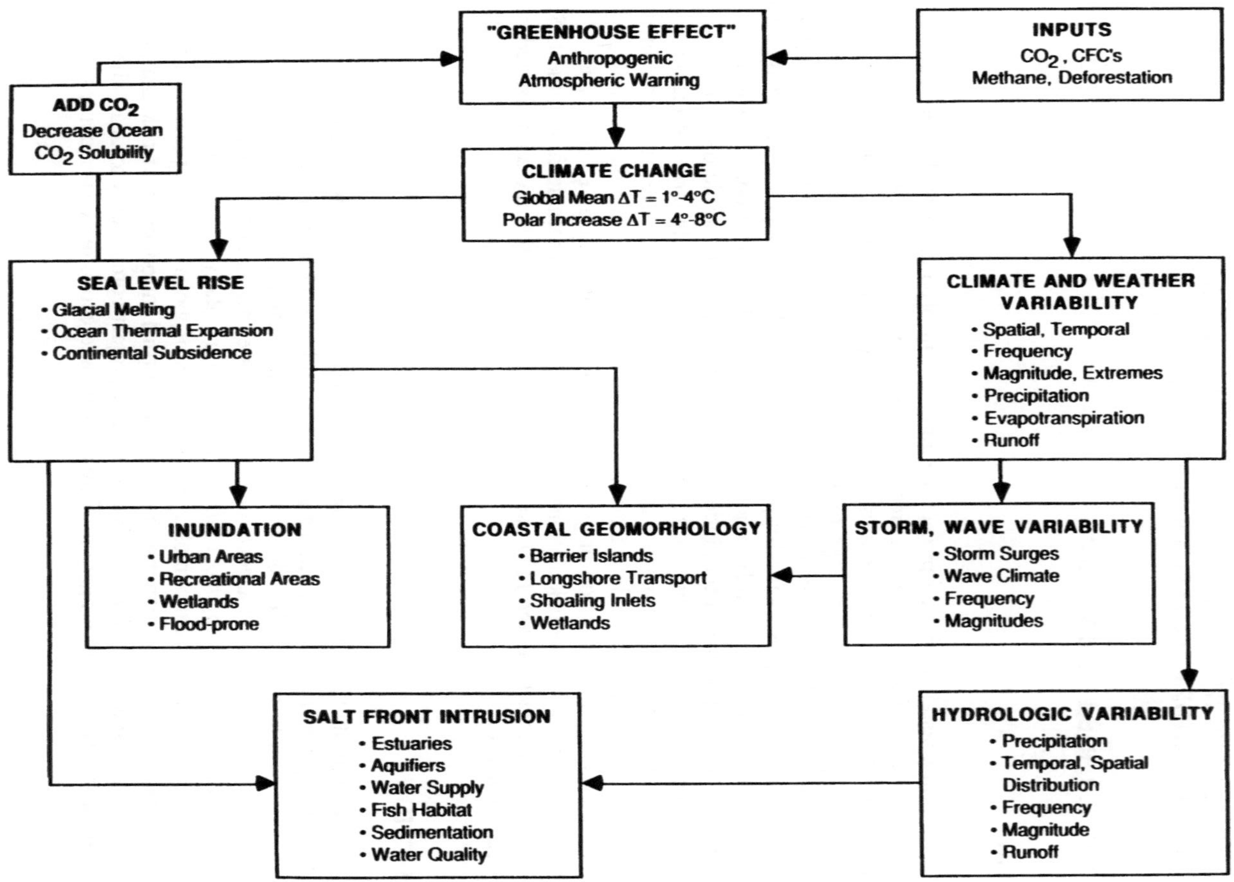

FIGURE 1. Physical Consequences of Climate Change

Source: Moser and Stakhiv, 1989

analysis has been transformed from an ad hoc set of procedures to an integral part of existing multiobjective evaluation procedures that guide Federal water resources development. These procedures are based on a body of social, economic, environmental, planning and decision theory literature that has been developed over the last 50 years.

It should be understood, however, that even without sea level rise or changes in the climate/weather regime, the prediction of future physical shorelines changes is in itself a problematic and highly uncertain enterprise. Two classes of predictive models commonly in use today, the longshore and cross-shore transport models, are probably accurate to no more than $\pm$ 30-60 percent, "providing there is no storm-induced interruption to the assumed prevailing conditions" (National Research Council, 1990; pg 130). Added to this is the large degree of uncertainty in the economic development sector, upon which the basis for justifying alternative coastal protection measures exists. In other words the benefits of shoreline protection depends on the economic damages prevented from a combination of (a) flood inundation due to higher water tables and storm surges; (b) wave damage; (c) wind damage and (d) beach erosion.

There are many uncertainties and unknowns associated with the physical effects of greenhouse warming and anticipated sea level rise. These compound the existing difficulties of planning under conditions of uncertainty regarding future growth, economic development and environmental effects. Sea level rise alone, even with the present weather regime, will logically cause the landward retreat of the shoreline following the Bruun rule (Figure 2) (Schwartz, 1967). Any change in the intensity and frequency of storm events, the direction of which is presently largely speculative, will either accentuate or diminish the physical consequences of sea level rise.

Storms produce damage to property in coastal areas from several causes. In addition to direct wind related damage, which is ignored here, a storm typically produces a surge that raises the water surface elevation well above the mean high tide level. This wind driven surge may be sufficient, even in the absence of waves, to produce flooding in low lying areas from standing water. In addition to the surge, storms also produce larger waves. Property subject to direct wave attack can suffer extensive damage to the structure and contents as well as foundation erosion threatening the stability of the entire structure. Storms also produce at least temporary physical changes at the land-water boundary by eroding the natural beach and dune that serve to buffer and protect the shore and property from the effects of storms. Increased wave energy during storms erodes the beach and carries the sand offshore.

Several components of coastal project evaluation are stochastic so that the evaluation can be computationally complicated. For instance, the damages from storms are dependent on characteristics, described in probabilistic terms, such as intensity, duration, wind direction, and diurnal tide level. Since these characteristics, in turn, influence the level of storm surge and significant wave height, these two direct factors in storm damage are also stochastic. Sea level rise can be considered as a shift in the base elevation for measuring storm surge and wave height.

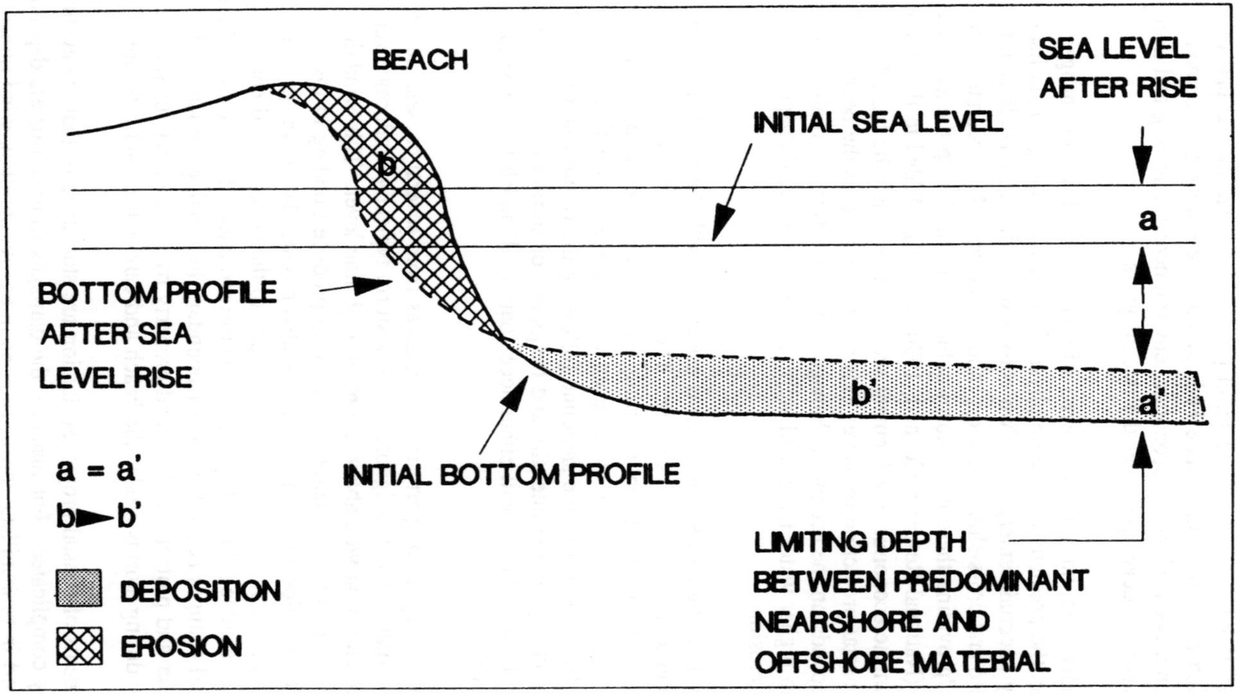

FIGURE 2. Brunn Rule

Source: Adapted from M.L. Schwartz (1967).

AN OVERVIEW OF ECONOMIC EVALUATION PRINCIPLES

The Federal government has a long history of planning coastal protection projects. By providing protection against the hazard, efficiency gains can be achieved resulting in an increase in the national output of goods and services, termed National Economic Development (NED) benefits. There are also additional regional and local economic gains that result from the transfer of economic activity from some other location. The identification and measurement of the national efficiency gains follows benefit-cost analysis (BCA) procedures developed, to a significant extent, to evaluate the national economic implications of Federal investments in what are inherently local water resources projects. These procedures are codified in the Economic and Environmental Principles and Guidelines for Water and Related Land Resources Implementation Studies (P&G), (WRC, 1983).

A fundamental question that climate change and sea level rise poses for society is how to effectively cope with the changes that are slow and appear irreversible. Many Federal, state, and local institutions are currently debating the possible strategies and specific measures for anticipating the most severe consequences and adapting to the inevitable changes. The Corps of Engineers, as one of these institutions, can effectively deal only with protective measures. This paper deals with how the Corps' economic evaluation principles and decision rules influence the choice of a particular shore protection measure in a risk analysis framework. There are many other effective alternative management measures, residing within the responsibilities of the states and local communities, that should be considered strongly in adapting to sea level rise. The range of public measures to mitigate the potential hazards to life and property from sea level rise and climate change will be the same as those available today under "normal" conditions. The probable difference will be that the emphasis on alternative management strategies will change to reflect the reality that the baseline condition is changing. Thus, it is likely that shore protection strategies would shift from protective measures such as groins, bulkheads, sea walls, and beach nourishment to adaptive land use modification measures. These adaptive strategies would consist of limiting investments in, and subsidies to hazard prone areas through regulation and disinvestment strategies such as transferable development rights, the use of financial incentives and tax deductions and flood and storm damage insurance programs.

Preventive and adaptive responses to sea level rise are generally the same as those considered for existing coastal erosion problems. These can be classified into four approaches or options:

hard engineering options--bulkheads, groin fields, sea walls revetments, and the elevation of the shoreline and structures,

soft engineering options--beach nourishment and dune stabilization,

management options--set-back requirements, development restrictions, land use controls, flood insurance, and

passive options--no systematic response, allowing coastal erosion to occur with private attempts to protect individual property.

Coastal protection projects, like all investments, involve spending money today to gain predicted benefits in the future. In addition, many types of projects, especially beach nourishment and maintenance, require a commitment to future spending to maintain the project. This future aspect requires that the current and future dollar costs and benefits must be commensurate or compared in a common unit of measurement; typically in terms of their present values or the average annual equivalent of the present values. Therefore, the discount rate used to determine the present values influences the economic feasibility of alternative projects. As is well known, large discount rates reduce the influence of future benefits and costs on present values. High interest rates generally favor the selection of projects with low first costs but relatively high planned future maintenance expenditures over those with high first cost but low future maintenance expenditures.

The standard for the identification and measurement of economic benefits from a water resources investment project is individual willingness to pay (WTP). For coastal protection projects, this value can be generated by a reduction in the cost to a current land use activity or the increase in net income possible at a given site. A project generates these values by reducing the risk of storm damage to coastal development. Conceptually, the risk from storms can be viewed as incurring a cost to development, i.e. capital investment, at hazardous locations. Thus, the cost per unit of capital, invested at risky locations, is higher than at risk free locations. Economic theory predicts that the risk of storm damage results in less intensive development and lower land values compared to the same location with a lower risk or otherwise equivalent, risk free locations. The risk component of the marginal cost of capital is composed of the expected value of the per unit storm damages plus a premium for risk. This risk premium results from the attitudes or preferences of the individual decision-maker toward risk. If the individual is risk averse, the risk premium is positive indicating that capital must earn a return to cover not only expected storm damages but also to compensate the capital owner for the risk.

THE EVALUATION FRAMEWORK

The establishment of a baseline or "without" project condition for incorporating physical effects and socioeconomic impacts into coastal protection measures is crucial in any evaluation. In the deterministic approach, a single forecast defines physical, developmental, cultural, environmental, and other changes. These changes are considered to occur with certainty in the absence of any systematic adaptive measure. This approach does allow, however, for individual property owners to respond to storm and erosion threats through the construction of protective measures or by property abandonment. The baseline requires assumptions to determine at what future time period these responses would occur. In the risk analysis approach this simplistic determination of the "without" condition is modified to incorporate uncertainties about storm frequencies, the distribution of wave heights, and the geomorphologic changes and property losses produced by storms and waves.

The variable sea state is measured as the sum of the level of storm surge plus the significant wave height. The level of storm surge is a function of the storm characteristics so that the annual probability of storm surge exceeding some level depends on the annual probability of storms that can generate a surge of that level or greater. The distribution of wave heights from a storm is not independent of the level of storm surge. (Bakker and

Vrijling, 1981) One can consider the storm surge to shift the probability density function for significant wave heights. The final component for incorporating classical risk-analysis techniques within the benefit evaluation framework for storm protection, specified by the P&G, is to estimate the "with" and "without" project future economic development and land values.

Without a public coastal protection project, property owners are presumed to repair structural losses with the damages from storms presumed to be capitalized into the value of the land. In addition, property owners are assumed to construct individual protection structures when the costs are less than the value of the preserved property and the avoided expected damages to improvements. "With" the project, landowners realize increases in economic rental values of land at protected locations. This rental value increase is typically considered to be equivalent to the annualized expected present value of avoided property losses with the project or the avoided costs of individual protection structures. The increase in rental value of land is location based resulting from a reduction in the external costs imposed by storms and it represents a national economic development (NED) benefit as required under P&G. It is this type of economic benefit that is compared to project costs to determine the economic feasibility of any proposed Federal project. The time stream of these benefits will reflect the stochastic nature of storm events. An important additional consideration stems from the chronological order of storms and damages. A large storm may result in damages that are so extensive that the buildings are not or cannot be rebuilt. Therefore, succeeding storms will inflict smaller losses if preceded by large storms.

The general description of the evaluation framework does not explicitly incorporate long term shoreline erosion. In many situations, the observed shoreline retreat is simply the by-product of the storm history at a particular location perhaps in combination with relative sea level rise. In other special case, coastal structures such as groins and jetties may induce sand starvation in down-drift areas. Typically, these are incorporated in project evaluation based on the average rate of historical shoreline retreat. For purposes here, any shoreline retreat is treated as storm induced.

Benefits produced by a project depend on the type and scale of the protection project. Even where two alternative projects have the same scale, as defined by the design level of storm protection, (e.g. 100-year storm or probable maximum hurricane), the impact on benefits will differ depending on the magnitude of residual losses from storms that exceed the level of protection. Consider two types of projects; one a structural type, such as a sea wall, and the other a maintenance type, such as beach and dune restoration, stabilization, and periodic nourishment. For a given level of protection, the sea wall is likely to result in different residual storm losses compared to beach and dune restoration, stabilization, and nourishment.

Once the alternative formulated plans are evaluated in economic terms, the expected net benefits can be calculated. Following the project selection criteria in P&G, the recommended type and scale of plan should be the one which "reasonably maximizes" net NED benefits. This is a key conceptual point in risk analysis: the net benefits decision rule for selecting the economically optimal project simultaneously selects the degree of protection and level of residual risk bearing. Thus, by varying the scale of project for each type, a benefit function can be derived for each type of project. Deviations from the NED plan can be recommended to incorporate risk and uncertainty considerations in addition to the explicit risk analysis used in the economic evaluation. These could be considerations for human health and safety or non-monetized environmental concerns (Moser and Stakhiv, 1989).

CLIMATE CHANGE AND SEA LEVEL RISE

Thus far the evaluation and selection of Federal coastal protection investments has assumed stationarity of climate and mean sea level. The underlying physical parameters and relationships yielding the historically observed distribution of sea states have been assumed to be constant. Most forecasts for sea level rise suggest that it is not an immediate problem for coastal development. In addition, there is a wide variation in the estimates of the rate of sea level rise. For instance, a recent National Research Council report (NRC, 1987) proposed three scenarios for sea level rise, with the absolute increase rate of sea level rise varying from 0.5 to 2 meters by the year 2100. That estimate has been refined by the U.N. Intergovernmental Panel on Climate Change (IPCC, 1990). Their estimate ranges from 30 to 110 cm by the year 2100.

Sea level rise can be included in the evaluation of planning alternatives as a mean increasing, variance preserving shift in the probability density function of storm surge. This results in an increase in the site cost of capital "with" and "without" each alternative. Note that a rise in sea level will likely have different impacts on the site cost of capital for different types of planning alternatives. The incorporation of higher future sea levels in project evaluation will favor the recommendation of larger, structural type projects over maintenance type. Two additional considerations temper this conclusion, however. First, building higher levels of protection than is economically efficient given the current mean sea level implies that current net benefits are sacrificed. The larger levels of protection are economically efficient only at higher mean sea levels that may or may not occur in the future. Second, since the increase in net benefits for a larger scale project occur in the future, the discounting process necessary to determine the present values of benefits and costs will reduce the influence of these future benefits on the determination of the appropriate project scale.

One way of presenting the economic tradeoffs between design project scales that have different time streams of future net benefits is to determine the breakeven discount rate for the projects. The breakeven discount rate is the interest rate that equates the present values of two streams of future net benefits. The present value of net benefits as a function of the discount rate is shown in Figure 3 for two alternative projects A and B. Project A provides the economically efficient level of protection today ignoring sea level rise while B provides a higher level of protection in anticipation of climate change and sea level rise. Notice that the present value of future net benefits for project B exceeds the present value of future net benefits for project A if the discount rate is less than approximately 3.2 %. This compares to the 1989 U.S. Federal discount rate of 8 7/8% used for project evaluation. In general, because sea level rise and its effects occur relatively far in the future, incorporating even a high forecast of future sea levels in the evaluation of project scale will have little impact on the economically efficient project design. Additionally, the uncertain prospect of the amount of sea level rise may support projects that are more flexible and that can easily incorporate staging of project increments as sea levels change.

Similar to the above analysis, project evaluation could incorporate the effects of forecasted climate change, expressed as a change in the frequency of storm events, through the calculation of expected values and sensitivity analysis. One hypothesis about the effect of climate change is that in many locations the frequency of severe storms will increase over time. Since recurring maintenance expenditures depend primarily on the frequency of storms, climate change that increases storm frequency will shorten the time period between these expenditures. This would tend to favor structural type projects since they have lower maintenance costs. Again, a perhaps overriding consideration for Federal projects is the impact of discounting on these future costs and their influence on project type and scale. Thus, even though climate change may result in a dramatic increase in total lifetime project costs, most of the increase occurs beyond the first 15 to 20 year of project life which has little influence on the present value of net benefits (Moser and Stakhiv, 1989).

In summary, the key issues affecting evaluation in order of relative importance, are the following:

(i) the dominant effect of the discount rate in benefit-cost analysis and computation of expected annual damages,

(ii) uncertainty of future economic development and storm damage assessment,

(iii) establishing the proper baseline for evaluating physical effects and socioeconomic impacts,

(iv) uncertainty of storm frequency and wave regime under climate change, and

(v) the rate and magnitude of sea level rise.

AN IDEALIZED CASE STUDY APPROACH

Beyond the conceptual issues discussed thus far, is the possibility of adapting elements of a feasible method for an integrated uncertainty analysis of sea level rise associated flooding that was developed for local flood damage alleviation projects by the Corps of Engineers (Ratick and Du, 1988). This method still requires considerable adaptation and refinement for the complex problem of sea level rise, but it serves as a useful starting point for combining uncertainties in the physical properties associated with sea level rise along with economic analysis uncertainties.

The integrated uncertainty analysis procedure is comprised of three basic steps. The first consists of separating the types of uncertainties confronted in determining flood damages into three independent areas: hydrologic uncertainty, represented by flood frequency-discharge relationships; hydraulic uncertainty, measured by the stage-flow relationship and economic uncertainty, represented by flood stage-damage relationships. Second, using established Corps procedures and analytical models for determining Expected Annual Flood Damages (EAD) (HEC 1984), specific discrete changes are made to the mathematical relationships in the areas that represent the effects of uncertainty on the levels of precision for inputs to these models.

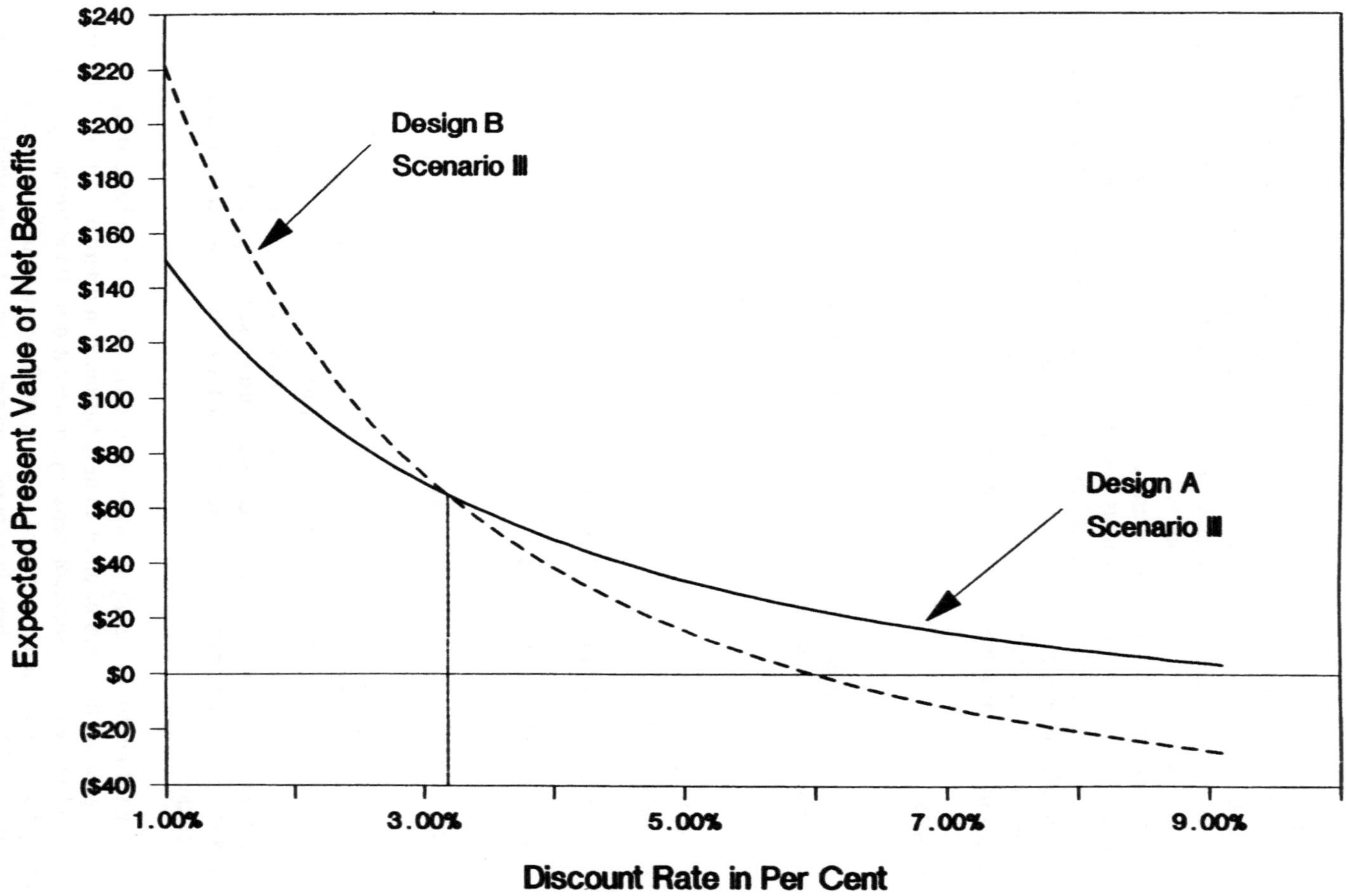

FIGURE 3. Expected Present Value of Net Benefits as a Function of Discount Rate

The EAD values that result from these calculations are used to develop mathematical response surfaces that represent the EAD values as polynomial functions of discrete changes to model input values (levels of imprecision). The response surfaces will , to a degree, closely replicate the results of running the sequence of HEC models for any input value that are within the limits of the discrete changes. Third, the response surfaces are used in Monte Carlo type simulations with probability distributions for input changes that yield a probability distribution of EAD values.

Assessing the effects of uncertainty on coastal flooding, however, represents a more complex undertaking. In addition to the uncertainties in forecasting damages for inland flooding, other elements such as: storm frequencies and intensities; tidal effects; salt-water effects, and subsidence and marsh deterioration, among others. This is made even more complex when the potential effects of global warming are introduced into the analysis. ["National Economic Development Procedures Manual - Coastal Storm Damage and Erosion" (IWR, 1990)].

The major steps in calculating Expected Annual Flood Damages (EAD) for conventional inland flooding can be separated into three independent analytical tasks: (1) determining the frequencies (return periods) for floods of different magnitudes and expressing this in the flood frequency discharge relationship (F-Q); (2) determining the flood stage at different flood discharge levels and expressing this in the flood stage-discharge (S-Q) relationship, and (3) determining damages at different flood stages and expressing this as the flood stage-damage (S-D) relationship. These three components are analytically combined through the use of the HEC-EAD computational frequencies to yield the EAD values.

The uncertainty/sensitivity analysis procedure involves making simultaneous discrete changes to the values in each of the relationships, using the corresponding EAD values to develop a response surface - a mathematical equation that will closely replicate the output of the EAD computational models - and using this equation in a Monte Carlo type simulation. These steps are detailed below.

(1) Initially the effects of discrete measures of uncertainty (for example, modifying model input values by -10%, -5%, +5% and +10%) on "expected annual flood damages" (EAD) in the flood frequency-discharge (F-Q) relationship, the flood stage-discharge (S-Q) relationship, and the flood stage-damage (S-D) relationships are analyzed independently. This is accomplished by running the HEC flood damage computational models a number of times with these fixed changes to the input data.

(2) The combinatorial effects of these dimensions of discrete uncertainty measures on the EAD are then obtained. (Figure 4) Each of the relationships (F-Q, S-Q and S-D) are logically combined to yield the flood frequency-damage curve that, when integrated over all frequencies, produces the EAD measure (see "Expected Annual Flood Damage Computation: User's Manual," The Hydrological Engineering Center and Water Resources Support Center, 1984, for a detailed description of this process).

(3) Response surfaces, mathematical equations with the EAD value as the dependent variable and the specific imprecision measures for each of the dimensions as the independent variables, are obtained by fitting a non- linear multi-dimensional surfaces to the results of step 2. This response surface estimates the EAD output values that would be expected if the HEC computational models were run for any combination of input imprecision values in the F-Q, S-Q and S-D relationships within the limits of the discrete changes that were made.

(4) The derivation of the response surface allows the use of Monte Carlo simulation to assess the affect of uncertainties in each of the relationships on EAD values. The equations for the EAD values as a function of levels of imprecision are input into a Lotus spreadsheet, and the Monte Carlo simulation is performed using an add-on software called "@ Risk". Levels of imprecision are expressed with probability density functions and used in the uncertainty simulation (Figure 5).

(5) Implementing steps 1 through 4 will yield a derived probability of EAD values resulting from the uncertainty assumptions.

Discrete Changes to the Frequency - Discharge (F-Q) and Stage - Discharge (S-Q) Relationships

The following example described below involves uncertainty analysis for two dimensions considered simultaneously, the frequency-discharged (F-Q) and the stage-damage (S-Q) relationship, the stage-damage relationship is kept as reported in the HEC manual (HEC, 1984). For this example five discrete changes to the F-Q and S-Q relationships were tested, -10%, -5%, normal, +5%, and +10%. Table 1 shows the results of running HEC-EAD; models with those combined changes for the base year (1976). Modifications to the stage- discharge (S-Q) curve resulted in very large changes to expected annual flood damages. The calculated EAD value for 1976 resulting from the normal input curves for both relationships (0%) change) is $579,810, with a range from a low of $95,860 when both relationships are decreased by 10%, to a high of $1,889,290 when both are increased by 10%.

The results were used to estimate the mathematical equation of the response surface, i.e. EAD values as a function of F-Q and S-Q imprecision (variables FQI and SQI respectively). For the Monte Carlo simulation the imprecision level variables (FQI and SQI) were each was sampled 3000 times. Figure 6 shows the resultant probability distribution of response surface. EAD values appear to follow a non-symmetric distribution for EAD values, reflecting the skewed nature of response surface in the S-Q dimension having exponentially increasing EAD values related to linearly increasing SQI values. The expected value for EAD is $608,980 and the standard deviation is $51,840. The 95% confidence interval has EAD values ranging from $510,860 to $685,720. EAD values and Figure 7 show the cumulative pdf for S-Q and F-Q uncertainties using the

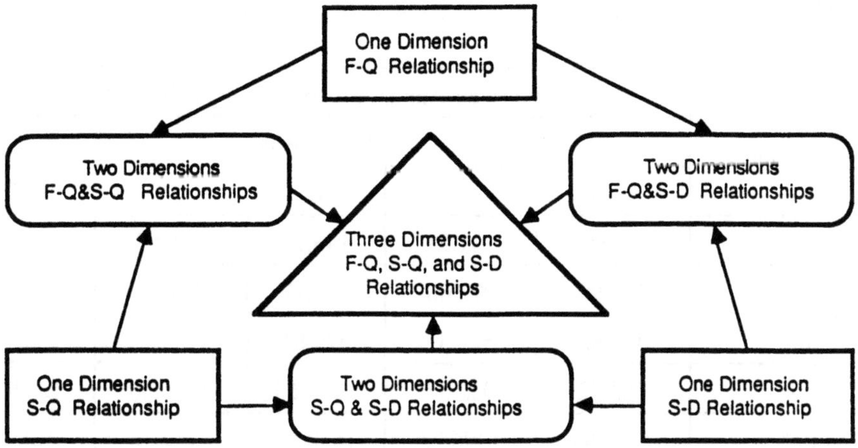

FIGURE 4. Combinatorial Structure of Discrete Uncertainties in the F-Q, S-Q and S-D Relationships

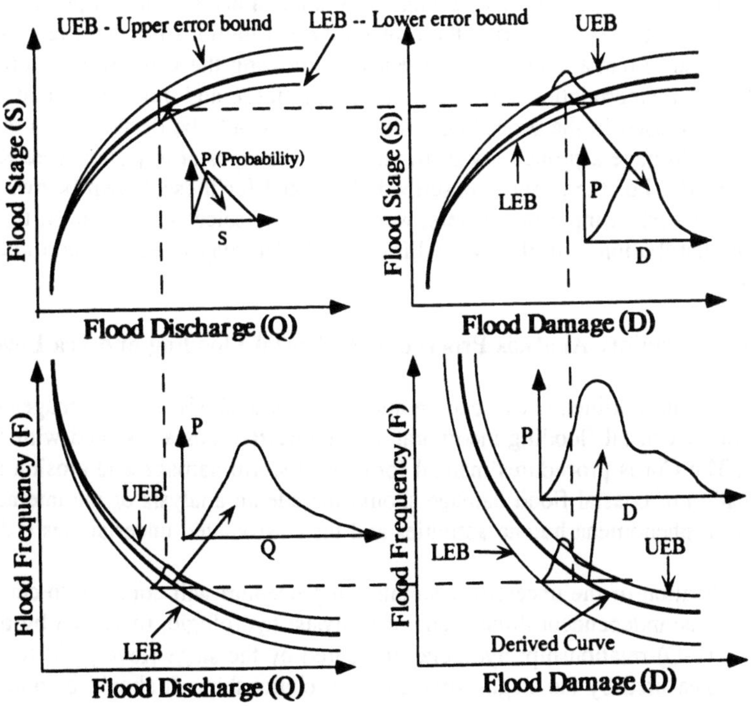

FIGURE 5. Uncertainty and Flood Damage Calculation

Table 1. The EAD Values Under Discrete Values of F-Q and S-Q Imprecision

S-Q Imprecision	Expected Annual Flood Damage ($000 1974)				
	F-Q Imprecision				
	-10%	-5%	0	+5%	+10%
+10%	95.86	131.93	145.16	158.68	172.54
+5%	245.90	272.08	299.17	327.08	355.72
0	476.62	527.37	579.81	633.51	688.46
+5%	863.42	952.57	1043.65	1136.30	12430.14
+10%	1363.98	1496.23	1630.51	1752.18	1889.29

Each of the relationships needed to assess expected annual flood damages are themselves the result of a number of analytical steps, each of which may be characterized by a variety of sources and types of uncertainties. The uncertainty simulation procedure described above may be used to derive the probability distributions for these relationships. For example, the stage-damage relationship relies on knowing the value of structures and contents at different uncertainties inherent in calculating these and other important values. The stage-damage curve used in the procedure outlined above would be derived from these estimates and would have a probability distribution for damages at any stage (see Ratick and Du, 1988, pp 90-96). For the analysis described above and for those shown below, probability distributions for uncertain input parameters, were arbitrarily selected. However, the choice of the appropriate distribution is itself a very difficult task that may involve expert knowledge and judgement.

Applying the Uncertainty Analysis Procedure to Coastal Flooding and Sea Level Rise

The procedure outlined above can be applied to an analysis of the damages due to slow inundation or coastal flooding incorporating the uncertainties associated with forecasting sea level rise. The Corps procedures manual for coastal storm damage and erosion suggests that the evaluation of coastal flood damage should include an analysis of the interaction of multiple physical phenomena human activities and their associated uncertainties (IWR, 1990).

The application of the uncertainty simulation procedure will continue to center on the assumption of three independent dimensions of analysis: hydrologic, measured by the frequency-flow (F-Q) relationship; hydraulic, measured by the stage-flow (S-Q) relationship, and economic, measured by the stage-damage (S-D) relationship. It should be noted that hydrologic, hydraulic, and economic dimensions of analysis for coastal regions are often very different from those used for inland flood damage assessment. They are generally derived through the analysis of complex, interactive and sometimes synergistic physical processes in coastal regions: coastal storms, coastal tides, hurricane surges, oceanic currents, erosion,

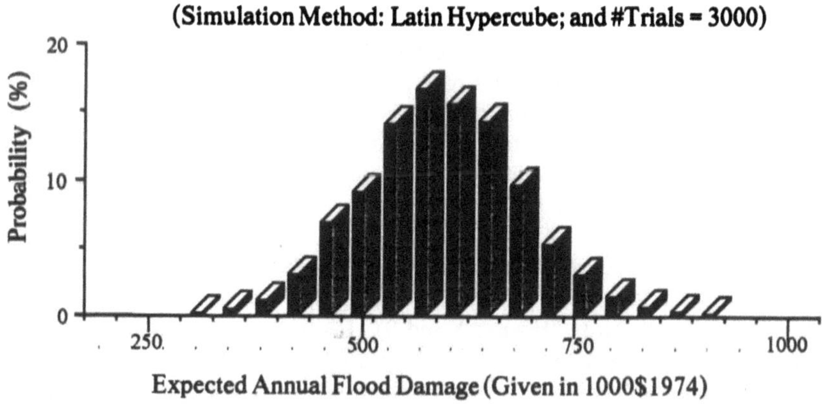

FIGURE 6. Probability Density Distribution of EAD Values

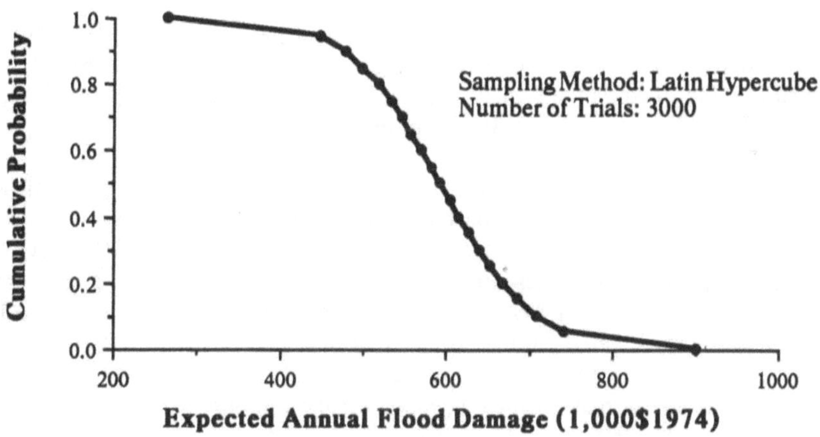

FIGURE 7. Cumulative Probability Distribution of EAD Values

among others. For example, sea level rise, coastal storms, hurricane surges, and coastal tides may substantially increase the flood stage and modify flood frequencies, and erosion and salt-water intrusion may significantly contribute to flood damage estimates in coastal areas. Those hydrologic, hydraulic and economic relationships are often obtained through use of causal or statistical models, or survey techniques. As an example, the Surge III model can provide information on the flood frequency - stage relationships that result from storm surges (US Army Corps of Engineers, 1989).

For the analysis the increase in mean sea level rise (SLR), has been assumed to be a random variable having a lognormal probability distribution with a mean value of 3 ft.(1 m) and a standard deviation of 0.6 ft. (Figures 8 and 9). Clearly this is a significant increase that most studies suggest may take approximately 100 years to occur. This large an increase has been chosen to illustrate both the type of impact increased sea levels can have on flood damage estimates and to enable us to distinguish the contribution of this phenomena and its associated uncertainties to coastal flood damage assessments. This does not mean to imply that such a large change will occur, or will occur rapidly enough to affect damage calculations in the manner presented below. A gradual increase in mean sea level will allow for adaptive responses by society that can change the stage-damage relationship significantly.

Case A: Coastal Flooding Combined with Sea Level Rise

Sea level rise can significantly affect the flood stage-discharge relationship. An increased sea level would result in an increased flood stage at given levels of flood discharge values (Figure 9). This, in effect, will shift the S-Q curve upwards to reflect the increase in flood stage associated with an increase in sea level. The actual size of this shift is dependent upon the forecast of sea level rise (SLR), which is considered here as a lognormally distributed random variable (the probability side graph in Figure 10). This shift is incorporated into the flood damage analysis procedure described previously. Again, the data provided in HEC-EAD manual (HEC 1984) are used for the analysis.

Table 2 shows the calculated EAD values in response to <u>discrete</u> changes (from -10% to +10%) input values in the F-Q relationship, discrete estimates of the magnitude of the upward shift in the F-Q curve (SLR values) when modified by ±5% and 10%, and modifying the S-Q input values by -10%. These discrete values are used to fit an EAD response surface as given by the equation. Here EAD values are a function of the imprecision in the F-Q, S-Q, S-D relationships as well as in the estimates of SLR.

As before, a mathematical equation for the response surface is estimated using results such as these, and the Monte Carlo uncertainty simulation is undertaken with the independent variables FQI, SQI, SDI and SLR are represented as random variables with the following probability distributions: -a normal distribution with a 10% standard deviation for F-Q imprecision a normal 10 distribution with a 1 % standard deviation for S-Q imprecision; a normal distribution with a 5% standard deviation for the S-D imprecision; and SLR represented as a lognormal distribution with a mean of 3 ft and a standard deviation of 0.6 ft., as mentioned previously.

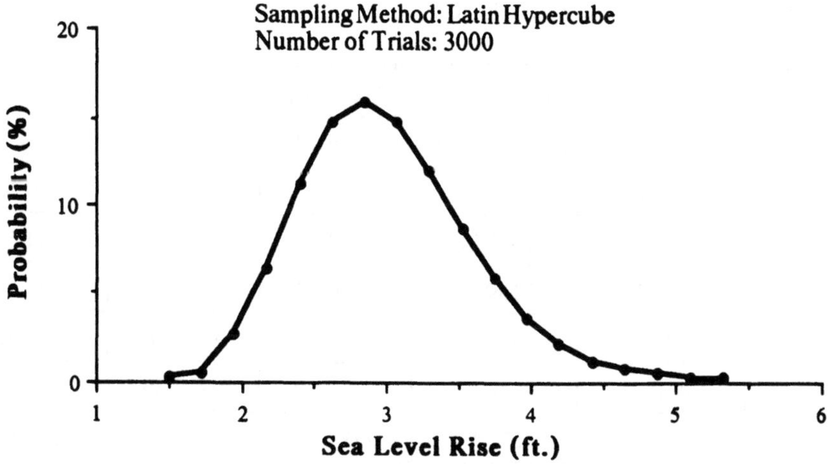

FIGURE 8. Probability Density Distribution of Assumed Sea Level Rise

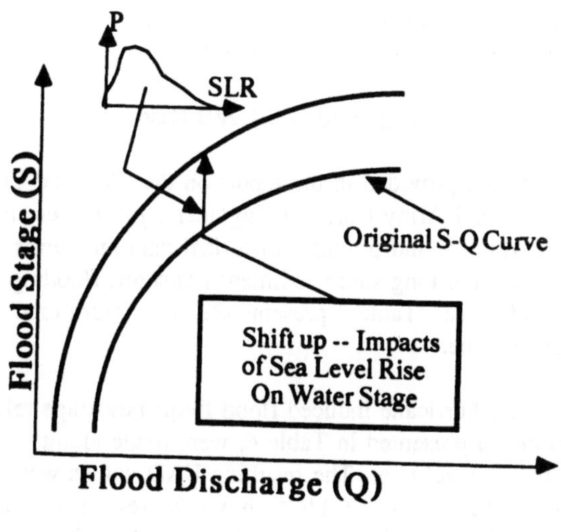

FIGURE 9. Influence of SLR on Stage-Discharge Relationship

Table 2. Discrete Sensitivity Analysis on Uncertainties for a Normal S-Q Curve

Condition: F-Q Imprecision	Expected Annual Flood Damage ($000s) Imprecision in Forecasts of Sea Level Rise				
	-10%	-5%	0	+5%	+10%
-10%	1907.82	1995.78	2082.84	2168.90	2254.69
-5%	2077.15	2169.06	2260.13	2350.40	2439.57
0	2244.68	2340.77	2435.75	2529.90	2623.36
+5%	2410.05	2510.31	2609.89	2708.56	2806.87
+10%	2574.92	2679.89	2784.25	2888.05	2991.67

The result of the Monte Carlo simulations are shown in Figures 11 and 12 representing the probability distribution and cumulative probability density functions for the resultant EAD values under other assumptions about uncertainty and sea level rise. The resultant distribution of EAD values had a mean of $2,421,000 and a standard deviation of $519,000, with a possible range of value of $3,530,298 from $1,260,525 to $4,790,824. The shape of the probability distribution for EAD values reflects the lognormal assumption for SLR, having a longer "tail" for EAD values larger then the mean. The 95% confidence interval for expected annual flood damages can be determined from the cumulative probability density function and ranges from $1,669,620 to $3,333,820.

Case B: Coastal Flooding, Storm Surge and Sea Level Rise

For this example the data provided in the report on the "Atlantic Coast of Maryland Hurricane Protection Project,"(US Army Corps of Engineers, 1989) were used. The project report describes the site conditions, and provides historical data on storms and their effects on Ocean City, Maryland erosion and long shore sediment transport, flooding and flood damage, and sea level rise in the study area. Table 3 presents the "maximum total water level" - frequency relationship for the area.

Discrete changes to the hurricane induced flood frequency-stage relationships and flood stage-damage relationship presented in Table 4, were made along with the discrete changes to our forecasts of sea level rise. The resulting EAD values were computed as before using the HEC-EAD model (HEC 1984). Table 5 shows the resultant EAD values for discrete changes to the stage-damage relationship an forecasted sea level rise, assuming no change to the frequency-stage relationship.

For this case, the range of discrete changes in mean sea level rise (SLR) were increased to ±25% and ±45%. Since coastal flood damage analysis usually focuses on sea level changes (flood stage) rather than flood discharge, the uncertainty procedure had to be

Table 3. Frequency - sea level rise relationship caused by a hurricane

Frequency (%)	20	10	5	2	1	0.4
Maximum total water level, ft.	8.8	9.6	10.6	11.2	11.9	12.8

Source: US Army Corps of Engineers, 1989. "Atlantic Coast of Maryland Hurricane Protection Project", Feasibility Report, Baltimore District

modified to consider only frequency-stage (F-S) and stage drainage (S-D) relationships. This is done in the HEC-EAD model by using a dummy default value for flow (Q) in the frequency flow (F-Q) and stage-flow (S-Q) relationships.

Using the same probability distributions for uncertainties given previously yields the derived probability density distribution and cumulative probability function for EAD values shown in Figure 13 and 14. The derived probability distribution of EAD values has an expected value of $3,408,653 compared to the original figure of $3,052,340 (Table 5), a standard deviation of $1,728,564 with a possible range of values from $786,230 to $12,080,080. The shape of the probability distribution for EAD values again reflects the lognormal assumption for SLR. The cumulative probability density function shows the 95% confidence interval for expected annual flood damages to range from $1,426,310 to $6,694,570.

CONCLUSIONS

Coastal protection projects can incorporate forecasts of sea level rise and storm frequency changes due to climate change through the application of risk and uncertainty analysis techniques. The incorporation of these forecasts is not a trivial matter, but well within the probabilistic analyses currently employed to estimate project benefits and costs for coastal projects. When the effects of sea level rise and climate change occur in the future, structural projects of larger scale than that warranted under current sea level and storm frequencies would offer greater benefits than those designed for the present conditions. In addition, sea level rise and climate change which increase recurring project maintenance costs tend to favor structural type projects. However, given the rather high discount rate used by the U.S. Federal agencies, it is not likely that the benefits will outweigh the dramatically increased costs for what is proposed as an anticipatory protection strategy.

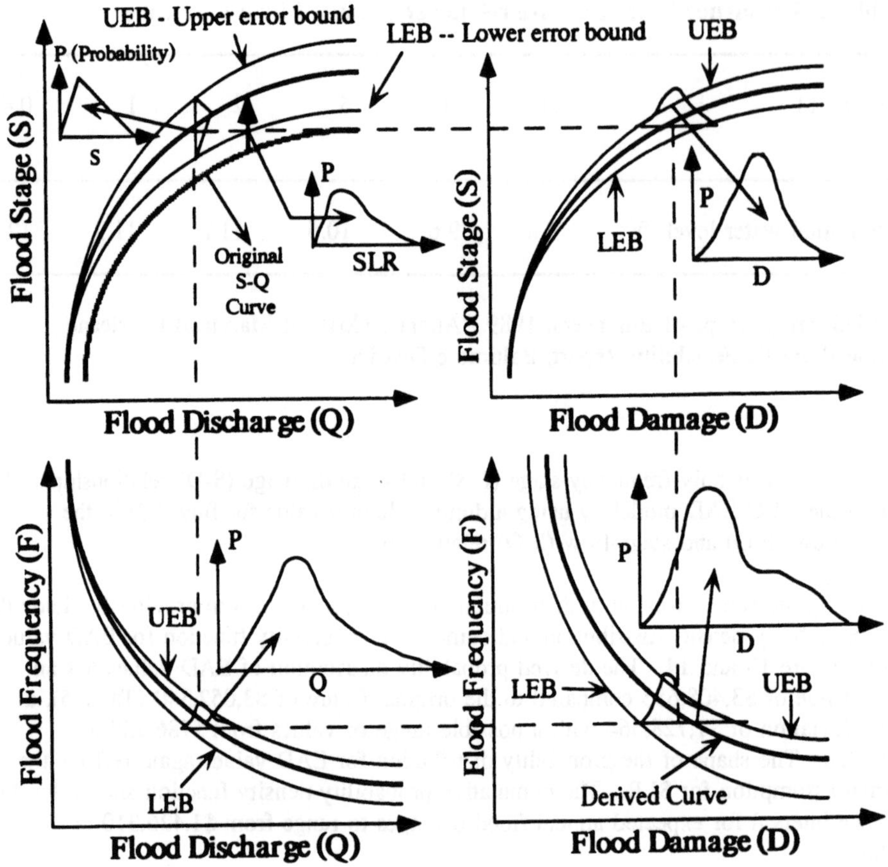

FIGURE 10. Uncertainty and Coastal Flood Damage Calculation

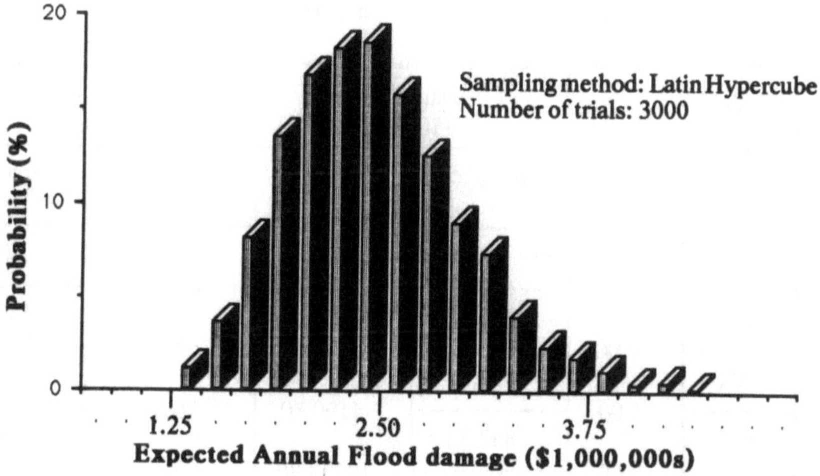

FIGURE 11. Probability Density Distribution of EAD Values Incorporating the F-Q, S-Q, S-D, and SLR Uncertainties

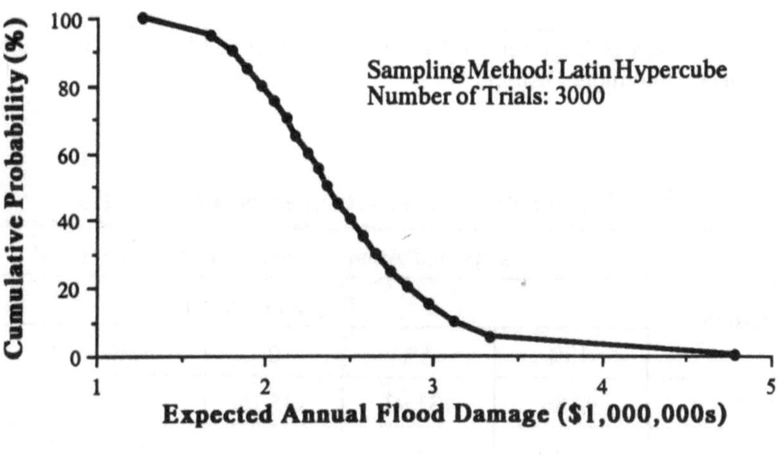

FIGURE 12. Cumulative Probability Distribution of EAD Values Incorporating the F-Q, S-Q, S-D, and SLR Uncertainties

Table 4. Average annual damage for existing conditions at reach 1.

Stage (ft)*	Frequency		Damage ($000)		Annual Damages ($000)	
	(%)	% change	at stage	average	Incremental EAD	Cumulative EAD
11.89	33.00		0.00			0.00
		13.00		1175.40	152.80	
11.89	20.00		2350.77			152.80
		10.00		2442.40	244.20	
11.89	10.00		2534.07			397.00
		5.00		4190.70	209.45	
12.38	5.00		5847.34			606.60
		3.00		10339.30	310.20	
13.22	2.00		1438.17			916.80
		1.00		26709.40	267.10	
14.38	1.00		38587.61			1183.80
		0.60		70510.60	423.10	
16.30	0.40		102433.54			1606.90
		0.40		102433.50	409.70	
16.30	0.00		1216.80			2016.60
		0.00		51216.80	0.00	
	0.00		51216.80			0.0

Source: US Army Corps of Engineers, 1989. "Atlantic Coast of Maryland Hurricane Protection Project", Baltimore District.

Table 5. Discrete Sensitivity Analysis on Uncertainties for a Normal F-S

S-Q Imprecision	Expected Annual Flood Damage (000$ 1974)				
	F-Q Imprecision				
	-10%	-5%	0	+5%	+10%
+10%	95.86	131.93	145.16	158.68	172.54
+5%	245.90	272.08	299.17	327.08	355.72
0	476.62	527.37	579.81	633.51	688.46
+5%	863.42	952.57	1043.65	1136.30	12430.14
+10%	1363.98	1496.23	1630.51	1752.18	1889.29

Hence, risk-cost analysis is not likely to yield definitive answers to the problem of choosing adaptive measures to cope with the risk of sea level rise, although it is a valuable method for assessing the relative effects that a range of economic and physical uncertainties might have on recommended options. It has been demonstrated through the use of a simplified model, that uncertainties in flood-stage economic computations dominate any uncertainties in physical phenomena (sea level rise or storm frequency) in the final computation of expected annual damages (EAD). Other considerations which incorporate cultural, social or environmental aspects related to sea level rise may be more important in choosing adaptive measures. Risk based may also approaches remind the analyst that hard engineering options may exacerbate losses by encouraging development and fostering a false sense of security. Hard engineering adaptations to sea level rise, particularly the barrier type, increase the potential for disaster should natural events exceed their designed level of protection. Therefore, decision makers should be wary of engineering solutions with high residual risks.

At the present time, there is considerable disagreement on the amount of sea level rise and the impact of climate change on storm frequency. More importantly, the adverse impacts on storm damages occur too far into the future, given the nature of discounting and the level of the Federal discount rate, to have much influence on the economically efficient type and scale of project recommended today. There is likely to be a greater reliance on non-structural, land use management solutions that require state and local regulatory controls. The uncertainties about the magnitude and rate of change in sea level rise emphasize the need to maintain flexibility and emphasizes the adoption of an incremental approach that preserves options.

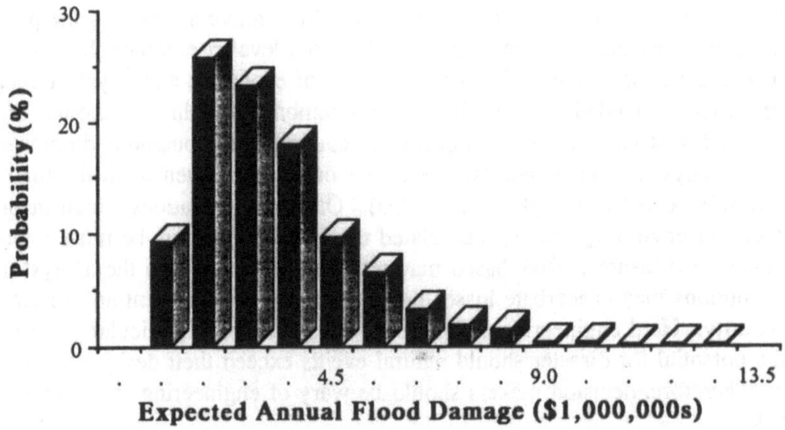

FIGURE 13. Probability Density Distribution of EAD Values Incorporating the F-S, S-D, and SLR Uncertainties

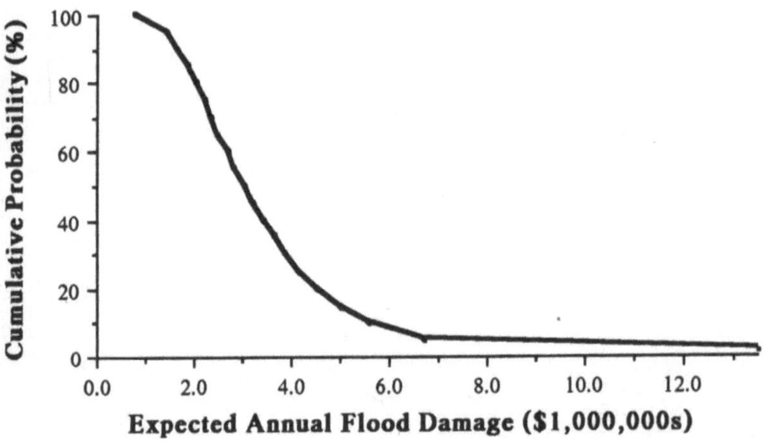

Figure 14. Cumulative Probability Distribution of EAD Values Incorporating the F-S, S-D, and SLR Uncertainties

REFERENCES

Bakker, W.T. and J.K. Vrijling. 1981. Probabilistic design of sea defenses. In: Proceedings of the Seventeenth Coastal Engineering Conference, Vol II. New York: ASCE.

Bruun, P. 1962. Sea level rise as a cause of shore erosion. Journal of Waterways and Harbors Division. 1:116-30.

Hydrologic Engineering Center (HEC), 1984. Expected Annual Flood Damage Computation, User's Manual. U. S. Army Corps of Engineers, Davis, CA.

Institute for Water Resources (IWR), 1990. National Economic Development Procedures Manual: Coastal Storm Damage and Erosion. Draft IWR Report, US Army Corps of Engineers, Fort Belvoir, VA.

Moser, D.A. and E.Z. Stakhiv, 1989. Risk-Cost Evaluation of Coastal Protection Projects With Sea Level Rise and Climate Change. In: Y.Y. Haimes and E.Z. Stakhiv (eds.) Risk Analysis and Management of Natural and Man-Made Hazards. Proc. Engineering Foundation conference, American Society of Civil Engineers. New York, NY. p.222-239.

Intergovernmental Panel on Climate Change (IPCC), 1990. Report of Working Group I. Chapter 9: Sea Level Rise. World Meteorological Organization/United Nations Environment Program.

National Research Council. 1987. Responding to Changes in Sea Level: Engineering Implications. Washington, D.C.: National Academy Press.

National Research Council, 1990. Managing Coastal Erosion. Washington, D.C. National Academy Press. 182p.

Ratick, S. J. and W. Du, 1988. Guidelines for Evaluating Risk and Uncertainty in Local Flood Damage Protection Projects: Draft Report prepared by U. S. Army Corps of Engineers Institute for Water Resources. Center for Technology, Environment and Development, Clark University .

Schwartz, M.L. 1967. The Bruun theory of sea level rise as a cause of shore erosion. Journal of Geology. 75:76-92.

US Army Corps of Engineers 1989. Atlantic Coast of Maryland Hurricane Protection Project. Feasibility Report, Baltimore District.

U.S. Water Resources Council. 1983. Economic and Environmental Principles and Guidelines for Water and Related Land Resources Implementation Studies. Washington, D.C.: U.S. Government Printing Office.

OPTIMAL FLOOD WARNING THRESHOLD AND A CASE STUDY IN MILTON, PENNSYLVANIA

Yacov Y. Haimes and Duan Li
Center for Risk Management of Engineering Systems
University of Virginia
Charlottesville, Virginia, USA
Eugene Z. Stakhiv
Policy and Strategic Studies Division
The Institute of Water Resources
U.S. Army Corps of Engineers
Fort Belvoir, Virginia, USA

Abstract

Flood warning systems are studied in a two-level hierarchical system framework. The interactions between the forecast subsystem and the response subsystem are investigated. The probabilistic evaluation of a forecast system coupled with a stochastic dynamic model of the evolvement of the response fraction in a community reveals that the desire for high present flood-loss reduction must be balanced with the possibility of high future flood loss. Multiobjective dynamic programming is used to select the optimal flood warning threshold. The proposed methodology is applied to the case study in Milton, Pennsylvania.

1. Introduction

Flood control can be provided by either structural or nonstructural measures or a combination of both. Structural flood control measures, such as an increase in dam height, affect the flood-frequency relationship. Nonstructural measures, such as a flood warning system, do not have an impact on the flood-frequency relationship; however, they modify the flood-damage relationship.

The benefits of flood forecasts have been studied and systems approaches to flood forecasting have been pursued by many research scholars for more than twenty years (NACOA 1972; Bhavnagri and Bugliarello 1965; Bock and Hendrick 1966; Day and Lee 1976; Lee et al. 1975; Sniedovich et al. 1974; Sniedovich and Davis 1977). Curtis and Schaake (1988) evaluated flood warning benefits both on a national (or regional) scale and on a specific site problem. Prediction models for loss of life

NATO ASI Series, Vol. G 29
Water Resources Engineering Risk Assessment
Edited by J. Ganoulis

from floods were studied by Lee et al. (1986) and Shabman (1987). Barrett et al. (1988) developed categories for flood warning systems based on types of flood forecasting systems and flood response systems.

Predicting the future behavior of a time-dependent random variable $\{X(t), t \leq T\}$ is a major research task in the theory and applications of stochastic processes. Critical events occur when the level of X(t) crosses a given high level (e.g., flooding level). An alarm is set off when X(t) exceeds a specified threshold level. An alarm system is considered optimal if it detects catastrophes with an acceptable level of probability and at the same time yields a minimum expected number of false alarms (Lindgren 1979, 1980, 1985; and de Maré 1980). The paper by de Maré (1980) indicates that when judging the performance of an alarm system, it is not very interesting to know, in the mean, how close the prediction is to the process X(t); however, it is important for a system to be able to detect catastrophes without causing too many false alarms.

In a series of papers, Krzysztofowicz and his colleagues (Alexandridis and Krzysztofowicz 1985; Ferrell and Krzysztofowicz 1983; Krzysztofowicz 1983a, b, and 1985; Krzysztofowicz and Davis 1983a, b, c, d, and 1984) conceptualized the flood forecast-response process in the form of a total system. This system is defined as a cascade coupling of two components: (1) the forecasting system, which includes data collection, flood forecasting, and forecast dissemination; and (2) the response system, which encompasses decisionmaking and action implementation. Based on the above mathematical description of the physical flood forecast-response process, Krzysztofowicz et al. establish performance measures of flood warning systems.

Paté-Cornell (1986) presents a method for assessing the performance of the forecasting system and human response, given the memory that people have kept on the quality of previous alerts. The trade-off between the rate of false alerts and the length of the lead time is studied to account for the long-term effects of "crying wolf." An explicit formulation of benefits from warning systems is derived under the above considerations.

In this paper flood warning systems are studied in a two-level hierarchical system framework. The interactions between the forecast subsystem and the response subsystem are investigated. Emphasis is placed on exploring the impact of the current selected flood warning threshold on the future response fraction of a flood warning. The probabilistic evaluation of a forecast system coupled with a stochastic dynamic model of the evolvement of the response fraction in a community reveals that the desire for high present flood-loss reduction must be balanced with the possibility of high future flood loss. Multiobjective dynamic programming is used to select the optimal flood warning threshold. The proposed methodology is applied to the case study in Milton, Pennsylvania.

2. The Overall Structure of a Flood Warning System

In general, the overall flood warning system can be viewed as and modeled in a two-level hierarchical system framework (see Fig. 1). There are two subsystems at the lower level. One is the forecasting subsystem, which issues a flood forecast based on hydrological and climatic information. The other is the response subsystem, which includes decisionmaking and action implementation of the community in response to flood warnings. At the upper level, we assume that a regional agency exists whose functions are to set a warning threshold, disseminate a flood warning to the community (e.g., by radio, TV), provide transportation during the evacuation process, and collect statistical data on the degree of the people's response to the warning system in the community. We should emphasize here that the main feature of this two-level model is that the upper level is in charge of issuing warnings based on the multistage

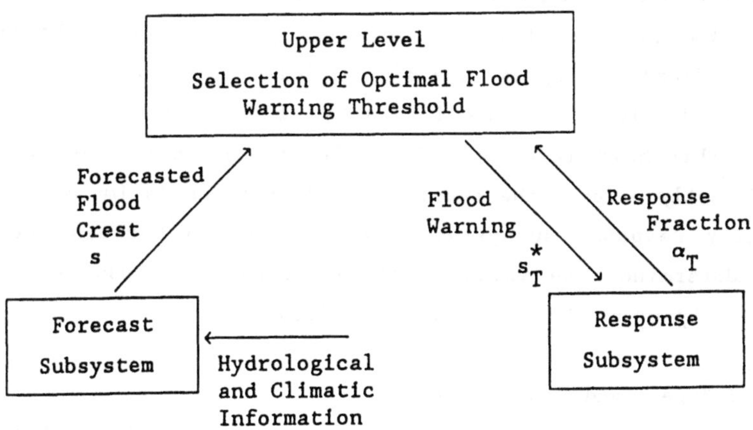

Fig. 1. Multilevel Structure of Flood Warning Systems

decision logic, which encompasses both forecast and response subsystems, while the function of the forecast subsystem is to prepare the forecast for the incoming flood crest. In this chapter, we will focus on the topic of the strategic selection of a flood warning threshold at the upper level based on the investigation of the interaction between the forecast subsystem and the community response subsystem in a multiobjective framework.

Define H to be a random variable which represents the actual flood crest and S to be a random variable which represents the forecasted flood crest. If the prior probability density function of the flood crest is denoted by g(h) and the conditional probability density function of s, given h, is denoted by f(s|h), then the posterior probability density function of h, given forecast s, is

$$f(h|s) = f(s|h)g(h)/k(s), \qquad (2.1)$$

where k(s) is the marginal probability density function of forecast s,

$$k(s) = \int_0^\infty f(s|h)g(h)dh. \qquad (2.2)$$

We assume in this paper that the flood loss is a function of the actual flood crest h. Based on hydrological and climatic information, both observed and historic, the forecast subsystem provides the forecasted value of the flood crest s. From the Bayesian formula given in Eq. (2.2), the prior representation of the uncertainty associated with the flood crest is then updated to a posterior form f(h|s).

The flood warning threshold s^* is defined by the fact that the domain for issuing a flood warning is $\{s \geq s^*\}$. In other words, a flood warning will be issued only when the forecasted flood crest s exceeds a preassigned threshold level s^*. The selection of the flood warning threshold directly affects the evacuation action which is implemented by the community response subsystem. The interaction between the two subsystems determines the value of the expected flood-loss reduction, which is the subject of the sections that follow.

3. Impacts of Type I and Type II Errors

There is no perfect forecasting system, and thus the statistical characteristics of a forecasting system must be analyzed. For a given physical forecast system, the performance of the system can be evaluated by the four probabilistic measures. Assume that the elevation of the floodplain zone under consideration is y; the probability then that this zone will be flooded, conditioned on the forecast s, is $P(h \geq y|s)$. The decision rule is to issue a flood warning if the forecasted flood crest exceeds a preassigned warning threshold s^*, and not to issue a flood warning otherwise. There exist four possible outcomes that follow a flood warning decision: a correct warning, a false warning, a missed warning, and a correct quiet (the decision not to issue a warning). A correct warning is a warning followed by a flood. The probability of a correct warning is

$$P_{11}(s^*,y) = \int_{s^*}^{\infty} P(h \geq y|s)k(s)ds. \qquad (3.1)$$

A false warning is a warning not followed by a flood. The probability of a false warning is

$$P_{10}(s^*,y) = \int_{s^*}^{\infty} P(h < y|s)k(s)ds. \qquad (3.2)$$

A missed forecast is a flood event which is not preceded by a warning. The probability of a missed flood warning is

$$P_{01}(s^*,y) = \int_{0}^{s^*} P(h \geq y|s)k(s)ds. \qquad (3.3)$$

A correct quiet is the case of no warning and no flood $(h < y)$. The probability of acting correctly in not issuing a flood warning is

$$P_{00}(s^*,y) = \int_{0}^{s^*} P(h < y|s)k(s)ds. \qquad (3.4)$$

These four probabilistic measures are related to each other. Knowing one of them and the prior flood probability $P(h \geq y)$ and the probability to issue a warning, $P(s \geq s^*)$, the other three can be calculated.

There are two types of prediction errors of a forecast system-- Type I and Type II errors. Type I errors are those of missed predictions. Type II errors are those of false alerts. It is clear from Eqs. (3.1) - (3.4) that the value of the selected threshold s^* plays a key role in determining the probabilities of Type I and Type II errors. If the threshold s^* is set lower, the forecast will have a lower probability value of a Type I error and a higher probability value of a Type II error. If the threshold s^* is set higher, the forecast will have a higher probability value of a Type I error and a lower probability value of a Type II error.

Type I and Type II errors have different impacts on flood-loss reduction. A Type I error will result in an immediate flood loss. Thus, it has a short-term impact. On the other hand, a Type II error will reduce the credibility of the forecast system. This "cry wolf" consequence has a long-term impact. Such errors do not cause a flood loss at the present stage, but will discourage the response to flood warnings for future flood events. The present fraction of people who respond to a flood warning is certainly an indispensable factor in constructing the flood-loss function for a community. It thus affects the selection of the flood warning threshold. On the other hand, the response fraction fluctuates as time evolves, based on the past performance of the warning system. The coupling between successive flood events is carried by dynamic evolvement of the fraction of people who respond to a flood warning. The past performance of a flood warning system affects the present fraction of people who respond to the warning system. Therefore, we need to investigate the operation of flood warning systems in a dynamic framework. Fig. 2 presents the interconnection between the flood warning threshold and the response fraction in the decision logic.

4. A Model of the Response System

In general, the response of a community to a flood warning system is affected by people's experience of flood and their subjective evaluation of the past performance of the forecasting system.

The general interaction between a forecasting system and a response system can be described from the following considerations. The

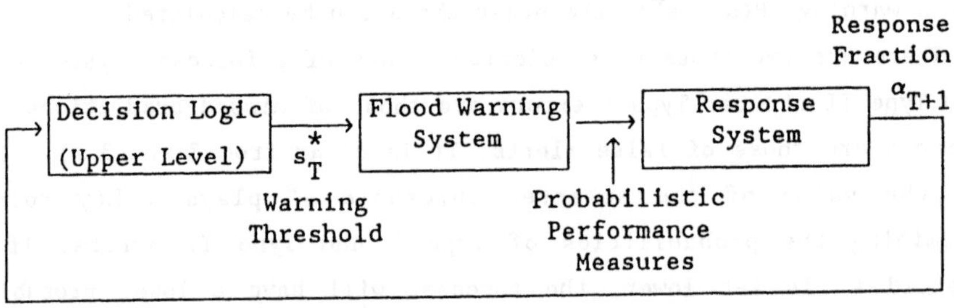

Fig. 2. Interaction between Forecast and Response Subsystems

effectiveness of a forecasting system can be judged from its performance measures of Type I and Type II errors. The response of a community to a flood warning can be described by a state variable, that is, the fraction of people in the community who respond to a call of evacuation when warned. This fraction is denoted by α_T for the Tth flood event. If a past flood event has been predicted, then the confidence in the flood forecasting system will increase, and thus, future rates of response will also increase. On the other hand, the "cry-wolf" (Type II error) event will decrease confidence in the flood forecasting system, thereby decreasing future rates of response. People tend to have decreased confidence in a flood warning system when they experienced a missed warning. However, the experience of flooding will increase people's alertness to the possibilities of future floods. For simplicity, it is reasonable to assume that the response fraction will remain unchanged after a missed warning has been experienced. We also assume that a correct quiet does not change the response fraction in the future. In view of the above discussion, the fraction α_T can be assumed to evolve dynamically, being governed by the following equation:

$$\alpha_{T+1} = \begin{cases} \alpha_T + c_1(1 - \alpha_T) & \text{with prob. } P_{11}(s_T^*, y) \\ \alpha_T & \text{with prob. } P_{00}(s_T^*, y) + P_{01}(s_T^*, y) \\ c_2\alpha_T & \text{with prob. } P_{10}(s_T^*, y) \end{cases} \tag{4.1}$$

where c_1 and c_2 are constants or functions of α_T. Their range is in (0,1) and they can be determined using identification methods based on historic data.

The response fraction in the community is described here as a controlled stochastic process. Knowing the present value of the response fraction, three possible transitions exist with given probabilities. The actual transition depends on the real outcome associated with the present warning decision. This stochastic system can be controlled in the sense of the expected value.

We note here that the feedback loop that encompasses the forecast and the response subsystems is closed only when the next flood event occurs. The present performance of a forecast system does not affect the fraction α_T at the present flood event, and it does affect the fraction α_{T+1} at the next flood event.

5. Multiobjective Multistage Optimization Model

A key aspect of flood warning systems is that the selection of the flood warning threshold cannot be viewed in isolation at each stage since the decisionmaker must balance the desire for high present flood-loss reduction with the possibility of high future flood loss. A multiobjective multistage optimization model is proposed here for finding the best value for the flood warning threshold of a flood warning system. Evaluating the trade-off between short- and long-term effects yields to an acceptable balance between the expected loss reduction at the current stage and the fraction of people who respond at the next flood event.

Assume that there are N successive flood events in the time horizon under consideration. At each flood event, the maximization of three noncommensurate/conflicting objective functions is considered in this study. At the Tth flood event, one objective is to maximize the expected property-loss reduction; the second objective is to maximize the expected life-loss reduction; and the third objective is to increase the system's credibility by reducing the "cry-wolf" effect (i.e., increasing the fraction of people who respond in the future).

At the Tth flood event, assume that the expected fraction of people that respond to the warning system is α_T. The decision logic for property loss is described in Fig. 3 for a given forecast s, where F is the event of flooding, w is the action of flood warning, the notation "$^-$" denotes

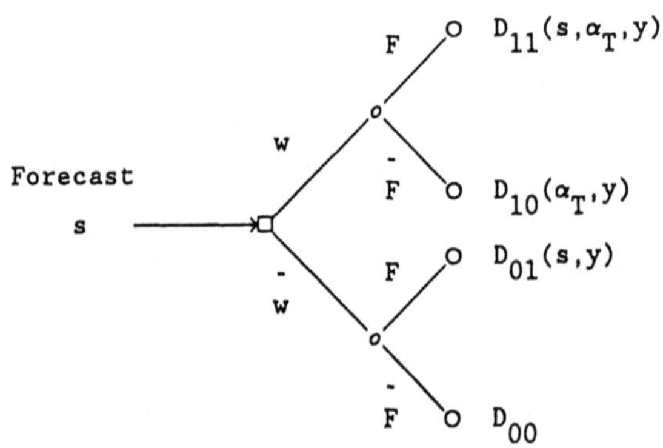

Fig. 3. Decision Logic at Stage T

the negation of an event, D_{00} (which is assumed to be zero) is the expected loss when no warning is given and no flood occurs, $D_{01}(s,y)$ is the expected community property loss without a warning, $D_{10}(\alpha_T,y)$ is the cost of the evacuation in the community, and $D_{11}(s,\alpha_T,y)$ is the expected community property loss with a warning. The difference between this decision logic and the one presented in Krzysztofowicz (1990) is that we are taking into account the response fraction in the construction of the flood-loss functions. The response fraction affects loss functions D_{10} and D_{11}. For a specific value of forecast s, the exceedance probability (or the probability that the zone with elevation y is flooded) is $q(s,y) = P(h > y|s)$. The expected property loss with no warning issued is $q(s,y)D_{01}(s,y)$, and the expected property loss with a warning issued is $q(s,y)D_{11}(s,\alpha_T,y) + [1 - q(s,y)]D_{10}(\alpha_T,y)$. Assume that the flood warning is issued when the forecast flood crest s exceeds a preassigned threshold level s_T^*. In other words, the range for issuing a flood warning is $\{s \geq s_T^*\}$. The expected property loss with a warning system then becomes

$$R(s_T^*,\alpha_T,y) = \int_0^{s_T^*}\{q(s,y)D_{01}(s,y) + [1 - q(s,y)]D_{00}\}k(s)ds$$

$$+ \int_{s_T^*}^{\infty}\{q(s,y)D_{11}(s,\alpha_T,y) + [1 - q(s,y)]D_{10}(\alpha_T,y)\}k(s)ds. \qquad (5.1)$$

Recall that the expected property loss without a warning system is

$$R_0(\alpha_T,y) = \int_0^{\infty} q(s,y)D_{01}(s,y)k(s)ds. \qquad (5.2)$$

Thus, the expected property-loss reduction, f_1^T, is

$$f_1^T = R_0(\alpha_T,y) - R(s_T^*,\alpha_T,y)$$

$$= \int_{s_T^*}^{\infty}\{q(s,y)D_{01}(s,y) - q(s,y)D_{11}(s,\alpha_T,y) - [1 - q(s,y)]D_{10}(\alpha_T,y)\}k(s)ds$$

$$(5.3)$$

It is easy to see that $f_1{}^T$ attains its maximum at s_T^*, which satisfies $q(s,y) = D_{10}(s,y)/[D_{01}(s,y) + D_{10}(\alpha_T,y) - D_{11}(s,\alpha_T,y)]$.

A similar derivation applies to the expected life-loss reduction $f_2{}^T$ at the Tth flood event, except that we do not count nonmonetary cost for evacuation,

$$f_2^T = \int_{s_T^*}^{\infty} q(s,y)\{D_{01}^L(s,y) - D_{11}^L(s,\alpha_T,y)\}k(s)ds, \tag{5.4}$$

where D_{01}^L and D_{11}^L are expected life losses without and with warning, respectively. It is easy to see that $f_2{}^T$ attains its maximum at $s_T^* = 0$ provided that D_{01}^L is greater than D_{11}^L (which is always the case).

The N-stage multiobjective optimization problem of flood warning systems is to maximize the sum of the expected property-loss reductions of all flood events over the time horizon under consideration; to maximize the sum of the expected life-loss reductions of all flood events over the time horizon under consideration; and to maximize the forecast system's credibility, which is implicitly expressed by $E(\alpha_{N+1})$, the expected fraction of people who respond to the warning beyond the time horizon under consideration. Mathematically, this overall multiobjective optimization problem can be posed as follows:

$$\max_{s^*} \begin{bmatrix} f_1 = E\{\sum_{T=1}^{N} f_1^T\} \\ f_2 = E\{\sum_{T=1}^{N} f_2^T\} \\ f_3 = f_3^N = E\{\alpha_{N+1}\} \end{bmatrix} \tag{5.5a}$$

$$\text{s.t. } \alpha_{T+1} = \begin{cases} \alpha_T + c_1(1 - \alpha_T) & \text{with prob. } P_{11}(s_T^*,y) \\ \alpha_T & \text{with prob. } P_{00}(s_T^*,y) + P_{01}(s_T^*,y), \\ c_2\alpha_T & \text{with prob. } P_{10}(s_T^*,y) \end{cases} \tag{5.5b}$$

where E is the operator of the expected value, and $f_1{}^T$ and $f_2{}^T$ are given by Eqs. (5.3) and (5.4), respectively.

Solving the multiobjective multistage optimization problem in Eq. (5.5) yields the set of noninferior solutions. A decision sequence $s^* = \{s_1^*, s_2^*, \cdots, s_N^*\}$ is said to be noninferior if there does not exist another decision sequence $\hat{s}^* = \{\hat{s}_1^*, \hat{s}_2^*, \cdots, \hat{s}_N^*\}$ such that $f_i(\hat{s}^*) \geq f_i(s^*)$, $i = 1,2,3$, with at least one strict inequality. The multistage multiobjective optimization problem in (5.5) can be effectively solved by multiobjective dynamic programming approaches, such as the envelope approach by Li and Haimes (1987 and 1988).

6. Application to Milton, Pennsylvania

One system in Milton, Pennsylvania, is selected as the illustrative application of the methodology described in the above sections. In this case, the flood crest H is of a normal distribution $N(24.88, 4.8216387^2)$ and the conditional probability density function of the forecasted flood crest S given h is of a normal $N(0.4503h + 12.1044, 1.8973^2)$. It can be shown that (1) the marginal probability density function of the forecast k(s) is of $N(23.307864, 2.8833638^2)$ and (2) the posterior distribution density function of h given s, $f(h|s)$, is $N(1.2591933s - 4.4691054, 3.1727162^2)$.

The probability of flooding given forecast s is obtained by

$$q(s,y) = \int_y^{\infty} f(h|s)dh$$

$$= 1 - \Phi[(y - 1.2591933s + 4.4691054)/3.1727162], \qquad (6.1)$$

where $\Phi(\cdot)$ is the cumulative distribution function of the standard normal. In our computer program, the cumulative distribution function $\Phi(\cdot)$ is calculated through its relationship with the error function,

$$\Phi(x) = 0.5(1 + \mathrm{erf}(x/\sqrt{2}), \qquad (6.2)$$

where the error function is defined by

$$\mathrm{erf}(x) = \frac{2}{\sqrt{\pi}} \int_0^x e^{-u^2} du. \qquad (6.3)$$

Substituting k(s) and q(s,y) into Eqs. (3.1) - (3.4), the four probabilistic measures of a forecasting system can be calculated for given warning threshold s^* and zone elevation y. Table 1 shows those measures for two values of y and various values of s^*. A trade-off between Type I and II errors can be clearly seen from Fig. 4. Different values of s^* associate different values of the probabilistic measures, $P_{11}(s^*)$, $P_{10}(s^*)$, $P_{01}(s^*)$, and $P_{00}(s^*)$. They thus yield different impacts on the respond fraction at the subsequent stage.

For simplicity, we only consider property losses in this illustrative example. To construct a reasonable loss function, the concept of category-unit loss functions by Krzysztofowicz and Davis (1983d) is adopted in the following. The main modification is that the notation α, which was originally used in Krzysztofowicz and Davis (1983d) as the response degree of an individual, is used here to represent the fraction of people who respond to the warning. The cost function is assumed to be of the form

Table 1. Probabilistic Measures of the Warning System

y = 19

s^*	$P_{11}(s^*,y)$	$P_{10}(s^*,y)$	$P_{01}(s^*,y)$	$P_{00}(s^*,y)$
15.0	0.88859	0.10943	0.0000862	0.0018939
15.5	0.88846	0.10816	0.0002140	0.0031715
16.0	0.88817	0.10620	0.0005011	0.0051294
16.5	0.88757	0.10332	0.0011061	0.0080047
17.0	0.88637	0.099286	0.0023056	0.012041
17.5	0.88413	0.093883	0.0045469	0.017444
18.0	0.88018	0.087000	0.0084949	0.024326
18.5	0.87361	0.078674	0.015060	0.032653
19.0	0.86329	0.069126	0.025381	0.042201
19.5	0.84792	0.058765	0.040750	0.052562

y = 22

s^*	$P_{11}(s^*,y)$	$P_{10}(s^*,y)$	$P_{01}(s^*,y)$	$P_{00}(s^*,y)$
19.0	0.71712	0.21529	0.0077255	0.059857
19.5	0.71088	0.19581	0.013971	0.079341
20.0	0.70090	0.17345	0.023947	0.10170
20.5	0.68587	0.14905	0.038977	0.12610
21.0	0.66449	0.12378	0.060363	0.15137
21.5	0.63571	0.098964	0.089143	0.17619
22.0	0.59901	0.075925	0.12584	0.19923
22.5	0.55461	0.055721	0.17024	0.21943
23.0	0.50351	0.039009	0.22134	0.23614
23.5	0.44745	0.025986	0.27740	0.24916

$$d_{10} = MC\alpha, \tag{6.4}$$

where MC = 4 (ten thousand dollars) is the evacuation cost for the community.

The flood-loss function without a warning is given by

$$d_{01} = MD\delta(h - y), \tag{6.5}$$

where MD = 100 (ten thousand dollars) is the maximum possible damage due to flood of the highest magnitude when no response is made, and $\delta(h - y)$ is the unit damage function specifying the fraction of MD which occurs

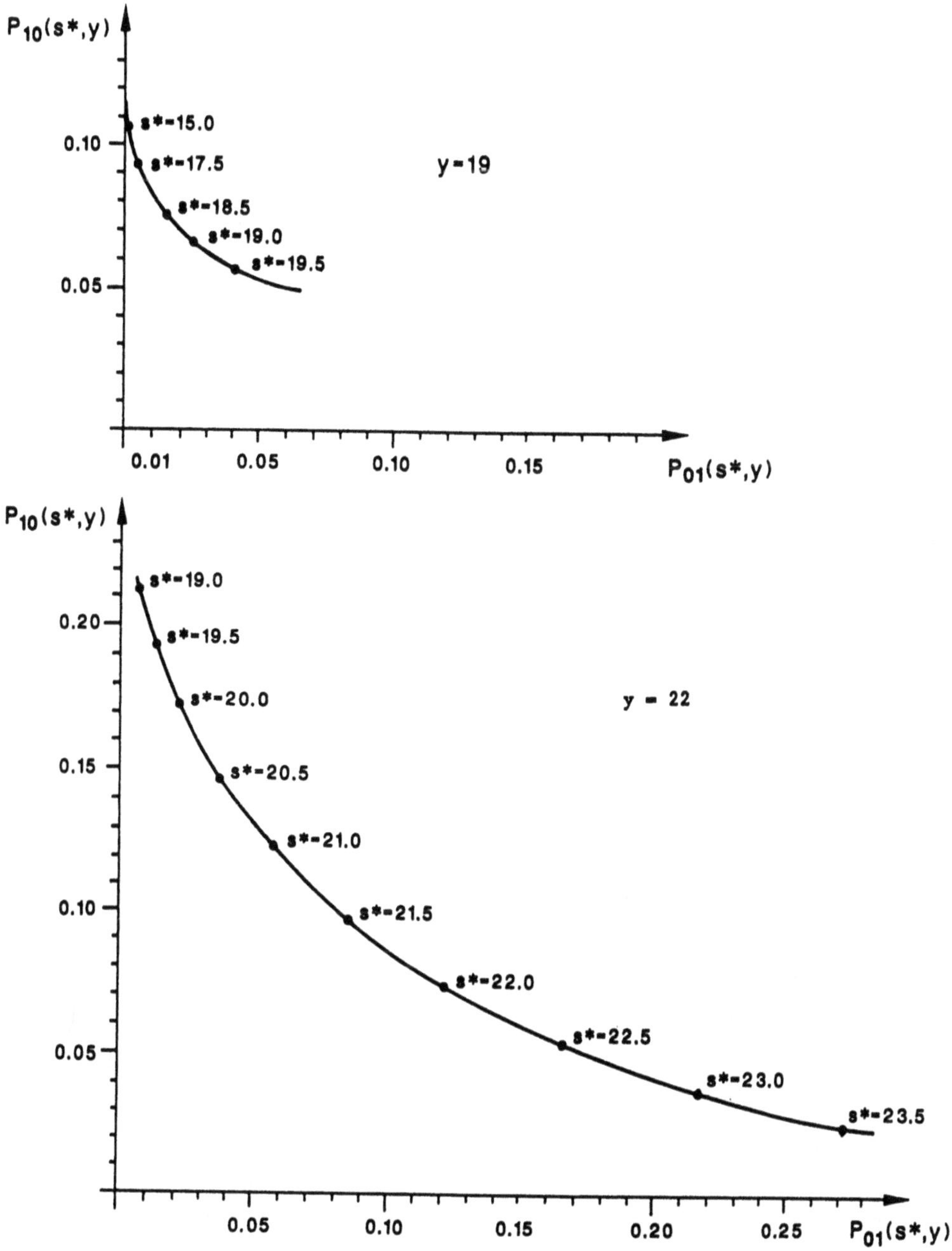

Fig. 4. Trade-off Between Type I and Type II Errors

when the depth of flooding is (h - y). We assume $\delta(h - y)$ equal to 0.4(h - y) when $h - y \le 2.5$ and equal to 1 when $h - y > 2.5$ (h and y are in the unit of 10 feet).

The loss function with a warning is

$$d_{11} = MC\alpha + MD[1 - \alpha MR(h - y)]\delta(h - y), \tag{6.6}$$

where MR(h - y) is the unit reduction function specifying the reduction of MD when the depth of flooding is (h - y) and full response of community is made, $\alpha = 1$. We assume MR(h - y) equal to $0.25 + 0.4(h - y) - 0.333(h - y)^2$ when $h - y \le 1.2$ and equal to 0.25 when $h - y > 1.2$.

The expected cost of the community evacuation conditioned on a given forecast s is

$$D_{10} = d_{10} = MC\alpha. \tag{6.7}$$

The expected property loss without a warning conditioned on a given forecast s is

$$D_{01} = \frac{1}{q(s,y)}\int_y^\infty d_{01}(h)f(h|s)dh$$

$$= \frac{MD}{q(s,y)}[\int_y^{y+2.5} 0.4(h - y)f(h|s)dh + \int_{y+2.5}^\infty f(h|s)dh]. \tag{6.8}$$

The expected property-loss with a warning conditioned on a given forecast s is

$$D_{11} = \frac{1}{q(s,y)}\int_y^\infty d_{11}(\alpha,h)f(h|s)dh$$

$$= MC\alpha + D_{01}$$

$$- \frac{\alpha MD}{q(s,y)}[\int_y^{y+1.2} [0.25 + 0.4(h - y) - 0.333(h - y)^2]0.4(h - y)f(h|s)dh$$

$$+ \int_{y+1.2}^{y+2.5} 0.25\cdot 0.4(h - y)f(h|s)dh + \int_{y+2.5}^\infty 0.25f(h|s)dh]. \tag{6.9}$$

Since f(h|s) is of normal distribution N(1.2591933s - 4.4691054, 3.1727162*), the above integrations can be performed using the following equations:

$$\int_a^b f(h|s)dh = \Phi(\frac{b - 1.2591933s + 4.4691054}{3.1727162})$$

$$- \Phi(\frac{a - 1.2591933s + 4.4691054}{3.1727162}) \tag{6.10}$$

$$\int_a^b hf(h|s)dh = \frac{3.1727162}{\sqrt{2\pi}} \{Exp[-\frac{(a - 1.2591933s + 4.4691054)^2}{2\cdot 3.1727162^2}]$$

$$- Exp[-\frac{(b - 1.2591933s + 4.4691054)^2}{2\cdot 3.1727162^2}]\}$$

$$+ (1.2591933s + 4.4691054)\int_a^b f(h|s)dh \quad (6.11)$$

$$\int_a^b h^2 f(h|s)dh = \frac{3.1727162}{\sqrt{2\pi}} \{aExp[-\frac{(a - 1.2591933s + 4.4691054)^2}{2\cdot 3.1727162^2}]$$

$$- bExp[-\frac{(b - 1.2591933s + 4.4691054)^2}{2\cdot 3.1727162^2}]\}$$

$$+ 3.1727162^2 \int_a^b f(h|s)dh + (1.2591933s + 4.4691054) \int_a^b hf(h|s)dh \quad (6.12)$$

$$\int_a^b h^3 f(h|s)dh = \frac{3.1727162}{\sqrt{2\pi}} \{a^2 Exp[-\frac{(a - 1.2591933s + 4.4691054)^2}{2\cdot 3.1727162^2}]$$

$$- b^2 Exp[-\frac{(b - 1.2591933s + 4.4691054)^2}{2\cdot 3.1727162^2}]\}$$

$$+ 2\cdot 3.1727162^2 \int_a^b hf(h|s)dh + (1.2591933s + 4.4691054)\int_a^b h^2 f(h|s)dh \quad (6.13)$$

The objective function f_1^T can be now posed as

$$f_1^T = \int_{s_T^*}^{\infty} \{q(s,y)D_{01}(s,y)-q(s,y)D_{11}(s,\alpha_T,y)-(1-q(s,y))D_{10}(\alpha_T,y)\}k(s)ds \quad (6.14)$$

Table 2 gives the calculated values of the expected flood-loss reduction for various response fractions and preselected warning thresholds where y is equal to 19 or 22. The relationship between the expected flood-loss reduction and the warning threshold is also depicted in Fig. 5 for various response fractions and elevation levels.

In this illustrative example, we assume that the parameters in the response fraction dynamic model, Eq. (4.1), take values of $c_1 = 0.1$ and $c_2 = 0.9$. We consider a five-stage problem with the initial responding

Table 2. Expected Flood-Loss Reduction

y = 19

s^* \ α	0.50	0.60	0.70	0.80
16.5	6.0133	7.2160	8.4186	9.6213
17.0	6.0144	7.2173	8.4202	9.6230
17.5	6.0116	7.2139	8.4162	9.6185
18.0	5.9989	7.1987	8.3985	9.5983
18.4	5.9751	7.1701	8.3651	9.5601

y = 22

s^* \ α	0.50	0.60	0.70	0.80
19.0	3.5921	4.3106	5.0290	5.7474
19.5	3.5956	4.3148	5.0339	5.7530
20.0	3.5744	4.2893	5.0041	5.7190
20.5	3.5119	4.2142	4.9166	5.6190
20.9	3.4207	4.1049	4.7890	5.4732

fraction α_1 equal to 0.7. Two values of the elevation y are considered, y = 19 (feet) and y = 22 (feet).

The overall problem can now be posed as follows:

$$\max_{s^*} \begin{bmatrix} f_1 = E(\sum_{T=1}^{5} f_1^T) \\ \\ f_3 = E(\alpha_6) \end{bmatrix} \tag{6.15a}$$

$$\text{s.t. } \alpha_{T+1} = \begin{cases} \alpha_T + 0.1(1 - \alpha_T) & \text{with prob. } P_{11}(s_T^*, y) \\ \alpha_T & \text{with prob. } P_{00}(s_T^*, y) + P_{01}(s_T^*, y) \\ 0.9\alpha_T & \text{with prob. } P_{10}(s_T^*, y) \end{cases} \tag{6.15b}$$

$$\alpha_1 = 0.7 \qquad T = 1,2,3,4,5$$

Equation (6.15) can be solved by the weighting method and dynamic programming.

At the fifth stage, the following optimization problem is solved for given weighting coefficient θ and given value of α_5:

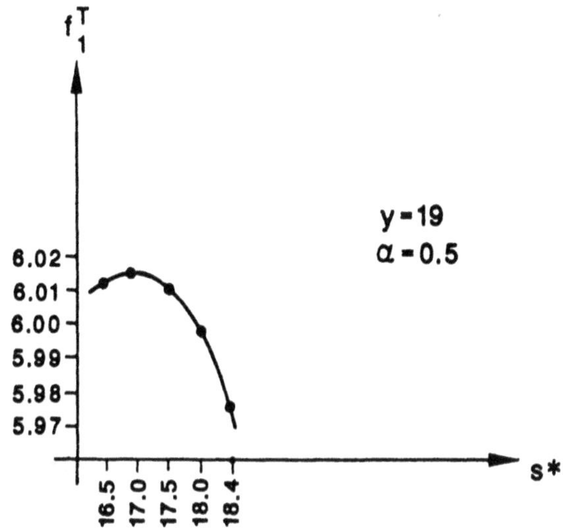

5(a)

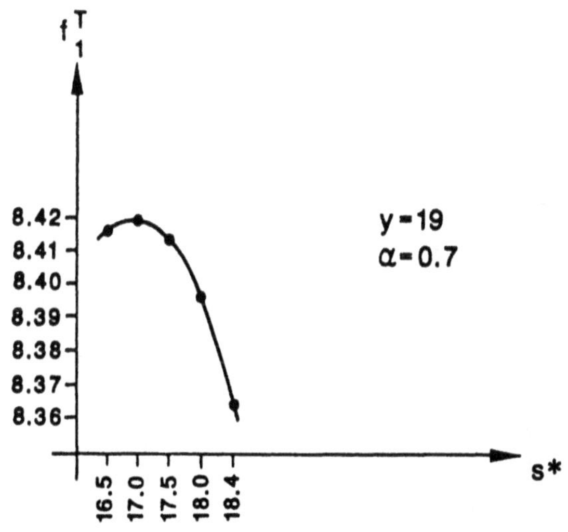

5(b)

Fig. 5. Relationship Between the Expected Flood-Loss Reduction and the Warning Threshold for Various Response Fraction and Elevation Levels

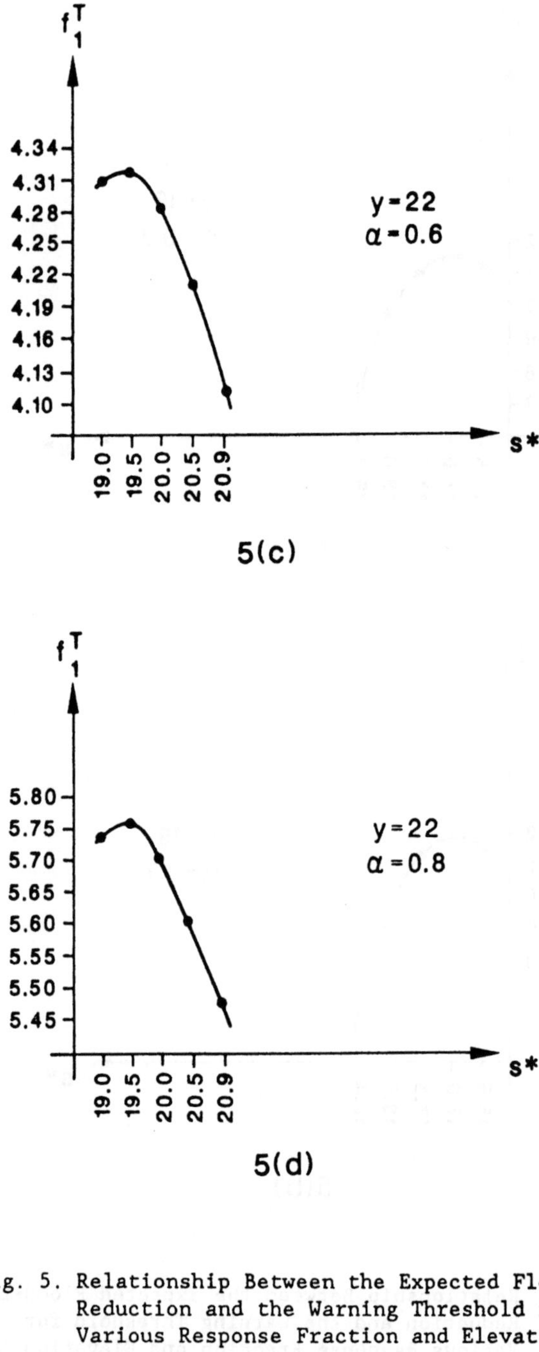

Fig. 5. Relationship Between the Expected Flood-Loss Reduction and the Warning Threshold for Various Response Fraction and Elevation Levels

$$\phi_5(\theta,\alpha_5) = \max_{s_5^*} \theta[f_1^5|\alpha_5] + (1 - \theta) E\{\alpha_6|\alpha_5\}, \tag{6.16}$$

subject to Eq. (6.15b) with T = 5.

At stage T, T = 4,3,2,1, the following optimization problem is solved for given value of the weighting coefficient θ and given value of α_T:

$$\phi\ (\theta,\alpha_T) = \max_{s_T^*} \theta[f_1^T|\alpha_T] + E\{\phi_{T+1}(\theta,\alpha_{T+1})|\alpha_T\}, \tag{6.17}$$

where α_T satisfies Eq. (6.15b), and $\alpha_1 = 0.7$.

To make the computational procedure feasible, the state space of α_T and the control space of s_T are quantized by the grid sizes 0.01 and 0.1, respectively. Table 3 provides noninferior solutions at various stages and various α_t's for four different weighting coefficients and two different values of zone elevation. It can be seen from Table 3 that (1) the lower the weighting coefficient θ associated with the first objective (loss reduction), the higher the value of flood warning threshold will be set to avoid possible high Type II errors; (2) it can be observed that in order to select a decision that maximizes the sum of flood-loss reductions, the flood warning threshold is set higher at the earlier stage than at the later stage (with respect to the same value of the response fraction) in order to reduce the probability of high loss at the later stages; and (3) the higher the present response fraction, the more cautious the selection of the threshold is. That means that a higher value of threshold is set for a higher value of the present response fraction in order to avoid losing a larger number of the response population. We should note here that the third conclusion may be model-specific.

Figures 6, 7, and 8 show the optimal flood warning threshold as functions of the weighting coefficient, the stage, and the response fraction, respectively.

Table 3a. Optimal Flood Warning Thresholds
Y = 19 θ = 0.02

α	Stage 1	Stage 2	Stage 3	Stage 4	Stage 5
0.45		17.6 1.0840	17.6 0.91482	17.5 0.75087	17.5 0.59272
0.50		17.7 1.1580	17.6 0.98348	17.6 0.81363	17.6 0.64895
0.55		17.8 1.2320	17.7 1.0522	17.7 0.87641	17.6 0.70519
0.60		17.8 1.3061	17.8 1.1209	17.7 0.93919	17.7 0.76143
0.65		17.9 1.3801	17.8 1.1896	17.8 1.0020	17.7 0.81768
0.70	18.0 1.6522	17.9 1.4542	17.9 1.2583	17.8 1.0648	17.8 0.87392
0.75		18.0 1.5282	17.9 1.3270	17.9 1.1276	17.8 0.93017
0.80		18.0 1.6023	18.0 1.3957	17.9 1.1904	17.8 0.98642

*)The upper number denotes the selected flood warning threshold; and the lower number denotes the expected loss reduction from the present stage to the final stage.

Table 3b. Optimal Flood Warning Thresholds
Y = 19 θ = 0.06

α	Stage 1	Stage 2	Stage 3	Stage 4	Stage 5
0.45		17.5 2.0455	17.4 1.6033	17.3 1.1838	17.2 0.78943
0.50		17.5 2.2011	17.5 1.7357	17.4 1.2906	17.2 0.86789
0.55		17.6 2.3567	17.5 1.8682	17.4 1.3975	17.3 0.94636
0.60		17.6 2.5123	17.6 2.0007	17.4 1.5043	17.3 1.0248
0.65		17.7 2.6680	17.6 2.1332	17.5 1.6112	17.3 1.1033
0.70	17.8 3.3909	17.7 2.8236	17.6 2.2657	17.5 1.7180	17.3 1.1818
0.75		17.8 2.9793	17.6 2.3982	17.5 1.8249	17.3 1.2602
0.80		17.8 3.1349	17.7 2.5307	17.5 1.9318	17.4 1.3387

*)The upper number denotes the selected flood warning threshold; and the lower number denotes the expected loss reduction from the present stage to the final stage.

Table 3c. Optimal Flood Warning Thresholds
Y = 19 θ = 0.1

α	Stage 1	Stage 2	Stage 3	Stage 4	Stage 5
0.45		17.4 3.0072	17.4 2.2919	17.3 1.6167	17.1 0.98616
0.50		17.5 3.2444	17.4 2.4882	17.3 1.7677	17.1 1.0869
0.55		17.5 3.4817	17.5 2.6845	17.3 1.9187	17.2 1.1876
0.60		17.6 3.7189	17.5 2.8808	17.4 2.0696	17.2 1.2883
0.65		17.6 3.9562	17.5 3.0771	17.4 2.2206	17.2 1.3890
0.70	17.8 5.1300	17.7 4.1935	17.6 3.2734	17.4 2.3716	17.2 1.4897
0.75		17.7 4.4308	17.6 3.4698	17.4 2.5225	17.2 1.5905
0.80		17.7 4.6681	17.6 3.6661	17.4 2.6735	17.2 1.6912

*)The upper number denotes the selected flood warning threshold; and the lower number denotes the expected loss reduction from the present stage to the final stage.

Table 3d. Optimal Flood Warning Thresholds
Y = 19 θ = 0.5

α	Stage 1	Stage 2	Stage 3	Stage 4	Stage 5
0.45		17.4 12.625	17.3 9.1779	17.2 5.9465	17.0 2.9536
0.50		17.4 13.678	17.3 10.013	17.2 6.5387	17.0 3.2769
0.55		17.5 14.732	17.4 10.847	17.2 7.1309	17.0 3.6001
0.60		17.5 15.786	17.4 11.682	17.2 7.7231	17.0 3.9234
0.65		17.6 16.839	17.4 12.517	17.3 8.3153	17.0 4.2466
0.70	17.7 22.522	17.6 17.893	17.5 13.352	17.3 8.9075	17.0 4.5699
0.75		17.6 18.947	17.5 14.187	17.3 9.4997	17.0 4.8932
0.80		17.6 20.001	17.5 15.022	17.3 10.092	17.0 5.2164

*)The upper number denotes the selected flood warning threshold; and the lower number denotes the expected loss reduction from the present stage to the final stage.

Table 3e. Optimal Flood Warning Thresholds
Y = 22 θ = 0.02

α	Stage 1	Stage 2	Stage 3	Stage 4	Stage 5
0.45		19.8 0.82894	19.8 0.72964	19.8 0.63179	19.8 0.53553
0.50		19.9 0.88792	19.9 0.78644	19.9 0.68621	19.9 0.58736
0.55		20.0 0.94693	20.0 0.84327	20.0 0.74066	19.9 0.63920
0.60		20.1 1.0060	20.0 0.90013	20.0 0.79513	20.0 0.69106
0.65		20.1 1.0651	20.1 0.95702	20.1 0.84962	20.0 0.74293
0.70	20.2 1.2349	20.1 1.1242	20.1 1.0139	20.1 0.90413	20.1 0.79480
0.75		20.2 1.1833	20.2 1.0709	20.1 0.95864	20.1 0.84668
0.80		20.2 1.2425	20.2 1.1278	20.2 1.0132	20.1 0.89856

*)The upper number denotes the selected flood warning threshold; and the lower number denotes the expected loss reduction from the present stage to the final stage.

Table 3f. Optimal Flood Warning Thresholds
Y = 22 θ = 0.06

α	Stage 1	Stage 2	Stage 3	Stage 4	Stage 5
0.45		19.7 1.3731	19.6 1.1215	19.6 0.87880	19.5 0.64563
0.50		19.7 1.4807	19.7 1.2160	19.6 0.95890	19.6 0.70996
0.55		19.8 1.5883	19.7 1.3105	19.7 1.0390	19.6 0.77429
0.60		19.8 1.6959	19.8 1.4050	19.7 1.1191	19.6 0.83862
0.65		19.8 1.8036	19.8 1.4996	19.7 1.1992	19.6 0.90295
0.70	19.9 2.2307	19.9 1.9112	19.8 1.5941	19.7 1.2794	19.6 0.96728
0.75		19.9 2.0189	19.8 1.6886	19.7 1.3595	19.7 1.0316
0.80		19.9 2.1266	19.8 1.7832	19.8 1.4396	19.7 1.0960

*)The upper number denotes the selected flood warning threshold; and the lower number denotes the expected loss reduction from the present stage to the final stage.

Table 3g. Optimal Flood Warning Thresholds
Y = 22 θ = 0.1

α	Stage 1	Stage 2	Stage 3	Stage 4	Stage 5
0.45		19.6 1.9177	19.6 1.5138	19.5 1.1260	19.5 0.75587
0.50		19.7 2.0741	19.6 1.6461	19.6 1.2319	19.5 0.83274
0.55		19.7 2.2305	19.7 1.7784	19.6 1.3378	19.5 0.90962
0.60		19.8 2.3869	19.7 1.9108	19.6 1.4437	19.5 0.98649
0.65		19.8 2.5433	19.7 2.0431	19.6 1.5496	19.5 1.0634
0.70	19.9 3.2281	19.8 2.6997	19.7 2.1754	19.6 1.6555	19.5 1.1402
0.75		19.8 2.8562	19.7 2.3078	19.7 1.7614	19.5 1.2171
0.80		19.8 3.0127	19.8 2.4401	19.7 1.8673	19.5 1.2940

*)The upper number denotes the selected flood warning threshold; and the lower number denotes the expected loss reduction from the present stage to the final stage.

Table 3h. Optimal Flood Warning Thresholds
Y = 22 θ = 0.5

α	Stage 1	Stage 2	Stage 3	Stage 4	Stage 5
0.45		19.6 7.3654	19.5 5.4377	19.5 3.5992	19.4 1.8585
0.50		19.6 8.0099	19.6 5.9485	19.5 3.9632	19.4 2.0610
0.55		19.7 8.6545	19.6 6.4594	19.5 4.3271	19.4 2.2636
0.60		19.7 9.2992	19.6 6.9702	19.5 4.6911	19.4 2.4661
0.65		19.7 9.9439	19.6 7.4811	19.5 5.0550	19.4 2.6686
0.70	19.8 13.207	19.7 10.589	19.6 7.9920	19.5 5.4190	19.4 2.8712
0.75		19.7 11.233	19.6 8.5028	19.5 5.7830	19.4 3.0737
0.80		19.7 11.878	19.6 9.0137	19.5 6.1469	19.4 3.2763

*)The upper number denotes the selected flood warning threshold; and the lower number denotes the expected loss reduction from the present stage to the final stage.

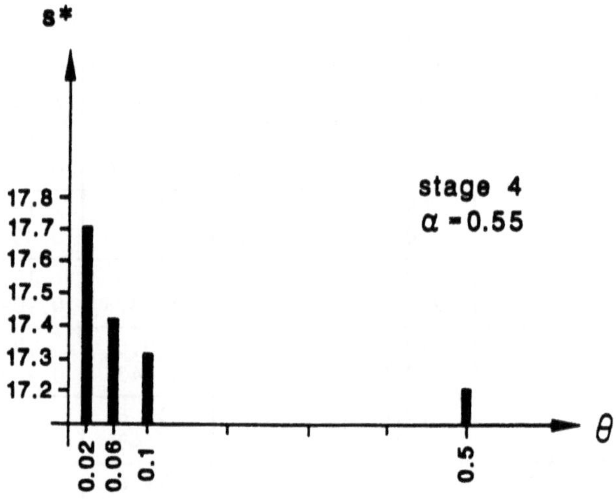

Fig. 6. Relationship Between the Optimal Warning Threshold and the Weighting Coefficient for α = 0.55 at Stage 4

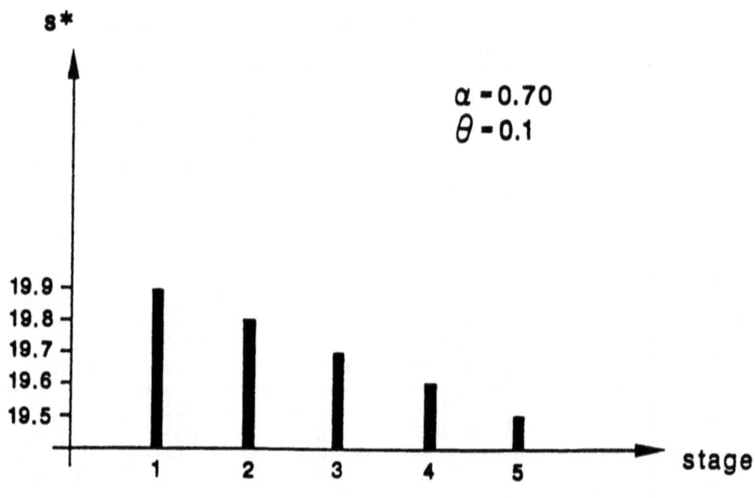

Fig. 7. Relationship Between the Optimal Warning Threshold and the Stages for α = 0.70 and θ = 0.1

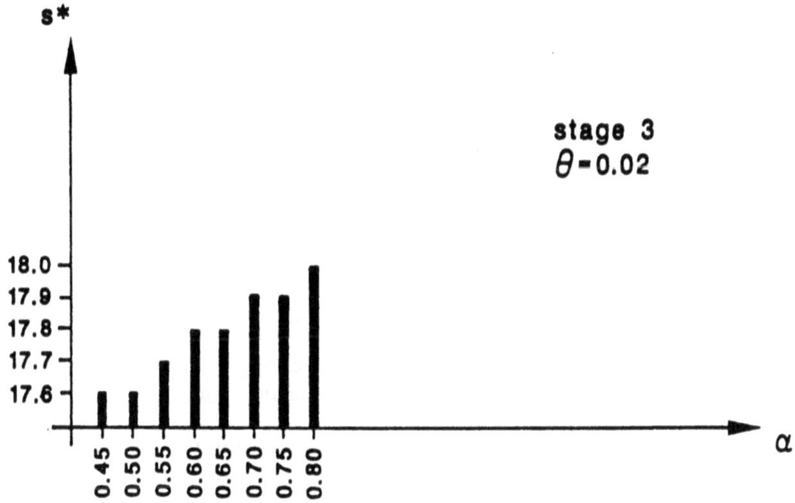

Fig. 8. Relationship Between the Optimal Warning Threshold and the Response Fraction for $\theta = 0.02$ at Stage 3

References

Alexandridis M G, Krzysztofowicz R (1985) Decision models for categorical and probabilistic weather forecasts. Applied Mathematics and Computation 17: 241-66.

Barrett C, Neal D M, Wright S, Lee R, Mojib R. (1988) Primer on flood forecast and response systems. Report to U.S. Army Corps of Engineers, Ft. Belvoir, Virginia.

Berger J O (1980) Statistical Decision Theory. New York: Springer-Verlag.

Bhavnagri V S, Bugliarello G (1965) Mathematical representation of an urban flood plain. J. Hydraul. Div. Am. Soc. Civ. Eng. 91(HY2): 149-73.

Bock P, Hendrick R L (1966) Benefits of river forecast information. Final Rep. 7229-229, Travelers Res. Center, Inc., Hartford, Connecticut.

Curtis D C, Schaake J C Jr (1988) Economic benefit estimation for flood warning systems. Water Resources Bulletin.

Day H J, Lee K K (1976) Flood damage reduction potential of river forecast. J. Water Resour. Plann. Manage. Div. Am. Soc. Civ. Eng. 102(WR1): 77-87.

de Maré J (1980) Optimal prediction of catastrophes with applications to Gaussian processes. The Annals of Probability 8(4): 841-50.

Ferrell W, Krzysztofowicz R (1983) A model of human response to flood warnings for systems evaluation. Water Resources Research, 19(6): 1467-75.

Krzysztofowicz R (1983a) Why should a forecaster and a decision maker uses Bayes theorem. Water Resources Research 19(2): 327-36.

Krzysztofowicz R (1983b) A Bayesian model of the flood forecast process. Water Resources Research 19(6): 1455-65.

Krzysztofowicz R (1985) Choice of action based on the probability of a flash flood. Report to Office of Hydrology, National Weather Service, National Oceanic and Atmospheric Administration, and U.S. Department of Commerce, Department of Systems Engineering, University of Virginia.

Krzysztofowicz R (1990) Performance characteristics under forecast uncertainty. Department of Systems Engineering, University of Vitginia, Charlottesville, Virginia.

Krzysztofowicz R, Davis D (1983a) A methodology for evaluation of flood forecast-response systems 1. Analysis and concepts. Water Resources Research 19(6): 1423-29.

Krzysztofowicz R, Davis D (1983b) A methodology for evaluation of flood forecast-response systems 2. Theory. Water Resources Research 19(6): 1431-40.

Krzysztofowicz R, Davis D (1983c) A methodology for evaluation of flood forecast-response systems 3. Case studies. Water Resources Research 19(6): 1441-54.

Krzysztofowicz R, Davis D (1983d) Category-unit functions for flood forecast-response system evaluation. Water Resources Research 19(6): 1476-80.

Krzysztofowicz R, Davis D (1984) Toward improving flood forecast-response systems. Interfaces 14(3): 1-14.

Lee K K, Day H J, Cheney P B (1975) Progress in simulation of integrated floodplain analysis and management. Paper presented at Second World Congress, Int. Water Resour. Assoc.. New Delhi, India, December.

Lee R, Hu P S, Neal D M, Ogles M R, Sorensen J H, Trumble D A (1986) Predicting loss of life from floods. Report to the Institute for Water Resources, U.S. Army Corps of Engineers.

Li D, Haimes Y Y (1987) The envelope approach for multiobjective optimization problems. IEEE-SMC 17(6): 1026-38.

Li D, Haimes Y Y (1988) Decomposition techniques in multiobjective discrete-time dynamic problems. In Control and Dynamic Systems. Ed. C. T. Leondes. Vol. 28, 109-80. San Diego, California: Academic Press.

Lindgren G (1979) Prediction of level crossing for normal processes containing deterministic components. Adv. Appl. Prob. 11: 93-117.

Lindgren G (1980) Model processes in nonlinear prediction with application to detection and alarm. The Annals of Probability. 8(4):,775-92.

Lindgren G (1985) Optimal prediction of level crossing in Gaussian processes and sequences. The Annals of Probability 13(3): 804-24.

National Advisory Committee on Oceans and Atmosphere (1972) The Agnes floods. Report for the Administrator of NOAA, Washington, D.C., November.

Paté-Cornell M E (1986) Warning systems in risk management. Risk Analysis 6(2): 223-34.

Press W H, Flannery B P, Teukolsky S A, Vetterling W T (1986) Numerical Recipes. Cambridge: Cambridge University Press.

Shabman L (1987) Measuring non-property residential flood protection benefits: A policy review. Report to The U.S. Army Corps of Engineers, Institute for Water Resources, Fort Belvoir, Virginia.

Sniedovich M, Davis D R (1977) Evaluation of flood forecasting-response systems. Journal of the Water Resources Planning and Management Division, 103(WR1): 83-97.

Sniedovich M, Davis D R, Duckstein L, Bogardi I (1974) A decision theoretic evaluation of hydrologic data for short-term forecasts of floods. Paper presented at the AGU National Meeting, Washington, D.C.

MULTICRITERION RISK AND RELIABILITY ANALYSIS IN HYDROLOGIC SYSTEM DESIGN AND OPERATION

Lucien Duckstein and Bijaya P. Shrestha
Department of Systems and Industrial Engineering
University of Arizona, Tucson, AZ 85721

Eugene Z. Stakhiv
IWR, Research Division, U.S. Corps of Engineers
Casey Building, Fort Belvoir, VA 22060

ABSTRACT In the first part of this study, reliability in hydrologic design and operation is defined within a general discrete state discrete time system framework. The system definition is composed of time scale, state, input, output, system function and readout or output function. The state includes performance indices PI (level of service, quality of service, ratio of resistance to load) and the output includes figures of merit FM combining the PI's. The reliability PI is the ratio of the number of times the system is not in a failure mode over the number of time periods. The definition of other uncertainty - related criteria (repairability or resilience, vulnerability, engineering risk) within the same system framework is also presented.

In the second part of this study, three simple but realistic design examples are presented, namely:

1. a bridge pier subject to scouring of the river bed
2. a concrete foundation subject to corrosion hazard
3. a mine dewatering scheme to prevent groundwater inrushes.

In each example, the system model leads naturally to a multicriterion analysis.

The third part of this paper consists of the development of scheme to cope with water supply under developing drought conditions. A decision rule combining long-term forecasts, existing and historical parameters is defined using a Bayes procedure. Multiple conflicting criteria are pooled by means of compromise programming and the operation policy is shifted form "normal" to "alert" to "emergency" policy as a function of the value of the present value of the Lp-norm. The case study of the Tenkiller Lake in Oklahoma illustrates the methodology.

NATO ASI Series, Vol. G 29
Water Resources Engineering Risk Assessment
Edited by J. Ganoulis

INTRODUCTION

The purposes of this paper are:

1. to provide a system framework for defining and trading off criteria related to risk and reliability in hydrologic systems design and operation with emphasis on extreme events.

2. to illustrate the concepts by means of simple, but realistic examples.

3. to provide along the same line of thought a dynamic model for switching operation policy of a supply system when drought occurs.

In order to relate the notion of incident and failure event in the physical (structure-related) sense to the parallel notion in a dynamic (target-related) sense, it appears to be useful to define a generalized "load" λ, a generalized "resistance" ρ, an "incident mode" μ and the proposition:

$$\text{Incident } \mu \text{ occurs if and only if } \lambda > \rho \tag{1}$$

For example, in the case of water being supplied from reservoirs to satisy users' requirements, the "load" λ would be the total requirement and the "resistance" ρ would be supply capacity. Either one of these quantitites may be probabilistic, fuzzy or both. Other possible interpretations for λ and ρ are presented in Duckstein et al.(1987). The incident mode μ refers to the type of incident that may occur. For example, for a structure, an incident may correspond to failure of all or part of the structure; for a supply system, an incident may be a shortage or excess of water. A prolonged shortage then leads to "failure."

The paper is organized as follows: The mathematical system framework is described in the next section, leading to a description of reliability, risk and other incident-related criteria as performance indices, resource indices and figures of merit. Section 4 consists of three examples of design under risk: a bridge pier, a concrete foundation and a mine dewatering scheme so as to demonstrate how multicriterion analysis is a natural sequel to system modeling under risk. Section 5 develops a procedure to switch water supply policy from "normal" to "alert" to "emergency" mode under developing drought conditions. A case study in the Illinois River in Oklahoma illustrates the methodology; the approach includes embedding forecasts into the decision by means of a Bayes procedure and the joint consideration of conflicting criteria by means of a distance-based multicriterion technique called compromise programming.

SYSTEM FRAMEWORK

System Elements Consider a discrete-time dynamic system Z as defined in Wymore (1976, 1991). For the sake of convenience, let this system be also discrete-state. Example of such systems may be found in Bogardi and Bardossy (1987), and Duckstein et al. (1990). Consider, here, a single multi-purpose reservoir whose system components are (T^*, S, X, Y, F, G) defined respectively as follows:

T^* : a time scale $(t = 1, \cdots, T, \cdots, J)$; the horizon J may be infinite.

S : a discrete state vector describing the system at time t. The elements of this vector belong to set S, which may be infinite and may include storage volume, water quality indices, cumulative shortage and flood volumes as well as physical characteristics of a dam, such as resistance or fatigue. Generally speaking, let the state set S include *performance indices* $PI^k, k = 1, \cdots, K$, which measure how well the system performs. Let $PI(t)$ be the performance index vector at time t. The state variable vector at time t is denoted $s(t) \in S$ and the state trajectory at time t is the sequence of vectors:

$$s(t) = (s(0), s(1), \cdots, s(t))$$

The performance trajectory is defined as the matrix:

$$P(t) = (PI(1), \cdots, PI(t)); \quad PI(t) \in S. \tag{2}$$

$X = (U, W)$: input set or alphabet composed of a set of controllable elements U and a set of non-controllable elements W. Controllable or decision vector elements $u(t) \in U$ may include both those parameters which are fixed by the design, such as the height of a dam, and adjustable quantities, such as the opening of a spillway gate. Non-controllable elements $w(t) \in W$ include such quantities as the total net inflow into a reservoir or the deterioration of a structure.

A particular subset of U is the decision trajectory or control policy

$$u(t) = (u(0), u(1), \cdots, u(t))$$

and a subset of W may be the historical random inflow sequence.

Y = an output set or alphabet comprising physical variables, such as water volume delivered and generalized loads. At every time t, the output usually includes an updated value of performance indices $PI(t)$ which were defined as elements of the state set S, such as losses due to shortage or excess spills.

The output also includes figures of merit (FM), which are criterion functions defined over the components of the performance matrix $P(J)$ equation (2) where J is the horizon. For example, the mean value of $P(J)$, the maximum or minimum component of $P(J)$, may be FM's. Generally speaking, the outputs $y(t) \in Y$, $t = 0, \cdots, T$, define the output trajectory

$$y(T) = (y(1), \cdots, y(T))$$

a subset of which may be the performance index trajectory $P(T)$.

F : a state transition function which calculates the state at time $t+1$ as a function of the previous states $s(t)$ and the previous input $x(t)$:

$$s(t+1) = F(s(t), x(t)) \tag{3}$$

Note that by recurrence, $s(t+1)$ depends on both state trajectory $x(t)$ and input trajectory $s(t)$; also function F in equation (3) includes the algorithm to calculate the performance index vector $PI(t+1)$.

In the case of reservoir operation, one component of vector function F is frequently the mass balance equations, as it is the case in the last example in this paper:

$$v(t+1) = v(t) - u(t) + w(t) \tag{4}$$

$$\begin{aligned} v(t) &= \text{water volume in storage at time t} \\ u(t) &= \text{controlled outflow, including release} \\ w(t) &= \text{net random inflow;} \end{aligned}$$

Other components of F may include design parameters, sediment movement equations, trophic state transformation.

G: an output function which provides the algorithm for calculating the present output $y(t)$ as a function of the present state $s(t)$:

$$y(t) = G(s(t)) \tag{5}$$

A simple example of output function is the transformation of a shortage incident into an economic loss.

System Experiment To operate, for example, a water supply system, the input vector or trajectory is manipulated to provide a desirable output vector or trajectory. The desirability of an output is measured by means of PI's and then FM's, calculated on the basis of a so-called *system experiment* $c(T) = (x(T), s(0), T)$ with $t = 0, 1, \cdots, T$, composed of an input trajectory (inflow) $x(T)$, an initial state (storage) $s(0)$, and a time horizon T. The state transition function $F(.)$ and the output function $G(.)$ determine, respectively, the state trajectory $s(T)$ and the output trajectory $y(T)$, given any experiment $c(T)$, as described in Table 1. Such a table may be used to describe the results of any digital simulation run, with decision trajectory $u(T)$ and noncontrollable input trajectory inflow $w(T)$.

Although in principle a performance index PI, such as grade of service (defined below), may be calculated at every time t, one usually only evaluates a single performance index

Table 1: Behavior of a system under experiment $c(T) = x(T), s(0), T$

TIME	INPUT	STATE	PERFORMANCE	OUTPUT
$j \in J$	$x(j) \in X = (U, W)$	$s(j) \in S$		$y(j) \in Y$
0	$x(0) = (u(0), w(0))$	$s(0)$ (given)		$y(1)$
1	$x(1) = (u(1), w(1))$	$s(1)$	$PI(1) \in S(1)$	$y(1)$
$\vdots$	$\vdots$	$\vdots$	$\vdots$	$\vdots$
T	$x(T) = (u(T), w(T))$	$s(T)$	$PI(T) \in S(T)$	$y(T)$

value $PI(T)$ per system experiment and then one figure of merit value per ensemble of experiments. The figure of merit thus appears to be a "super criterion".

Performance Indices Nine incident-related PI's which will be used to assess the desirability of an input/output trajectory pair are now defined, namely:

PI^1 : grade of service
PI^2 : quality of service
PI^3 : speed of response
PI^4 : reliability
PI^5 : incident period
PI^6 : mission reliability
PI^7 : availability
PI^8 : repairability or resilience
PI^9 : vulnerability

These nine performance indices are to be traded off in a multicriterion procedure. Further one defines resource indices such as economic indicators, land usage, manpower required, which are also traded off and then performance and resource figures of merit are traded off.

The first three indices can be calculated solely from the system characteristics given so far and are respectively defined as:

PI^1: the grade of service is the relative frequency of providing a service, such as supplying water, when it is required, that is, when the demand arises. Shamir (1987) defines this PI as the "fraction of the demanded water which is supplied, computed over a specific period of time, such as a day, a month, a year".

PI^2: the quality of service or of response, measures for example the percentage of requirement satisfied. Shamir (1987) defines a related but somewhat different quantity, namely, the "fraction of time the consumer is connected to a source and has some supply (even if less than demanded)". Alternatively, a negative quality of service PI^2 may be "the number of customers who have at least one failure event, greater than some value, once or more during the time period".

PI^3: the speed of response is the time elapsed between the occurrence of the demand for a service and the delivery of that service.

The other PI's can be calculated only after the incident mode , which includes failure modes, has been specified as in equation (1). The decision maker must make a subjective judgement to define the range, $(\lambda - \rho)$ values for which an incident is minor, major, or constitutes a failure. Such a judgment may be encoded as a fuzzy number (Kaufman and Gupta, 1985, 1988).

Examples of incidents usually classified as minor incidents are:

1. the inundation of a low-yield agricultural floodplain because of high reservoir release.
2. a 10% agricultural water shortage occurring during a non-critical crop growth phase.
3. a 5% exceedance of the nitrate standard for one day

Examples of incidents which would generally be classified as failures are:

1. a dam break causing heavy downstream damage
2. an extreme drought causing 80% crop losses.

There appears to be a continuum between a small incident (a few per cent shortage) and large failure (total shortage). The main point is that incident or failure modes μ must be properly defined. Note that the definition may involve a Boolean combination of simple events, such as: $\{\mu : \text{incident occurs if and only if} (PI^1 < 0.7 \text{ and } PI^3 > 1)\}$.

PI^4: The realiability performance index, PI^4, under the condition that has been specified, is defined as an estimate of the relative frequency that the system is not in incident mode μ during the experiment $c(T)$. To be precise, let $\delta(\mu, t)$ be an indicator function such that

$$\delta(\mu, t) = \begin{cases} 1 & \text{if } Z \text{ is in mode } \mu \text{ at time } t \\ 0 & \text{otherwise} \end{cases} \tag{6}$$

Then the reliability PI, with respect to mode μ, is by definition

$$PI^4\left(C(T), \mu\right) = \frac{T + 1 - \sum_{t=0}^{T} \delta(\mu, t)}{T + 1} \tag{7}$$

This definition embeds both structural reliability and target-related reliability (Plate and Duckstein, 1987). An operational definition of the remaining PI's is now given:

PI^5: the incident period is the mean interarrival time between entries into mode μ; it is also called the average recurrence time. If $d_5(\mu, n), n \geq 1$, denotes the duration of the n-th interarrival time, then

$$PI^5\left(C(T),\mu\right)=\frac{1}{N-1}\sum_{n=1}^{N-1}d_5(\mu,n) \tag{8}$$

where $N = N(\mu)$ is the number of incidents of mode during time horizon t. Shamir (1987) defines a similar PI as the "number of failure events per unit time", or else, as the "number of failures per unit time which are longer than some minimum". The descriptions of drought given in Dracup and Kendall (1991) make use of similar ideas.

PI^6: The mission reliability is an estimate of the probability that the system will not have an incident between the time that a non-null input requirement arrives and the corresponding service (water supply) is delivered. For a flood control problem, PI^6 may count the number of floods crested below a given level by the reservoir with respect to the total number of incoming floods. Vogel (1987) gives real world examples of calculations of PI's.

PI^7: The availiability is the probability that the system will not be in mode when the demand for a service occurs, for example, that the reservoir will not be above a critical control level when a flood arrives. PI^7 differs from PI^6 in that a large flood may still result in substantial spillway overflow and damage downstream. PI^6 and PI^7 are precise restatements of PI^7.

PI^8: The repairability is the average length of time that a system stays in mode during the experiment $c(T)$:

$$PI^8\left(C(T),\mu\right)=\frac{\sum_{t=0}^{T}\delta(\mu,t)}{T+1}=\frac{\sum_{n=1}^{N}d(\mu,n)}{N} \tag{9}$$

where $d_8(\mu,n)$ is the duration of the n-th mode μ incident ($n = 1,2,\cdots,N$). Index PI^8 is closely related to resilience which broadly speaking, measures the average recovery time from the occurrence of an incident mode given that one has occurred. For example, Hashimoto et al. (1982) state that resilience measures "how quickly the system returns to a satisfactory state" and define an index which is the limit of PI^8 as $N \to \infty$. A large resilience indicates a quick return to normal state or short times in incident mode; a small resilience indicates the opposite phenomenon.

PI^9: The vulnerability measures the average severity of an incident event. Shamir (1987) defines a similar PI as the "magnitude of the largest shortfall during the unit time". Here, mode μ must take on more than one value. Let be a random μ variable such as shortage with known distribution function $F(\mu)$, then PI^9 may be the expected value:

$$PI^9\left(c(T),\mu\right)=E(\mu) \tag{10}$$

or some other statistic.

Next, the resource indices are grouped under a 10th vector PI:

PI^{10}: The economic PI is a vector whose components may include expected costs, losses and benefits, rates of return, and cash flows.

Figures of Merit The PI's are usually calculated only at the end of each system experiment $\{c(T)\}$ although, as stated before, they could be calculated at every time period if so desired. The FM's are then defined as a function of the PI's calculated over the ensemble of system experiments $\{c(T)\}$.

If, for a given initial state $s(0), c(T)$ is taken as the realization of a stochastic process with known distribution $F_{\mathbf{c}}(c)$, then a set of possible FM may be defined by

$$FM^i(\mu) = \int PI^i(c,\mu)dF_{\mathbf{c}}(c) \quad \text{for } i = 1, \cdots, I \tag{11}$$

An alternative definition may also be given in terms of a percentile or another statistic; furthermore, an FM may consist of a combination of PI's (for μ given):

$$FM = FM\left(PI^1, PI^2, \cdots, PI^I\right) \tag{12}$$

Two particular FM's, implementation reliability FIR and engineering risk RI, are of special interest in the present study.

If a system is composed of a moderate number of independent elements, with known reliability, placed in a series-parallel configuration, then a so-called *implementation reliability* FIR may be calculated as in classical reliability theory. However, it should be realized that FIR represents only an upper or lower bound to the actual reliability because of the simplifying assumptions which have to be made in order to keep the calculations at a manageable level. Independence and lag one Markov dependence are examples of such assumptions, as illustrated in the third example presented in this paper.

Thus, if water is being supplied from a set of I independent sources, each with a reliability $RE(i), i = 1, 2, \cdots, I$, then the implementation reliability of the sources placed in series and in parallel are given in Eqs. (13) and (14) respectively:

$$FIR = \prod_i RE(i) \tag{13}$$

$$FIR = 1 - \prod_i (1 - RE(i)) \tag{14}$$

The reliability of series-parallel combinations can be calculated in a similar manner, but how realistic would such a model be? This point should be investigated on a case-by-case basis.

The other FM to be introduced at this point is engineering risk RI. The qualification "engineering" is needed to limit the scope of the word "risk" which could encompass concepts of social and psychological sciences that are quite important but outside of the scope of the present study. The comparison of a set of I alternative systems Z_i is reduced to the comparison of I vectors of FM's may then be (Wymore, 1991):

$$FM(Z_i) = \left(FM^1(Z_i), \cdots, FM^I(Z_i)\right) \quad i = 1, \cdots, I$$

where I is the number of FM's selected. This comparison constitutes a multicriterion decision problem whose "cost-effectiveness" table (Goicoechea et al., 1982) or decision matrix (Hiessl et al., 1985) is an array of alternative systems versus FM's. Next, to each choice of system, associate a measure of the consequences of uncertainty in the system elements, such as uncertainty in the inflow sequence or in the requirements. Engineering risk is then defined as a joint probability distribution function over the set of all possible values of the FM's. As an example, the *engineering risk* of a supply scheme Z may be:

$$RI\,(Z, a(4), a(5), a(8)) = Pr\left[FM^4(Z) < a(4); FM^5(Z) < a(5); FM^8(Z) > a(8)\right] \quad (15)$$

If we let reliability limit be $a(4) = 0.95$, mean recurrence time limit, $a(5) = 30$ months and proportion of down time threshold, $a(8) = 0.10$, then $RI(Z)$ is the probability that the system Z will have a reliability below 0.95, a mean time between μ-events under 30 months and be in mode more than 10% of the time.

Note that system reliability $RE = FM^4$ is embedded into the definition of engineering risk, as shown for example in equation (15). Also, the particular case of that equation:

$$RI\,(Z, a(4)) = Pr\left(PI^4(Z) < a(4)\right)$$

expresses that the simplest measure of risk, the probability of failure, is the complement of reliability.

MULTICRITERION DESIGN UNDER RISK

Levels of Design The design of water related structures which by definition involve hydrologic inputs, is usually seen as a problem in which the hydrologist is selecting the design information, and then an engineer uses it to find the dimensions of the appropriate structure. In reality, however, hydrologic input and structure design must be seen as integral and interacting parts of the general design procedure.

This procedure can be formulated in terms of a general systems approach presented in Section 2 above, as well as in Plate and Duckstein (1988). Depending on the performance index which has been selected, one can distinguish a number of different design approaches. Four different performance indices are used, which define different design levels: where the understanding is that the higher the level the more data are required for its application:

1. Level I uses the exceedance probability PE as performance index,

2. Level II uses probability of failure PF as a performance index, under the assumption that all probability densities are Gaussian (or transformable into Gaussian distributions)

3. Level III also uses the probability of failure PF as a performance index, but probability distributions keep their true shape.

4. Level IV uses the generalized risk RI as a performance index, with special cases included.

5. Level V uses reliability RE as a performance index relating incident (or failure) probabilities to the life expectancy of the system; uncertainties should be considered in the estimation of RE.

The above concepts are illustrated by the three examples in the next section and the case study in Section 5.

EXAMPLES OF DESIGN UNDER RISK

Bridge Pier Design Consider the bridge design problem described in Laursen (1970) and again in Davis and Dvoranchik (1971). A 500-foot bridge is to rest on four piers, each containing 25 piles with a sinking cost of \$4 per foot of pile. Cost of the bridge portion that may be lost because of pier scour due to flood is \$150,000. An empirical relationship $x = x(Q(t))$ between scour depth $x(t)$ and extreme streamflow $Q(t)$ at time t, is employed. It is desired to minimize cost of sinking the piles plus expected cost of bridge replacement, that is,

$$\text{minimize } E(L(h, \mathbf{Q}); \quad \mathbf{Q} = \{Q(t); t = 1, 2, \cdots, T\} \tag{16}$$

In the case investigated by Davis and Dvoranchik (1971), $\mathbf{Q}(t)$ is the peak annual flow of the Rillito river near Tucson, Arizona, USA, and the DF $F_{\mathbf{Q}}(q|\theta)$ of $\mathbf{Q}(t)$ is lognormal with parameter vector $\theta = (\mu, \sigma^2)$; μ and σ^2 are, respectively, the true mean and variance of the logarithm of flows Q assumed to be unknown because of short sample record. The expected loss is calculated as:

$$E\left(L(h, Q)\right) = \int_{\mathbf{Q}} L(h, q) dF_{\mathbf{Q}}(q|\theta) = g(h, \theta) \tag{17}$$

The design h is sought which minimizes the expected value $R(h)$ of $g(h, \theta)$, called goal function, which respect to a prior $pdf\ \pi(\theta)$. $R(h)$ is the Bayes risk. The problem is thus summarized as follows:

Find h that minimizes

$$R(h) = \int_{\theta} g(h, \theta) \pi(\theta) d\theta \tag{18}$$

In terms of the general systems model, the following elements are defined for bridge model Z:

1. time t: year $t = 0, 1, \cdots, T$, where T is the design horizon.

2. input: trajectory $f > 0$ given by the sequence of annual peak floods $Q(t)$ obtained from model $F_{\mathbf{Q}}(q|\theta)$; also, prior $\pi(\theta)$ based on data D and decision h at time 0.

3. state: $x(t) > 0$, the depth of scour in year t, assumed to be independent of antecedent conditions.

4. state transition function Z: the empirical relationship between $x(t)$ and $Q(t)$.

5. output: failure of the bridge, and cost (Bayes risk).

6. output function Z: cost function such as given in equations (16), (17) or (18).

For the purpose of calculating the reliability performance index PI^4 and system reliability $RE(Z)$, let the output function be rewritten as the following performance index PI, in 10^3 dollars:

$$L(h,Q) = \begin{cases} 150 + 0.4h & \text{if } x(Q) > h \text{ (bridge fails)} \\ 0.4h & \text{if } x(Q) < h \text{ (bridge does not fail)} \end{cases} \tag{19}$$

Let $x^{-1}(.)$ be the inverse function of $x(Q)$. Then the Bayes risk component of the output is calculated as

$$BR(h) = \int_H \left[\int_0^{x^{-1}(h)} 0.4h \; dF_{\mathbf{Q}}(q|\theta) + \int_{x^{-1}(h)}^{\infty} (150 + 0.4h) \; dF_{\mathbf{Q}}(q|\theta) \right] \pi(\theta) d\theta \tag{20}$$

It may be noted that $BR(h)$ is indeed a figure of merit $FM(Z)$ obtained by taking the expectations with respect to Q and θ of the performance index represented by equation (16).

Equation (20) may be integrated to yield

$$BR(h) = 0.4hM_{\mathbf{Q}}\left(x^{-1}(h)\right) + (150 + 0.4h)\left[1 - M\left(x^{-1}(h)\right)\right] \tag{21}$$

where $M_{\mathbf{Q}}(q)$ is the predictive, unconditional or Bayesian DF of input Q (Berger, 1985, Bernier, 1987):

$$M_{\mathbf{Q}}(q) = Pr(Q < q) = \int_{\theta}^{q} \int_H f_q(q|\theta)\pi(\theta) d\theta dq \tag{22}$$

$M_{\mathbf{Q}}(q)$ may be considered as a reliability figure of merit, that is, the complement of the Bayesian probability of failure obtained by taking the expectation of equation (19) with respect to the set of possible input trajectories $(\mathbf{Q}, \pi(\theta))$. These input trajectories are generated on the basis of a given data set D. To express this dependence, equation (22) may be rearranged as:

$$FM(h|D) = M_{\mathbf{Q}}\left(x^{-1}(h)\right) = 1 - P_F(h) = \frac{150 + 0.4h - BR(h|D)}{150} \tag{23}$$

Thus, to every system design h and data set D, there corresponds a reliability figure of merit. Consider the Bayes design h^* and data sets as given in Table 1 of Davis and Dvoranchik (1971). The quantities of equation (23) are given in Table 2. If the same calculations are done on the basis of 5-year data sets (Table 2 in reference cited), then the results of Table 3 are obtained, which show a substantial variation of reliability as a function of D: between 0.8285 and 0.9304. In a hypothetical case when the set of all input trajectories would be the 5-year ones shown in Table 3, with equal probability, the engineering risk would be the average of the complements of $FM^4(h^*|D)$, namely:

$$RI(h^*) = 1 - \sum FM^4(h^*|D(j))P(D(j)) = 1 - .8693 = 0.1307$$

The average Bayes risk, which includes the sample model uncertainty (Bernier, 1991) may be calculated in an analogous manner.

Alternatively, a design with fixed reliability may be sought, leading to various optimum pier depths as a function of the data set considered. For example, if a reliability of $PI^4(h|D) = .95$ is desired with the data set D of 1920-1929, the calculations in Davis and Dvoranchik (1971) show that the pier depth should be 22 ft. and the Bayes risk, 9.55. More generally, a bicriterion analysis corresponding to the trade off between Bayes risk (a cost figure) and reliability may be done.

Concrete Foundation Design In Bogardi and Duckstein (1978), a foundation is to be designed so as to be protected against corrosion by sulfurous groundwater, the level of which fluctuates randomly. The groundwater level stochastic process (input trajectory) is described as a partial duration series, where the jth event is defined by the maximum exceedance $Z(j)$ above the foundation floor, the exceedance duration $D(j)$, and the time $W(j-1)$ elapsed between the beginning of event $j-1$ and that of event j (Figure 1). Four decisions may be made (or four alternative systems may be chosen) to protect the structure against possible damage by high levels of groundwater with substantial sulfate concentration; these four decisions correspond to taking or not taking two protection measures: $e(1)$ = to waterproof the floor against water pressure; $e(2)$ = to use special concrete resisting to sulfate. The four actions and corresponding costs and losses (or outputs), as calculated in Bogardi and Duckstein (1978), are shown in Table 4, for the specific set of Data D with 37 events used in that study. In particular, consider the loss corresponding to action $a(2)$, calculated by the output or readout function

$$L\left(a(2)\right) = K\ Pr(SL \leq \beta) \tag{24}$$

Table 2: Optimum Bayes Risk and Reliability for Data Sets D of Varying Lengths

Data D	1920 - 1924	1920 - 1929	1920 - 1939	1920 - 1939
Optimum Pier depth h^*	21.2	16.5	11.7	11.0
Optimum Bayes Risk $BR(h^*)$	12.045	7.845	5.059	4.680
Bayes Reliability Figure of Merit: $1 - P_F = FM^4(h * \|D)$	0.8632	0.9060	0.9351	0.9395

$$K = \text{cost of replacing the foundation}$$
$$SL = \text{strength or resistance loss of concrete}$$
$$\beta = \text{threshold defining ``complete failure''}$$

Let $T(\beta)$ be the expected time to a complete failure, or failure period PI^5. A first order analysis in the cited study yield the relationship:

$$PI^5(\beta) = T(\beta) = -0.59 + 3.62\beta^{\frac{1}{2}} \tag{25}$$

in which β is measured in percentage points. The probability of failure is the reciprocal of $T(\beta)$:

$$P_F = \left[-0.59 + 3.62\beta^{\frac{1}{2}}\right]^{-1} \tag{26}$$

The threshold β may be a decision variable in terms of type or thickness of the concrete. A reliability performance index may thus be defined as the complement of P_F in equation (26). For example, if β takes on the values 10, 20 and 30, one finds a reliability $PI^4(a(2), D)$ of, respectively, 0.908, 0.936 and 0.948.

Next, when considering action a(3) - use of special concrete - another type of failure mode must be defined, namely, flooding of the basement whenever $Z(j) \geq Z(0)$. Then a vulnerability performance index PI^9 may be defined as the average flooding damage per event. Other performance indices, such as availability PI^7 and repairability PI^8, can also be defined, especially if a dewatering system is installed. Note that action (or system) $a(1)$ is obtained by coupling $a(2)$ and $a(3)$, while $a(4)$ is the "complement" of $a(1)$.

System reliability and engineering risk may be defined by simulating various input trajectories (Figure 1) and taking the expected value of the various performance indices, as it was explained for the bridge pier example. A multicriterion analysis can then be performed to trade off the PI's thus calculated.

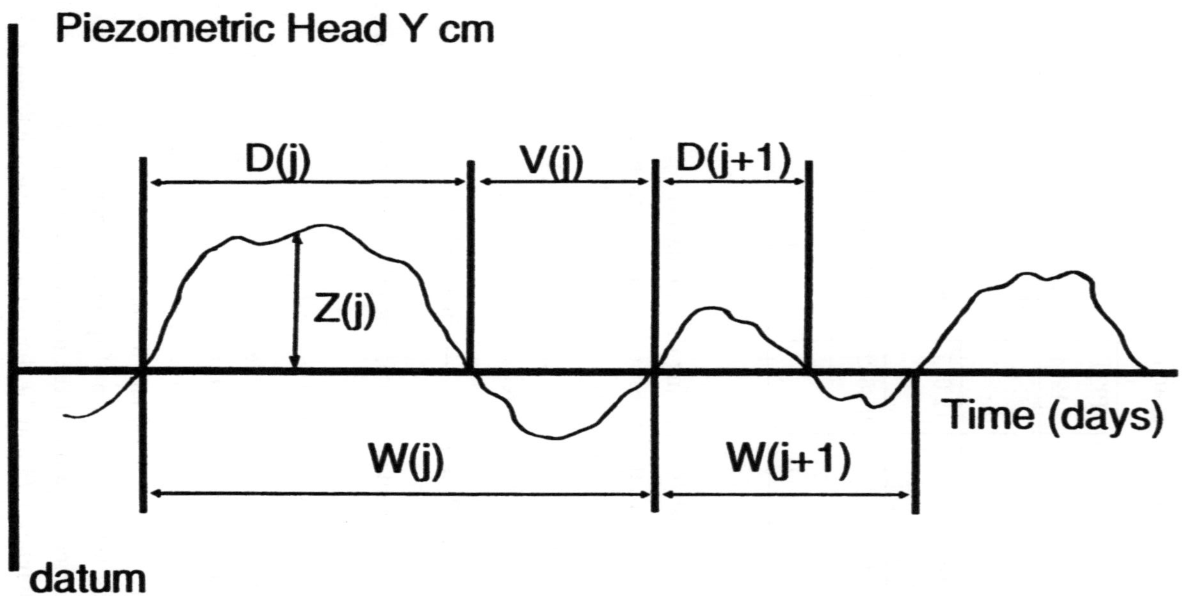

Figure 1: Definition sketch of hydrologic input trajectory for the foundation design example

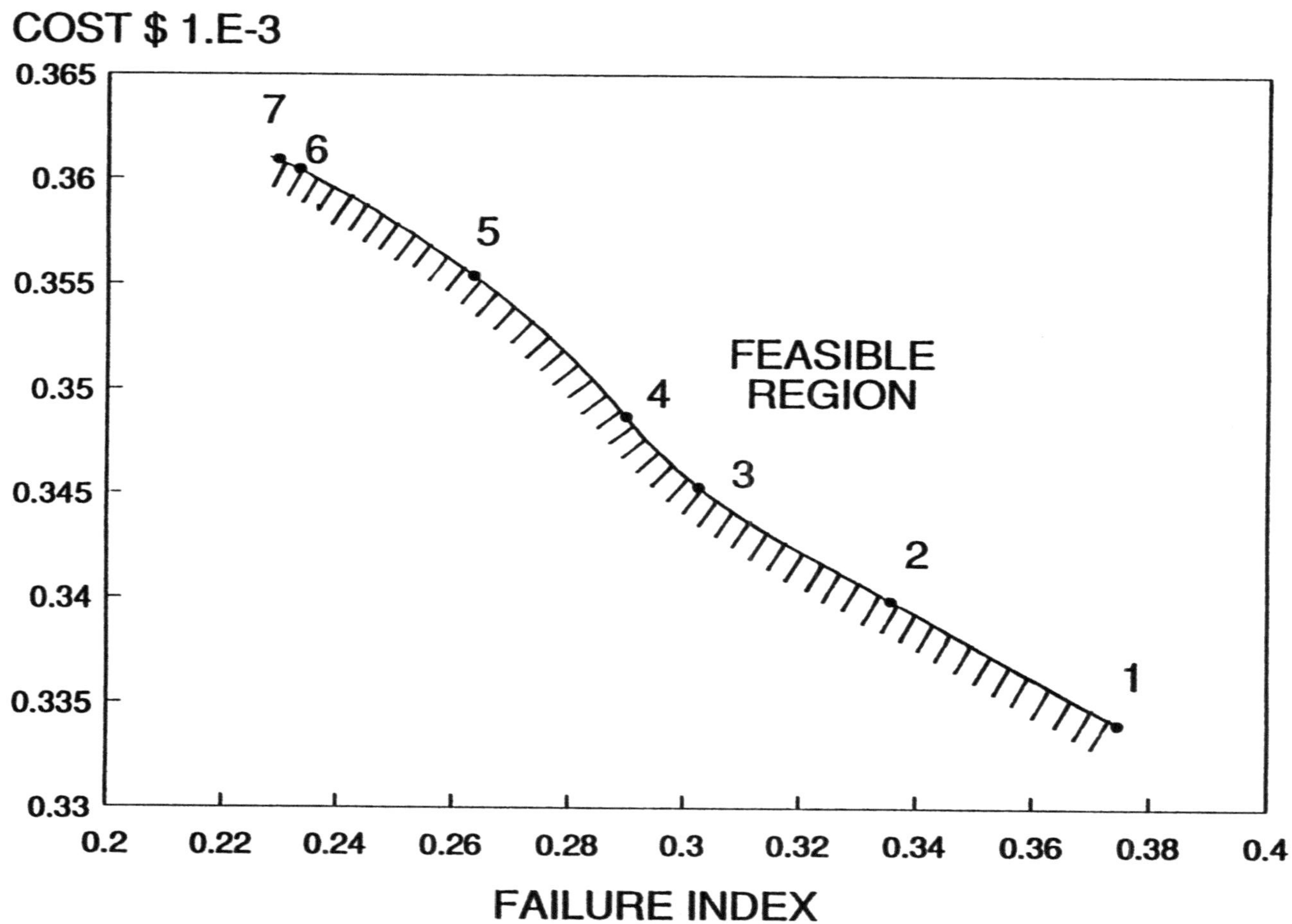

Figure 2: Tradeoff between failure probability and cost for the dewatering system example

Table 3: Optimum Bayes Risk and Reliability for 5-year Data Sets

Data Set D	1920 - 1924	1925 - 1929	1930 - 1934	1935 - 1939	1940 - 1944	1945 - 1949	1950 - 1954
h^*	21.2	26.2	11.7	18.8	21.6	22.8	20.0
$BR(h^*)$	12.405	15.327	5.766	10.322	12.841	13.575	11.164
$FM^4(h^*\|D)$	0.8632	0.8285	0.9304	0.8811	0.8568	0.8487	0.8762

Table 4: Action Versus Cost and Loss

Action	e(1) Waterproof	e(2) Special Concrete	Annual Cost $\$10^6$ Ft.	Expected loss $\$10^6$ Ft(1)
a(1)	Yes	Yes	3.4	0
a(2)	Yes	No	1.5	$K\ Pr[SL \geq \beta]^2$
a(3)	No	Yes	1.9	$76E[\sum_j Z(j) - Z(0))^2]$ $Z(j) \geq Z(0)$
a(4)	No	No	0	$K\ Pr[SL \geq \beta]$ $+76E[(Z(j) - Z(0))^2]$ $Z(j) \geq Z(0)$

1Ft. = \$0.05 K = Cost of replacing the foundation

Table 5: Efficient set trading off reliability $FMR(Z)$ versus cost $PI^{10}(Z)$

Design Z	Failure Index $100(1 - FMR(Z))$	Cost $10^3 PI^{10}(Z)$
1	0.376	0.3317
2	0.334	0.3398
3	0.350	0.3446
4	0.289	0.3492
5	0.263	0.3552
6	0.232	0.3607
7	0.230	0.3642

Dewatering System Design In this example, the system implementation reliability or figure of merit FMR is calculated. Consider an underground structure subject to flooding, such as the basement in the preceding example under designs a(3) and a(4), or a mining tunnel, as in Bogardi et al. (1987). A pumping system is designed to remove the water whenever a flooding event occurs. This system is composed of N independent subsystems in series, such as booster stations to remove water from deep mines. The system fails if any of the subsystems fail. For this reason, each subsystem j built with $n(j)$ identical units or pumps in parallel. In the example problem investigated in Sakawa (1978), a trade-off is to be found between system implementation reliability FMR and cost. Using Sakawa's example with $N = 4$, $n(1) = n(3) = 7$, $n(2) = n(4) = 8$, cost functions and constraints as given in that study, the results shown in Table 5 are obtained, in which $100(1 - FMR(Z))$ is the percentage point of failure and $PI^{10}(Z)$ is the cost along the non-dominated or efficient set. The curve representing this efficient set is shown in Figure 2.

To reach a compromise or tradeoff solution, the surrogate worth tradeoff method (Haimes et al, 1975) is used in Sakawa (1978) yielding alternative $Z = 3$. As an alternative technique, compromise programming (Zeleny, 1973; 1982; Duckstein and Opricovic, 1978) may be used. For example, if the numbers in Table 5 are considered to be normalized with weight taken as unity and the ideal point is taken as the orign, then the compromise solutions for $p = 1$ is obtained by minimizing the sum of failure and cost indices, and the solution for $p = 2$, by minimizing the sum of the squares of those indices. In both cases, the compromise solution is found for alternative design $Z = 6$. The main point in all these design examples is that the system model leads, in a natural way, to a multiple criteria analysis.

MULTICRITERION OPERATION UNDER DROUGHT RISK

In this section a dynamic problem with multiple objectives is developed. The case study of the Tenkiller reservoir described in the Appendix. Whenever drought conditions begin developing, the question arises to switch the operating procedure of the reservoir from "normal" to "alert" mode and then may be to "emergency" mode. As in the Appendix, multiple conflicting objectives must be taken into account in this approach which consists of the following steps.

1. Stochastic dynamic programming (SDP) is used to establish "normal", "drought alert" and "drought emergency" modes of operating (operating rules of the reservoir and water allocation). For this purpose, three different inflows, derived from the record with normal, low and very low means respectively, are generated and the multiobjective SDP algorithm is applied.

2. Let $\theta = (\mu, \sigma^2)$ be the historical parameter vector of inflow distribution. The National Weather Service and Soil Conservation Service, jointly, issue a long-range forecast of several month (basically for a season), yielding a prior parameter vector *pdf* $f'(\theta)$. Bayes theorem is used to obtain an updated parameter *pdf* $f''(\theta)$.

3. Using the existing storage (at time 0 of the analysis), and future flows generated on the basis of $f''(\theta)$, each of the three operating policies is applied in turn, and the policy yielding a minimum present value of the L_p norm is selected.

DISCUSSIONS AND CONCLUSIONS

While the application of the system model to the examples has not been done in an exhaustive manner, it is hoped that the usefulness of such a general system model, in which various performance indices and figures of merit related to system failure are defined precisely, has been demonstrated. Clearly, the system framework may not be necessary in every case; for example, if the reliability of a levee reach is sought, then the approach used in Szidarovszky et al. (1976) and Duckstein and Bogardi (1981) may be sufficient. Or else, dam safety may be studied as in Plate (1984) or Kara and Krzysztofowicz (1983).

The following concluding points may be made:

1. A mathematical system framework that embeds both structural reliability and target-related reliability has been defined.

2. An incident is said to occur when a generalized load exceeds a generalized resistance.

3. The discrete-time discrete-state model is composed of five sets. The initial state and input trajectory uniquely determine the behavior of the system over time horizon t.

4. The performance of the system is measured by means of a set of running criteria called performance indices PI, and a set of "super-criteria" called figures of merit FM. The FM's are combinations of PI's calculated over a set of observation sequences "experiments", such as simulation runs.

5. Most performance indices can only be calculated after the incident modes have been defined.

6. Reliability of a system component is a PI that can usually be calculated, but reliability of a complex system cannot in general be calculated from the reliability of the components. Furthermore, reliability is embedded in engineering risk.

7. When sample uncertainty is present, as in the bridge pier design example, a Bayesian reliability may be defined and traded off against Bayes risk (cost).

8. The foundation design example illustrates the case when different actions may lead to different modes of failure.

9. The dewatering system example, which is based on a tradeoff between implementation reliability and cost, illustrates how the application of the system framework leads quite naturally to a multiple criteria analysis.

10. The case study of the Tenkiller lake in Oklahoma illustrates how multicrterion analysis under risk may be performed in a dynamic setting.

ACKNOWLEDGEMENTS

Research results reported in this paper have been supported in part by NSF Grant #ECS 8802920 and #BCS 9016462.

REFERENCES

Bardossy, A., and Duckstein, L., Analysis of a Karstic Aquifer Management Problem by Fuzzy Composite Programming, Working Paper, 90-17, 1990.

Berger, J.O., "Statistical Decision Theory and Bayesian Analysis". Springer Verlag, New York, 583 p., 1985.

Bernier, J., Elements of Bayesian Analysis of Uncertainty in Hydrological Reliability and Risk Models (in) Engineering Reliability and Risk in Water Resources, L. Duckstein and E.J. Plate (eds.) *NATO ASI Series*, M. Nijhoff, Dordrecht, The Netherlands, 1987.

Bogardi, I., Duckstein, L. and Szidarovsky, F., Reliability Estimation of Underground Water Control Systems Under Natural and Sample Uncertainty (in) Engineering Reliability and Risk in Water Resources, L. Duckstein and E.J. Plate (eds.) *NATO ASI Series*, M. Nijhoff, Dordrecht, The Netherlands, 1987.

Bogardi, I. and Bardossy, A., Reliability Aspects of Multicriterion Watershed Management (in) Engineering Reliability and Risk in Water Resources, L. Duckstein and E.J. Plate (eds.) *NATO ASI Series*, M. Nijhoff, Dordrecht, The Netherlands, 1987.

Bogardi, I. and Duckstein, L., Design Under Stochastic Groundwater Fluctuation, *Journal of the Hydraulics Division*, January, HY1, pp. 59-74, 1978.

Cunha, R., Galler, W. S. and Hodgson, T. J., Multiple Objective Analysis in the Scheduling of Reservoir Releases, *Proceedings* (to appear), IX-International Conference on Multiple-Criteria Decision Making, Washington, D. C., 1990.

Davis, D. and Dvoranchik, W., Evaluation of the Worth of Additional Data, *Water Resources Bulletin*, Vol. 7, No. 4, August, pp. 700-707, 1971.

Dracup, J.A. and Kendall, D. R., Risk Analysis of Water Supply Network of Southern California, Risk and Reliability in Water Resources and Environmental Engineering, *NATO ASI*, Greece, 1991.

Duckstein, L. and Bogardi, I., Application of Reliability Theory to Hydraulic Engineering Design, *Proc. ASCE*, Vol. 107, J. Hydr. Div., pp. 799-815, 1981.

Duckstein, L., Head, L., and Bogardi, I., A mathematical system model of nitrate contamination, *Proceedings* (to appear), NATO Workshop: "Nitrate Contamination: Exposure, Consequence and Control," Univ. of Nebraska, Lincoln, September, 1990.

Duckstein, L., Plate, E. J., and Benedini, M., Water Engineering Reliability and Risk: A System Framework (in) Engineering Reliability and Risk in Water Resources, L. Duckstein and E.J. Plate (eds.), *NATO ASI Series*, M. Nijhoff, Dordrecht, The Netherlands, 1987.

Duckstein, L. and Opricovic, S., Multiobjective Optimization in River Basin Development, *Water Resources Research*, Vol. 16, No. 1, pp. 14-20, 1978.

Goicoechea, A. M., Hansen, D. R., and Duckstein, L., "Multiobjective Decision Analysis with Engineering and Business Applications", John Wiley & Sons, New York, 1982.

Haimes, Y., Hall, W., and Freedman, H., "Multiobjective Optimization in Water Resources Systems: The Surrogate Worth Trade-Off Method", Amsterdam-Oxford-New York: Elsevier Scientific Publishing Company, 1975.

Hiessl, H., Duckstein, L. and Plate, E. J., Multiobjective Q- Analysis with Concordance and Discordance Concepts, *Applied Mathematics and Computation*, Vol. 17 pp. 107-122, 1985.

Kara, F. and Krzysztofowicz, R., Bayesian Decision Analysis of Dam Safety, Working paper, Department of Engineering Science and Systems, University of Virginia, Charlottesville, Virginia 22901, 1983.

Karamouz, M. and Houck, M. H., Comparison of Stochastic and Deterministic Dynamic Programming for Reservoir Operating Rule Generation, *Water Resour. Bull.*, 23(1), pp. 1-8, 1987.

Kaufmann, A. and Gupta, M. M., "Introduction of Fuzzy Arithmetic: Theory and Applications", Van Nostrand Reinhold, New York, 351p, 1985.

Kaufmann, A. and Gupta, M. M., "Fuzzy Mathematical Models in Engineering and Management Science", North Holland, Amsterdam, 338 p, 1988.

Laabs, H., and Harboe, R., Generation of Operating Rules with Stochastic Dynamic Programming and Multiple objectives, *Water Resources Management*, 1988.

Laursen, M., Bridge Design Considering Scour and Risk, *Transportation Engineering Journal*, TE2, May, pp. 149-164, 1970.

Plate, E. J., Reliability Analysis of Dam Safety (in) "Frontiers in Hydrology" W.H.C. Maxwell and L.R. Beard (eds.), *Water Resources Publications.* Littleton, Colorado, U.S.A. pp. 288-304, 1984.

Plate, E. J. and Duckstein, L., Reliability in Hydraulic Design (in) Engineering Reliability and Risk in Water Resources, L. Duckstein and E.J. Plate (eds.) *NATO ASI Series*, M. Nijhoff, Dordrecht, The Netherlands, 1987.

Plate, E. J. and Duckstein, L., Reliability-Based Design Concepts in Hydraulic Engineering, In press, Water Resources Bulletin, 24(2): 235-245 April, 1988.

Plate, E. J. and Duckstein, L., Stochastic aspects of water quality modeling for non-point sources. (in) DeCoursey, G., editor. *Proceedings*, International Symposium on Water

Quality Modeling of Agricultural Non-Point Sources, Part 2. U.S. Department of Agriculture, Agricultural Research Service, ARS-81, pp. 6531- 654, 1990.

Sakawa, M., Multiobjective Optimization by the Surrogate Worth Trade-off Method, *IEEE Transactions on Reliability*, Vol. R-27, No. 5, December, pp. 311-314, 1978.

Shamir, U., Reliability of Water Supply Systems (in) Engineering Reliability and Risk in Water Resources, L. Duckstein and E.J. Plate (eds.) *NATO ASI Series*, M. Nijhoff, Dordrecht, The Netherlands, 1987.

Shrestha, B. P. and Bogardi, J. J., Optimal Operation of a Multipurpose Reservoir System with Stochastic Inflows and Irrigation Demand, *Proceedings*, International Symposium on Application of Water Resources Systems, Canada, 1990.

Shrestha, B. P., Sen, S. and Duckstein, L., Deriving Reservoir Operation Rules by Multicriterion Stochastic Dynamic Programming, Working paper, 90-11, SIE, Univ. of Arizona, Tucson, AZ 85721, presented in part at TIMS-ORSA Fall Meeting, Philadelphia, 1990.

Szidarovszky, F., Duckstein, L., and Bogardi, I., Levee System Reliability Along a Confluence Reach, *Journal of the Engineering Mechanics Division, ASCE*, Vol. 101, No. EM5, Proc. Paper 11631, October, pp. 609-622, 1976.

Szidarovszky, F. and Duckstein, L., Dynamic multiobjective Optimization: A Framework with Application to Regional Water and Mining Management, *European Jour. of Oper. Res.*, 24, 305-317, 1986.

Szidarovszky, F., Gershon, M. E. and Duckstein, L., "Techniques for Multiobjective Decision Making in Systems Management", Elsevier, Amsterdam 506p., 1986.

Tai, F. K. and Goulter, I. C., A Stochastic Dynamic Programming base Approach to the Operation of a Multi-reservoir System, *Water Resour. Bull.*, 23(3), pp. 371-377, 1987.

Tauxe, G., Imman, R. and Mades, D., Multiobjective Dynamic Programming with Applications to a Reservoir, *Water Resour. Res.*, 15(6), pp. 1403-1408, 1979.

Tecle, A., Fogel, M. M. and Duckstein, L., Multicriterion Analysis of Forest Watershed Management Alternatives, *Water Resour. Bull.*, 24(6), pp. 1169-1178, 1988.

US Army Corps of Engineers, Southwestern Division, Tulsa District, Restudy of Tenkiller Ferry Lake Illinois River, Oklahoma, *Draft Survey Report and Environmental Impact Statement*, 1982.

Vogel, R. M., Reliability Indices for Use in the Design of Water Supply Reservoir (in) Engineering Reliability and Risk in Water Resources, L. Duckstein and E.J. Plate (eds.) *NATO ASI Series*, M. Nijhoff, Dordrecht, The Netherlands, 1987.

Wymore, A. W., "Systems Engineering Methodology for Interdisciplinary Teams", Wiley, Inc., New York, p. 362, 1976.

Wymore, A. W., "A Mathematical Theory of Systems Engineering" to appear, CRC Press and SANDS, Tucson, 1991.

Yakowitz, S., Dynamic Programming Applications in Water Sources, *Water Resour. Res.*, 18(4), pp. 1089-1099, 1982.

Yeh, W. W-G., Reservoir Management and Operation Model: A State-of-art Review, *Water Resour. Res.*, 21(12), pp. 1797-1818., 1985.

Zeleny, M., Compromise Programming, in "Multiple Criteria Decision Making", J. L. Cochrane and M. Zeleny (Eds.), University of South Carolina Press, Columbia, pp. 262-301, 1973.

Zeleny, M., "Multiple Criteria Decision Making", McGraw-Hill Book Company, New York, 563 p., 1982.

APPENDIX

A Multicriterion Stochastic Dynamic Programming Model for Designing Reservoir Operation Policy

As explained in Shrestha et al. (1990) stochastic dynamic programming (SDP) algorithm is one of the suitable techniques to derive optimal operation policies of reservoir systems when the multiple conflicting objectives and stochastic characteristic of the inputs are jointly considered. Yakowitz (1982) and Yeh (1985) in their reviews of dynamic programming applications in water resources discussed the development of the SDP applications during the three preceding decades summarizing all essential work untill 1985. Recent contributions to explicit SDP in reservoir management as found in Karamouz and Houck (1987), Tai and Goulter (1987) and Shrestha and Bogardi (1990) specify the potential use of the SDP algorithm in solving different types of water resources management problems. Multiobjective control policies have been examined under deterministic inflow hypotheses in Tauxe et al. (1979) and Cunha et al. (1990). Laabs and Harboe (1988) have been taken random inflows into account but their model handled only for local problem.

In this study, we attempt to develop a model that takes into consideration both randomness of inputs and the multiobjective nature of water resources system, by coupling SDP technique with a distance-based multiobjective technique known as compromise programming (Zeleny, 1973, 1982; Duckstein and Opricovic, 1978; Tecle et al., 1988).

The model may be described as follows.

There are six objectives related to flood control, energy generation, water supply, water quality, recreation and streamflow regulation. All six are formulated as deviation from a target to be minimized. Consider for example energy production, where the expected deviation from target firm energy is to be minimized, the formulation is as follows:

$$Min\ E\left[\sum_{t=1}^{T} Z_{1,t}\right] \quad Z_{1,t} = \begin{cases} (\text{Energy}_t - TFE)^2 & \text{if Energy}_t < TFE \\ 0 & \text{if Energy}_t \geq TFE \end{cases} \tag{A1}$$

where,

$$\begin{aligned} \text{Energy}_t &= \eta\gamma R_t H_t \Delta t \\ \eta &= \text{overall efficiency} \\ \gamma &= \text{unit weight of water} \\ R_t &= \text{release at t} \\ H_t &= \text{net average head during } \Delta t \end{aligned}$$

Δt	=	time duration
T	=	total number of time steps in a year
$E(.)$	=	Expectation
TFE	=	target firm energy

Some of the objectives or targets may be imprecise, in which case it may be appropriate to introduce a fuzzy set description (Bardossy and Duckstein, 1990). The SDP algorithm is implemented using a discrete lag one Markov chain. Thus during any given time period, the inflows can be described by a conditional probability $P(I_{t+1}|I_t)$. The recursive equation, performance function and mass balance equation are written as follows:

Recursive Equation:

$$f(S_t, I_t) = Opt\left(B(S_t, I_t, S_{t+1}) + \sum_{I_{t+1}=0}^{I_{t+1,max}} P(I_{t+1}|I_t)\, f_{t+1}(S_{t+1}, I_{t+1})\right) \quad \forall\, S_t, I_t;\, S_{t+1} \text{ feasible} \tag{A2}$$

Performance Function:

$$B(S_t, I_t, S_{t+1}) = \sum_{i=1}^{I} \alpha_i Z_{i,t} \qquad \begin{array}{c} 0 \leq \alpha_i \leq 1 \\ \sum \alpha_i = 1 \end{array} \tag{A3}$$

Mass Balance Equation:

$$S_{t+1} = S_t + I_t - R_t - E_t \tag{A4}$$

$S_t,\ I_t,\ R_t$	=	storage, inflow and release at time t
E_t	=	evaporation loss at t
$f(S_t, I_t)$	=	expected return from the optimal operation of the system from t^{th}time period to the end
$B(S_t, I_t, S_{t+1})$	=	return obtained from t to t+1
$P(I_{t+1}\|I_t)$	=	one-step transition probability
α_i	=	weight assigned to objective i

The boundary conditions for the model are

Boundary Conditions:

$$\begin{array}{c} S_{t,min} \leq S_t \leq S_{t,max} \\ R_{t,min} \leq R_t \leq R_{t,max} \end{array} \tag{A5}$$

$S_{t,min}$	=	min. storage level at time t
$S_{t,max}$	=	max. storage level at time t
$R_{t,min}$	=	min. release at time t
$R_{t,max}$	=	max. release at time t (capacity of d/s channel)

As stated earlier, the SDP model is used in conjunction with compromise programming.

$$Min\ L_p$$
$$L_p = \left[\sum_{i=1}^{I} \left(\theta_i \frac{Z_i^* - Z_i}{Z_i^* - Z_i^{**}}\right)^p\right]^{\frac{1}{p}} \quad \forall\, 1 \leq p < \infty \tag{A6}$$

where,

L_p = deviation from ideal solution
Z_i = value of objective function i
Z_i^* = most preferred value of objective i
Z_i^{**} = least preferred value of objective i
θ_i = weight assigned to objective i by decision maker
I = total number of objectives
p = parameter of L_p as described earlier

A fuzzy version of this technique has been developed in Bardossy and Duckstein (1990) may be used in further investigations. The methodology is applied to the Tenkiller Lake dam in Illinois river basin (Oklahoma) using only the four objectives in the example case given here namely energy generation, water supply, recreation and water quality. The dam is located at river mile 12.8 on the Illinois river about 7 miles northeast of Gore, and about 25 miles south of Tahlequah, as shown in Fig. A1. The reservoir is built to serve two main purposes flood control and hydropower generation. The project also provides water supply, recreation facilities and fish and wildlife conservation.

The top of Tenkiller lake dam is at elevation 677.2 which is about 197 feet above the streambed. The top of flood control pool is at elevation 667.0, the top of conservation pool is at elevation 632.0, the top of inactive pool is at elevation of 594.5. The lake area at the top of conservation pool is about 12,900 acres. The total conservation storage is about 371,000 acre feet. The reservoir stretches about 34 miles with 130 miles of shoreline. The other pertinent data can be found in US Army Corps of Engineers (1982). This report also provides the hydrological and climatological data that are required for the study. The US Army Corps of Engineers have also studied two additional extension plans (National Economic Development Plan, NED, and Environmental Quality Plan, QE).

Using a discrete lag-one Markov chain model to describe the stochastic nature of streamflow into the reservoir, conditional probabilities are constructed from the available 55 years of historical data (1923-1977). Three inflow states, namely, high, medium and low are distinguished on the basis of distribution of the inflow for a particular period (say a month). This means for inflow mean $\bar{X}$ and standard deviation of S the high, medium and low inflow states represent the inflows which are greater than $\bar{X} + S/2$, in between $\bar{X} - S/2$ and $\bar{X} + S/2$ and less than $\bar{X} - S/2$, respectively. The reservoir storage is discretized into N equal volume intervals, each describing a storage state (N is to be chosen according to available computer memory and CPU; here N=50 has been selected).

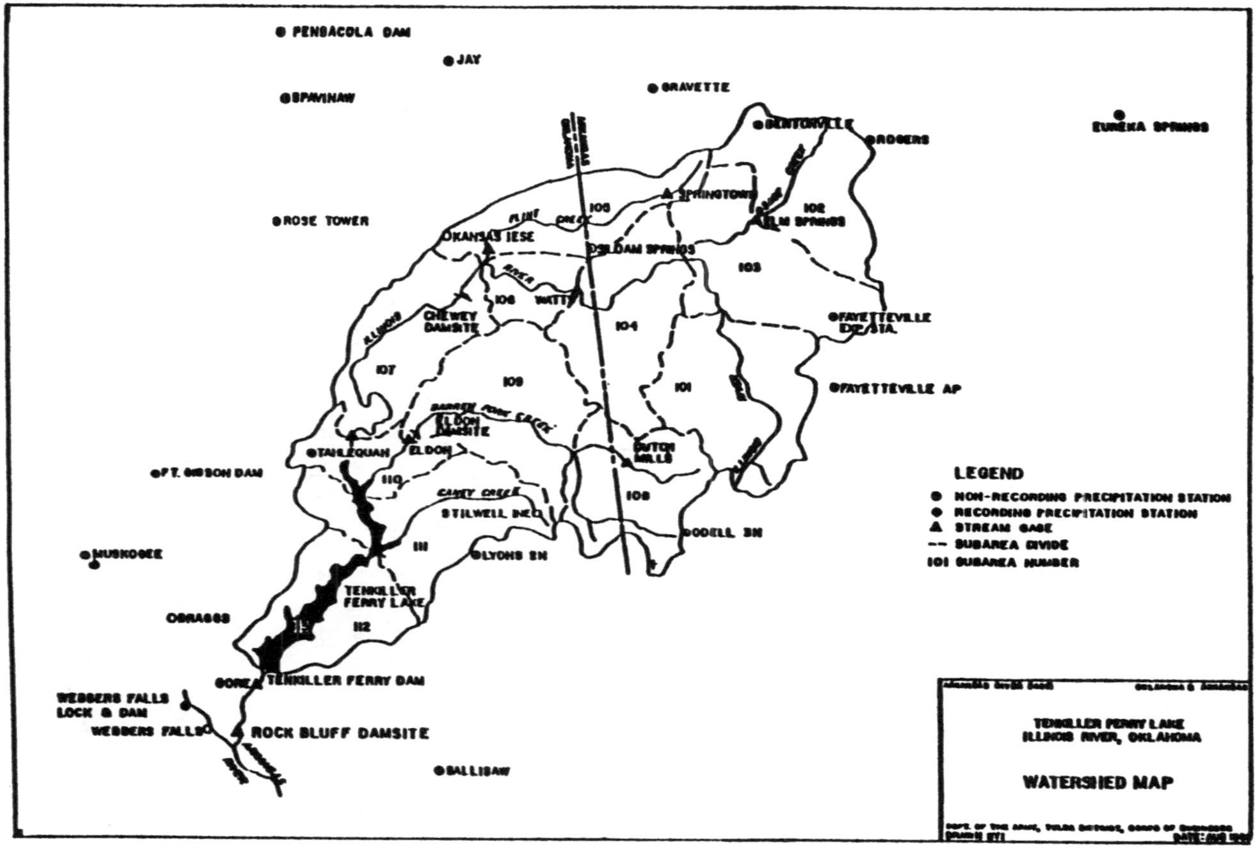

Figure A1: The Tenkiller Lake and its Watershed

Combination of SDP with a distance-based MODM technique creates dimensionality problem (inherent to dynamic programming) due to the fact that each additional objective function requires an additional state variable (Szidarovzsky and Duckstein, 1986). This in turn causes an exponential increase in computer memory (RAM) and computer time (CPU) requirements. Further, the non-commensurate objectives have to be converted into one commensurate objective. Here, the so called weighting technique (Szidarovzsky et al., 1986) with reliability-based objective is used to tackle these problems. The non-dominated solutions are thus obtained for different sets of weights assigned to different objective functions. These non-dominated solutions are the alternative reservoir operating policies. The criteria are weighted so as to reflect the importance that the decision maker places upon each objective function.

Fig. A2 shows the operating policies for the alternative with equal weights for the month of January obtained by compromise programming with $p = \infty$.

The policies obtained by SDP are compared on the basis of reliabilities of the four (rather than six) objectives under historical inflows. This alleviates the problem of non-commensurate objectives. Another approach could have been to transform the deviations into value functions or fuzzy set membership functions. The results are displayed in Table A1. The generated policies operate on the basis of incoming flows, so that higher returns can be expected with higher expected inflows.

The compromise solution set indicates that the alternative (1) generated by using equal weights in the composite objective function is preferable in terms of L_2 and L_∞ (Shrestha et al., 1990). The ELECTRE II and MCQA2 which have been applied to the same problem indicate that this alternative stays ranked high. The three techniques also show alternative with equal weights of 0.35 assigned to power and water supply and equal weights of 0.15 assigned to the other two objectives may be a preferable to the alternative with equal weights.

A summary of the results obtained so far is as follows:

1. SDP accounts for inflow randomness and dependence between two consecutive monthly inflows.

2. The policies obtained by SDP have been compared on the basis of reliabilities of four objectives evaluated under historical inflows.

3. The operating policies show that higher inflows in the present month lead to higher storage (and higher releases) in the coming month due to higher expected flows.

4. The compromise solution set indicates that alternative # 1 (all weights equal) is preferable both in $p = 2$ and $p = \infty$. Alternative # 15 is preferable in $p = 1$.

5. Ranking vary considerably as a function of balancing factor p in L_p.

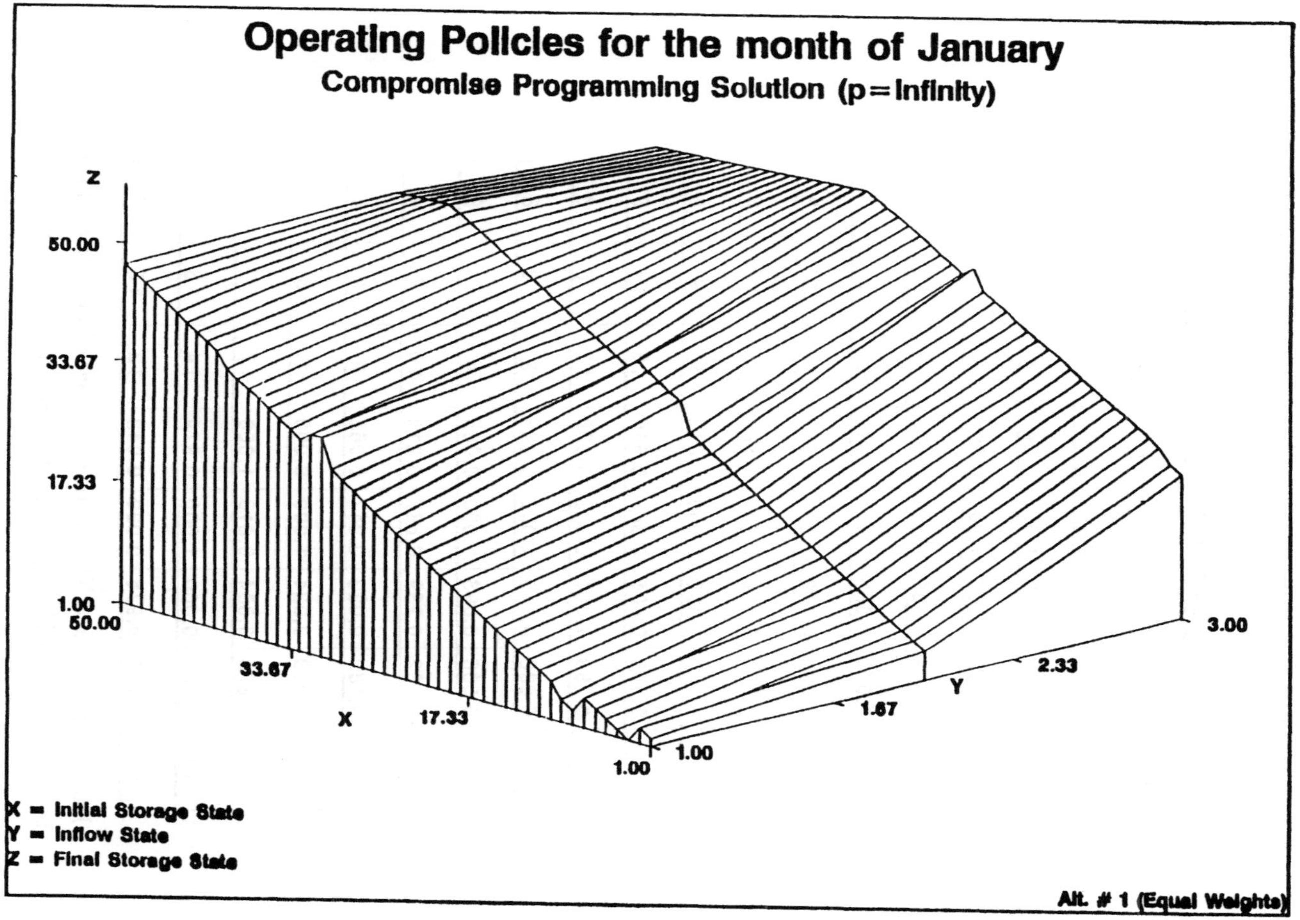

Figure A2: Operating Policy for January

Table A1: Set of the alternatives and its ranking using L_p norms

	Objectives				Ranking by Lp		
Alt. #	Z1	Z2	Z3	Z4	L1	L2	Linf
1	0.364	0.262	0.761	0.256	11	1	1
2	0.060	0.049	1.000	0.042	3	6	8
3	0.152	0.000	1.000	0.021	6	11	8
4	0.484	0.309	0.834	0.235	13	12	8
5	0.072	0.025	0.941	0.042	2	3	5
6	0.030	0.072	1.000	0.857	7	8	8
7	1.000	0.452	0.497	0.660	16	15	8
8	0.120	0.000	1.000	0.042	5	9	8
9	0.848	0.405	0.543	0.596	15	14	3
10	0.758	1.000	0.033	0.958	18	18	8
11	0.000	0.119	1.000	0.086	8	10	8
12	0.000	0.096	1.000	0.065	4	7	8
13	0.182	0.239	0.867	0.213	10	2	4
14	0.182	0.023	0.947	0.065	9	4	6
15	0.000	0.024	1.000	0.000	1	5	8
16	0.878	0.834	0.019	1.000	17	17	8
17	0.090	0.000	1.000	0.979	14	16	8
18	0.726	0.976	0.000	0.042	12	13	7

Each alternative is a NDS defined by a set of weights assigned to objectives by DM (e.g. alternative # 1 is for equal weights).

RISK MANAGEMENT FOR NITRATE-CONTAMINATED GROUNDWATER UNDER IMPRECISE CONDITIONS

M.F. Dahab and Y.W. Lee
Department of Civil Engineering
University of Nebraska-Lincoln
W348 Nebraska Hall
Lincoln, NE 68588-0531 USA

Abstract

Many areas throughout U.S. and Europe have experienced problems with increased nitrate contamination of groundwater supplies. The concern over nitrate contamination stems from the fact that nitrates have been associated with the genesis of infant methemoglobinemia and linked to the possible formation of nitrosoamines and nitrosoamides which are known etiologic agents for human gastric cancer. This paper describes research which was directed toward the development of risk management techniques for nitrate-contaminated groundwater supplies. To reduce nitrate risk from groundwater supplies, several strategies can be developed based on an acceptable level of human health risk, the reasonableness of nitrate control cost, and the technical feasibility of nitrate control methods. However, the objectives of the risk reduction and cost are in conflict with each other, and the objectives may be of varying degrees of importance. In addition, each phase of the risk and cost analysis, and the evaluation of technical feasibility are associated with imprecision. In this paper, a multicriterion decision-making methodology is presented to assist decision makers in determining, with imprecise information, a nitrate risk management strategy that best satisfies the objectives of input variables (reducing the risk and cost, and assuring the technical feasibility).

1 Introduction

This paper describes research which was directed toward the development of risk management methodology for nitrate-contaminated groundwater supplies. The methodology developed addresses specific, yet common, cases of groundwater contamination in the midwestern U.S. as well as many areas throughout Europe. The risk management methodology considers both health risk and the cost of control with special attention to nitrate treatment, and imprecision associated with the assessment of the risk and cost which are characterized by using fuzzy set

NATO ASI Series, Vol. G 29
Water Resources Engineering Risk Assessment
Edited by J. Ganoulis

approach. Various nitrate inputs as well as different control strategies, including source control and a variety of treatment systems, were considered in the development of this methodology.

Nitrates are proven precursors to methemoglobinemia a condition more commonly known as the "blue-baby syndrome." This syndrome usually occurs in infants up to 6 months old with those under 3 months of age being particularly susceptible [WHO, 1978]. Nitrate intake also appears to be a major contributor of gastric nitrite which in turn has been linked to the production of nitrosoamines and nitrosoamides. This is more troublesome since the latter compounds are well known etiologic agents for human gastric cancer [Mirvish, 1990; Crespi and Ramazzotti, 1990].

To reduce nitrate risk from groundwater supplies, several strategies can be developed. Final decisions on the appropriate strategy may be based on: 1) an acceptable level of human health risk, 2) the reasonableness of nitrate control cost, and 3) the technical feasibility of nitrate control methods. A discrete set of nitrate control methods include source control, development of new water supplies, blending of two or more existing water supplies, and direct treatment for nitrate removal. Strategies with high cost may provide a high degree of human health risk protection, while strategies with low cost may not provide adequate protection. In other words, the objectives of risk reduction and cost are in conflict. In addition, the objectives may be of varying degrees of importance. Thus, the ultimate goal of nitrate risk management is to determine which strategy "best" satisfies the reduction of both human health risk and nitrate control cost (a risk versus cost trade-off analysis).

Human health risk analysis, nitrate control cost analysis, and assessment of the technical feasibility of each control method introduce uncertainty (or imprecision) in the risk management and decision making. Sources of uncertainty (or imprecision) include nitrate exposure assessment, interspecies conversion, dose-response assessment, and inaccuracy of data used [Dahab and Bogardi, 1990]. In this paper, the uncertainty in problem variables (risk, cost and technical feasibility) and its impact on results are represented by using fuzzy set theory. The fuzzy set approach, pioneered by Zadeh [1965], has been widely applied to solve, under imprecise conditions, decision-making problems [Watson et al., 1979;

Zimmermann, 1985] and, specifically, multicriterion decision-making problems [Yager, 1977; Bogardi and Bardossy, 1983b; Anandalingam and Westfall, 1988].

2 Methodology For Nitrate Risk Management

A methodology is devised to assist decision makers in determining a nitrate risk management strategy that "best" satisfies the objectives of problem variables (reducing the risk and cost, and assuring technical feasibility). In formulating this methodology, a fuzzy composite programming method [Lee et al., 1991] is employed. As such the methodology is organized into the following sequential format:

1. Define nitrate risk management strategies.
2. Define basic indicators.
3. Group basic indicators into progressively fewer and more general groups.
4. Evaluate and rank the management strategies.

2.1 Selection of Basic Indicators

To represent both potential benefits and problems of each nitrate risk management strategy, basic indicators are selected. These indicators are then used as the input variables for evaluation of each strategy. Because the selection of the basic indicators tends to be case specific, it is difficult to generalize. In other words, basic indicators vary with nitrate control methods used, and the number of basic indicators depends on the level of analysis desired (i.e., preliminary or detailed).

2.2 The Composite Procedure

The composite procedure involves a step by step regrouping of a set of various basic indicators to form a single indicator [Bogardi and Bardossy, 1983a]. Fig. 1 shows an example of the indicator grouping. Basic (first-level) indicators are grouped into a smaller subset of second-level indicators. For example, basic indicators such as risks of methemoglobinemia and human cancer can be grouped into human health risk, an element of the subset of second-level indicators. The status of technology and the ease (simplicity) of operation are grouped into the technical feasibility, the other element of the subset of second-level indicators. Further, second-level indicators such as human health risk and the adverse impact of performance are grouped into environmental risk, an element of the subset of third-level indicators. Similarly, the final (system) indicator can be formed by composing environmental risk, technical feasibility and cost.

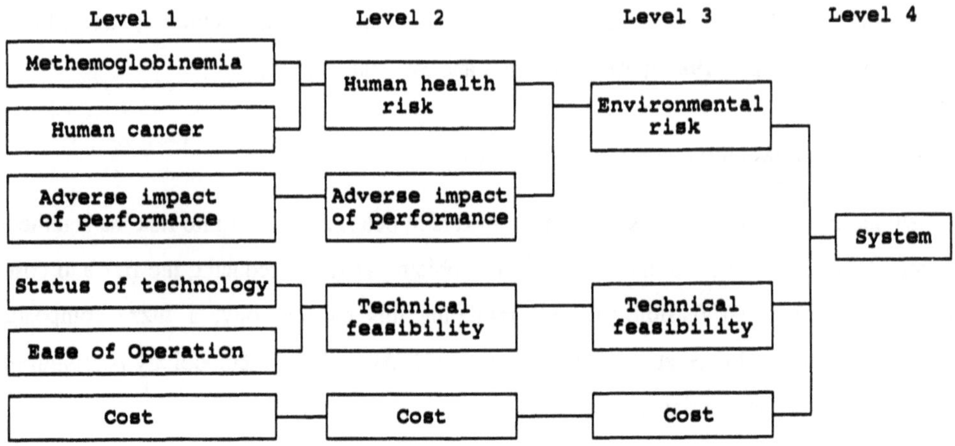

Fig. 1. Example of composite structure

2.3 Trade-off Analysis

The values of the basic indicators for nitrate risk management are estimated as fuzzy numbers to characterize their imprecision. The fuzzy numbers are values which belong to a given set (interval) with a degree of membership. In order to evaluate the various nitrate risk management strategies under uncertainty, let $Z_i(x)$ be a fuzzy value for the i-th basic indicator, and let its membership function $\mu(Z_i(x))$ be a trapezoid (Fig. 2), where x is one element of the discrete set of nitrate risk management strategies. Note that if the trapezoid is reduced to a vertical line, it represents a "crisp" number. A level-cut concept [Dong and Shah, 1987] can be used to define the interval of each basic indicator at various levels of "confidence." The confidence level for an uncertain value can be determined using the "expert judgement" based on experience and observed measurement variability [Stansbury et al., 1991]. As shown in Fig. 2, $Z_{i,h}(x)$ is the interval value of the i-th basic indicator at level-cut h (i.e., $a \leq Z_{i,h}(x) \leq b$).

Since the units of basic indicators such as risk and cost are different and thus difficult to compare them directly, the actual value of each basic indicator ($Z_{i,h}(x)$) should be transformed into an index. Using the best value ($BESZ_i$) of Z_i and the worst value ($WORZ_i$) of Z_i for the i-th basic indicator, the index value, $S_{i,h}(x)$, resulting from the actual value $Z_{i,h}(x)$ can be

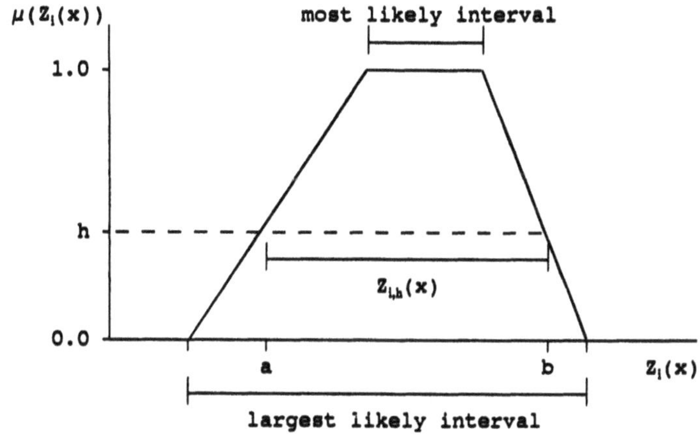

Fig. 2. Fuzzy estimate of the i-th basic indicator

represented by the piece-wise linear function:

(1) If $BESZ_i > WORZ_i$, index value $S_{i,h}(x)$ is defined by:

$$S_{i,h}(x) = \begin{cases} 1 & , Z_{i,h}(x) \geq BESZ_i \\ \dfrac{Z_{i,h}(x) - WORZ_i}{BESZ_i - WORZ_i} & , WORZ_i < Z_{i,h}(x) < BESZ_i \\ 0 & , Z_{i,h}(x) \leq WORZ_i \end{cases} \quad (1)$$

(2) If $BESZ_i < WORZ_i$, index value $S_{i,h}(x)$ is defined by:

$$S_{i,h}(x) = \begin{cases} 1 & , Z_{i,h}(x) \leq BESZ_i \\ \dfrac{Z_{i,h}(x) - WORZ_i}{BESZ_i - WORZ_i} & , BESZ_i < Z_{i,h}(x) < WORZ_i \\ 0 & , Z_{i,h}(x) \geq WORZ_i \end{cases} \quad (2)$$

To assign the best and worst values (i.e., $BESZ_i$ and $WORZ_i$) of the i-th basic indicator, one of two options can be used. The first is to assign the best and worst values of the i-th basic indicator according to the overall best and worst values of the i-th basic indicator, among the given strategies. The second is to assign the best and worst values of the i-th basic indicator according to expert judgment. Since the actual value $Z_{i,h}(x)$ is an interval with lower bound a and upper bound b (Fig. 2), the index value $S_{i,h}(x)$ resulting from $Z_{i,h}(x)$ is also an interval.

Next, index values, $L_{j,h}(x)$, for second-level composite indicators can be calculated by using the index values of basic indicators, or:

$$L_{j,h}(x) = \left[\sum_{i=1}^{n_j} w_{i,j} \, (S_{i,h,j}(x))^p \right]^{(1/p)} \quad (3)$$

where:

n_j = the number of element in the second-level group j,

$S_{i,h,j}$ = the index value for the i-th basic indicator in the second-level group j of basic indicators,

$w_{i,j}$ = the weight reflecting the importance of each of basic indicators in group j; $\Sigma\, w_{i,j} = 1$, and

p = the balancing factor among indicators for group j.

Weights represent the relative importance between indicators in a group. The greater the importance of an indicator, the greater should be the weight assigned to it. The balancing factors are assigned for each group of indicators. The balancing factor p ($p \geq 1$) reflects the importance of the maximal deviations of the indicators, where maximal deviation means the maximum difference between an indicator value and the best value for that indicator. The processes available for determining the weights and balancing factors are given in detail elsewhere [Lee et al., 1991].

Trade-off index values for the other levels are calculated in similar fashion until a final index value is determined for each management strategy.

2.4 Ranking Nitrate Risk Management Strategies

As the first step in determining the ranking of various nitrate risk management strategies, let L(x) be the fuzzy number representing the final composite (system) indicator of strategy x. The membership function, $\mu(L(x))$, of the fuzzy number L(x) can then be defined as follows [Lee et al., 1991]:

$$\mu(L(x)) = \begin{cases} 1 , & r_{min} \leq L(x) \leq r_{max} \\ \dfrac{L(x) - R_{min}}{r_{min} - R_{min}} , & R_{min} \leq L(x) < r_{min} \\ \dfrac{L(X) - R_{max}}{r_{max} - R_{max}} , & r_{max} < L(x) \leq R_{max} \\ 0 , & \text{otherwise} \end{cases} \tag{4}$$

where r_{min} and r_{max} = lower and upper bound values, respectively, of the index value of the final composite indicator obtained by using $Z_{i,h=1}(x)$, and R_{min} and R_{max} = lower and upper bound values, respectively, of the index value of the final composite indicator calculated by using $Z_{i,h=0}(x)$.

If there are n management strategies, there would be n fuzzy numbers (L(x), x=1,...,n) which are restricted by the membership function of Equation 4. The ranking of these n fuzzy numbers, which corresponds to the ranking of strategies, can be determined by using the ranking method given in Appendix I.

3 Illustrative Example

A community with a nitrate water quality problem is chosen as an illustrative example. This community consumes about 3000 cubic meters of water per day. The existing water source is groundwater that has nitrate level of 20 mg-NO_3-N/L, sulfate level of 120 mg/L, and total dissolved solids of 800 mg/L.

Table 1. Nitrate risk management strategies for the example community

Strategy	Nitrate Control Method	Nitrate in finished Water mg-NO_3-N/L
1	Bio-denitrification	1 - 4
2	Ion-Exchange	4 - 7
3	Bio-denitrification	4 - 7
4	Ion-Exchange	7 - 10
5	Blending two supplies	8 - 12

Table 1 shows five strategies available to the community to reduce the nitrate level in the water supply. Strategies 1 and 2 treat all of the raw water by using bio-denitrification and ion-exchange, respectively. Strategies 3 and 4 treat a fraction of the raw water by using the same technologies. The treated water is then blended with the remainder of the raw water. The fraction treated is determined to achieve the target level of nitrate in the finished water (i.e., acceptable human health risk level). To calculate the treatment rates, average nitrate removal efficiencies of 90 % for bio-denitrification and 70 % for ion-exchange were assumed [Dahab, 1990; Rogalla et al., 1990]. Strategy 5 is to blend the existing water with groundwater of an alternate source. Up to 60 % of the water demand can be satisfied using the alternate source. The alternate water supply has a nitrate concentration in the range of 0 to 6 mg-NO_3-N/L.

Table 2 contains data of the basic indicators for each of the strategies. Cancer risk was determined using a relationship based on the work of Lee et al. [1991a]. Cost data were obtained from the literatures [Dahab, 1987; Guter and Kartinen, 1989; Rogalla et al., 1990]. The cost data assigned to strategy 5 were hypothetical. Data on the adverse impact of performance corresponded to the volume (m^3/day) of waste produced by each of the processes used to remove nitrate. Data on methemoglobinemia, the status of technology and the simplicity of operation were assigned by transferring qualitative definitions (such as bad, good and excellent) into numerical values, as was the case in Lee [1991].

As shown in Table 2, the value of each basic indicator is represented by two intervals (i.e., most likely interval and largest likely interval) to reflect its imprecision. These intervals are used to construct the membership function (Fig. 2) for each basic indicator. Table 3 displays the best and worst values of the six basic indicators. Weights and balancing factors are used to express preferences for certain variables and control substitution of inferior alternatives, respectively. The values assigned to each indicator are shown in Table 4.

With Tables 2, 3 and 4, a fuzzy composite programming software developed for use on a personal computer [Lee et al., 1990] was used to determine the "best" of the five nitrate risk management strategies. Accordingly, the ordering value for each of the five strategies was calculated as shown in Table 5.

Fig. 3 shows an example of the output from the software relating to environmental risk and cost for each nitrate risk management strategy. The risk-cost boxes are produced as the result of trade-off between indicators under the third level, and the widths of the boxes represent the range of uncertainty in trade-offs. With the boxes formed between the risk and cost, one can identify the status of the trade-off analysis. That is, Strategy 1 provides the lowest risk, but it requires very a large cost. Strategy 4 provides the lowest cost, but it does so at a relatively high risk. Based on the widths of the risk-cost boxes, Strategies 4 and 5 include relatively high uncertainty. The "ideal point" located at (1,1) in Fig. 3 represents the lowest cost and environmental risk.

Table 2. Basic indicator values for five different strategies

Strategies	Basic* Indicators	Most Likely Interval Low	Most Likely Interval High	Largest Likely Interval Low	Largest Likely Interval High
1	(1)	1.0	1.0	1.0	1.0
	(2)	$5.5x10^{-10}$	$9.2x10^{-8}$	$1.7x10^{-11}$	$3.7x10^{-7}$
	(3)	11.4	12.6	9.6	14.4
	(4)	0.4	0.6	0.3	0.7
	(5)	0.2	0.4	0.1	0.5
	(6)	950	1050	800	1200
2	(1)	0.9	1.0	0.8	1.0
	(2)	$1.1x10^{-8}$	$1.0x10^{-6}$	$3.4x10^{-9}$	$2.0x10^{-6}$
	(3)	4200	4640	3540	5300
	(4)	0.6	0.8	0.5	0.9
	(5)	0.5	0.7	0.4	0.8
	(6)	570	630	480	720
3	(1)	0.9	1.0	0.8	1.0
	(2)	$1.1x10^{-8}$	$1.0x10^{-6}$	$3.4x10^{-9}$	$2.0x10^{-6}$
	(3)	9.1	10.1	7.7	11.5
	(4)	0.4	0.6	0.3	0.7
	(5)	0.2	0.4	0.1	0.5
	(6)	760	840	640	960
4	(1)	0.8	0.9	0.6	1.0
	(2)	$6.2x10^{-8}$	$3.5x10^{-6}$	$3.0x10^{-8}$	$5.7x10^{-6}$
	(3)	3500	3870	2950	4400
	(4)	0.6	0.8	0.5	0.9
	(5)	0.5	0.7	0.4	0.8
	(6)	456	504	384	576
5	(1)	0.2	0.4	0.0	0.5
	(2)	$1.2x10^{-7}$	$5.7x10^{-6}$	$4.9x10^{-8}$	$9.9x10^{-6}$
	(3)	0.0	0.0	0.0	0.0
	(4)	0.9	1.0	0.8	1.0
	(5)	0.8	0.9	0.7	1.0
	(6)	523	578	440	660

*(1); methemoglobinemia, (2); cancer risk, (3); adverse impact of performance, (4); status of technology, (5); ease of operation, (6); the total cost required for capital and O & M (dollar/day).

Table 3. Best and worst values of basic indicators

Basic Indicators	Best value	Worst value
Methemoglobinemia	1.0	0.0
Human cancer	1.7×10^{-11}	9.9×10^{-6}
Adverse impact	0	5300
Status of technology	1.0	0.3
Ease of operation	1.0	0.1
Cost	384	1200

Table 4. Weights (w) and balancing factors (p)

<table>
<tr><th>Basic Indicators</th><th>w</th><th>p</th><th>Second-level Indicators</th><th>w</th><th>p</th><th>Third-level Indicators</th><th>w</th><th>p</th><th>Final Indicator</th></tr>
<tr><td>Methemoglobinemia</td><td>0.6</td><td rowspan="2">1</td><td rowspan="2">Human health risk</td><td rowspan="2">0.9</td><td rowspan="3">2</td><td rowspan="3">Environmental impact</td><td rowspan="3">0.4</td><td rowspan="6">2</td><td rowspan="6">System</td></tr>
<tr><td>Human cancer</td><td>0.4</td></tr>
<tr><td>Adverse impact of performance</td><td>1.0</td><td>1</td><td>Adverse impact of performance</td><td>0.1</td></tr>
<tr><td>Status of technology</td><td>0.5</td><td rowspan="2">2</td><td rowspan="2">Technical feasibility</td><td rowspan="2">1.0</td><td rowspan="2">1</td><td rowspan="2">Technical feasibility</td><td rowspan="2">0.3</td></tr>
<tr><td>Ease of operation</td><td>0.5</td></tr>
<tr><td>Cost</td><td>1.0</td><td>1</td><td>Cost</td><td>1.0</td><td>1</td><td>Cost</td><td>0.3</td></tr>
</table>

Table 5. Ranking of nitrate risk management strategies

Ranking	Strategies	Ordering Values
1	Strategy 2	0.548
2	Strategy 4	0.547
3	Strategy 5	0.494
4	Strategy 3	0.362
5	Strategy 1	0.345

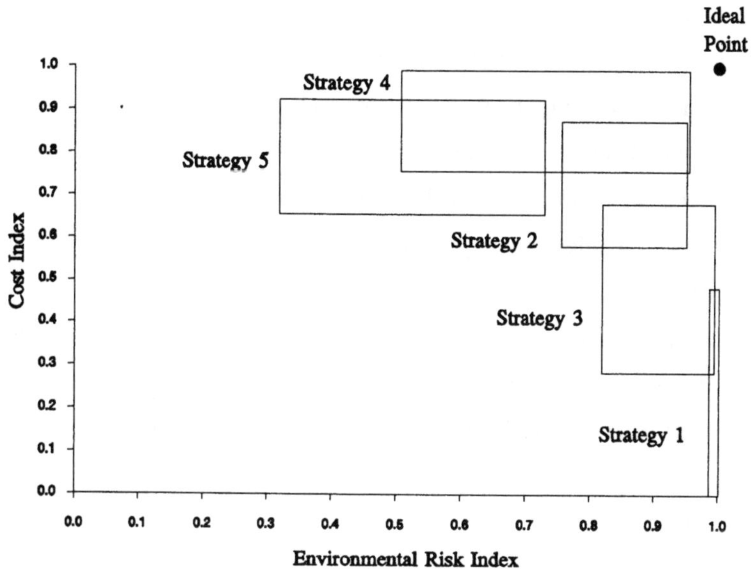

Fig. 3. Risk versus cost boxes shown at level-cut h=0

4 Discussion and Conclusions

In this paper, a multicriterion decision-making methodology to assist decision makers in solving nitrate risk management problems under uncertainty (or imprecision) was described. In the methodology, the basic indicator values were transformed into fuzzy numbers to represent their imprecision. Using these fuzzy numbers, the imprecision in basic indicator values can be incorporated into the trade-off analysis.

The methodology was illustrated using an example community with nitrate-contaminated groundwater supply. In this example, Strategy 2 was ranked as the best of five strategies because it had the highest ordering value. Strategy 2 appeared to be the most cost-effective alternative. This can be demonstrated by comparing it with the "next best" strategy for the cost. Since the ordering value of Strategy 4 is close to that of Strategy 2, it might be reasonable to consider a combined use of the Strategies 2 and 4. Combining the two strategies means changing nitrate level in the finished water because both of them use ion-exchange to remove nitrate in the water supply (Table 1). Note that the final result (ranking of the strategies) may vary with the weights and balancing factors assigned to each indicator and group.

The following conclusions can be made from this study:

1. Imprecision in the values of basic indicators selected for nitrate risk management can be represented using the fuzzy set approach.
2. The methodology presented in this paper can be a useful tool for solving nitrate risk management problem where there are conflicting objectives; values of the basic indicator variables are imprecise; and the objectives are of varying degrees of importance.
3. Under the situation of the illustrative example, ion exchange was the most effective nitrate removal method since this process showed the "best" compromise between environmental risk, technical feasibility, and nitrate control cost.

Acknowledgements

The work described in this paper was supported, in part, by the University of Nebraska-Lincoln Water Center and Center for Infrastructure Research, and in part by the U.S. EPA National Network for Environmental Management Studies (NNEMS) program (Contract No. U-913265-01-0).

References

Anandalingam, G., and M. Westfall, 1988, "Selection of hazardous waste disposal alternative using multi-attribute utility theory and fuzzy set analysis," Journal of Environmental Systems, 18(1), 65-85.

Bogardi, I., and A. Bardossy, 1983a, "Application of MCDM to geological exploration," In Essays and Surveys on Multiple Criterion Decision Making, P. Hansen, Editor, Springer-Verlag.

Bogardi, I., and A. Bardossy, 1983b, "Regional management of an aquifer for mining under fuzzy environmental objectives," Water Resources Research, 19(6), 1394-1402.

Chen, S.H., 1985, "Ranking fuzzy numbers with maximizing set and minimizing set," Fuzzy Sets and Systems, 17, 113-129.

Crespi, M., and V. Ramazzotti, 1990, "Evidence that n-nitroso compounds contribute to the causation of certain human cancers," NATO/ARW on Nitrate Contamination: Exposure, Consequences and Control, Lincoln, NE, Sep. 9-14.

Dahab, M.F., 1987, "Treatment alternatives for nitrate contaminated groundwater supplies," Journal of Environmental Systems, 17, 65-75.

Dahab, M.F., and I. Bogardi, 1990, "Risk management for nitrate-contaminated groundwater supplies," Final Report submitted to the U.S. Geological Survey, Water Resources Division, Reston, VA 22092.

Dahab, M.F., 1990, "Nitrate treatment methods: an overview," NATO/ARW on Nitrate Contamination: Exposure, Consequences and Control, Lincoln, NE, Sep. 9-14
Dong, W. and H.C. Shah, 1987, "Vertex method for computing functions of fuzzy variables," Fuzzy Sets and Systems, 24, 65-78.
Guter, G.A., and E.O. Kartinen, 1989, "Alternatives for reducing nitrate in municipal water supplies," Presented at the Nitrates in Drinking Water Symposium, Dallas, Texas.
Lee, Y.W., I. Bogardi, and A. Bardossy, 1990, "Fuzzy composite programming software and manual," Department of Civil Engineering, University of Nebraska, Lincoln, NE.
Lee, Y.W., 1991, "Nitrate risk management under uncertainty," EPA 101/F-91/047, Office of the Administrator (A-101F6), U.S. Environmental Protection Agency.
Lee, Y.W., I. Bogardi, and J. Stansbury, 1991, "Fuzzy decision-making in dredged material management," Accepted for publication by Journal of Environmental Engineering, ASCE, March.
Lee, Y.W., I. Bogardi, and A. Bardossy, 1991a, "Nitrate risk analysis under uncertainty using expert knowledge-based regression method," Submitted for possible publication to the Journal of Environmental Systems, April.
Mirvish, S.S., 1990, "The significance for human health of nitrate, nitrite and n-nitroso compounds," NATO/ARW on Nitrate Contamination: Exposure, Consequences and Control, Lincoln, NE, Sept. 9-14
Rogalla, F., G. de Larminat, J. Coutelle, and H. Godart, 1990, "Experience with nitrate removal methods from drinking water," NATO/ARW on Nitrate Contamination: Exposure, Consequences and Control, Lincoln, NE, Sept. 9-14
Stansbury, J., I. Bogardi, Y.W. Lee, and W. Woldt, 1991, "Multiobjective decision making under uncertainty," IX-th International MCDM Conference, Fairfax, Virginia, Aug. 5-8.
Watson, S.R., J.J. Weiss, and M.L. Donnell, 1979, "Fuzzy decision analysis," IEEE Transactions on Systems, Man, and Cybernetics, SMC-9(1), 1-9.
WHO, 1978, "Task group on environmental health criteria for nitrates, nitrites, and n-nitroso compounds," Nitrates, Nitrites, and N-Nitroso Compounds, World Health Organization, Geneva, 7-84.
Yager, R.R., 1977, "Multiple objective decision-making using fuzzy sets," International Journal of Man-Machine Studies, 9, 375-382.
Zadeh, L.A. (1965). "Fuzzy sets," Information and Control, 8, 338-353.
Zimmermann, H.J., 1985, "Fuzzy set theory and its application," Kluwer-Nijhoff Publishing, Hingham, MA 02043.

Appendix I: Ranking Method

If there are n management strategies, there are n fuzzy numbers (L(x), x=1,...,n). To determine the ranking of these n fuzzy numbers, the ranking method developed by Chen [1985] can be applied. This method determines the ranking of n fuzzy numbers by using a maximizing set and a minimizing set.

The maximizing set M is a fuzzy subset with membership function μ_M given as:

$$\mu_M(L) = \begin{cases} \dfrac{L - L_{min}}{L_{max} - L_{min}}, & L_{min} \le L \le L_{max} \\ 0, & \text{otherwise} \end{cases} \tag{5}$$

where for x = 1, ..., n, $L_{min} = \min[\min L_{h=0}(x)]$ and $L_{max} = \max[\max L_{h=0}(x)]$. Then, the right utility value, $U_R(x)$, for strategy x is defined as:

$$U_R(X) = \max\{\min[\mu_M(L), \mu(L(x))]\} \tag{6}$$

The minimizing set G is a fuzzy subset with membership function μ_G given as:

$$\mu_G(L) = \begin{cases} \dfrac{L - L_{max}}{L_{min} - L_{max}}, & L_{min} \le L \le L_{max} \\ 0, & \text{otherwise.} \end{cases} \tag{7}$$

The left utility value, $U_L(x)$, for strategy x is defined as:

$$U_L(x) = \max\{\min[\mu_G(L), \mu(L(x))]\} \tag{8}$$

The total utility or ordering value for strategy x is:

$$U(x) = (U_R(x) + 1 - U_L(x))/2 \tag{9}$$

The strategy that has the highest ordering value in the discrete set of strategies is then selected as the "best" strategy.

FACING THE RISK OF UNCONTROLLED DENITRIFICATION IN SEWAGE TREATMENT PLANTS

Anastasios I.Stamou
Department of Civil Engineering
National Technical University of Athens
Iroon Polytechniou 5, 15773 Zografou, Athens
Greece

ABSTRACT

Operational measures are presented to face occasionally the uncontrolled denitrification in settling tanks. For a more permanent control, certain methods are also described and compared. Among these methods, it seems that the "Bio-Denitro" is the most appropriate for application in medium size (10000-300000 PE) sewage treatment plants in Greece, due mainly to its stable operation and operational flexibility to deal with wide variations in sewage temperatures and BOD loads. For treatment plants with relatively constant sewage characteristics, the "recirculation" is probably the simplest method. For small treatment plants, where a consistent N-removal efficiency is not required, the simultaneous denitrification method is expected to be the cheapest alternative.

1.INTRODUCTION

The problem of <u>rising sludge</u> in settling tanks is well-known. Pieces of sludge rise to the tank surface carried by fine bubbles and form dark clumps. The cause of this phenomenon is denitrification. For denitrification to occur three conditions must be present in the bottom of the tanks : (a)low DO (<0.5ppm), (b) high nitrate (>5ppm) and (c) sufficient BOD (>10ppm) concentrations. Under these conditions ((a) and (b) are called anoxic), certain bacteria use the nitrates as acceptors of electrons to form nitrogen gas, which becomes trapped in the sludge mass. If enough nitrogen is produced, portions of the sludge mass become buoyant, rise to the surface and escape with the effluent mass over the outlet weir, thus leading to the deterioriation of the effluent quality.

NATO ASI Series, Vol. G 29
Water Resources Engineering Risk Assessment
Edited by J. Ganoulis

In the present paper various operational measures and methods to face this problem are presented and discussed.

2. MEASURES AND METHODS TO AVOID UNCONTROLLED DENITRIFICATION

To face the problem of uncontrolled denitrification there are certain operational measures (WPCF,1987), that can be tried to an existing sewage treatment plant with the following scopes: (a) to increase the DO concentration, (b) to reduce the nitrates drastically and (c) to reduce the BOD concentration in the sludge. DO concentration can be increased by using a higher scraper speed or by reducing the detention time in the settling tanks, so the microorganisms have not enough time to consume the DO. The latter can be achieved by reducing the number of settling tanks in use (which however leads to high values of overflow rates) or by increasing the return flow rate (up to a certain value controlled by the sludge blanket depth). Nitrate concentrations can be reduced by inhibiting the process of nitrification. For low sewage temperatures nitrification can be inhibited by reducing below a certain value the sludge age. For relatively high sewage temperatures,however,only chlorination of the return sludge can be an effective solution. BOD concentration in sludge is reduced, when the BOD removal efficiency is increased (e.g. by reducing the surplus sludge wasting rate,so the sludge age increases to ensure almost complete BOD removal).

These measures can be tried by the operator of the plant to face the problem of uncontrolled denitrification occasionally and mainly for sewage treatment plants designed with small sludge ages and for low sewage temperatures. For extended aeration treatment plants or for plants with high sewage temperatures (such as these in Greece) the only safe solution is to perform the denitrification process in a "controllable" environment. This is usually performed in the design phase of the plant with the so-called nitrification-denitrification (N-DN) process. Alternatively, any other well-known nitrogen removal method may be applied (ammonia air stripping, ion exchange, breakpoint chlorination or root zone treatment).

The N-DN process (MetCalf and Eddy,1979 and ATV,1985) is perhaps the best , because of its high potential N-removal efficiency,high process stability and reliability, easy process control, low land-area requirements and moderate cost. N-DN is carried out in two steps. In the first step (N=Nitrification) the ammonia of the sewage is aerobically converted to nitrate by autotrophic bacteria. In the second step (DN=Denitrification) heterotrophic bacteria utilize the nitrates (instead of oxygen) for the oxidation of BOD converting them to nitrogen. For denitrification to occur organic matter must be present either as soluble BOD or as external C-source (methanol). Because of the high cost of most external C-sources, a number of processes have been developed in which N-DN is performed into a single process without addition of any external C-source. Among these processes the most well-known are the following: (a) the recirculation (Rec-DN) or "Bardenpho", (b) the simultaneous-DN (Sim-DN) and (c) the alternating N-DN or "Bio-Denitro" method. These methods are shown schematically in Fig.1.

In the Rec-DN (Fig.1a) the required for the denitrification process nitrates are returned to the first anoxic tank directly from the second aerobic tank. In the Sim-DN (Fig.1b), which is normally performed in tanks, where sewage circulates (e.g. oxidation ditches), both nitrification and denitrification are performed in the same tank. Immediately downstream of the entrance of the incoming sewage (and return sludge) anoxic conditions prevail (DO=0.1-0.3ppm) and denitrification occurs. Nitrification is performed downstream of the aerators (aerobic conditions) together with CBOD-oxidation. In the Bio-Denitro two tanks are normally employed, as in Kavala's sewage treatment plant in Northern Greece (see Loukas,1991). In these tanks denitrification and nitrification are performed alternately according to the operational cycle shown in Fig.1c. In phases A and C sewage enters to either of the two tanks through a mechanical inlet distributor to perform denitrification, passes to the other tank to perform nitrification and CBOD-oxidation and exits to a common settling tank. Phases B and D are intermediate serving for further nitrification to increase the nitrate concentration.

By comparing these methods the following remarks can be made:

(1) The required volumes for nitrification and denitrification are generally independent of the method applied and thus equal for all three methods. Especially for Rec-DN a separate anoxic tank is required resulting to generally higher civil engineering costs.

(2) For denitrification to be effective a certain degree of mixing is required. In Rec-DN additional mixing devices should be installed in the anoxic tank. In the other two methods, the aerators used for nitrification, can be used operating at lower speeds.

(3) Bio-Denitro and Rec-DN need two tanks to operate. Thus, for very small plants, where only one tank is involved, the Sim-DN method is expected to be the cheapest alternative.

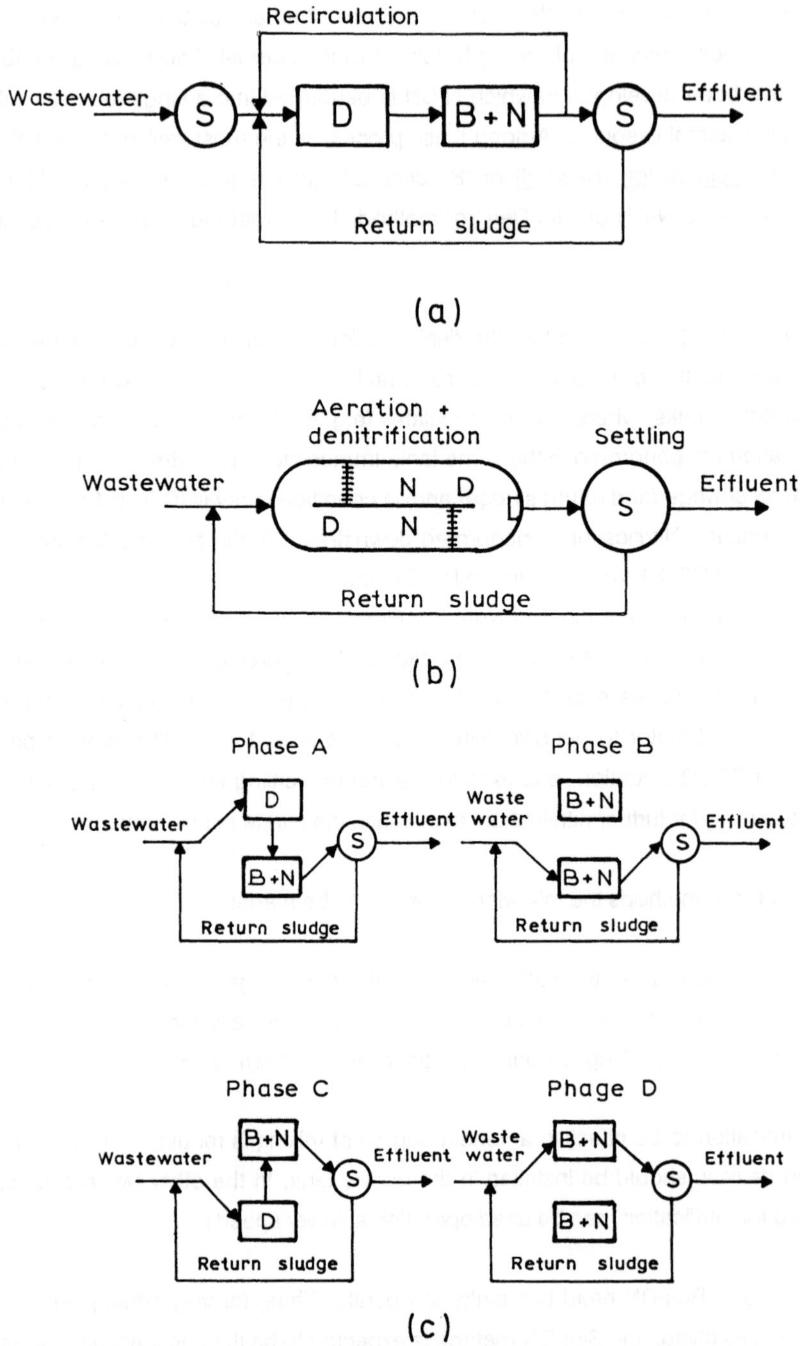

Fig.1. Nitrification-Denitrification methods:
(a) Recirculation, (b) Simultaneous,
(c) Bio-Denitro

(4) The Rec-DN method requires a very large amount of recirculation flow. Normally, for a N-removal of 80% a recirculation rate of 700% is required, which results to a significant increase of the initial and running costs. The other two methods require the normal activated sludge return rates.

(5) The Sim-DN can be efficiently applied only to plug flow reactors, where a certain portion of the volume of the tank upstream of the aerators can be used as anoxic. For completely mixed reactors, where the total volume of the tank at any instant is either aerobic or anoxic, Sim-DN can be applied with limited operational capability and thus efficiency.

(6) In the Rec-DN the anoxic and aerobic volumes are always fixed and thus Rec-DN offers no operational flexibility, when BOD load and sewage temperature variations are expected. However, if such variations are not expected Rec-DN is probably the simplest and most appropriate method.

(7) The Bio-Denitro is the only method, in which fluctuating BOD loads and sewage temperatures can be efficiently controlled by varying the duration of aerobic and anoxic phases, and consequently the capacity for nitrification and denitrification. This phase variation can be performed within rather wide limits with the help of a computer program and thus the operation of the plant can be adjusted for maximum nitrogen removal, even for sewage compositions very different from design. With the tanks operating at any instant as aerobic or anoxic a full and independent control of the anoxic and aerobic volumes is allowed, leading to a great flexibility in process control to face seasonal temperature changes. The Rec-DN has almost no operational flexibility to face such load variations, since the anoxic and aerobic volumes are always fixed. The Sim-DN has a certain operational flexibility, especially if a significant number of DO meters is installed in the tanks to control the process. However, the possibility of e.g. increasing the nitrification rate in winter by increasing the oxygen concentration in the aerobic zones cannot be utilized in plants with Sim-DN without trading in some of the denitrification capacity.

(8) Bio-Denitro can be operated as a standard activated sludge plant, if required. For the Sim-DN to operate in the standard method extra aeration equipment should be added, while for the Rec-DN extra volume is required.

(9) Due to the discontinuous use of the aerators an increased installed capacity for aeration is theoretically required for the Bio-Denitro. In practice, however, it has been found (Loukas,1991) that Bio-Denitro can operate with relatively low DO concentrations (1ppm) in the aeration period. If this behavior is due to the specific operation of Bio-Denitro, then a sig-

nificant reduction of the oxygen and aeration requirements is allowed, which balances the theoretical need for further installed aeration capacity. Significant energy savings and less wearing of the aerators are further advantages of the Bio-Denitro, because of the discontinuous operation of the aerators.

Thus, the Bio-Denitro method seems to be an efficient method to control the problem of denitrification in settling tanks and to remove nitrogen from the sewage, especially for medium size (10000-300000 PE) plants in Greece due to its capability to face the wide variations of sewage temperature (normally from 12 to 30°C). Indeed, the majority of the medium size treatment plants in Greece, either in operation (Karditsa,Kavala and Agios Nicolaos-Kriti) or in construction (Kos and Arta) have been designed according to the Bio-Denitro method.

3.CONCLUSIONS

Certain operational measures are presented to face occasionally the problem of uncontrolled denitrification in settling tanks. A more permanent control can be achieved by the biological removal of Nitrogen using various methods based on the nitrification-denitrification process. Among these methods, the Bio-Denitro seems to be the most appropriate method for medium size (10000-300000 PE) sewage treatment plants in Greece, due mainly to its stable operation and operational flexibility to deal with wide variations in sewage temperatures and BOD loads. For small treatment plants, however, where a consistent N-removal efficiency is not required, the simultaneous denitrification method is expected to be the cheapest solution. For plants with relatively constant sewage characteristics the recirculation method is probably the simplest alternative.

4.REFERENCES

ATV(1985): "Nitrifikation und Denitrifikation" in "Lehr- und Handbuch der Abwassertechnik", Band IV, p.313, Vol.IV, Ed.: Ernst und Sohn.

Loukas,N.(1991): Examining the Kavala's Sewage Treatment Plant, Diploma Thesis, Dept. of Civil Eng., National Technical University of Athens, Athens, Greece.

MetCalf and Eddy (1979): Wastewater Engineering, Treatment, Disposal, Reuse,McGraw Hill, Inc.New York, U.S.A.

WPCF (1987): Activated Sludge, Manual of Practice No. OM-9, Alexandria, U.S.A.

Part 4.
CASE STUDIES

STOCHASTIC MODELING OF A WATER RESOURCE SYSTEM: ANALYTICAL TECHNIQUES VERSUS SYNTHETIC APPROACHES

E. Parent,
Department of Applied Mathematics and Computer Science
ENGREF, 19 av du Maine, 75015 Paris, France

F. Lebdi,
Department of Rural Engineering
INAT, 43 ave C. Nicolle, Tunis, Tunisia.

P. Hurand,
Neste Operationnal Center,
CACG, Chemin de l'Alette, 65001 Tarbes, France

ABSTRACT

This paper deals with bicriterion (irrigation & water quality) weekly operation of a water resource system during dry periods. Two ways of handling the problem are proposed and compared for a real case study:

-a stochastic dynamic programming model using system analysis language;

-a more "synthetic" model using direct hydrological and water resources concepts.

Numerical results based on historical series are derived in both cases The insight provided by the water resources engineering approach is pointed out and guidelines for model choice are suggested.

1. INTRODUCTION

Most of stochastic modeling methods in water resources are based on a set of analytical model equations, mainly continuity and flow-routing ones that includes the description of the stochastic behavior of model variables such as in Bodo & Unny (1990). These techniques are particularly designed for simulation to compute indirectly reliability indices via Monte Carlo randomizations and are well fit for sequential optimization. Indeed they can be considered as a "quintessential" branch of state space modeling and control system theory applied to the specific field of water resources. A quasi exhaustive list of potential engineering applications of these methods can be found in TAPIERO(1988). It is then sometimes tempting to take the attitude

NATO ASI Series, Vol. G 29
Water Resources Engineering Risk Assessment
Edited by J. Ganoulis

that, by adopting such state space representation coupled to a stochastic dynamic programming optimization method , one has in hand a mathematical panacea for the operation of water resources systems. Since the works of MASSE (1959), dynamic programming methods have been widely used in water resource operation and planning. YAKOWITZ (1982) and YEH (1985) have set up extensive state-of-art reviews of these sequential optimization techniques.

However, at time the automatic application of such system analysis techniques might appear to be somewhat improper , resulting in heavy computing tasks and oversophisticated optimization devices. This paper argues that stochastic modeling of a water resources system through sequential decision process is not the only way and that much attention should be paid to simple and comprehensive methods based on a good sense of hydrology. In addition, these can give more direct values to output variables of interest, specially for reliability and risk analysis.

These two formulations are developed and illustrated by a French case study, the Neste system, where two conflicting objectives have to be satisfied, namely, irrigation and water quality demand. The first approach is a typical system analyst 's stochastic dynamic model solved by discrete stochastic dynamic programming. It is then compared to a synthetic model built on the basis water resources engineering considerations where the state is defined in terms of the statistical frequency of consumption of all present water resources in the near future. A practical decision rule is derived without an optimization scheme.

Numerical results based on historical series reveal that the two types of methods give very similar operational results. We carefully review all the implicit and explicit assumptions inherent to each method. Our main conclusion is that systems analysts have much to learn from water resources engineers.

2 THE NESTE WATER RESOURCE SYSTEM

The Lannemezan Plateau in the Gascogne basin is bordered by the Pyrenees Mountains in the South and the Gascogne and Adour rivers in the Est and West . Consequently, the Gascogne basin rivers are real french "oueds" that used to dry out in summer when no precipitation occurs. As early as 1850, a 29 km long canal (the Neste canal) was built in order to convey water from the mountain river Neste to the top of the Gascogne watershed. From now on, the "Neste system" will refer to the so-called Association of the Neste to Gascogne Canal and the Watershed Basin including its irrigation zones and domestic water users. For more than a century, this abundant new facility has allowed for the development of 30,000 ha of intensively irrigated corn agriculture . Figure 1 shows this extraordinarily increasing development during the last two decades (each bar represents the annual total authorized pumping discharge for irrigation in l/s) .

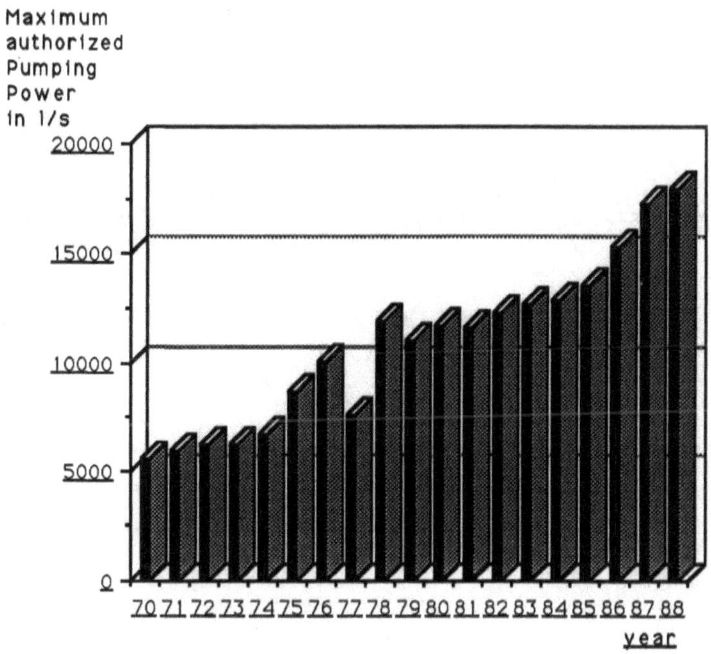

Figure 1 : Irrigation Development on the Neste System

A second purpose of the Neste canal is to supply continuous minimum levels in the Gascogne rivers for the purposes of water quality control and domestic supply. In order to fulfill increasing irrigation demand, new dams of a total capacity of 42 hm^3 have been built in the hills of the Gascogne basin in addition to the 48 hm^3 reservoir in the Pyrenees Mountains that already supplies the river Neste. Nowadays, water supply and irrigation demand compete for the same limited resource. The aim of our study is to develop multi-purpose models so that the decision maker can reach a rational allocation of the Neste water resource.

3 FROM REAL TIME OPERATION TO STRATEGIC MANAGEMENT

For the main Neste system reservoirs, there already exists a real time operation device based on control theory: every three hours, the levels in the rivers are checked, parameters for irrigation demand model are filtered and readjusted, and a feedback loop taking into account propagation delays along the rivers modifies the gate opening so as to obtain a predetermined level at downstream checkpoints . As indicated in Figure 2, this self governing regulation loop works perfectly well but is of no use without a good strategy to choose which level should be targeted downstream.

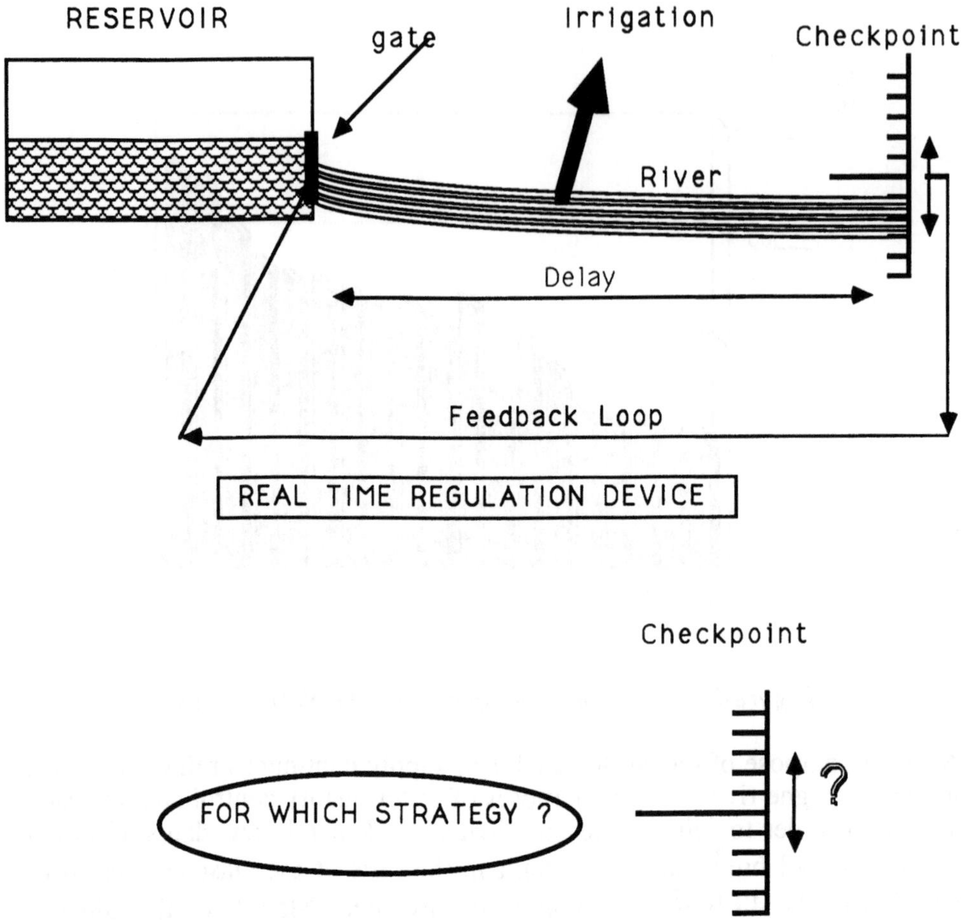

Figure 2: Tactical VS Strategic Point of view

Clearly enough , a strategic rule should avoid lowering too much this target output requirement because then, the reservoirs would remain nearly full until the operation horizon, which means that the system manager would have continuously, uselessly and dangerously threatened ecological balance and domestic water use . On the other hand, if the target is set too high a shortage in the near future may occur due to this present "waste" of water resources.
Strategic rules belong to a higher level of decision process than real time operation and are designed to work with a longer time step (a weekly one for the Neste system). They also provide a decision making tool to suggest new

alternatives and integrate reliability and risk concepts into water resources allocation. They must take into account both risk analysis and reliability as defined for example in ASKEW (1974) or SIMONOVIC and MARINO (1984).

4 SYSTEM DESCRIPTION

In case of water supply failure, the whole system is involved: this is why, in order to find a satisfying water resources allocation strategy, we will study the weekly operation of the Neste system from a global point of view. This system can be described as follows:

- a single reservoir S(t) at the beginning of week t including both mountains and Gascogne storages.

- a lumped release L(t) that is adjusted in real time by the tactical regulation loop so as to maintain the output value as close as possible to a predetermined target.

- the river Neste inflow N(t). It is the main natural water supply for the system, flowing directly from the Pyrenees mountains.

- this water is diverted into two valleys:
  * the natural Neste river valley,
  * the Gascogne watershed via the Neste canal.

- the total irrigation demand D(t) that occurs in week t in the Gascogne plateau.

- the natural inflow G(t) of the lumped Gascogne river.

River Neste flow r_n after the canal diversion and water supply volume r_g are the outputs of the system; here target output requirements for quality control and domestic use must be achieved. r_g is determined for a complete week before knowing what the inflows and demands will be during that week. Figure 3 represents the Neste system with identification of these variables. The study period begins on the 1st of June when all reservoirs are full and lasts 30 weeks.
r_n must stay above a minimum constraint $q_{minN} = 4$ m^3/s that must be kept in the river NESTE. r_g may be represented as a contract between the system manager and the water users. The system manager assumes the responsibility to set r_g as high as possible , for water quality and aquatic life purposes. In

any case r_g should not be dropped below $q_{minG} = 5\ m^3/s$ which is the tenth of the mean annual runoff required by law.

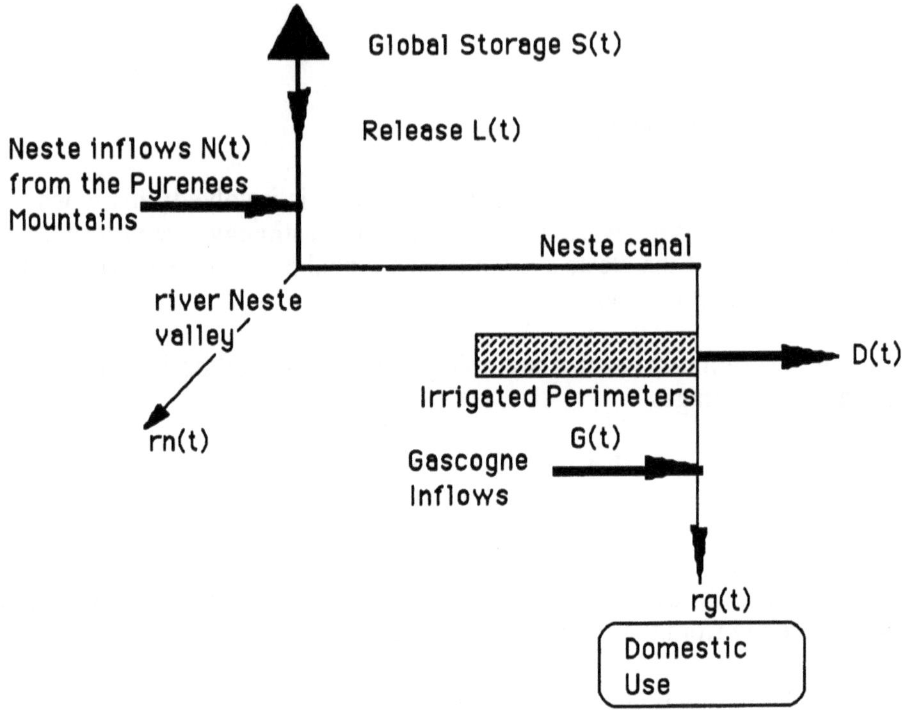

Figure 3: System description

In normal periods , the reservoirs release enough water to satisfy irrigation demand and other uses , and there is a general agreement that it is of no use that r_g be greater than q_{objG}= 12 m3/s. Considering the climatological and agronomical conditions, irrigation is a necessary means to get maximum corn yields. Nevertheless, in case of water shortage, domestic use and water quality become higher priorities. In this case, the farmers and the system manager agree on a certain amount (referred as $(1-\gamma)\ D(t)$ in the text) that will not be furnished to crops. When the system enters this restrictive operation mode, it practically cannot switch back because of administrative inertia.

5 SYSTEM ANALYST'S APPROACH

A first way of dealing with the above case study is to reformulate this problem using state space methods. In classical control theory terms, one would describe the system as follows:

- each week t would be a stage in the decision process ,
- the decision vector $\vec{u}$ contains the two control variables r_g and γ ,
- N(t), G(t), D(t) are the results of stochastic inputs ω into the system,
- the state $\vec{x}$ is what we have in hand to make our decision at time t. It includes S(t), the reservoir storage; N(t-1) the previous flow in the river Neste and γ (t-1) to know if a restriction mode has been initialized for irrigation.

With $\vec{x}(t) = \begin{pmatrix} S(t) \\ N(t-1) \\ \gamma(t-1) \end{pmatrix}$ and $\vec{u}(t) = \begin{pmatrix} r_g(t) \\ \gamma(t) \end{pmatrix}$ the dynamic aspect of the Neste system has now to be expressed in the form of a transition function :

$$\vec{x}(t+1) = \vec{f}(t, \vec{u}(t), \vec{x}(t), \omega) \qquad (1)$$

5-1 Equations, state set and transition functions

In order to modelize the behavior of the system during week t with Equation (1) many difficulties must be overcome especially in view of the stochastic input ω. Hydrological assumptions are needed: for instance, D(t) and G(t) are assumed to be statistically independent variables. In a former hydrological study based on 15 years of data, it was found that Log N(t) could be represented as a Markov chain with lag 1. The set of equations that governs the dynamic behavior of the system can now be written as follows:

$$r_n(t) = q_{min}N + \mathrm{Sup}(0, N(t) - Tcap - q_{min}N) \qquad (2)$$

$$\mathrm{Log}\, N(t) = a(t)\mathrm{Log}\, N(t-1) + b(t) + \sigma(t)\ \varepsilon(t) \qquad (3)$$

$$S(t+1) = S(t) - r_g(t) + N(t) - r_n(t) - \gamma(t)\, D(t) + G(t) \qquad (4)$$

$$0 \le \gamma(t+1) \le \gamma(t) \le 1 \qquad (5)$$

Tcap stands for the Neste canal transfer capacity value which is presently 14 m^3/s. Equation (2) reads that, due to the limited canal capacity, the diverted flow entering the Neste system is at best Tcap and the remaining part r_n flows towards the natural Neste channel.

Equation (3) describes the stochastic evolution of the natural river Neste inflows. a(t), b(t) and $\sigma(t)$ are deterministic yearly periodic function of t which were assessed during the hydrological study. $\varepsilon(t)$ is a discrete white noise with unit variance describing the main pertubation of the system. D(t) and G(t) are tabulated independent stochastic functions.

The mass balance Equation (4) indicates that the reservoir releases and the Neste and Gascogne rivers are the only resources used to satisfy all the needs in the Neste system.
Equation (5) is a constraint referring to administrative inertia when entering a restriction mode.

A operation rule is a mapping from the state space $\vec{x(t)} = \begin{pmatrix} S(t) \\ N(t-1) \\ \gamma(t-1) \end{pmatrix}$ to the decision space $\vec{u(t)} = \begin{pmatrix} r_g(t) \\ \gamma(t) \end{pmatrix}$ which generates a feedback decision law:

$$\vec{u(t)} = \vec{u}(t, \vec{x(t)}) \qquad (6)$$

In other words, using this rule in a feedback loop, the system manager makes his decision at time t on the basis of his knowledge from the system at time t.

5-2 "Optimum" Strategic Rule

Following the system analyst's approach we need some criteria to choose among all the possible decision vectors $\vec{u(t)} = \vec{u}(t, \vec{x(t)})$. Perhaps the most difficult function of systems engineering is to define the objectives to be attained through operation of a water resource system. Although no monetary benefits can be defined as in the case of hydro-electric systems, the system manager wants an "optimum" allocation with no clear definition of what this optimum may be. This may explain why hydroelectric dam operation has been extensively studied by many authors with various algorithms (COLLETER et al , 1978; DODU et al 1981; TURGEON(1981) but still there is a lot to be done when one cannot define an energy cost to be minimized. Optimal operating procedures for water allocation have been derived for several hydro-agricultural engineering problems (BURAS 1972, MERGOS 1987, HOUGHTALEN and LOFTI 1988; FANG et al , 1989). We could have used a multiattribute utility approach which has already been applied to such agro-hydraulics engineering studies (see for instance KRZYSZTOFOWICZ, 1982) but we prefer to help the system manager to formulate which incidents should be avoided by a good operation rule (HASHIMOTO et all 1982).
It is often easier to think in terms of incidents and failures than performance objectives to be maximized. We will focus on incidents (ie events that occur beyond a prespecified range of bounds). In other words, we strive to answer the question : What does the system manager really want? First of all no trouble with irrigation but at the same time a smooth variation of r_g during the whole study horizon.

5-2-1 Water quality index

From a system analysis point of view a quadratic function $Q(r_g)$ can be used. This function, $Q(r_g)$ is a vulnerability index designed so as to get a value between 0 and 1 for week t :

$$Q(r_g) = 0 \text{ if } r_g < qminG \text{ and } Q(r_g) = 1 \text{ if } r_g > qobjG$$

$$\text{otherwise } Q(r_g) == \left(-\left(\frac{r_g - qminG}{qobjG - qminG}\right)^2 + 2\left(\frac{r_g - qminG}{qobjG - qminG}\right)\right) \qquad (7)$$

This quadratic function sketched in Figure 4 possesses a reasonnable shape: a marginal variation of r_g around q_{objG} is of little use, the same increase around q_{minG} results in great improvement of the criterion.

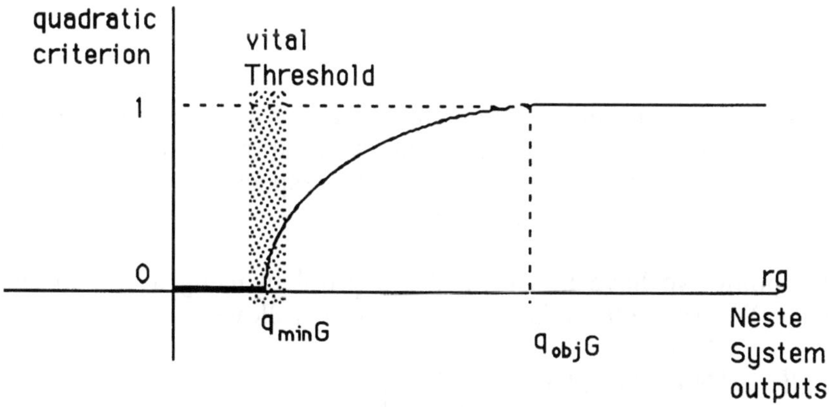

Figure 4 : Quadratic index for water quality and aquatic life

We suppose that the system manager behaves so as to maximize the cumulative value of the $Q(r_g(t))$:

$$\text{Max} \left\{ E \sum_{t=t_0}^{T} Q(r_g(t)) \right\} \qquad (8)$$

Notice that this function (8) would give, under deterministic conditions, a constant solution (r^*_g) for all t.

5-2-2 Irrigation criterion

At the same time the system manager is concerned about irrigation: he cannot afford to take r_g too close to q_{objN} because he wants to satisfy, as much as possible, irrigation demand. Let a performance index for irrigation be defined as follows:

$$\mathrm{Max}\left\{E\sum_{t=t_0}^{T}\gamma(t)\right\} \quad (9)$$

This objective function (9) may be interpreted as the average length of the irrigation season: ($\gamma(t) = 1$ means that irrigation continues and $\gamma = 0$ when it has stopped)

5-2-3 Compromise between irrigation and water quality

By introducing a positive penalty coefficient PEN, we obtain a one parameter family of objectives to be maximized over the whole period from t_0 to T :

$$\mathrm{Max}\left\{E\sum_{t=t_0}^{T}Q(r_g(t)) + \mathrm{PEN}\ \gamma(t)\right\} \quad (10)$$

PEN is a coefficient trading off quality objectives and the expected length of the irrigation period .This relative weight of irrigation and water quality is the model parameter to be adjusted so as to respect the system manager's preference function, assumed to be linear.

At this point, we have transformed our problem using a system analysis language. We have reformulated the problem as follows:

With $\vec{x(t)} = \begin{pmatrix} S(t) \\ N(t-1) \\ \gamma(t-1) \end{pmatrix}$ and $\vec{u(t)} = \begin{pmatrix} r_g(t) \\ \gamma(t) \end{pmatrix}$,

$$L(\vec{u(t)},\vec{x(t)},t) = Q(r_g(t)) + \mathrm{PEN}\ \gamma(t)) \quad (11)$$

The dynamic aspect of the Neste system is now expressed in the form of a transition function as in Equation (1); we look for an optimum feedback strategy $\vec{u(t)} = \vec{u}(t, \vec{x(t)})$ such that the sequence:

$\vec{u(t)}_{t\ \in [t_0,\ T]}$ maximizes the criterion:

$$J(u(.)) = \left\{E\sum_{t=t_0}^{T} L(\vec{u(t)},\vec{x(t)},t)\right\} \quad (12)$$

This criterion includes relative weight PEN for irrigation and water quality. One can use simulated or historical series to adjust the coefficient PEN so as to reflect the system manager's preference function.

5-3 Computational aspects

When the optimum feedback strategy $\vec{u(t)} = \vec{u}(t, \vec{x(t)})$ is applied the state process is then a Markov chain of order 1. To find this optimum strategy let:

$$V\left(\vec{x}(t_0)\right) = \text{Max } J(u(.)) = \text{Max}\left\{E \sum_{t=t_0}^{T} L(\vec{u(t)}, \vec{x(t)}, t)\right\} \quad \text{.......................... (13)}$$

The maximum is to be calculated for $\vec{u(t)}$, $t \in [t_0, T]$

Equation(13) can be reformulated by separating time t_0 and the future periods for which V given in equation (13) is supposed to have already been computed (from T to t_0+1):

$$V\left(\vec{x}(t_0)\right) = \text{Max } (E\left\{ L(\vec{u(t_0)}, \vec{x(t_0)}, t_0) + V\left(\vec{x}(t_0+1)\right)\right\} \quad \text{........................ (14)}$$

$$V\left(\vec{x}(t_0)\right) = \text{Max } (E\left\{ L(\vec{u(t_0)}, \vec{x(t_0)}, t_0) + V(\vec{f}(t_0, \vec{u(t_0)}, \vec{x(t_0)}, \omega))\right\} \quad \text{..(15)}$$

This time, the maximization operates only on $\vec{u}(t_0)$ that is expressed as a function of $\vec{x(t_0)}$. Equation (15) gives a backwards functionnal recurrence for V(.) known as the Massé-Bellman function(1957,1959).

Discrete stochastic dynamic programming techniques are used to solve the 30 stage sequential problem. Klemes (1977) has underlined the unsound consequences of unappropriate discretization for stochastic reservoir optimization. In the present case, we use 21 levels of storage ranging from O to 75 hm^3 . The logarithm of the Neste inflows has been discretized to 7 levels corresponding to the cumulated probabilities of 0.025, 0.16, 0.29, 0.5, 0.71, 0.84, 0.975 of the unconditioned stochastic variable. The mathematical expectations have been calculated using 5 levels for each variable N(t), D(t) and G(t). The Massé-Bellmann function has then been estimated backwards in time by linear interpolation on a grid with 147 points. Although the software mainly consists in two simple loops (one for the stages and the other for updating the states) , a complete run takes 12 hours on a micro computer PC-AT.

5-4 An example of Massé-Bellman sequential Optimization

Consider the rules of allocation for t = 9, assuming that $\gamma(8)= 1$. All our results are then a function of the preceding Neste inflow N(8) and the actual storage level S(9). The resulting surface obtained from a grid of 147 points can be plotted in three dimensions with abscissa X (storage), ordinate Y (Neste preceding inflow) and vertical axis Z (value of interest) .

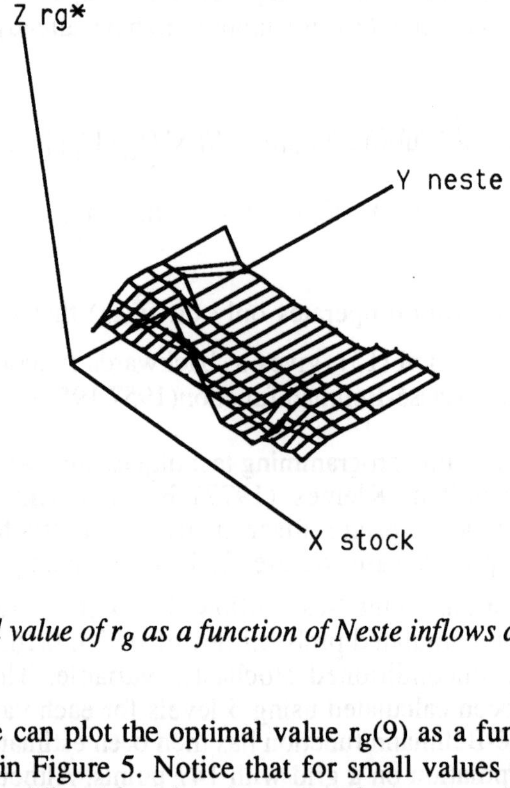

Figure5:Optimal value of r_g as a function of Neste inflows and reservoir levels

For instance one can plot the optimal value $r_g(9)$ as a function of N(8) and S(9) as showed in Figure 5. Notice that for small values of N(8) and S(9) , $r_g(9)$ increases rapidly. After that, it restarts from a very low value (q_{minG}) and then it increases very much with the storage and stands by when the Neste inflow is high.

Our explanation is as follows: at first, when there is a shortage of water, the optimal policy is to restrict irrigation. Consequently, more water can be used for water quality purposes . This fact is reflected in the first part of the surface corresponding to low values of N and S. When there is enough water, the reservoirs are used more efficiently than the natural Neste inflows (some reservoirs are directly on the sites of the irrigation perimeters) which are limited by the transfer capacity of the canal.

6 SYNTHETIC WATER RESOURCES ENGINEERING APPROACH

This other method focuses on 2 points:

- As explained in Figure 6, the interesting stochastic variable is the net cumulative difference between the natural inflows and the amount withdrawn for irrigation purposes and domestic needs.

Let:

$$VA(t)=\sum_{t=1}^{t} \left(\mathrm{Inf}\{N(t),Tcap + r_n(t)\} + G(t) - D(t)\right) \quad (16)$$

The term $\mathrm{Inf}\{N(t),Tcap + r_n(t)\}$ is due to the limited capacity ($Tcap = 14\ m^3/s$) of the Neste canal. For a fixed t, a normal probability law can be fitted to the available data.

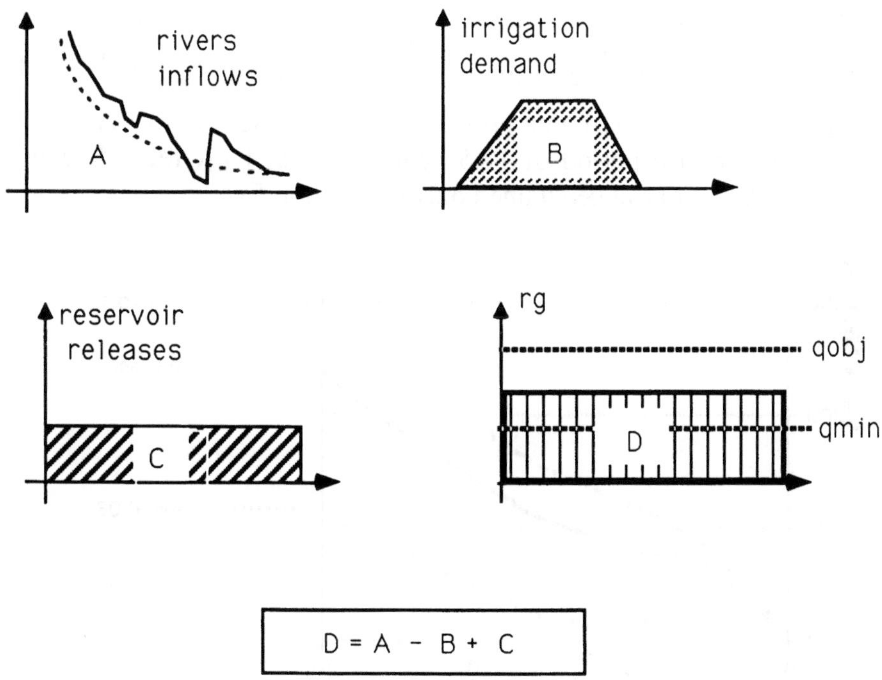

Figure 6: Synthetic engineering approach to the Neste System

-From a statistical viewpoint this quantity given in Equation (14) (see Figure 7) is relevant to the hydrological state of nature. In fact we search for a mapping from this hydrologic state to a value of the output target r_g at the exit of the Neste system.

Suppose that we have reached the end of the 30 weeks of operation (period T). We can compute VA(T) . By adding the storage volume VL(T) that has been released from the reservoirs we can determine what could have been the constant ideal value for r_g if we had known the future:

$VA(T) + VL(T) = r_g\ T$ ....(17)

Under drought conditions , we release all the volume that can be ($VL(T) = VL_0 = 74\ Mm^3$) to guarantee a minimum target level q_{minG}. The mass balance Equation (17) becomes:

$VA_{min}(T) + VL_0 = q_{minG}\ T$..(18)

In very wet situations , we choose a release policy (VL_1) between a mean attitude (VL_1=0) and an unnecessary spill of water resources ($VL_1 = VL_0$) :

$VA_{max}(T) + VL_1 = q_{objG}\ T$..(19)

By simulation based on the 15 years of available data with $VL_1 = \frac{VL_0}{2}$, it could be assessed that the tail area probability F_{min} associated to $VA_{min}(T)$ corresponds to a return period of 12.5 years and that F_{max} related to the critical value $VA_{max}(T)$ can be attained only one year in five.

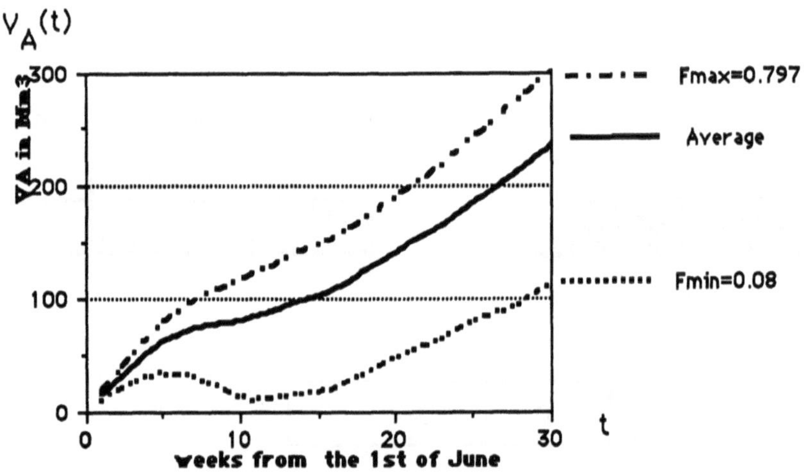

Figure 7: Law of VA(t) for $t \in [0,30]$

6-1 Strategic rule

Suppose that in week t , a value for the hydric state F has been determined:

- if $F < F_{min}$, the minimum level required q_{min} cannot be met for all remaining periods. A restriction of all irrigation should be enforced to preserve water quality and aquatic life.

- if $F_{min} < F < F_{max}$ we want a continuous monotonically increasing function $r_g(F)$ to define a policy so as to respect the following conditions: $r_g(F_{min}) = q_{min} < r_g(F) > q_{obj} = r_g(F_{max})$. Let U(F) be the Gaussian value of F

$$\text{such that : } F = \int_{-\infty}^{U(F)} \frac{\exp(-\frac{x^2}{2})}{\sqrt{2\pi}} dx .$$

A possible mapping could be :

$$r_g(F) = a + b\, U(F) \qquad (20)$$

where the coefficients (a,b) are defined by :

$$r_g(F_{min}) = a + b\, U(F_{min}) = q_{minG} \qquad (21)$$
$$r_g(F_{max}) = a + b\, U(F_{max}) = q_{objG} \qquad (22)$$

- if $F > F_{max}$ this means that we have abundant water facilities and we set $r_g(F) = q_{objG}$.

6-2 Determination of the state F

The idea is to extrapolate from stage t where VA(t) is known for the operation horizon T :

$$V_A(T) = VA(t) + \sum_{t=t}^{t+3} (\text{Inf}\{N(t), \text{Tcap} + r_n(t)\} + G(t))$$
$$+ \sum_{t=t+3}^{T} (\text{Inf}\{N(t), \text{Tcap} + r_n(t)\} + G(t)) - \sum_{t=1}^{T} D(t) \qquad (23)$$

The first term is theoretically known but can practically be re-adjusted as a function of the past releases to take into account the past perturbations around the strategic rule. The second term is estimated through hydrological models that can give reliable predictions for 3 weeks in the case of the river Neste. The last two terms belong to an unpredictable future . Practically , their values is assigned to a given frequency (10%) that is statistically adjusted so as to minimize the quadratic deviation between ideal r_g's that can be computed afterwards as in Equation (17) and r_g's given by strategic rules according to Equation (20).

7 COMPARISON OF THE TWO METHODS ON PAST SERIES

The strategies computed by the the systems analysis approach and the synthetic water resources engineering method have been compared using the 15 years of available observation data. The systems analysis approach generally gives slightly smoother policies. Nevertheless both strategies are very close as can be seen in Figures 8,9,10,11.

7-1 Normal year such as1980:

Figure 8 shows that the irrigation needs start late in the season and stay relatively low. Consequentely, reservoir releases are limited. Figure 9 indicates that the systems analysis approach is a little more cautious around t = 10 when the peak irrigation demand is reached.

7-2 Dry year such as 1985:

In 1985, there was practically no rain from July to September. The irrigation curve started to increase early and stayed at its maximum value for a long time (Figure 10). Both models would have dropped r_g near q_{minG} (Figure 11) and finally restricted irrigation to keep some resources for aquatic life and water quality needs until the end of the season before the high winter flows.

8 CONCLUDING COMMENTS

The operating rules of a hydro-system with irrigation and quality objectives have been defined within the framework of a classical stochastic dynamic programming model. Such rules can also be obtained from a more synthetic water resources engineering approach. From the present application to the Neste system, the conclusions are :

1) Both procedures allow the system manager to formulate operating strategies in a rational way:

- An operating rule can be derived to allocate water so as to meet a combination of the various objectives. It is expressed as a feedback law linking what we know from the system (F or S(t) and N(t-1)) to how we control its evolution (by setting r_g to predetermined values).
- Both methods need a parameter (VL_1 or PEN) to be set up by stochastic simulation.
- They give close results on the basis of the past data and can be conveniently proposed to system managers.

2) The 15 years of data are used to calibrate both systems analysis and water resources engineering models. In this case study, the same data are used to make a validation study and a comparisom of the two methods... which is of

course questionnable and which would certainly gives optimistic results and high bias. For practical reasons of making valid hydrological estimators, we couldn't divide this very short series of available data into a calibration period and a testing period: due to the weak number of dry years (1976 and 1985) this separation would have lead to ignoring drought phenomena !

3) The system analysis approach is based on stochastic dynamic programming. It can be efficiently used to derive optimal feedback rules of operation and can routinely deal with complex decisions such as limiting irrigation when a shortage is to occur or take the risk to keep going and decrease output targets for water quality management. At the same time, this procedure entails heavy computing time, uneasy interpretation of the weighting coefficient between irrigation and water quality objectives, and a rather artificial elicitation of the global compromise.
Such an approach is very well fit for simulation because it is composed of elementary blocks that are gathered in a transition relationship as Equation (1) to describe the system's dynamic evolution . This approach also provides a means to get an optimal policy as long as the system manager accepts the necessity to formulate an objective function consistentily with dynamic programming (i.e. stages are separable and additive). Of course this optimal allocation should be carefully examined because of model uncertainties influencing both the system response and the hydrological behavior. Klemes (1978) has proved that such disturbances could be of the same order of magnitude as the computed "optimized" results.

4) The synthetic method may appear more attractive from the engineering point of view for the following reasons:

- the state F is easily interpreted in terms of "dry year with a return period of 10, normal year , exceptionnally wet year "and so on. The trade-off coefficient is the volume one wants to keep in the reservoir s at time T for a wet year relative to the frequency F_{max} . Consequently if the parameter is chosen with "good sense" and no optimization scheme is needed.
- there is no computation except a mass balance equation and a normal probability law adjustment which is very easy because it deals with cumulative quantities.

5) Such models are designed to serve only as multicriterion decision making aids. In very dry days such as occured in summer 1976 or 1989 in France they cannot create additional water resources...still they can help the system manager by constant up dated multidimensional estimation of the risks that may be encountered when following different operation rules. In the case of the Neste system, a real-scale experiment began in 1989: in real time operation, both models worked on line as decision-making supports, and the system manager made a thorough study of the hydrological conditions when the two approaches did not agree on the same stragegy for the following week.

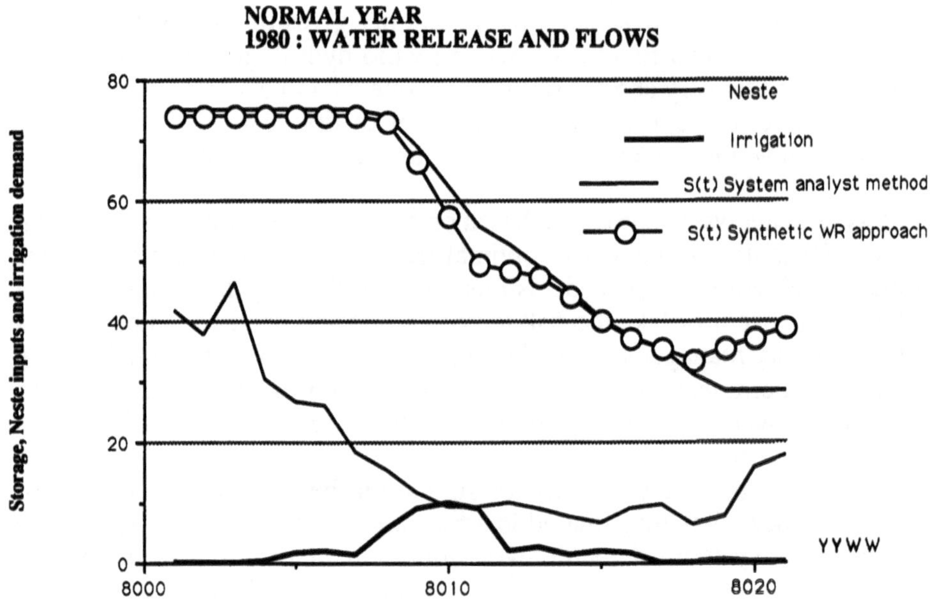

Figure 8 : S(t) for a normal year; the time scale t is in year-weeks: 8010 means 1980, week no 10

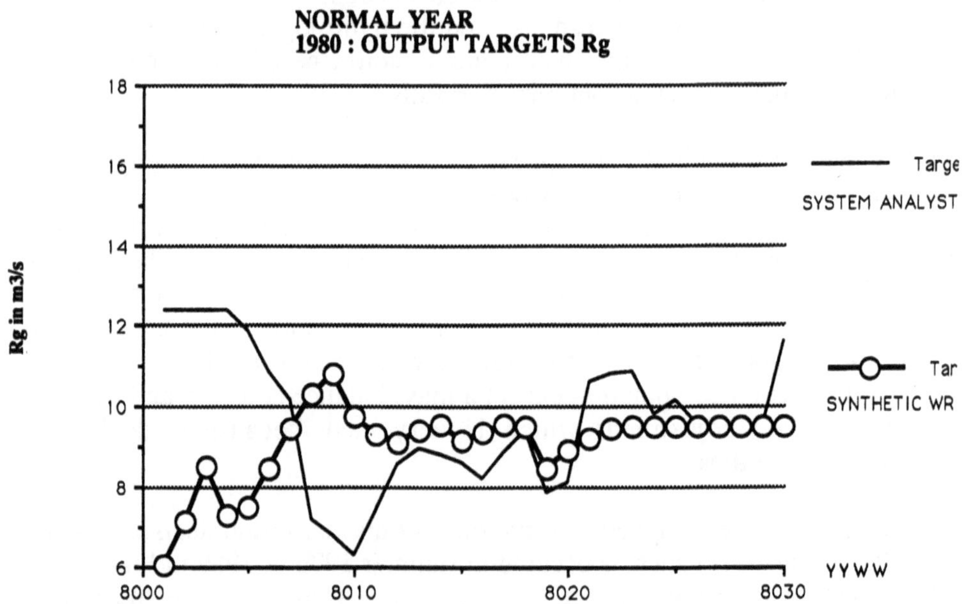

Figure 9 : $r_g(t)$ for a normal year

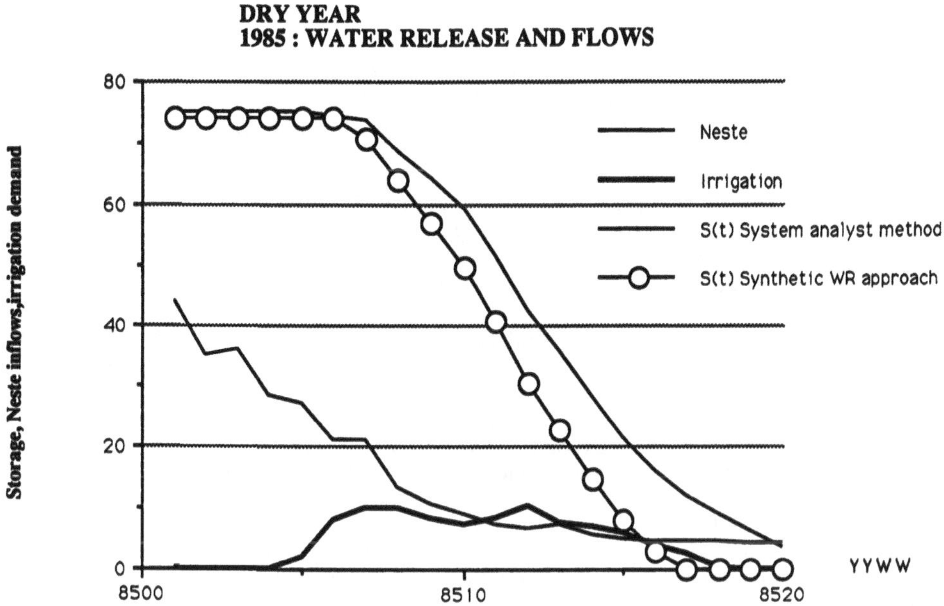

Figure 10: S(t) for a dry year

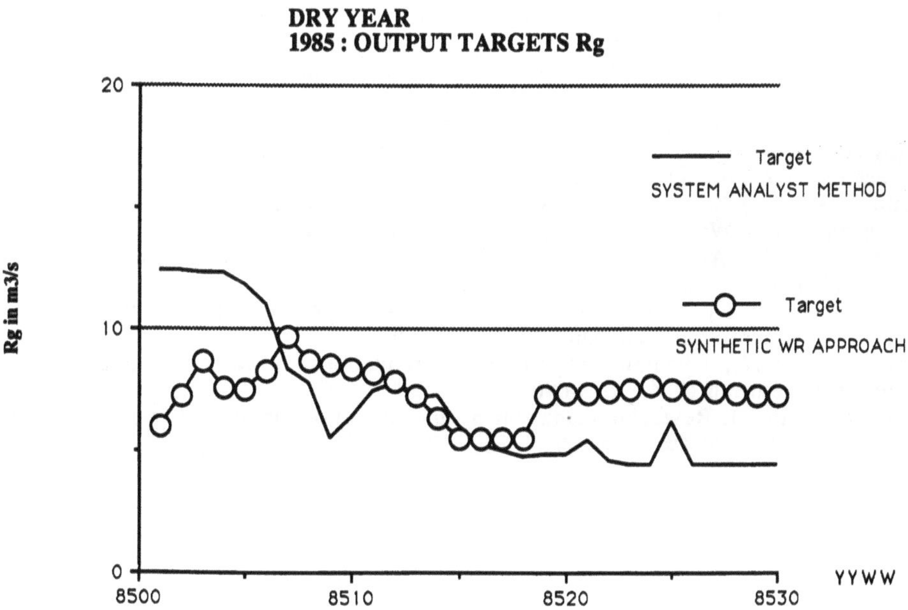

Figure 11 : $r_g(t)$ for a dry year

BIBLIOGRAPHY

Askew J.A. (1974), Optimum reservoir operating policies and the imposition of a reliability constraint. Water Resources Research 10, 1, p.51-56.

Bellman R.E. (1957), Dynamic programming. Princeton University Press,.

Bodo B.A., Unny T.E. (1990), Modeles lineaires stochastiques theoriques pour la reponse des petits bassins. Revue des sciences de l'eau 3, p.151-182.

Buras N. (1972), Scientific allocation of water resources. Elsevier, N.Y.

Colleter P., Delebecque F., Falgarone F. and Quadrat J.P. (1978), Application du controle stochastique a la gestion des moyens de production d'energie en nouvelle-Caledonie. In "E.D.F. Boulletin de la direction des etudes et rechereche, Serie C. Mathematiques-Informatique" (E.D.F.), p.1-28.

Dodu J.C., Goursat M., Hertz A., Quadrat J.P. and Viot M. (1981), Methodes de gradient stochastique pour l'optimisation des investissements dans un reseau electrique, E.D.F. 2, p.133-164.

Fang Z.X., Voron B., Bocquillon C. (1989), Programmation dynamique: Application a la gestion d'une retenue pour l'irrigation. Hydrolo.Sci. J. 34(4), p.415-424.

Hashimoto T., Stedinger J.R. and Loucks D.P. (1982), Reliability, resiliency, and vulnerability criteria for water resource system performance evaluation. Water resources research 18, 1, p.14-20.

Houghtalen R.J. and Loftis J.C. (1988), Irrigation water delivery system operation via aggregate state dynamic programming. Water resources bulletin 24, 2, p.427-434.

Klemes V. (1977), Discrete representation of storage for stochastic reservoir optimization. Water resources research 13, 1, p.149-158.

Klemes V. (1978), The unreliability of reliability estimates of storage reservoir performance based on short streamflow records. In "Water resource management, Water resources publications, Fort Collins, Colorado", p.193-205.

Krzysztofowicz R. (1982), Utility criterion for water supply: quantifying value of water and risk attitude. In "Decision-Making for hydrosystems-Forecasting and operation" (ED: T.Unny), 43-62.

Masse P. (1959), Le choix des investissements. Dunod.

Mergos G.J. (1987), Evaluation of irrigation projects under uncertainty: A symmetric quadratic programming approach. Water resources management 1, p.45-56.

Simonovic S.P. and Orlob T.G. (1984), Risk-reliability programming for optimal water quality control. Water resources research 20, 6, p.639-646.

Tapiero C.S. (1988), Applied stochastic models & control in management. In "Advanced series in management". V12. North-Holland.

Turgeon A. (1981), Optimal operation of multireservoir power systems with stochastic inflows. Water resources research 16, 2, p.275-283.

Yakowitz S. (1982), Dynamic programming applications in water resources. Water resources research 18, 4, p.673-696.

Yeh W.W.G. (1985), Reservoir management and operations models: S state-of-the-art review. Water resources research 21, 12, p.1797-1818.

RISK AND RELIABILITY IN PREDICTING DROUGHTS: THE USE OF PREHISTORIC TREE-RING DATA

John A. Dracup and Donald R. Kendall[1]
Civil Engineering Department
University of California, Los Angeles
Los Angeles, CA 90024-1593

Abstract

The climate variability scenarios of greatest interest to water resource managers are those of extreme high or low streamflows. Our study focused on low streamflow events in the Colorado River Basin. Drought analyses based on tree-ring reconstructed streamflows dating back to the year 1520 were compared with more recent data from the historical record. The analysis suggests that a repeat of the low streamflow hydrologic periods 1520-1599 or 1600-1679, both of which contained severe drought events, would have significant impacts on water availability and shortage allocations, particularly in the Lower Basin. For the thirty-year planning horizon considered, Lower Basin average shortages were on the order of 1.45 million acre-feet (maf) by the year 2020. These tree-ring reconstructed streamflows -- based on actual, prehistoric climatic events -- evidenced shortages far more severe than those in the recent historical record, suggesting that there are risks involved in relying on the historical record for prediction.

I. Introduction

Traditionally, water resources design, such as reservoir sizing, has been predicated on the assumption that the statistical parameters of historic streamflow series are stationary, that is, that the parameters do not change with time. This is not to say that climate is actually stationary, since changing climate has been accounted for in water resource system designs by considering the variance of streamflow (σ^2) about its mean ($\overline{Q}$). Stationarity assumes, however, that future variations in climate, as expressed in streamflow, will be similar to those observed in the past. Thus water resource managers look to the historical record to help preduct future events such as extreme high or low streamflows. They must take for granted that the

---

[1] Loyola Marymount University, Department of Civil Engineering and Environmental Science, Loyola Blvd at W. 80th St., Los Angeles, CA 90045

NATO ASI Series, Vol. G 29
Water Resources Engineering Risk Assessment
Edited by J. Ganoulis

historical record is sufficiently extensive to be representative of large-scale streamflow trends. In fact, though, since the historical record in many cases covers a relatively short period of years, doing this presents a risk. Do the conclusions evident from the historical record in fact correspond to those evident in macro-scale periods of hundreds of years? In looking into this question, our study focused on prehistoric data, namely low streamflow events evident from tree-ring studies in the Colorado River Basin. We found that the prehistoric data suggested a greater variation in streamflow events than that evidenced in the historical record. Historical climate analogies provide a different look into the future than that provided by large-scale computer models. Glantz (1988, 1989) posed four caveats for the use of such analogies. First, the use of the climate analogy must be made clear so that it is not misleading. Second, the climate analogy must not be overextended or expected to bear unreasonable weight. Third, limitations of the analogy due to "cultural or historical reasons" must be noted. Four, the researcher must be on the lookout for possible but inconsistent scenarios. The topic addressed here, namely, the operation of water resource systems under climate stress, is of course not new. One of the best discussions can be found in the National Research Council's study on climate, climate change, and water supply (Wallis, 1977). Here Wallis (1977) and Matalas and Fiering (1977) emphasized the calculus of economic regrets, noting that "unless the exact sequence of future flows can be predicted with certainty there may be little benefit to hydrologic system design...." Furthermore, Matalas and Fiering (1977) proposed designing water resource systems with robustness (a design that performs reasonably well under a variety of possible climates) and resilience (the property that allows a system designed for one climate and set of conditions to be modified in response to persistent new climates or other conditions). Climate analogies can be used to determine whether or not an existing water resource system exhibits the properties of robustness and resilience.

II. The Colorado River Basin

The Colorado River, which dominates water resource development in the seven southwestern United States, is one of the most carefully managed river systems in the world. Its multipurpose uses of water supply, hydroelectric power, and water-based recreation compete with its management priority of flood control. With the exception of the deserts of the Great Basin, this 243,000-square-mile basin has the greatest water deficiency (average precipitation less potential evapotranspiration) of any basin in the contiguous United States. Yet more water is exported from it than from any other basin in the United States. The Colorado River Compact of 1922 divided the basin into the Upper and Lower Colorado River Basins for purposes of interstate administration. The Upper Basin drainage includes those areas of Arizona, Colorado, New Mexico, Wyoming, and Utah that drain into the Colorado

River above Lee Ferry, Arizona. It is bounded on the east and north by mountains forming the Continental Divide, and on the south it opens to the Lower Colorado region. The Lower Basin drainage includes most of Arizona, parts of southeastern Nevada, southeastern Utah, southeastern California, and western New Mexico.

II.A Upper Basin Water Availability Estimates

A wide range of climate occurs in the Colorado River Basin because of differences in altitude, latitude, and topographic features. About 83% of the water that flows in the entire Colorado River Basin comes from the Upper Basin. The average annual precipitation throughout the Upper Basin is about 16 inches, which amounts to 93,440,000 acre-feet per year. Approximately 15% of the precipitation runs off, as most is lost to evapotranspiration. One of the most famous and controversial hydrologic records in the United States is that of the virgin flow of the Colorado River at Lee Ferry, Arizona (see Figure 1). Lee Ferry is defined as a point on the Colorado River one mile below the mouth of the Paria River. Estimates of virgin flow have been made there for the Upper Basin since 1896; however, runoff has been measured and recorded only since the first gauging station was established at Lee Ferry during the summer of 1921. (The Bureau of Reclamation now uses natural flow, not virgin flow, for its operation studies.)

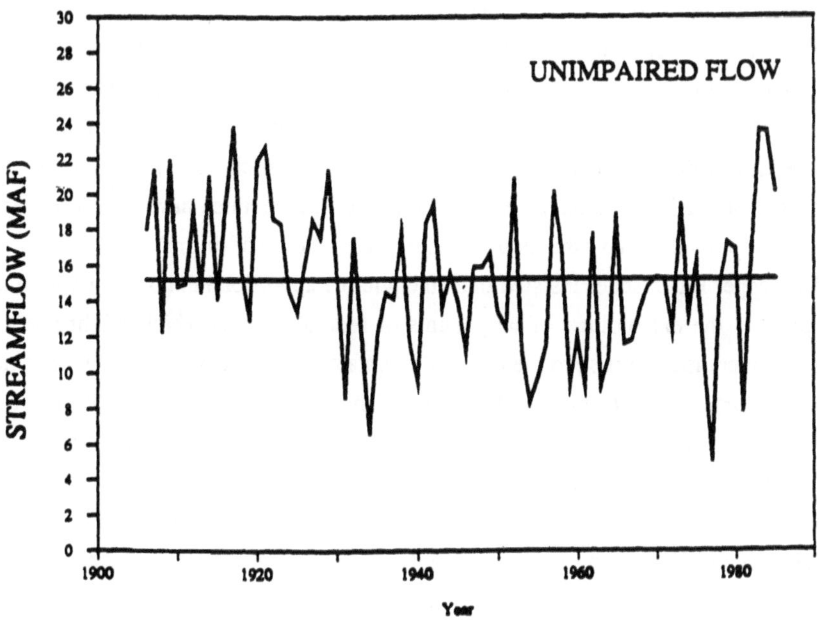

Figure 1. Colorado River Flow at Lee Ferry, Arizona

Estimates of the long-term annual average flow vary from 11.8 to 16.8 maf depending on the time period selected (Colorado River Board of California, 1990). Recent tree-ring analysis dating back to 1520 has shown the long-term mean to be approximately 13.5 maf (Stockton, 1977). Current estimates of available surface water supplies within the Upper Basin are less than those at the time the Colorado River Compact was negotiated. This is because of the abnormally wet period that occurred during the early part of this century. The range of annual natural flow at Lee Ferry has varied from a low of 5.0 maf in 1977 to a high of 24.0 maf in 1917.

II.B Normal Operation of the Colorado River Basin

The laws governing the Colorado River have been presented in detail by Meyers (1966) and Hundley (1975, 1983). Only a brief summary of the major treaties, laws, and compacts will be presented here. The allocation of Colorado River water is based on the concept of beneficial use. The allocation system operates at four levels: international, interregional, interstate, and intrastate (Weatherford and Jacoby, 1975). The international allocation was accomplished by the Mexican Treaty of 1944. Mexico was guaranteed an annual amount of 1.5 maf except in times of extreme shortage. However, this treaty contained no provision for water quality. Joint agreements in 1965 and 1973 called for a temporary agricultural drainage water bypass and eventually a desalting plant to improve the quality of water crossing the border. The interregional allocation was achieved when Congress approved the Colorado River Compact (United States, 1928 and 1983), which became effective in June 1929. Sectional rivalry has caused the states included in the drainage basin to agree to an equal apportionment in the use of the Colorado River system waters among the states of the Upper Basin and the states of the Lower Basin (Colorado River Compact, 1922, Article III (b) and Article III (d)). Traditionally, the fertile lowland valleys in the Lower Basin states have developed economically more rapidly than have the mountain headwater ''areas of origin'' in the Upper Basin states. The Upper Basin states insisted that an equitable apportionment of the river be made to them prior to the expenditure of large sums of federal money, which might result in a modification of equities adverse to the Upper Basin states. This is essence what was achieved in the 1922 Colorado River Compact. The intent of this landmark document was to give each basin the perpetual right to the ''exclusive beneficial use of 7.5 maf of water per annum....'' However, the Lower Basin was assured that depletion in the Upper Basin would allow at least 75 maf flow to the Lower Basin at Lee Ferry in each successive ten-year period. Thus, the Lower Basin received a guaranteed ten-year, not annual, minimum flow, and the Upper Basin assumed the burden of any deficiency caused by a hydrologic dry cycle. It is important to note that the division of the use of water between the Upper and Lower Basins is a fixed amount rather than a proportional amount (e.g., one-half of a ten-year moving average). Each basin currently possesses many storage facilities and a large linchpin reservoir, Lake Powell for the Upper Basin and Lake Mead for the Lower Basin. The storage in these two reservoirs totals 51.0 maf, or 85% of the total storage in the entire Colorado River Basin. The reservoir system now stores about four times the annual flow of the river. This volume of water in storage reflects the determination of the Basin states to conserve as much water as possible, providing a margin of safety in the event that a run of dry years occurs. The Upper Basin states prefer that releases to the Lower Basin be the absolute minimum required by law. Furthermore, the current operational regulation requires that the river's manager, the U.S. Bureau of Reclamation, maintain Lake Mead and Powell at or near equal volumes of water in storage at the end of each year (30 September) (Nathanson, 1978). Each basin is thus assured "equal ownership" of the river. The climate analogy utilized as a climate change scenario for this study was a severe and sustained drought on the Colorado River Basin. Conclusions drawn from the study are largely influenced by the joint operation of Lake Powell and Lake Mead which respond to the following criteria in the Law of the River (Nathanson, 1978):

1. A Lake Powell minimum objective release of 8.2 maf per year.

2. Additional releases from Lake Powell to equalize end-of-year active storages in Lakes Powell and Mead if Lake Powell would otherwise contain more water in storage.

3. A minimum required water in storage in the Upper Basin reservoirs combined, called 602(a) storage under the Colorado River Basin Project Act of 1968, and described in detail below.

II.C Severe and Sustained Drought in the Colorado River Basin

Perceptions of climate stability in the Colorado River Basin are not borne out by historical data. As demonstrated in Figure 1, the variability in the river's streamflow is substantial. As recently as 1977, the western United States was hit with a severe drought which resulted in a significant drop (to 5.0 maf) in the estimated virgin flow of the Colorado River (Upper Colorado River Commission, 1982). Since the Colorado River has a storage capacity of over four times the average annual natural runoff and an elaborate system of laws governing water allocation and system operation, some entities believe that the basin has been drought-proofed. However, a close examination of the hydrologic reconstructions of unimpaired flows at Lee Ferry, Arizona based on tree-ring analysis indicates that droughts are much more severe than those indicated from the conventional streamflows which have occurred during the period of historical measurement. A study of severe and sustained drought affecting the Southwest involves establishing streamflow characteristics and probabilities of drought events, establishing the effects of storage systems on water availability at key points in the systems over time given streamflow characteristics during drought, and establishing relationships between available supplies and projected demands at key points during the period of analysis.

III. The Analytical Model

Severe and sustained drought in the Colorado will be buffered by the massive storage provided in the Colorado River Basin. For this reason, this drought analysis focuses on annual operation of the two major reservoirs in the Upper and Lower Basins, Lake Powell and Lake Mead, and related effects on water availability relative to aggregate demand. Unimpaired measured streamflow at Lee Ferry, Arizona was taken to be the base-case hydrology, and compared against reconstructed sequences of equal length from the tree-ring data, dating back to 1520 (Meko, 1990). A Colorado River Annual System Regulation (CRASR) model (MWD, 1980) was used to simulate the basin operation. It was substantially modified and rewritten by the authors to reflect current operating regimes. The model simulates annual regulation of Lake Powell and Lake Mead. Seasonal operation is not critical in this study as in other river basins because of the large amount of available storage capacity compared to annual runoff. The average annual runoff of the Colorado River Basin is about 15 million acre-feet (maf) while the total reservoir storage available is almost 60 maf. The combined live capacity of Lake Powell and Lake Mead is over 51 maf. The reservoir system model operates in accordance with the Law of the River, and within certain physical limits. Maximum storage capacities of Lake Powell and Lake Mead are 25.00 maf and 25.88 maf, respectively. Typical minimum storage capacities are 4.13 maf and 10.02 maf, and represent the minimum power pool below which power generation at the projects would be interrupted. Shortage releases are made from Lake Mead when storage falls below 10.62 maf. The model also takes into account Section 602(a) of the Colorado River Basin Project Act, which allows for additional Upper Basin storage so as to assure future deliveries to the Lower Basin without impairing annual consumptive use in the Upper Basin. Factors considered in determining this storage include a minimum objective release of 8.23 maf per year from Lake Powell, Upper Basin demands, minimum power pool in the other Upper Basin reservoirs, and natural flow at Lee Ferry over the 12-year critical period (1953-1964). As input, the CRASR model uses Upper and Lower Basin demands, initial reservoir storages, virgin flow or reconstructed sequences at Lee Ferry, Arizona, and certain operating constraints. A strict accounting is maintained of all quantities of water involved. Each acre-foot of virgin flow to Lake Powell is accounted for as Upper Basin depletions (which includes reservoir evaporation), storage change in Lake Powell or Lake Mead, or Lake Mead outflow to meet water demands. Within the two reservoirs, flow continuity is maintained by a simple mass balance:

$$O = I \pm \Delta S \tag{1}$$

It is felt that the CRASR model, simplified though it is, captures the essential features of the operation of the Colorado River system.

IV. Using Tree-Ring Data

Streamflow is a useful hydrologic variable for drought studies at the basin scale, since it integrates a variety of processes over a watershed, such as runoff, soil moisture, and evapotranspiration. Changes in climate are thus reflected as changes in streamflow volumes over a specified time period. Typically in the western United States, the record consists of a period of streamflow measurement on the order of sixty to eighty years. At Lee Ferry, Arizona, the Colorado River streamflow records date back to 1906. This 80-plus-year period contrasts to over 450 years (from 1520) of reconstructed streamflow records at Lee Ferry based on tree-ring studies, which offer a record of greater variability than the measured record. Under conditions of climate change, the magnitude of the change in the statistical parameters is unknown. However, we do know that climate will cause a change not only in the mean ($\overline{Q}$), but also in the variance (σ^2) and the skew (γ). Rather than perform a sensitivity analysis with these parameters, and create a number of what-if scenarios, streamflow reconstructions provided by tree-ring analyses from the basin's recent past -- that is, tied to actual streamflow events -- are used as a basis for this climate analogy study of droughts. A comparison of the means of some reconstructed sequences with a hydrologic streamflow record of equal length are shown in Figure 2. What is most striking is the smaller mean annual flow for the reconstructed sequences, which ranges from 12.97 to 13.78 maf, compared to the 1906-1985 record of 15.06 maf. A standard t-test shows all of the means to be statistically different from the measured 1906-1985 period at a 95% level of significance, with the exception of the period 1680-1679. Stochastic flows generated from an autoregressive model of order 1 (AR(1)) were also considered in the drought study (Kendall and Dracup, 1991). It has been shown that hydrologic sequences provided by tree-ring reconstructions were the most appropriate choice for this type of analysis (Kendall and Dracup, 1991). Drought analyses are typically considered on a monthly or annual time scale. For this analysis, a time scale of one-year intervals was chosen, as this is consistent with the resolution provided by the tree-ring reconstructions. Threshold levels in drought analyses are typically taken to be a streamflow's mean or median. Other appropriate levels may include average annual demand. For this analysis, the threshold level was taken to be the mean annual measured unimpaired flow at Lee Ferry, important because comparisons using this threshold level will be made with reconstructed annual unimpaired flows at Lee Ferry from tree-ring data.

V. Drought Parameters for the Colorado River

A study of the Colorado River using streamflow indicates that there is a tendency for low flows to follow low flows and for high flows to follow high flows. One reason forwarded for this behavior is that the process which governs streamflow is non-stationary in its mean. Incorporation of tree-rings into this type of analysis gives a measure of support to this idea. An anthropogenically induced climate change would exacerbate this process also. Evidence of a nonstationarity in the mean is shown in Figure 2. Previous analyses performed by a panel on water and climate for National Academy of Sciences (1977) argued against a stationarity independent or short memory process for the Colorado River streamflows. Furthermore, the panel estimated that the long-term mean for flows at the Lee Ferry compact point was 13.5 maf (Dracup, 1977).

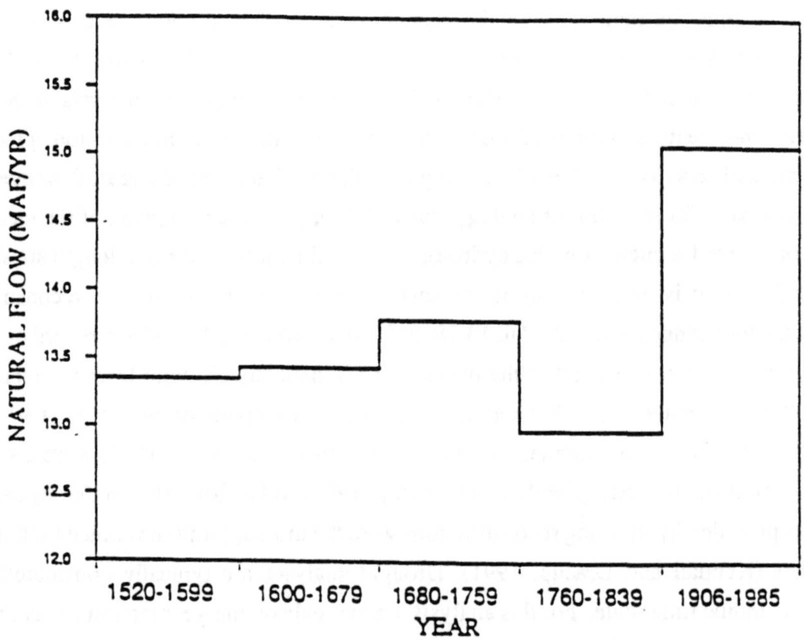

Figure 2. Colorado River at Lee Ferry - Nonstationary Mean

V.A Definition of Drought

Drought can be defined as a water shortage relative to a specified need for water within a conceptual supply-and-demand relationship. In particular, a hydrologic drought may be characterized as any year or series of consecutive years during which average annual streamflow is continually below some specified threshold level, typically taken to be the long-term mean ($\overline{Q}$) (Yevjevich, 1967; Dracup et al., 1980a). Comparative drought analyses at different threshold levels relative to the mean show the non-linear relationship to drought severity (Dracup and Kendall, 1990). A drought event (Figure 3) is considered to be composed of three defining attributes: duration, D, in years; severity, S, or the cumulative deficit; and magnitude, M, or the average water deficit below the threshold (Dracup and Kendall, 1987). More specifically,

$$S = \int_{o}^{D} Y(t)dt \tag{2}$$

where $Y(t)$ = deficit in year t of a D-year drought. In terms of streamflow, equation 2 may be written as

$$S = \int_{o}^{D} (\overline{Q} - Q(t))dt \tag{3}$$

where $Q(t)$ = unimpaired streamflow in year t of a D-year drought.

By definition, drought magnitude is written as $M = 1/D \int_{o}^{D} Y(t)dt$, such that $S = M \times D$. These drought parameters are interrelated, and only two of them are necessary and sufficient to completely define a single drought event (Dracup et al., 1980b). Duration and severity are the most closely correlated parameters, and may be considered to be the two primary parameters which depend directly on the streamflow values. In contrast, magnitude is taken to be a secondary parameter, owing to the fact that duration and magnitude are weakly correlated. Based on this definition of drought, we may write the severity of a D-year drought as

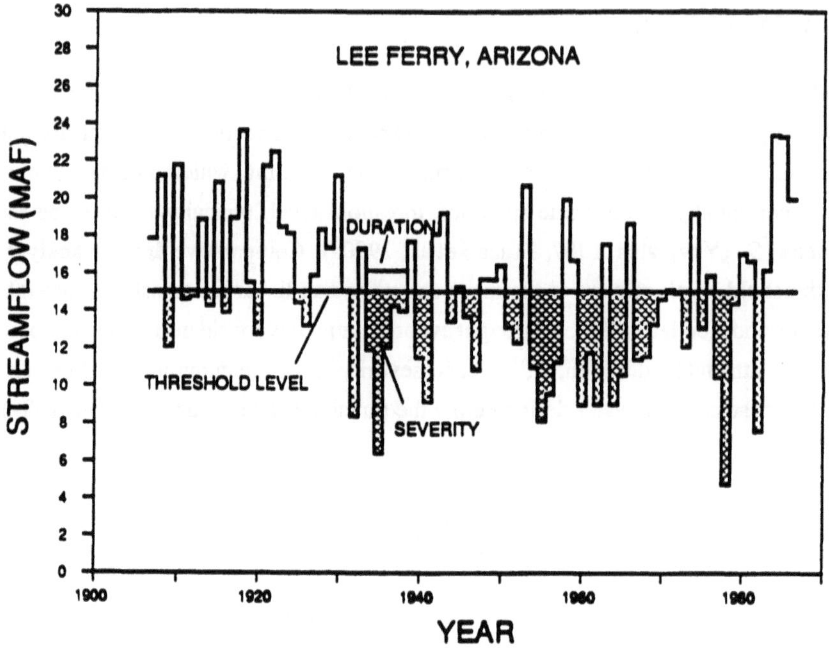

Figure 3. Definition of Hydrologic Drought

$$S_D = Y_1^D + Y_2^D + \ldots + Y_D^D \tag{4}$$

where:

D = drought duration (years)

Y_n^D = deficit in year n of a D-year drought

Assuming that the deficits are independent and identically distributed, we may write the expected value and the variance of drought severity as

$$E\ [S_D] = E\ [Y_1^D + Y_2^D + \ldots + Y_D^D] = D \cdot E\ [Y_i^D] \tag{5}$$

and

$$VAR\ [S_D] = E\ [D \cdot VAR\ [Y_i^D]] = VAR\ [D \cdot E\ [Y_i^D]] \tag{6}$$

Note that the relationships from these last two equations hold without any assumptions when viewed from a computational standpoint.

V.B Droughts on the Colorado River

Table 1 lists the twenty hydrologic droughts that occurred on1the Colorado River at Lee Ferry during the period of record from 1906 through 1985. Of these, four had a severity of over 15 maf, with a duration ranging from 3 to 5 years. It is apparent that a severe and sustained drought in the Colorado River Basin is not necessarily produced by a single continuous drought event, but may arise from a series of events, separated by one or two years. During the period of record, a series of droughts began in 1954 and continued through 1971.

Table 1: Historic Hydrologic Droughts on the Colorado River at Lee Ferry, Arizona 1906-1985.

Drought No.	Duration (Years)	Starting Year	Severity (maf)	Magnitude (maf/Year)
1	1	1909	2.954	2.954
2	2	1911	0.665	0.333
3	1	1914	0.769	0.769
4	1	1916	1.103	1.103
5	1	1920	2.290	2.290
6	2	1925	2.432	1.216
7	1	1932	6.609	6.609
8	5	1934	16.520	3.304
9	2	1940	9.384	4.692
10	1	1944	1.617	1.617
11	2	1946	5.506	2.753
12	2	1951	4.678	2.339
13	4	1954	20.092	5.023
14	3	1960	15.267	5.089
15	2	1964	10.422	5.211
16	4	1967	9.176	2.294
17	2	1972	2.971	1.486
18	1	1974	1.945	1.945
19	3	1977	15.351	5.117
20	1	1982	7.410	7.410

VI. Analogous Tree-Ring Sequences

Two sets of 80-year sequences from tree-ring reconstructed streamflows at Lee Ferry which exhibited different types of severe droughts were chosen for the purpose of comparison. The periods selected were 1520-1599 (Figure 4) and 1600-1679 (Figure 5). These periods display different types of droughts with regard to their number, duration, and severity. The sequence from 1520-99 displays a series of droughts which contribute collectively toward a severe and sustained drought condition, while the 1600-79 sequence displays a single severe drought.

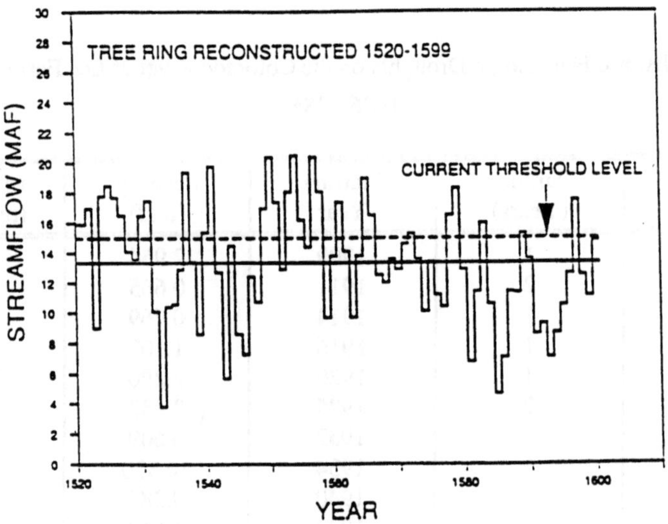

Figure 4. Streamflow Sequence at Lee Ferry, 1520-1599

Maintaining the same threshold level of 15 maf (the long term mean from 1906-85) for all sequences, fifteen drought events were identified for both tree-ring reconstructed streamflow records. The drought characteristics differ significantly. The period from 1520-1599 exhibits four droughts beginning in 1573 and extending through 1597. These are shown in Table 2. Conversely, the period from 1600-1679 displays a single severe drought of 15-year duration, as shown in Table 3.

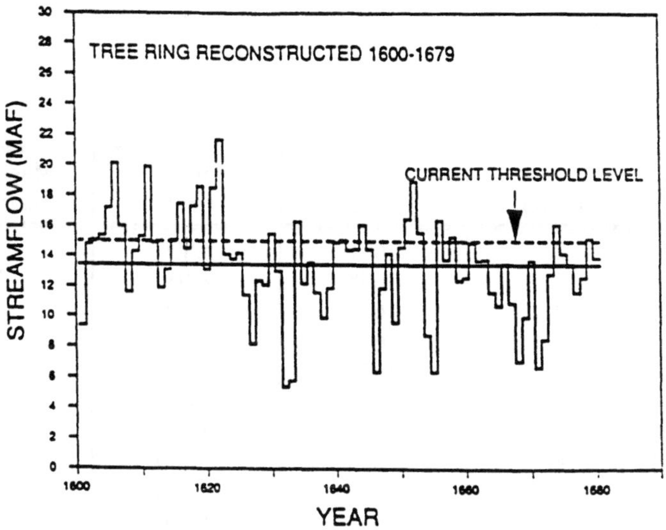

Figure 5. Streamflow Sequence at Lee Ferry, 1600-1679

Drought No.	Duration (Years)	Starting Year	Severity (maf)	Magnitude (maf/Year)
1	1	1523	6.000	6.000
2	2	1528	2.300	2.300
3	5	1532	27.20	5.440
4	2	1538	8.000	4.000
5	7	1542	33.10	4.729
6	1	1552	2.100	2.100
7	1	1556	0.600	0.600
8	2	1559	6.500	3.250
9	3	1562	7.400	2.467
10	5	1567	9.400	1.880
11	5	1573	16.30	3.260
12	3	1580	13.90	4.633
13	5	1584	29.90	5.980
14	7	1590	34.60	4.943
15	3	1598	6.600	2.200

Table 2: Tree-Ring Reconstructed Hydrologic Droughts on the Colorado River at Lee Ferry, Arizona 1520-1599

Table 3: Tree-Ring Reconstructed Hydrologic Droughts on the Colorado River at Lee Ferry, Arizona 1600-1679

Drought No.	Duration (Years)	Starting Year	Severity (maf)	Magnitude (maf/Year)
1	2	1601	5.800	2.900
2	2	1608	4.100	2.050
3	3	1612	5.100	1.700
4	1	1617	0.500	0.500
5	1	1620	1.900	1.900
6	7	1623	18.80	2.686
7	3	161	20.80	6.933
8	6	1635	15.90	2.650
9	2	1642	1.100	0.550
10	6	1645	18.80	3.133
11	2	1654	14.80	7.400
12	1	1657	1.200	1.200
13	15	1659	52.10	3.773
14	4	1675	8.100	2.025
15	1	1680	1.100	1.100

VII. Index-Sequential Hydrologic Sequences

In order to develop probability estimates of water supply, many agencies in California and the United States use the index-sequential method to create a set of realistically probable streamflows (Kendall and Dracup, 1991; United States, 1989; Metropolitan Water District, 1989). Each sequence differs from the next only by the fact that the first historic streamflow year is incremented by one. For example, in hydrologic sequence 1, the first historic streamflow year is assumed to occur in the first demand year of a study. In sequence 2, the second historic streamflow occurs in the first demand year, with the first historic streamflow being wrapped to the bottom of the deck, so to speak. Consequently, there are as many sequences as there are years of hydrologic data. In order to develop probability distribution functions, we used this approach to generate pseudo-likely hydrologic traces of both the measured and tree-ring reconstructed streamflows. Matalas and Fiering (1977) noted that a bias exists in geochronological records such as tree-rings when used to estimate streamflows. Long-term mean estimates based on streamflows constructed from tree-rings are reasonable, but tree-ring indices are more normally distributed and more highly correlated than recorded historical streamflow. However, it has never been determined whether these differences in statistical properties are significant enough to offset the utility of the reconstructed flows in water resources systems analyses.

VIII. Demand Assumptions

System regulation studies to determine future water supplies are performed over a series of years to accommodate growing demands. For this study, a demand period from 1989 to 2020, or 32 years, was simulated to accommodate the rather large drought periods observed in the tree-ring reconstructed hydrologic sequences. Demand data is taken from data used by the Bureau of Reclamation as the base case for its Colorado River System simulation (CRSS) model (United States, 1989). The model output is directly related to the hydrologic sequence input. For water system design purposes, streamflow is recognized as a random process, of which the historic hydrologic record is one expression. The hydrologic sequences comprise what can be taken to be year-to-year water supply. Coupled with this are year-to-year water demands. Water losses due to evaporation or seepage from the conveyance system can be regarded as an additional demand. An Upper and Lower Basin demand schedule used in the analysis is shown in Tables 4 and 5. These data are consistent with those used by the USBR's model of the Colorado River. Lower Basin demands include all uses from Lake Mead and downstream. The demands include those for California, Arizona, Nevada, and Mexico as well as net losses below Lake Mead.

Table 4: Normal Upper Basin Water Demands (maf)

Water Year	Moderate Demand
1988-89	3.540
1989-90	3.550
1990-91	4.040
1991-92	4.050
1992-93	4.060
1993-94	4.070
1994-95	4.080
1995-96	4.090
1996-97	4.200
1997-98	4.310
1998-99	4.410
1999-2000	4.520
2000-01	4.630
2001-02	4.640
2002-03	4.660
2003-04	4.680
2004-05	4.700
2005-06	4.720
2006-07	4.740
2007-08	4.760
2008-09	4.780
2009-10	4.800
2010-11	4.830
2011-12	4.840
2012-13	4.860
2013-14	4.870
2014-15	4.890
2015-16	4.900
2016-17	4.910
2017-18	4.930
2018-19	4.940
2019-20	4.960

Table 5: Normal Lower Basin Water Demands (maf)

Water Year	Normal Demands (inc. Mexico and losses)
1988-89	9.370
1989-90	9.460
1990-91	9.480
1991-92	9.490
1992-93	9.490
1993-94	9.500
1994-95	9.500
1995-96	9.510
1996-97	9.520
1997-98	9.530
1998-99	9.540
1999-2000	9.550
2000-01	9.550
2001-02	9.550
2002-03	9.550
2003-04	9.550
2004-05	9.550
2005-06	9.550
2006-07	9.550
2007-08	9.550
2008-09	9.550
2009-10	9.550
2010-11	9.550
2011-12	9.550
2012-13	9.560
2013-14	9.560
2014-15	9.560
2015-16	9.560
2016-17	9.570
2017-18	9.570
2018-19	9.570
2019-20	9.580

IX. Reservoir Results

Results were derived from simulating 80 index-sequential sequences for each of three 80-year hydrologic periods on the Colorado River: 1906-1985, the base-case years of record;

1520-1599; and 1600-1679. Each simulation included streamflow at Lee Ferry and reservoir storages, and releases, given water demands as specified in Tables 4 and 5 for periods beginning in water year 1988-89.

IX.A Lake Powell

Exceedance probabilities developed for Lake Powell indicate that a repeat of a hydrologic period like that of 1520-1599 or 1600-1679 would be significant. At about the 90% exceedance level (i.e., there is a 90% probability that the reservoir storage will be equal to or greater than the indicated amount), the difference between the base case period 1906-1985 and 1520-1599 is about 11 maf for year 2000. Conversely, there is a 10% probability that the reservoir storage will be equal to or less than the indicated amount. Furthermore, Lake Powell is at minimum power pool at an 85% exceedance probability. For the hydrologic period 1520-1599, the minimum power pool probability exceedance decreases to about 80% by year 2010, and to about 75% by year 2020. A comparison of the two drought scenarios shows that they have similar probability curves with maximum differences at the higher exceedance levels. While both sequences exhibit quite different types of droughts, preceding or subsequent high flow periods tended to produce an equalizing effect between them, when considering the entire 80-year sequence. Lake Powell average storage levels were calculated by taking the average from 80 wrapped sequences for each simulation year. Results are shown in Figure 6. Note that the average storage level difference under severe sustained drought conditions ranges from approximately 5 to 6 maf from 1998 to 2020.

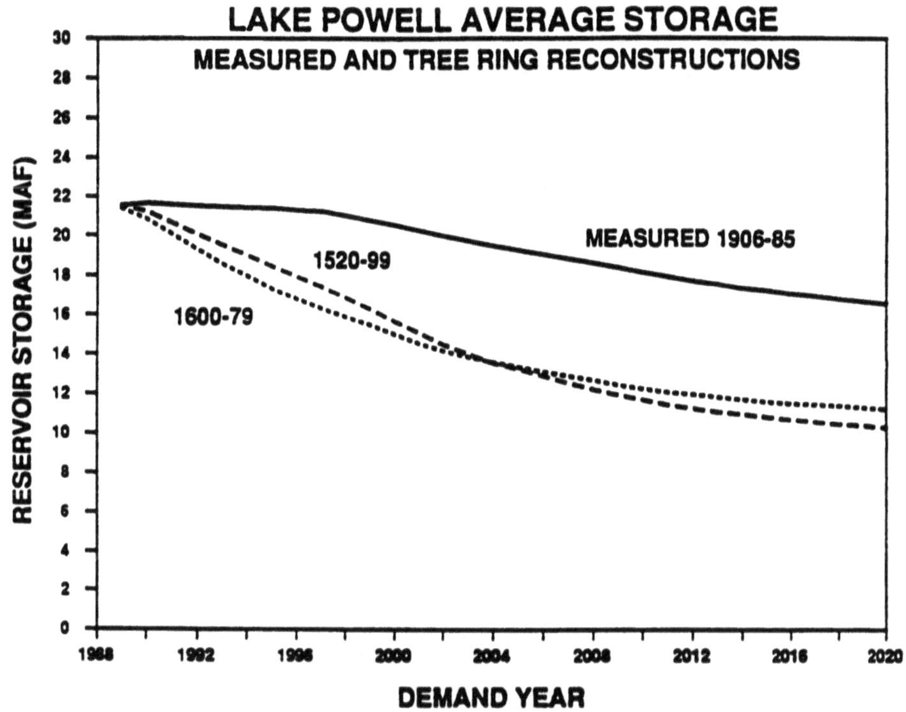

Figure 6. Lake Powell Average Storage Comparison

IX.B Lake Mead

Lake Mead simulation results are similar, with the difference between the base case and both drought sequences on the order of 6 maf for year 2000 at about 90% exceedance. Minimum power pool exceedance probabilities are about 80%. For the most severe drought scenario (1520-1599), minimum power pool exceedance levels decrease to about 50% by year 2010, while those for the sequence 1600-1679 decrease to about 70%. By simulation year 2020, these exceedance levels decrease to about 40% and 55%, respectively. Minimum power pool exceedance probability predicted by the base case was greater than 95%. Average storage levels for Lake Mead are shown in Figure 7, and indicate a maximum difference of about 6 maf between the base case and severe and sustained drought scenarios.

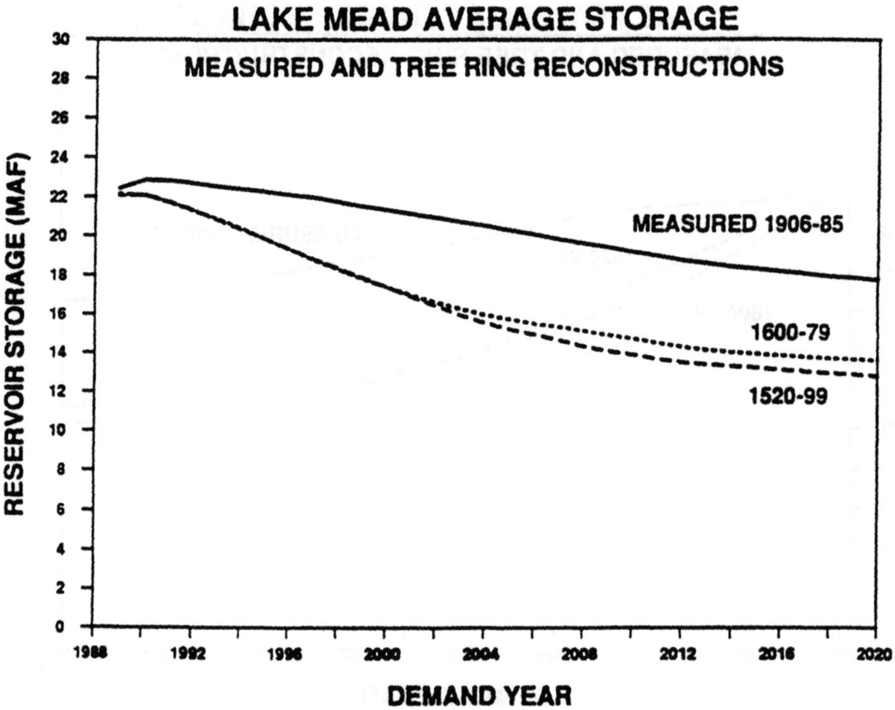

Figure 7. Lake Mead Average Storage Comparison

X. Shortages

Water shortages in the Upper Basin were allocated according to an interpretation of the Law of the River and storage provision requirements. Following the same procedure used for the development of exceedance probabilities for reservoir storages, similar empirical cumulative distribution functions were developed for Upper Basin shortages. The base case hydrologic sequence (1906-1985) indicates no shortages. The sequence 1520-1599 indicates shortages of about 300,000 acre-feet at the 75% exceedance level. The 1600-1679 sequence indicates Upper Basin shortages of about 300,000 acre-feet at about the 35% exceedance. Shortages of about 300,000 acre-feet at the 95% and 85% exceedance levels are indicated by Year 2010 for the two drought sequences. The base case sequence indicates no shortages for the Upper Basin. Similar results are shown for simulation year 2020 except that exceedance probabilities have increased in all cases. That is, shortages are present at the 98% and 85%

exceedance levels for the two drought scenarios, while the base case shortages are indicated at 50% exceedance. Average shortages for the 32-year simulation period are shown in Figure 8.

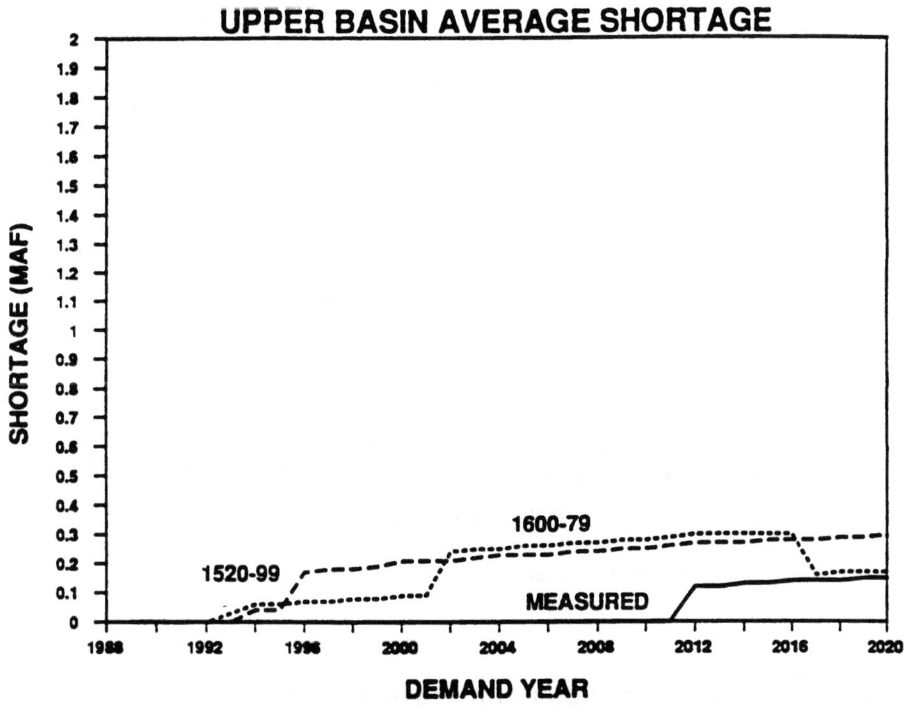

Figure 8. Upper Basin Average Shortage

Calculated Lower Basin shortages were much more severe than those for the Upper Basin. For simulation year 2000, no shortages were predicted by the base case hydrology, although 1.45 maf shortages were calculated at the 70% and 30% exceedance levels for the drought sequences 1520-1599 and 1600-1679 respectively. For year 2010, the base case hydrology yielded 1.45 maf shortages at about the 50% exceedance level, while the 1520-1599 drought sequence showed the same shortages at the 95% exceedance. The drought sequence 1600-1679 indicated 1.45 maf shortages at 85% exceedance. For simulation year 2020, the base case exceedance levels increase to about 60%. Note that for the hydrologic period 1520-1599 and 1600-1679, the exceedance probabilities stay approximately the same. Lower Basin average shortages are shown in Figure 9.

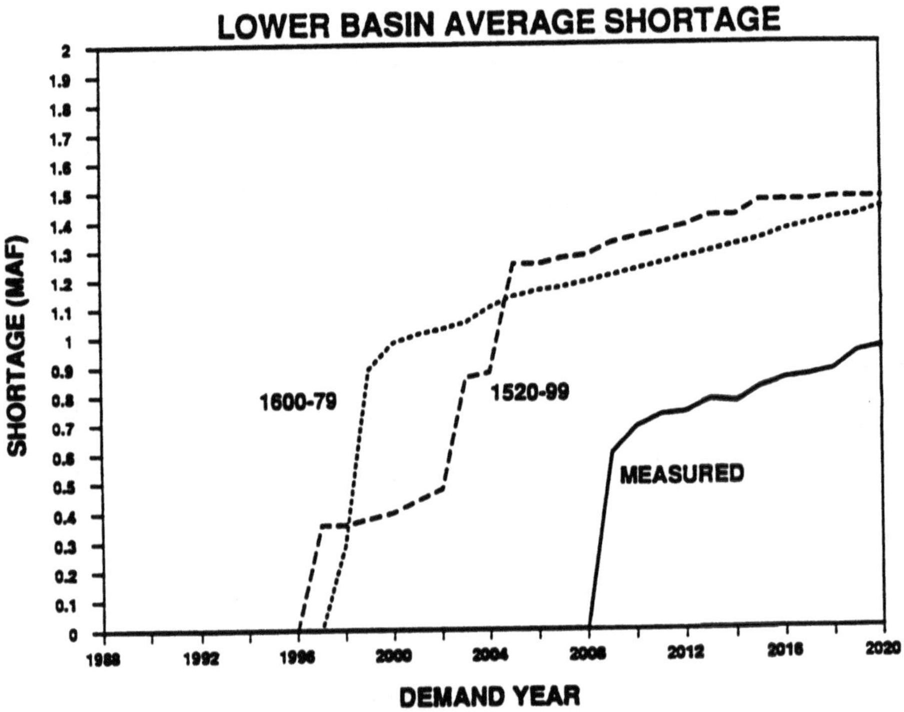

Figure 9. Lower Basin Average Shortage

XI. Conclusions and Implications for Further Study

A repeat of the hydrologic periods 1520-1599 or 1600-1679 on the Colorado River would have significant impacts on water availability and shortage allocations in the Upper and Lower Basins. It should be noted that results presented here are not speculative what-if climate scenarios. They represent outcomes of hydrologic sequences that have been reconstructed from the past. Observation of the data indicates that the river has had above-average flow of 15.06 maf over the past eighty years as opposed to its previous average of about 13.5 maf per year. The potential for anthropogenic climatic change effects could exacerbate a situation where the equilibrium level of the Colorado River over the long term (its mean flow) is lower than that measured over the past century. Shortages on the order of 300,000 acre-feet (under current system operation regimes) could occur in the Upper Basin as a result of hydrologic conditions derived from study of tree-ring data. Lower Basin shortages on the order of 1.5 maf have probabilities of occurrence of over 90% and about 85%, respectively, for the 1520-99 and 1600-79 droughts. Areas of future study need to focus on those activities that can help mitigate negative impacts in the Upper and Lower Colorado River Basins. These would include conservation practices, the conjunctive management of surface and groundwater, and systems for orderly and mutually agreeable movement of water to higher-valued uses, including temporary water transfers under drought conditions. The merits of the drought study lie in the use of data drawn from actual climatic events. While there is a degree of bias in the tree-ring reconstructed sequences, estimates of the streamflow mean are considered to be reliable. It is doubtful that Lake Powell and Lake Mead would be allowed to drop minimum power pool levels before other types of shortage procedures were invoked, since in real-time operation, there is no way to forecast whether a drought has ended. This analysis distributes shortages in the Upper and Lower Basin in a manner that would be consistent with existing laws and operating criteria. An area for future research is a critical assessment of potential shortages allocations and strategies that might be invoked by the Secretary of the Interior in the event of a severe and sustained drought. It is in the area of politics that decisions will be made about ways of responding to, mitigating, and avoiding drought effects. The extraordinary amount of storage on the Colorado River points to a robustness and a resilience in the past design process. The lesson of importance to policy makers is to develop a flexibility in the operation procedures that matches the variability of the resource being controlled, namely streamflow. For example, rigid operation procedures may need to be set aside and new operation procedures developed that allow for changing the current power pool and base flow requirements.

References

Colorado River Board of California (1990) Executive director's report. Los Angeles, October 9.

Dracup, J.A., (1977) Impact of the Colorado River Basin and Southwest water supply. In Wallis J (1977), Chap. 8.

Dracup, J.A. and Kendall, D.R., (1990) Floods and droughts. In Climate change and U.S. water resources, ed. P Waggoner and R Revelle, John Wiley and Sons, New York, Chap. 11.

Dracup, J.A. and Kendall, D.R., (1987) Frequency analysis of potential drought occurrences on the Rhone, Po, and Danube Rivers. *Proceedings, International symposium on natural disasters*, Rome, June.

Dracup, J.A., Lee, K.S., and Paulson E.G. (1980a) On the definition of droughts. *Water Resources Research* 16(2):297-302.

Dracup JA, Lee KS, Paulson EG (1980b) On the statistical characteristics of drought events. *Water Resources Research* 16(2):289-296.

Glantz MH (1989) Forecasting by analogy: societal responses to regional climate change. Environmental and Societal Impacts Group, (U.S.) National Center for Atmospheric Research, Boulder, Colorado.

Glantz MH (1988) Societal responses to regional climatic change: forecasting by analogy. Westview Press, Boulder, Colorado.

Hundley N (1983) The West against itself: the Colorado River in historical perspective. Colorado River working symposium: management options for the future, Santa Fe, New Mexico.

Hundley N (1975) Water and the West. University of California Press, Los Angeles.

Kendall DR, Dracup JA (1991) A comparison of index-sequential and AR(1) generated hydrologic sequences. *Journal of Hydrology* 131 (January).

Matalas NC, Fiering MB (1977) In Wallis J (1977).

Meko D (1990) Personal communication, University of Arizona.

Meyers CJ (1966) The Colorado River. Stanford Law Review.

Metropolitan Water District of Southern California (1980) Colorado River Annual System Regulation Model, Report 941, July.

Metropolitan Water District of Southern California (1989) Resources Division, State Water Project Simulation Model.

Nathanson MN (1978) Updating the Hoover Dam documents. U.S. Government Printing Office, Denver, Colorado.

Stockton CW (1977) Interpretations of past climatic variability from paleoenvironmental indicators. In Wallis J (1977).

United States. Colorado River Compact. Congressional Record (1928), documents 324 and 325; Congressional Record (1983), H. Doc. 605, 67 Cong., 4 Sess., 8-12.

United States (1989) Department of the Interior, Bureau of Reclamation. Colorado River Simulation System (CRSS).

Upper Colorado River Commission (1982) Thirty-fourth annual report, Salt Lake City, Utah.

Wallis J (ed) (1977) Climate, climatic change, and water supply. National Academy of Sciences, Washington, D.C.

Weatherford GD, Jacoby GC (1975) Impact of energy development on the law of the Colorado River. Univ. of New Mexico School of Law, Natural Resources Journal 171-213.

Yevjevich VM (1967) Objective approach to definitions and investigations of continental droughts. Hydrology Paper 23, Colorado State University, Fort Collins.

RELIABILITY IN REMOTE SENSING AND NUMERICAL MODELING OF COASTAL POLLUTION : THE CASE OF THE GAMBIA ESTUARY

B. Caussade
Institut de Mécanique des Fluides
Unité de Recherche Associée au CNRS
Avenue du Professeur Camille Soula
31400 Toulouse
France

INTRODUCTION

Research on measurable space parameters relevant to the study and monitoring of continental ecosystems necessitates :
-theoretical and experimental studies which allow the linking up of the quantities measured by satellites to the variables describing the observed systems and the parameters measured in-situ,
-the development of models which allow a description of their evolution
-the development of methods for incorporating relevant data.

Remote sensing is a relatively new technique, and satellites operating in the visible and near infrared part of the electromagnetic spectrum open new possibilities for the study of oceans, sea-shelves and continental biosphere. The main advantage is the possibility to study a very large surface, i.e.,the field view of SPOT is 60 by 60 km, the ground resolution being 20 by 20 m.

There are also disadvantages. An area on earth is only recorded when the satellite passes over it i.e., the time delay is 26 days for SPOT and may be much longer if the weather is not good (i.e.,when clouds cover the earth's surface). The instant picture is not sufficient to explain the observations and to determine their causes, if the observed phenomenon time scale is of the same order as or shorter than the orbital cycle (but there exist long time series of images which can allow the study of long term changes).

We have focused our study on the sediment transport in the river mouth of the Gambia River (West Africa). This site has been chosen because excellent images exist showing interesting phenomena which could be related to the transport of suspended matter.The results of this preliminary study have emphasized that the satellite remote sensing, because of its repetition and spatial resolution,is certainly a good way to obtain surface parameters which are

NATO ASI Series, Vol. G 29
Water Resources Engineering Risk Assessment
Edited by J. Ganoulis

necessary in digital simulation, and, in the near future, an indispensable database for the development of any management procedure.

We present the results of an experiment carried out at the Institut de Mécanique des Fluides in Toulouse (IMFT) in collaboration with the Centre d'Etudes Spatiales des Rayonnements (CESR) which concern the monitoring of concentrations of suspended matter in the estuary of the Gambia River.

REMOTE SENSING AND MATHEMATICAL MODELING OF SUSPENDED MATTER COASTAL DISPERSION

DESCRIPTION OF THE PROBLEM

Soil erosion observed in certain watersheds is generally connected to development carried out over centuries .These hazards have noticeably increased with the growth of agriculture at an industrial scale. Damage to crops and soil deterioration has also increased with surfaces left almost bare during the greater part of winter and spring, clearing of trees at a large scale .These hazards also weigh on the environment through water pollution (excessive muddiness, fixation of pesticides, phosphorus but also of heavy metals).

We do not plan to study here the mechanisms of erosion, certainly detectable in terrain radiometry, but rather to consider sediment transport in coastal zones and more especially in the estuary zones of rivers which are the outlets of the watersheds concerned and where the observed results are the clearest. Besides the problems of pollution already mentioned, the quantities, sometimes enormous, of transported sediment can modify the morphology of the river mouth and may also have catastrophic consequences on the biological cycles of the ecosystem.

The site chosen is the Gambia estuary, a river in West Africa, where interesting phenomena linked to heavy concentrations of suspended matter have been observed on the SPOT1 satellite images.The extraction of information contained in the spatial images in qualitative and especially quantitative terms poses many problems. Our method based on a hydrodynamic model and adjusted by data found in scientific litterature aims to compare results through modelization and images in order to be able to provide an interpretation. In this perspective, two studies have been carried out in parallel, one at the C.E.S.R. , the other at the I.M.F.T., in order to help interpret two SPOT 1 satellite images, taken vertically over the estuary of the Gambia River.

At the C.E.S.R. the radiometries recorded by the satellite were studied. These were treated in order to show on a map of the region observed, the isoradiometric contours corresponding to certain wave lengths. A first piece of research on this map allows some hypotheses to be made on the phenomena observed. To verify the validity of these hypotheses a joint study was undertaken with Sylvain Ouillon at the I.M.F.T.. First this consisted in presenting maps of currents, then of Suspended Matter Transport (SMT) in the coastal region near Gambia for atmospheric and meteorological conditions as close as possible to those which existed when the picture was taken. Next, a comparison was made between the map obtained by numerical modelling and the image treated. The analysis and the comparison of the results allowed us to check the hypotheses made and to draw lessons as to the modelling strategy ; it also pointed out the complementarity of modelling and satellite imaging.

SPACE IMAGING

General points

This study on solid transport consists in two parts: quantifying the characteristics of suspended particles and, then, their motion.The former consists of image treatment and physical parameter extraction from SPOT HSR (High Spatial Resolution) images which Martel and Flouzat (1989) and Martel (1990) carried out.The latter will be supported by the combination pattern recognition - hydrodynamics modeling.

HRV sensor and images

The SPOT HRV sensor gives a fine spatial resolution (10 to 20 m) well adapted to coastal phenomena but its temporal resolution does not allow us to study dynamics at this scale. The sensor sensibility and the weak instrumental gains, adapted to the luminance dynamics and to large gradients, make it an essentially Earth sensor because in spatial oceanography the luminance and their gradients are weak and because the noise is important. However the XS1 and XS2 channels have spectral ranges between 0,5 and 0,7 mm which give the best total quantity assessment of suspended matter. These channels are neither sufficiently numerous, nor sufficiently narrow to allow an accurate analysis of the complex mixing ; but the calibration based on *in situ* data improves the results.

Two images were treated. They came from two SPOT 22-323 scenes established in multistage mode on 09 May 1986 and 23 December 1987 (see Plates 4 and 5). Each one covers about 60 km to 60 km (see Figure1).

Image treatment

"On the Gambia estuary SPOT images, the range between adjacent radiometries (or more exactly the structures we could qualify as isoradiometric in the absence of noise) are of the same order of magnitude as the noise" explains Martel. That is the reason why it was necessary first to filter the images using several smoothing and iterative methods. Then, the segmentation was carried out using two methods, namely by thresholding and by outline extraction.

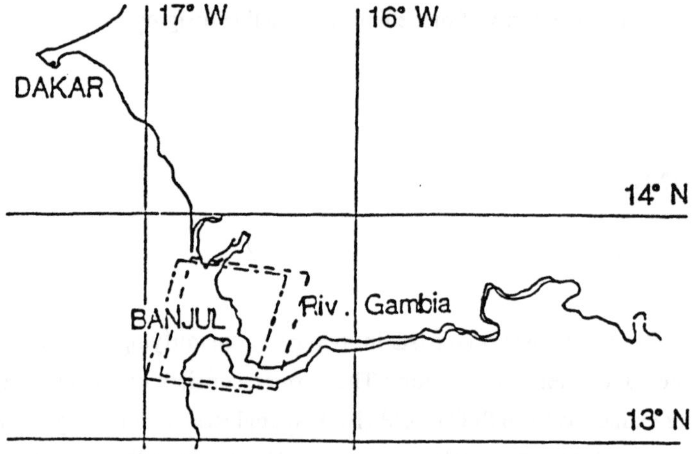

Figure 1 : SPOT scenes localization

Physical parameters extraction

The signal observed by a detector in the solid angle corresponding to a target depends on the target itself, on its environment, on the atmosphere, on the lighting conditions and on the sensor geometry. The assessment of the atmospheric effects, a general term which includes all the influences perturbing the useful signal coming from the target, is necessary to work out the absolute parameters characterizing an object (for example, the reflectance). The corrections are based on the atmospheric components effects modeling, taking the surface reflection into account before isolating the signal coming from the water column.

This signal depends on the substances present and water concentration. The study of diffusion and absorption phenomena in the marine environment has given theoretical explanations bounding spectral reflectances and components concentrations (Martel and Flouzat ,1989). These relations, calibrated thanks to *in situ* measurements allowed us to estimate the suspended particles contents in the shallow water. It must be underlined that they were obtained by new methods at a location where we possess only few *in situ* data.As an example we present a succession of images showing the different stages of the treatment on Plates 1 to 5.

This technique still remains imperfect in addition it is applied at a site where we have little in-situ data. This leads us to be circumspect about making any definite interpretation and also about comparing the simulation results with the image analysis results.

NUMERICAL MODELING

Available data

The most useful study on the Senegalese continental shelf was conducted by Rebert (1983) who focused on the Ocean. Howard Humphreys and Sons (1974) made a complementary study germane to the river.Gambia has two seasons : the dry season between November and May, and the wet season between June and October. During the dry season, the freshwater discharge is almost non-existent so that the river, whose slope is very small, becomes a "sea branch" on its downstream side (about 520 km). The two SPOT images were taken during the dry season, so we should consider the river 520 km inland (where there is a threshold), simulate the tidal propagation and not take freshwater discharge into account.

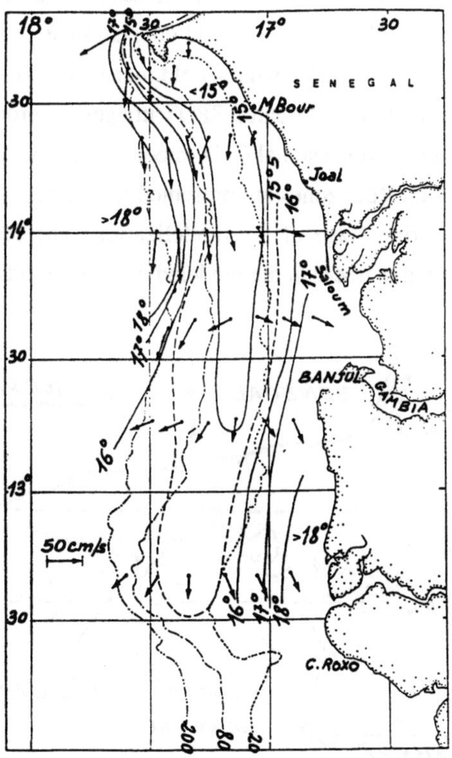

Figure 2 : Currents at a depth of 5 m during the period 20-31 March 1974
(From : Rebert, 1983)

In the ocean, the main currents flow N-S during the dry season (Figure 2). They are guided by two great circulations in the open sea off the West African coast and move simultaneously with the Inter Tropical Front whose crossing determines the season change. The main feature affecting the cross and vertical circulations is the existence of a current discrepancy at a depth of -35/-45 m which corresponds to a sea cliff at the boundary of the Senegalese continental shelf.

It is also to be noticed that the tide is propagating from the South to the North along the coast, and that the wind contributes to a great extent to mixing of the sea water by an increase of turbulence ; therefore, to simplify, we can consider that there is a relative vertical homogeneization.

HYDRODYNAMIC MODEL

Preliminary remarks

The phenomena complexity should require a three-dimensional (3-D) model. However the lack of useful data for such a model and the relative vertical homogeneity enables us to use a two-dimensional (2-D) plane model in the ocean and coastal areas, where the variables are integrated along the water depth. The results, which yield mean values close to the surface, are acceptable for a first comparison with the treated SPOT images. In order to simulate the tide propagation, we can only use a one-dimensional (1-D) model in the river, interactively coupled with the 2-D model at the mouth. Both zones are presented in Figure 3. The simulation of the hydrodynamics makes up the first stage of the modelization. The second one consists in injecting a passive and conservative tracer in the modeled velocity fields in order to follow its temporal evolution

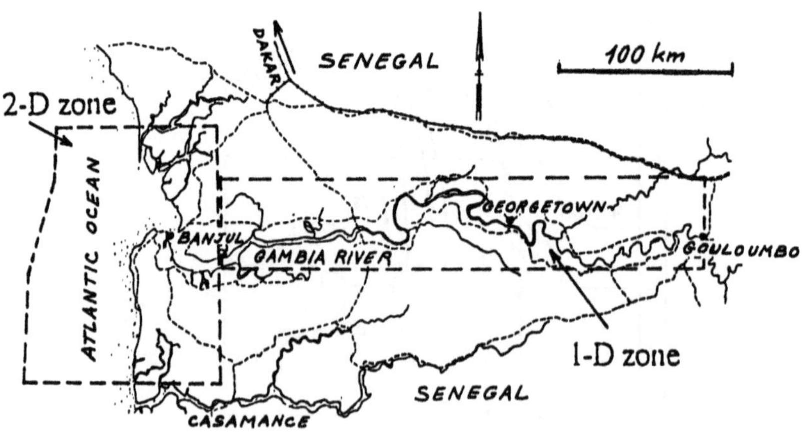

Figure 3 : 2-D and 1-D study zones

Equations

The basic equations of fluid mechanics express the conservation of mass, momentum and energy with a local instantaneous formulation. In our problem, the temperature is considered as uniform. Therefore,we have to solve only the partial differential equations called "Navier-Stokes equations". In fact, it is inconvenient to deal with the local instantaneous formulation.

To make the turbulence appear, each variable is broken down following the statistical average of the variable and the randomly fluctuating quantity. Then the local equations in statistical average are averaged in time (Reynolds equations) and in space to give the Barre de Saint-Venant equations (Abbott, 1979 ; Ouillon and Caussade, 1990).

For the 2-D model, time averaged flow quantities are integrated along the water depth : $<\overline{G}> = \frac{1}{Z_S - Z_B} \int_{Z_B}^{Z_S} \overline{G}\, dZ$. For the 1-D model, mean flow quantities at a cross-section of the river are defined by means of the following operator $\int_A dA$.

The 2-D model yields equations (1) to (3) below to determine water level and flow variation :

$$\frac{\partial H}{\partial t} + \frac{\partial HU}{\partial x} + \frac{\partial HV}{\partial y} = 0 \qquad (1)$$

$$\frac{\partial HU}{\partial t} + \frac{\partial HUU}{\partial x} + \frac{\partial HUV}{\partial y} = - g H \frac{\partial Z_S}{\partial x} + \frac{1}{r}(\tau_{Sx} - \tau_{Bx})$$
$$+ D\left(\frac{\partial^2 HU}{\partial x^2} + \frac{\partial^2 HU}{\partial y^2}\right) + f H V \qquad (2)$$

$$\frac{\partial HV}{\partial t} + \frac{\partial HUV}{\partial x} + \frac{\partial HVV}{\partial y} = - g H \frac{\partial Z_S}{\partial y} + \frac{1}{r}(\tau_{Sy} - \tau_{By})$$

$$+ D\left(\frac{\partial^2 HV}{\partial x^2} + \frac{\partial^2 HV}{\partial y^2}\right) - f H U \qquad (3)$$

Let us now define the 1-D model determining the discharge and the area at each cross-section :

$$\frac{\partial A}{\partial t} + \frac{\partial Q}{\partial x} = 0 \qquad (4)$$

$$\frac{\partial Q}{\partial t}+\frac{2\,Q}{A}\frac{\partial Q}{\partial x}-\frac{Q^2}{A^2}\frac{\partial A}{\partial x}=-g\,A\,\frac{\partial \frac{A}{b}}{\partial x}-\frac{g\,b}{c_z^2}\frac{Q|Q|}{A^2}+g\,A\,I \qquad (5)$$

Initial conditions

As the problem is unsteady, it is important to define good initial conditions. Therefore, several tidal cycles have been simulated corresponding to the real cycles during the preceding days of the one considered. At the beginning, velocities are zero and the water level corresponds to the mean level. When we obtain an established oscillating system, we store the values of the variables to be used as initial conditions.

Boundary conditions

The features previously discussed allow us to consider the following boundary conditions (see Figure 4).

a - The velocity components measured by Rebert (1983) in the season under consideration are selected.

b - This boundary corresponds to a current line. We lay down an impermeability condition.

c - The water level variations due to the tide are applied at the southern boundary.

d - The boundary in the river corresponds to the coupled area of the two models. The 2-D model input is the river discharge given by the 1-D model and, in a similar way, the 1-D model input is the water level calculated by the 2-D model. We set the discharge upstream of the 1-D model at zero during the dry season.

e - The coast-ocean and coast river boundaries correspond to impervious boundaries.

TRANSPORT MODEL

Equation

Dietrich et al (1982) specify that "sediments in motion in the river are overwhelmingly fine sediments eroded during the wet season (at the upper end of the river)". However, the lack of precise *in situ* measurements makes it impossible to define accurate erosion and sedimentation relationships. Moreover, we should take the turbulence and vertical velocities into account in order to develop a precise model. Our only possibility is to follow the evolution of a conservative and passive tracer in the calculated velocity field. Let us define the simple transport equation :

$$\frac{\partial C}{\partial t} + U\frac{\partial C}{\partial x} + V\frac{\partial C}{\partial y} = 0 \qquad (6)$$

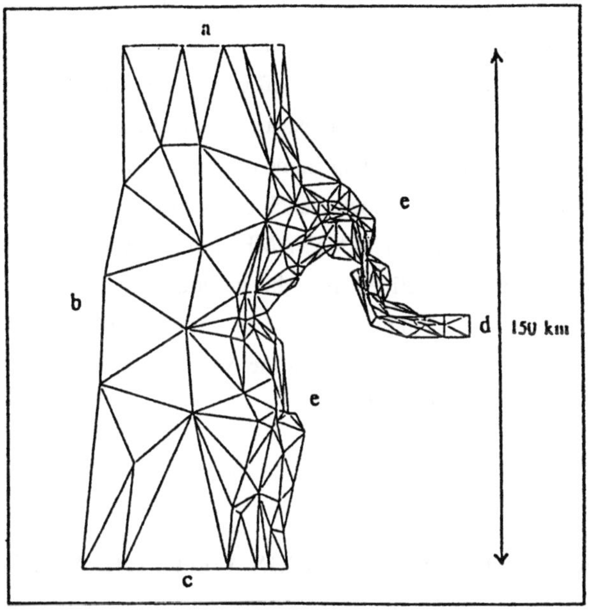

Figure 4 : Grid on the 2-D area

Initial and boundary conditions

The SPOT images give information to define boundary conditions. In order to obtain initial conditions we have considered that the sediments are brought to the river mouth during a flood at the wet season and that their distribution needs a long time to evolve under the tidal influence before becoming almost stable. This schematic hypothesis lead us to adopt the following approach : we simulate a flood in the river neglecting the tide with a zero tracer concentration everywhere at the beginning. Then, we use the distribution thus obtained as initial conditions to follow its evolution with floods (case of December) or without floods (case of May) in the presence of the tide. Indeed the December SPOT image shows a disturbed distribution making us think that there was or has just been a flood in the river. This approach enables us to highlight some trends while keeping up with system behavior.

NUMERICAL PROCEDURE

The advantages of the finite element method, which we prefer to the finite difference method are : a better definition of area boundaries, and grid flexibility. Its most important drawback is the need for a longer computing time. The equations are applied to each mesh introducing three parameters values at each node with the Galerkin method. The grid we use consists of 143 nodes and 230 triangular meshes for the 2-D area (see fig. 4), and in 12 nodes and 11 meshes for the 1-D area. The bottom stress coefficients were calculated using the water level measurements given in Howard (1974). We use a computer code developed in Liné et al (1987) at the I.M.F.T. adapted from a computing frame proposed in Dhatt and Thouzot (1981). It consists of a set of several functional blocks, each of them treating a special part of the problem with the finite element method. The MEFAME (Finite Element Method Applied to Environmental Modeling) program is defined from this computer code to solve the equations systems determined above. The numerical resolution of the unsteady nonlinear matrix system is based on the implicit Euler algorithm coupled with the Newton-Raphson method. For more details, see Ouillon and Caussade (1990 , 1991).

SIMULATION RESULTS AND COMPARISON WITH THE IMAGES

Case I - 09 may 1986

On Plate 6, we present : the SPOT image, a map of the instant velocities, a map of the residual currents which are calculated over the tidal cycle ending at the time the image has been taken (about 11 h 45 a.m.), two maps of the tracer distribution obtained after having laid down one tidal cycle on the initial conditions, and, a zoom of the mouth of the river to observe more important details and to make more accurate the analysis of the high concentration areas.

Concerning the currents, we observe that the residual currents are much greater in the ocean than in the river zones ; but, owing to the continuous oscillation of the flows, these observations are reversed on the map of instant velocities.

Concerning the tracer concentrations, the concentration fields (6.d and 6.e) emphasize the main features observed on the SPOT image (6.a). However one may observe some spurious features which can be explained as given below.Let us have a sharper look at the different zones located by numbers.

Zone 1. The reversal zone at the moment the image was taken presents weak residual velocities (6.c). Hence it is a sedimentation area that we observe on the SPOT image (6.a). The weak shallow concentrations observed in the simulation (6.d and 6.e) make us believe that the presence of the main current which carries the sediments farther out into the open sea may also explain the phenomenon.

Zone 2. The important velocity shear due to the N-S current (6.c) brings the suspended particles back to the coast. This explains the "tongue" shape observed in the digital simulation (6.d) in complete agreement with the SPOT image (6.a).

Zone 3. One may observe an anti-clockwise circulation due to the coastal morphology (6.c). The residual velocity decreases at its middle and explains that the sediments are deposited and that the observed shallow concentration is weak on image (6.a). This appears in the digital simulation but not at the exact place ; this zone is shifted (certainly by the main current) in the East direction, see (6.d) or (6.e).

Zone 4. It corresponds to the junction between the N-S oceanic current and the outgoing current from the river. It presents weak velocities (6.c) and therefore weak shallow concentrations that we observe as possessing a "corner" shape in image (6.a). The calculated fields (6.d) are different, which could be due to an overestimation of the oceanic current.

Zone 5. The marshy area in the coastal zone is not well represented,which explains the small concentrations observed on the digital simulation (6.d and e.)

Zone 6. The clockwise circulation observed in (6.c) traps the particles suspended in its middle. We thus observe a great concentration of sediments (6.a and e.) which cannot be deposited because of great instantaneous velocities, (see 6.b).

Zone 7. Concerning the currents the same phenomenon as in zone 6 occurs here, as shown in (6.c), but this is not observed in the digital simulation of tracer concentration (6.e) because the initial conditions in this zone were inaccurate due to our disregarding the tide .

Zone 8. The weak residual velocities (6.c) explain why sediments do not propagate to the ocean and a slight lack of concentration is observed in (6.e) in relation to the SPOT image. We should take the diffusion into account to remedy this problem.

Zone 9. The flow divides into two branches due to a separation phenomenon at the expansion of the river (6.c) These two branches correspond to shallow areas, which explains why we observe large concentrations in the image (6.a) and the calculated field (6.e)

Zone 10. The opposition between the circulation in zone 7 and the main current and a great depth explain the weakness in the residual velocities (6.c). The dilution and the diversion of the suspended particles cause weak concentrations, as it can be observed in image (6.a) and the digital simulation (6.e).

The consequences of these last two noticeable points can be clearly observed in (6.f) and (6.g) , as well as the good agreement with the SPOT image.Through this first study, the digital simulation of hydrodynamics appears to be useful for explaining main movement and some particular behavior of the sediments at the mouth. The explanation and interpretation are in fine agreement with the isoline plots of concentration observed on the SPOT image. The calculated field of concentration is less accurate for an immediate comparison because we have only considered a coarse transport law . However it enables us to understand sediment behavior in some zones and remains complementary to the hydrodynamic simulation.

Case II - 23 December 1987

On Plate 7 we present : the SPOT image (7.a) a map of the residual currents (7.b) which are calculated over the tidal cycle ending at the time the image had been taken (about 11h 45 a.m), three maps of the tracer distribution which were obtained respectively after one (7.e) three (7.f) and five (7.g) tidal cycles imposed upon the initial conditions. In order to have an accurate distribution of tracer in the river, the first distribution was calculated with a zero concentration gradient in the river as an upstream boundary condition. These zones in the SPOT image and the calculated field are marked for a better qualitative analysis, (7.c and 7.d).

We observed that there tracer concentration in the initial conditions was not sufficient to obtain a sufficiently realistic ocean distribution with regard to the SPOT image. Thus, afterwards, we changed the boundary conditions and imposed the concentration upon the river boundary. The corresponding fields are presented in Figures (7.f and 7.g).A global view shows that the residual velocity (7.b) is quite similar to the field of the previous case. The same particular circulations appear and the same outlines are emerging. On the other hand the concentration fields are different.

Let us examine local results pointing out when different features than in the preceding study arise.The "tongue" shape due to the N-S current (7.b) appears in our simulation (7.f and 7.g) However,the values of the concentrations are quite different from those calculated by the image treatment. Let us have now a closer look at the same zones indicated by the same numbers as in case I.

Zone 1. We observe a high concentration along the coast (7.f and 7.g) in agreement with the SPOT image.However,at a short distance from the coast, the comparison is not good. This second point can be explained by the lack of a sedimentation model.

Zone 2. Here the same phenomenon persists. This is certainly due to a possible overestimation of the river flood and of the N-S current, indicating an over-exportation of tracer out of the area.Above all a question arises as to the use of a 2-D model in deep water zones. Indeed in such zones the mean values become rather different from the shallow values. The limits of a 2-D model have probably been exceeded.

Zone 3. An anti-clockwise circulation is observed in Figure 7.b, but the weak velocities bring few sediments into the zone: therefore, the weak initial concentrations explains the weak calculated concentrations (7.f and 7.g).

Zone 4. The residual and instantaneous velocities are smaller than the upstream ones in the mouth. It causes an accumulation of tracer (7.f and 7.fg.) which undoubtedly explains the presence of a sand bank that has been noticed in scientific literature.

Zone 6. The simulation presents the same contrast in tracer distribution (7.f and 7.g) as in the SPOT image (7.a) The circulation we observe (7.b) causes an accumulation and gives rise to concentration.

Zone 10. The calculated distribution is in real good agreement with the concentration field calculated from the SPOT image (7.c and 7.d.). This is mainly due to the fact that the behavior of the suspended particles is controlled by the advective motions in this zone. The great instant velocities make it impossible for the sediments to settle. So that they stay permanently in suspension.

DISCUSSION AND CONCLUSIONS

It is now possible to draw some lessons from this first approach.The study emphasizes the fact that digital simulation is well-adapted to the zones where the main phenomena area are mastered at the modelization level. The evaluation of our simple transport model shows that it gives some good results in the shallow areas where we had accurate initial conditions. This makes clear the importance of having a good database in order to define accurate initial and boundary conditions, and also the need to take diffusion, erosion and sedimentation into account in further studies. These remarks must not hide the fact that the digital simulation of hydrodynamics has proved itself to be worthwhile in explaining the main motions and some peculiar behavior of the sediments, and that in some cases the transport model may give useful and more complete information.Beyond these remarks concerning uncertainties of numerical models, the main conclusion concerns the opportunity of the comparison of remote sensing and digital simulation : indisputably the complementarity of these techniques has been clearly demonstrated.

Efficient management of the naturel environment subject to man's actions rests on the monitoring of relevant parameters or indicators of the state of the medium coupled with good knowledge and observations of all of the processes involved.We have tried to demonstrate on a concrete example, that of transport of suspended matter in estuaries, how modern methods of observation and study i.e. satellite imaging and numerical modeling can help towards providing a solution to this problem.The qualitative results of a preliminary study concerning the transport of suspended matter in the estuary of the Gambia river allow us to draw some important conclusions.

Concerning satellite remote sensing images, it is important to remember that we stress the fact that the treatment of images and drawing up of physical characteristics are not easy operations and produce uncertainties in the nature of the observed phenomena.
Concerning the numerical approach, the maps of instant and residual currents show the role which tidal movements, influenced by the discharge of the river, local coastal and bottom geometry may play in the water circulation patterns ; in this manner a good physical explanation of some important details observed on the images has been given. On the other hand, some poor results in the deep zones make us wonder about the appropriateness of using a two dimensional model associated with very poor laws concerning suspended matter transport. The lack of laboratory or field information concerning the nature and physical characteristics of suspended matter destroys the capabilities of the transport model to predict adequately this phenomenon.
Beyond the gaps and uncertainties inherent to these techniques, the main conclusion concerns the opportunity of their comparison. Assuming the presence of suspended matter, expressed at the time, when the images have been treated and explained , we have confirmed the complementarity of the two approaches.Today, we can say that numerical simulation and remote sensing are two essential techniques in the study, monitoring and management of the environment , to the same extent as in-situ observation.Their intrinsic qualities must be highlighted if, in the long or short term, we want to set up efficient tools capable of managing, for the better, the hazards in our environment .This supposes, as far as remote sensing is concerned, the development of tools which allow a better spatial cover (resolution, frequency of passage, diversity of measured parameters etc...) and also a better interpretation ; as far as numerical modeling is concerned, we should be focus and systematically utilize advanced three dimensional codes taking turbulence into account.

ACKNOWLEDGMENTS

The author gratefully acknowledge his PhD student Sylvain Ouillon and colleagues Guy Flouzat and Jean-Philippe Gastellu-Etchegorry for their contribution to this work.

LIST OF SYMBOLS

A	Area of a cross section of the river
b	Width of the river
C	Concentration
C_z	Chézy coefficient

D	Diffusion and dispersion coefficient
f	Coriolis parameter
g	Gravity acceleration
H	Water level
I	Slope with respect to the horizontal plane
Q	Discharge
t	Time
U,V	Velocity components
x,y,z	Coordinates (rectangular Cartesian coordinate system)
Z_B	Bottom level
Z_S	Surface level
r	Water density
τ_{Bj}	Bottom shear stress components
τ_{Sj}	Surface shear stress components

REFERENCES

Abbot M.B., (1979) Computational Hydraulics (1) : Elements of the Theory of Free Surface Flows. Pitman (editor)

Dhatt G.,and Thouzot G., (1981) Une présentation de la méthode des éléments finis. Collection Universitaire de Compiègne - Malone S.A. Editeur.

Dietrich J., Hagstrom A.,and Navntoft E., (1982) Studies of the effect of a barrage on sedimentation. Danish Hydraulic Institute. Internal report.

Howard Humphreys and Sons (1974) Etudes hydrologiques et topograhiques du bassin du fleuve Gambie. Doc. n° 40360. Effectué pour le compte de l'O.N.U.

Line A., Masbernat L.,and Mocke G. (1987) Etude de l'hydrodynamique du bassin de Thau par modélisation mathématique. I.M.F.T. Rapport E.M.E. n° 332.

Martel C.et Flouzat G. (1989) Etude du panache des matières en suspension à l'estuaire de la Gambie. Méthodes de segmentation pour des images représentant des phénomènes à faible gradient. Bulletin de la SFPT ; n° 114.

Martel C. (1990) Analyse géométrique et physique d'images d'observation de la terre ; Eléments d'hydrodynamique côtière. Thèse de doctorat de l'Université Paul Sabatier de Toulouse.

Ouillon S. et Caussade B. (1990) Modélisation mathématique de l'hydrodynamique et du transport d'un constituant passif dans l'estuaire du fleuve Gambie : aide à l'interprétation d'images satellitaires. I.M.F.T. - Rapport Hydre n° 1.

Ouillon S. and Caussade B. (1991) "Numerical Simulation and remote sensing : two techniques applied on the Gambia Estuary" Computer Methods in Water Resources II Edited by C.A. Brebbia, D. Ouazar and D. Ben-Sari. Computational Mechanics Publications (Southampton)

Rebert J.P. (1983) Hydrologie et dynamique des eaux du plateau continental sénégalais. C.R.O.D.T. ; D.S. n° 89 Institut sénégalais de recherches agricoles.

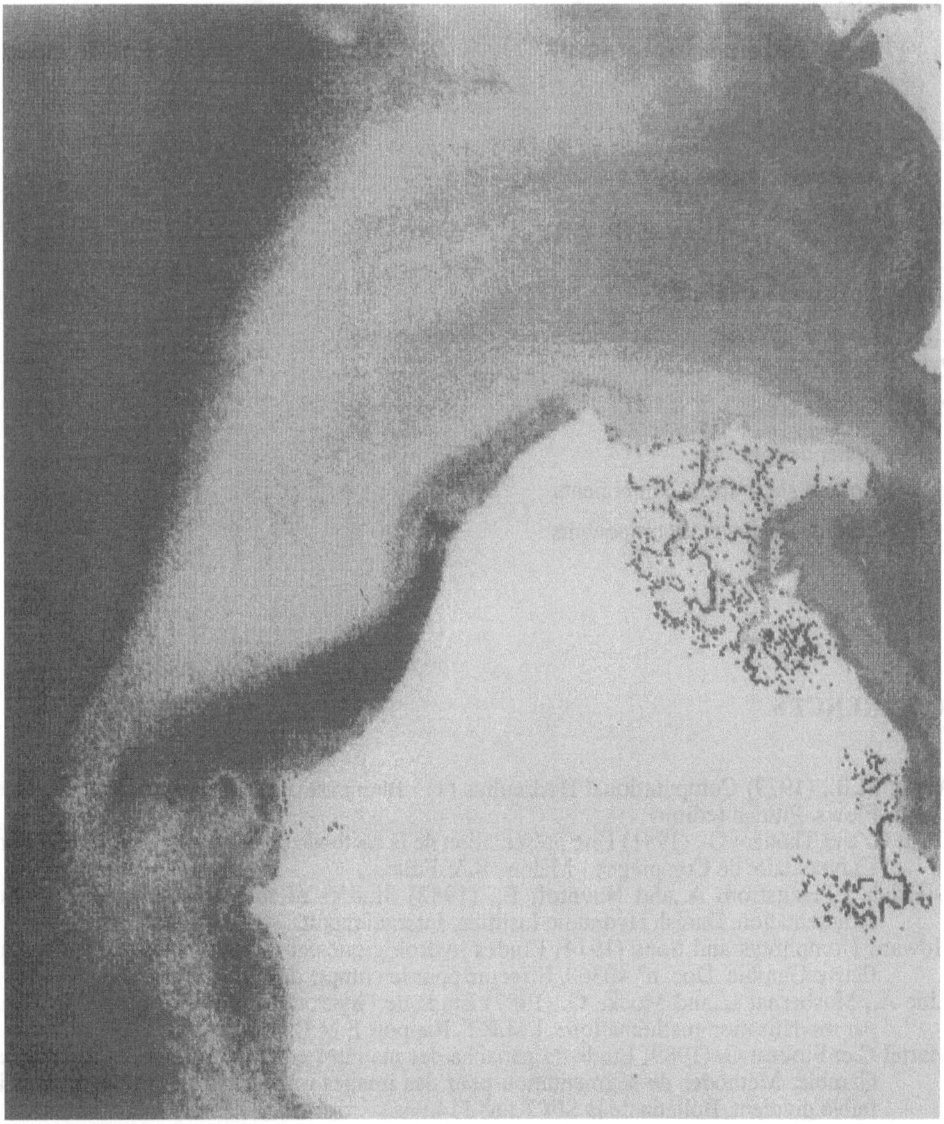

Plate 1 : Estuary of the Gambia River : rough image.

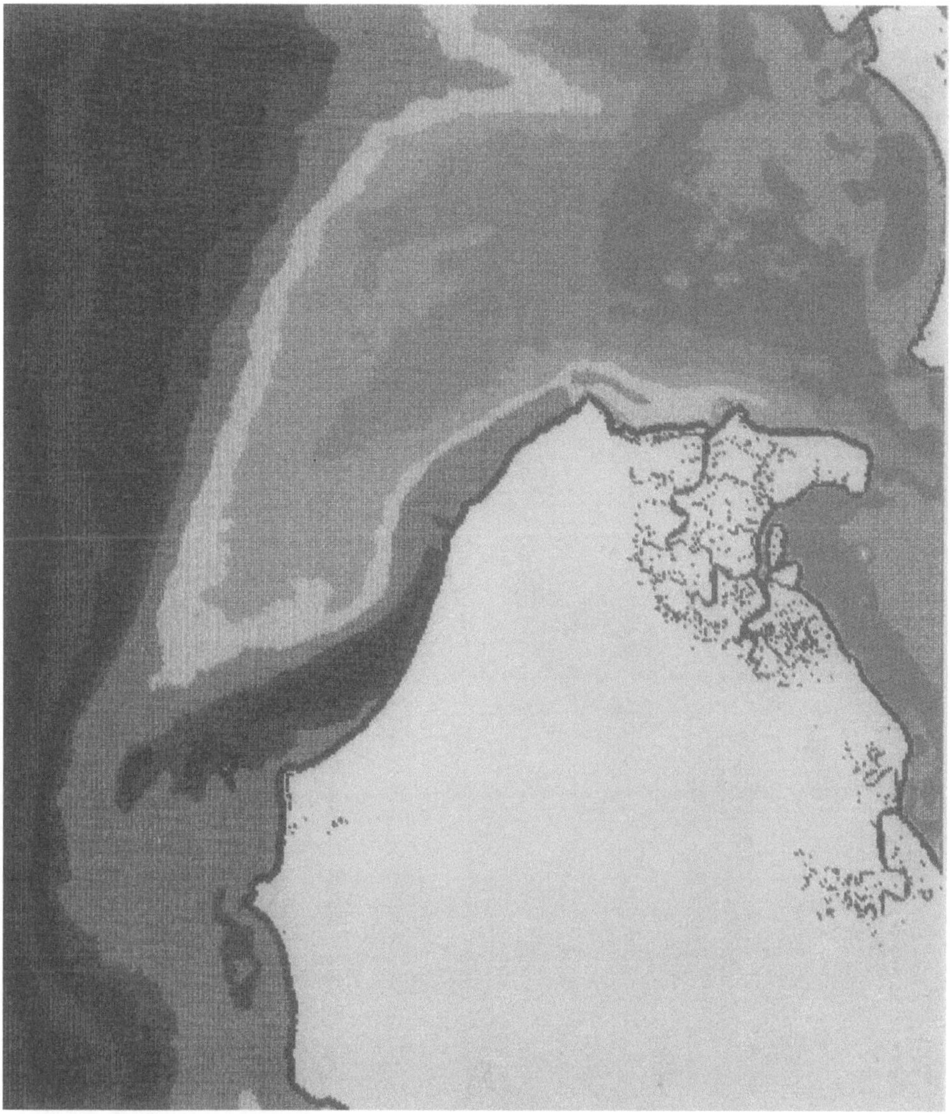

Plate 2 : Estuary of the Gambia River : filtered image.

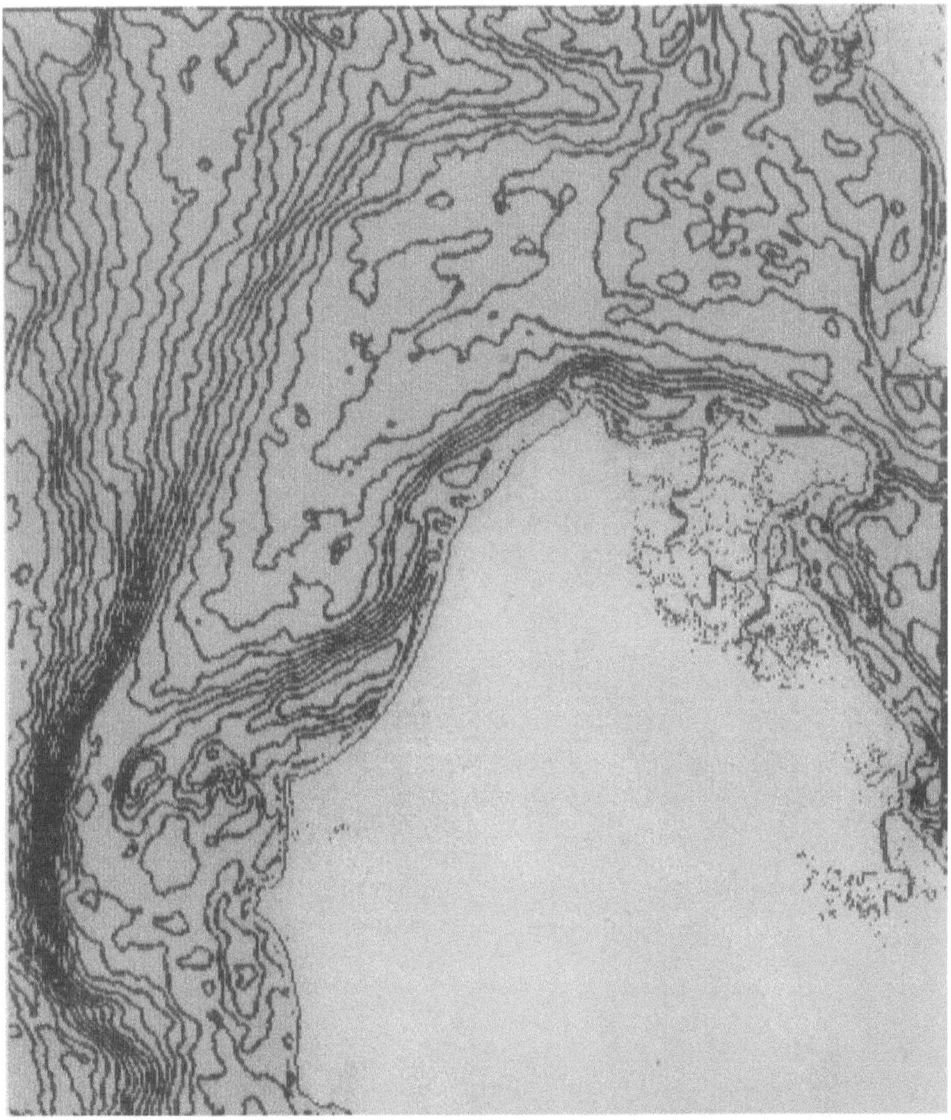

Plate 3 : Estuary of the Gambia River : isoradiometry lines.

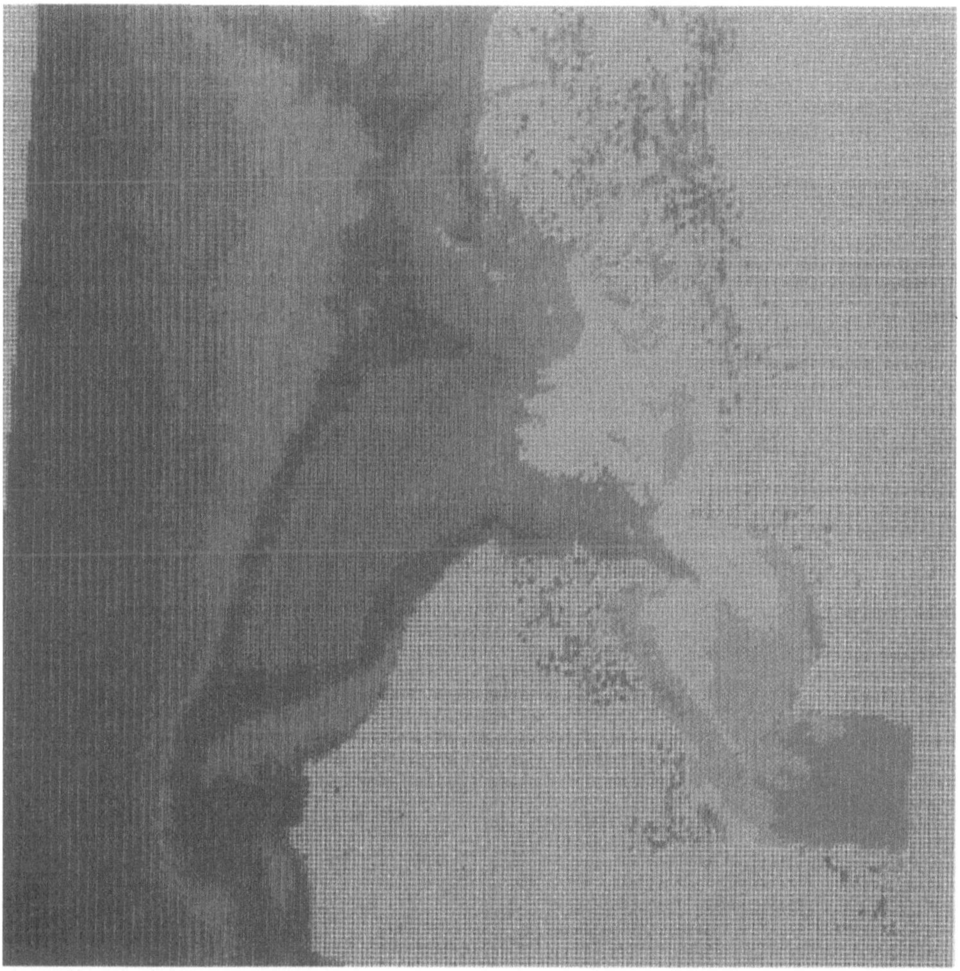

Plate 4 : Estuary of the Gambia River : image treated.
Map of the concentrations of suspended matter on 9th May 1986.
(isodensity by ranges of 10 mg/l).

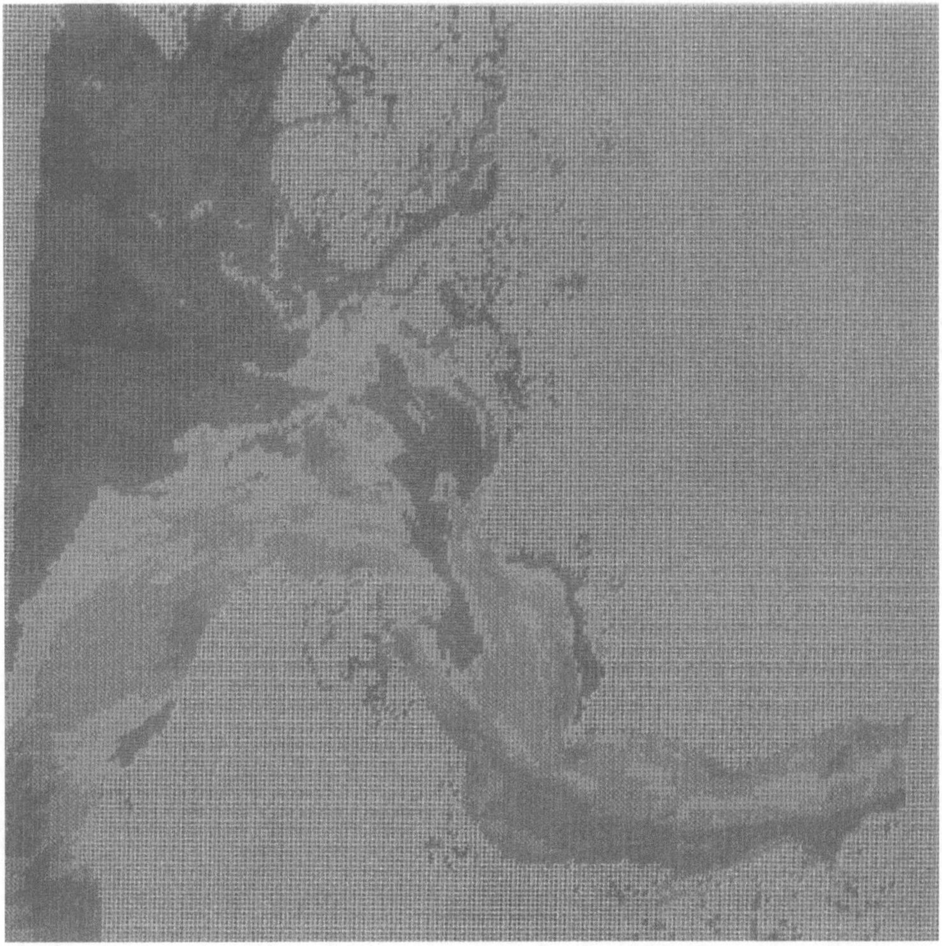

Plate 5 : Estuary of the Gambia River : image treated.
Map of the concentrations of suspended matter on 23rd December 1987.
(isodensity by ranges of 10 mg/l).

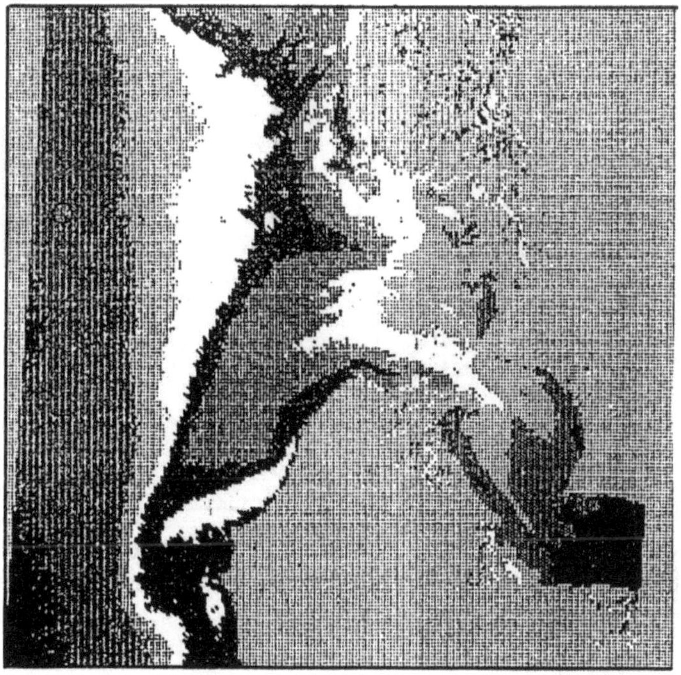

a - copy of the SPOT image

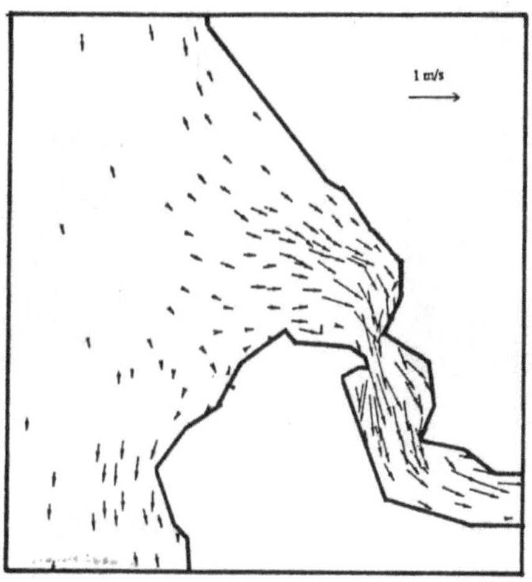

b - Instant currents

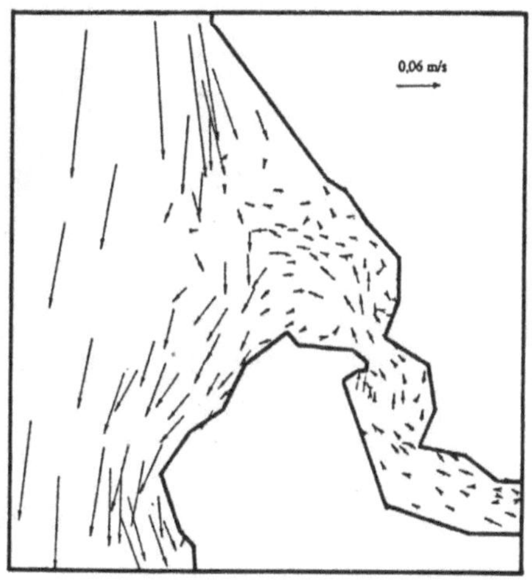

c - Residual currents

Plate 6 - Estuary of the Gambia River : numerical simulation on 09th May 1986

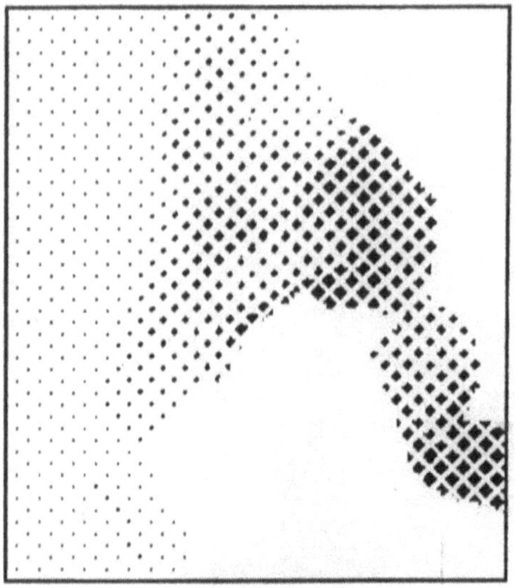

d - Concentration field
1st representation

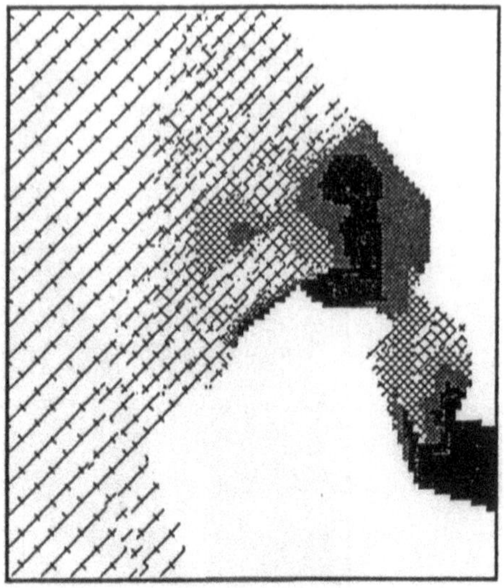

e - Concentration field
2nd representation

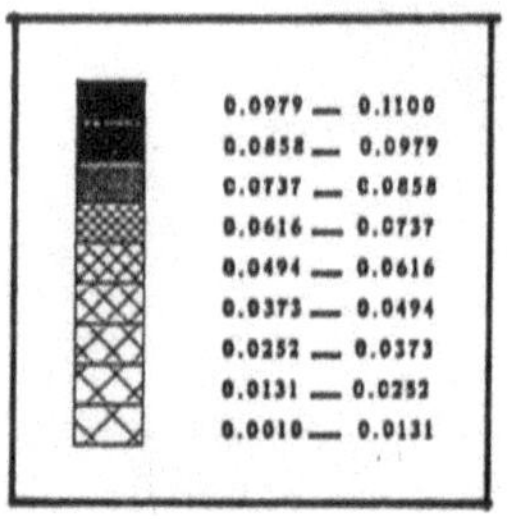

Scale (g/l)

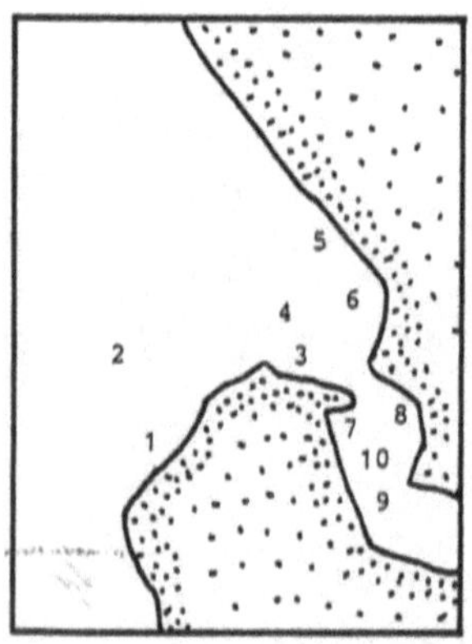

f - Focus from 6a

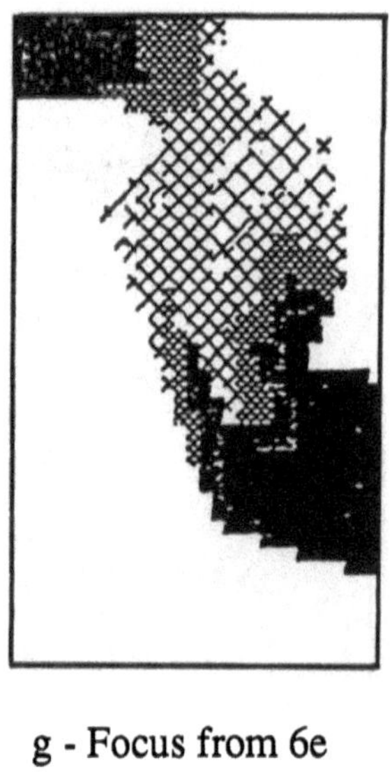

g - Focus from 6e

Plate 6 - (continuation) Estuary of the Gambia River : numerical simulation on 09th May 1986

a - copy of the SPOT image

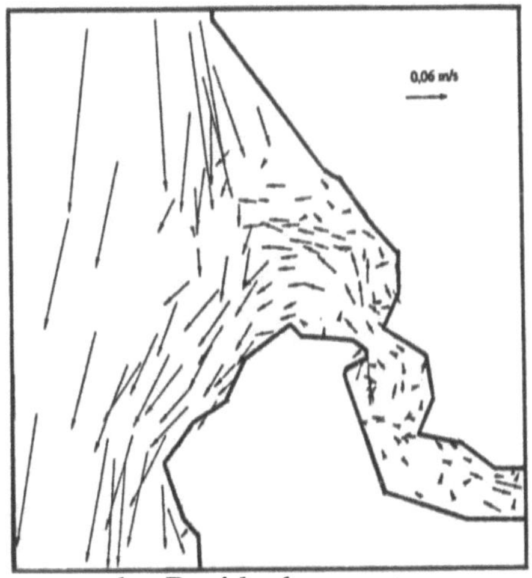

b - Residual currents

c - Focus from 7a

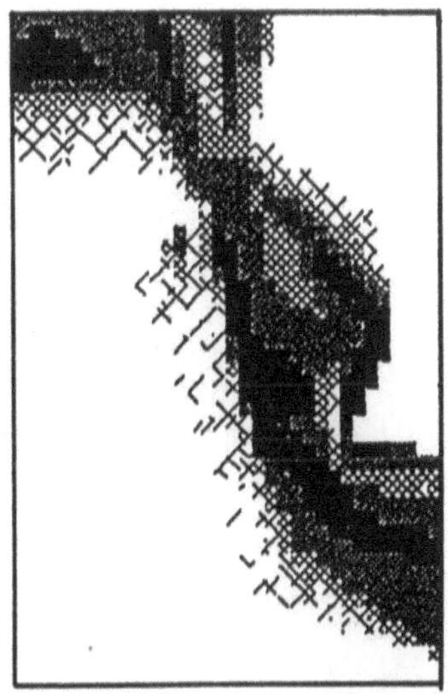

d - Focus from 7e

Plate 7 - Estuary of the Gambia River : numerical simulation on 23rd December 1987

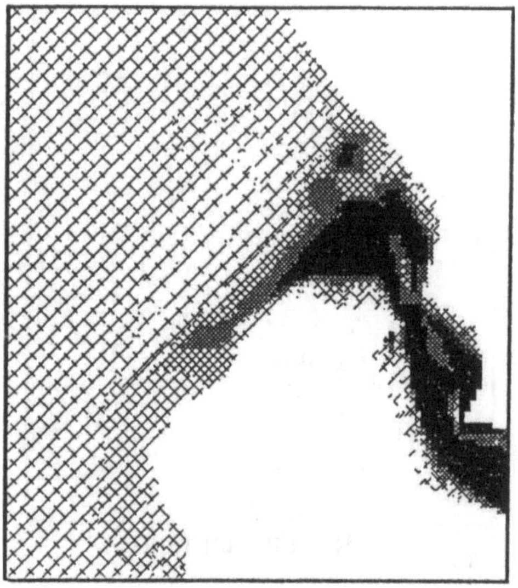

e - Concentration field after
one tidal cycle

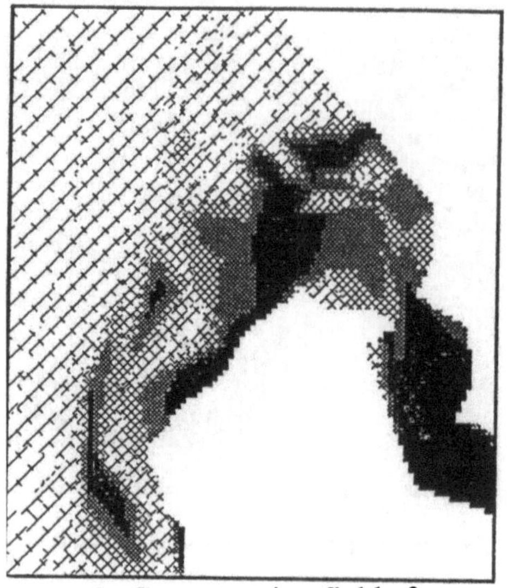

f - Concentration field after
three tidal cycles

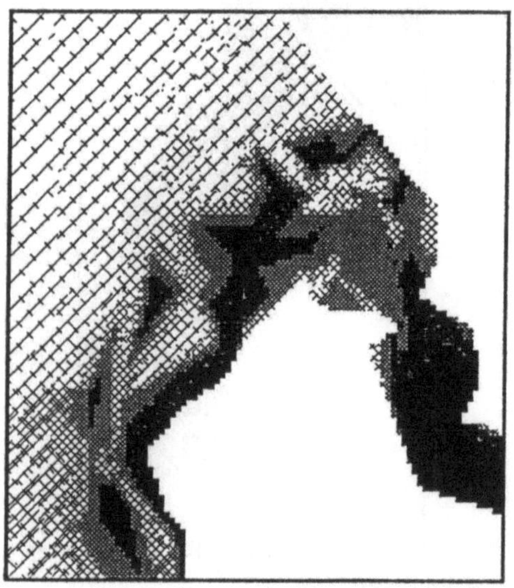

g - Concentration field after
five tidal cycles

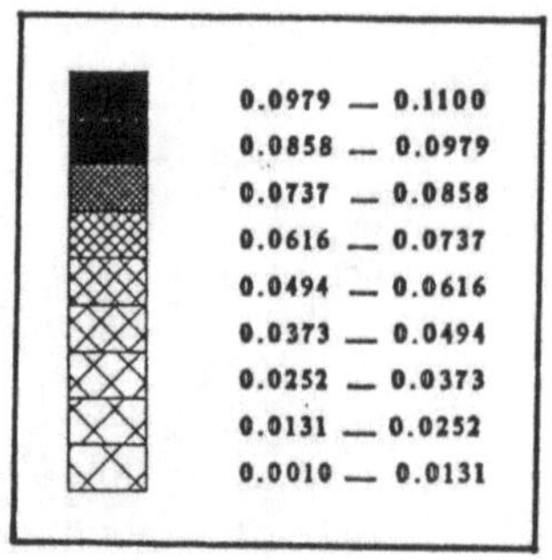

Scale(g/l)

Plate 7 - (continuation) Estuary of the Gambia River : numerical simulation on 23rd December 1987

WATER QUALITY RISKS IN ISTANBUL, PAST AND PRESENT STATUS

A. Samsunlu, B. Beler Baykal, G. Ubay
Istanbul Technical University
Environmental Engineering Department
80626 Maslak - Istanbul - Turkey

1. INTRODUCTION

Surrounded by the Black Sea the Marmara Sea and the Bosphorous, Istanbul is the greatest metropolitan city of Turkey. It experiences a continual increase both in terms of population and industrial activities. The population has increased from 2.8 million in 1970 to 4.6 million in 1980, and the industry has expanded from 1100 hectares in 1970 to 2200 hectares in 1980 bringing together with them significant environmental problems. Similar increases are observed today as well. Despite the vast enlargement in population and industrial activities the infrastructure was not sufficient to properly serve the needs of the growing population and industry. Although there had been some efforts to control environmental problems earlier in this century, rational and properly planned action was not started till 1980's. Early attempts include various projects such as the Master Plan by Prof.D. Kehr in 1966 and the Master Plan and Feasibility Study by DAMOC Consortium in 1970 , the latter being revised by CAMP-TEKSER Consortium in 1975. In 1981, the responsibility of water pollution control in Istanbul was delegated to a single local authority,the Istanbul Water and Sewerage Administration (ISKI). ISKI has also undertaken the duty and power to prepare and implement necessary regulations for the purpose of water pollution control and enforcement for compliance.

At that time one of the most obvious features of the environmental problems of Istanbul was the inadequacy of the sewerage system. One of ISKI's first major efforts was to manage this situation through the " Metropolitan Sewerage

NATO ASI Series, Vol. G 29
Water Resources Engineering Risk Assessment
Edited by J. Ganoulis

Project ", which was concerned with a complete network of sewers and interceptors along with relevant treatment systems and marine outfalls.

The purpose of this paper is to put forth the potential environmental risks in Istanbul, their changing status within years, with special emphasis on the effects of the Metropolitan Sewerage Project in alleviating these risks.

2. THE METROPOLITAN SEWERAGE PROJECT

Although Istanbul enjoys the advantage of being surrounded by the Black Sea, the Marmara Sea and the Bosphorous, with their capacity as receiving water bodies, it also has to watch for their welfare in order to maintain their recreational, touristic and aesthetic values along with fishing and water products potentials.

The Bosphorous is a good example of a well established current system with its definite bilayer stratification. The upper layer comprises of the low salinity (20-25 parts per thousand) Black Sea water and the lower layer arises from the high salinity (35-38 parts per thousand) Mediterranean water. The halocline lies at a depth of about 10-20 meters. The summer temperature in the upper layer is about 24 °C which drops to 8 °C during the winter. The bottom layer temperature is almost uniform all year round at around 14 °C (Meric, 1990).

In early eighties, 80% of Istanbul did not have a healthy sewerage system . The sewerage network that had existed at that time was old and inadequate. Some districts did not have any sewer lines at all and sewage was discharged into the sea randomly.

It was the aim of the Metropolitan Sewerage Project to construct a proper sewerage network and necessary treatment facilities for a healthy discharge of the sewage.

In early eighties, Istanbul had about 3000 km of old damaged sewer lines compared to a total street length of 12 000 km (Damalı et.al., 1989). Wastewater flow from Istanbul in 1990 was about 1.350 000 m^3/day. The estimated flow for 2020 is 5.230 000 m^3/day. Currently, the total planned sewer line is 8000 km , 6000 km of which has been constructed up to 1990 (Samsunlu, 1990a).

Within the scope of the Metropolitan Sewerage Project, the city of Istanbul has been divided into eleven sewage collection zones (Figure 1). Four zones will have biological treatment systems and the rest will have preliminary treatment comprising of screens and grit chambers. Characteristics of the eleven zones are shown in Table 1. Characteristics of discharges are presented in Table 2 (generated from Samsunlu,1990a).

The concept of preliminary treatment followed by marine outfalls was to exploit the well established current system and the bilayer nature of the Bosphorous. It was hoped that the fresh water flow from the neighboring seas will play a positive role in dilution. The discharges are assumed to be confined in the bottom layer with the sewage plume never reaching the surface and ultimately dissipating in the Black Sea. The Turkish Water Pollution Control Regulations permit a first dilution of 40 and recommend a first dilution of 100 for deep sea outfalls. The dilution ratio for the Yenikapı plant is calculated to be over 1000.

3. MARINE WATER QUALITY AND RELATED RISKS

A vast variety of pollutants are continuously discharged into the surrounding water bodies which carry together with them significant environmental risks. Among the major risks are, public health risks, eutrophication risks and risks for the aquatic organisms.

The significant water quality parameters concerning environmental risks in the marine environment of Istanbul are

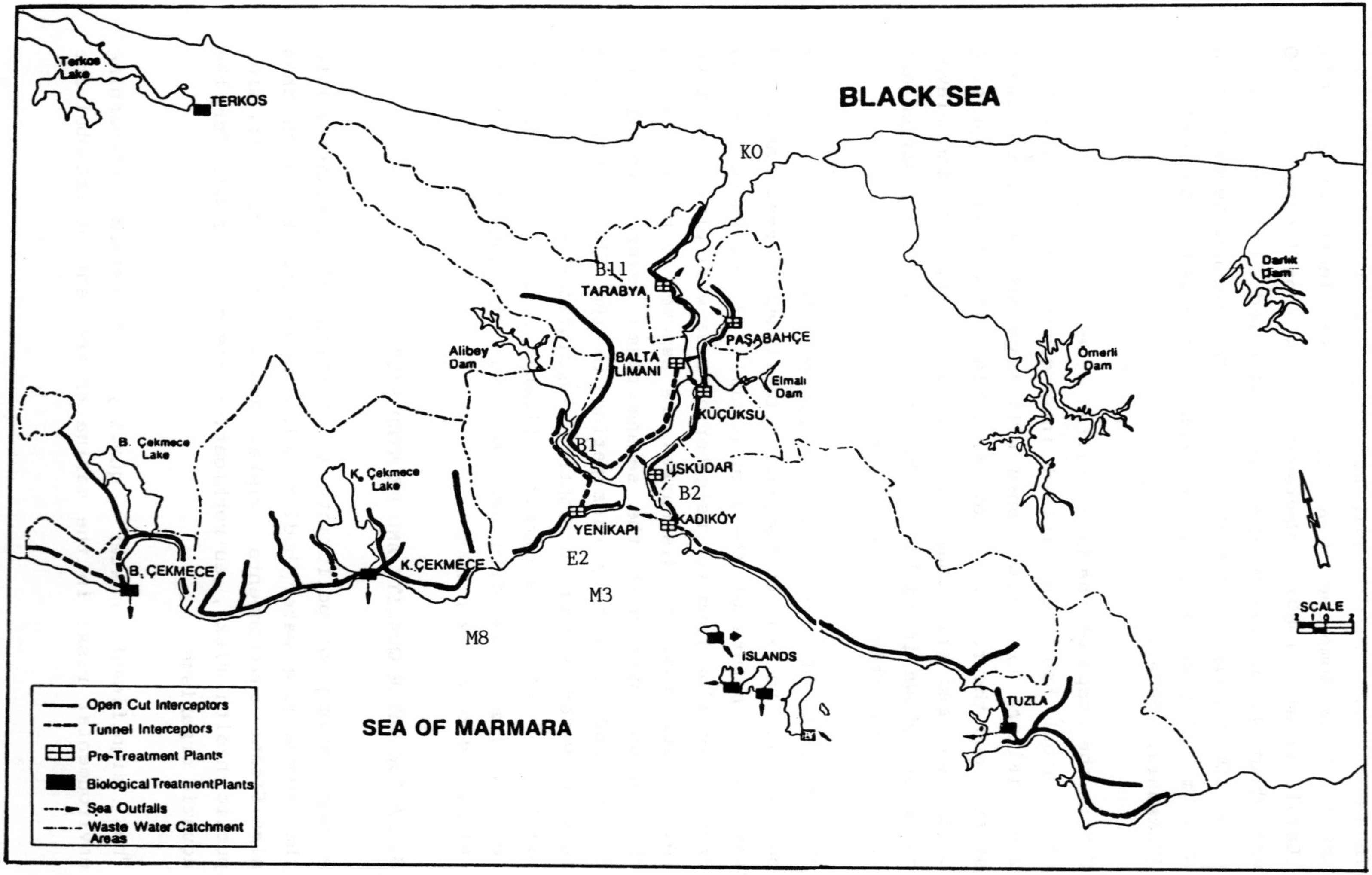

Figure 1. The Metropolitan Istanbul Drainage Catchment Areas and Trunk Sewerage Facilities

Table 1. The Eleven Zones Flowrates and Pollutant Loads.

Name of Zone	Flowrate			BOD_5	
	m^3/day	m^3/s	%	t/day	%
1. Kadikoy	319 400	3.69	23.68	60.1	18.35
2. Yenikapi	313 000	3.63	23.19	63.9	19.51
3. Kucukcekmece	204 500	2.37	15.15	44.1	13.46
4. Baltalimani	198 700	2.30	14.72	45.5	13.89
5. Tuzla	141 200	1.63	10.46	78.3	23.90
6. Buyukcekmece	49 000	0.57	3.63	10.5	3.20
7. Uskudar	47 300	0.54	3.50	9.3	2.84
8. Kucuksu	30 700	0.36	2.27	6.2	1.89
9. Tarabya	20 200	0.23	1.50	3.7	1.13
10. Adalar	12 900	0.15	0.96	2.8	0.85
11. Pasabahce	12 700	0.15	0.94	3.2	0.98
TOTAL	1 349 600	15.62	100	327.6	100

Name of Zone	SS		TKN		Total-P	
	t/day	%	t/day	%	t/day	%
1. Kadikoy	71.6	19.71	10.5	19.02	2.08	22.39
2. Yenikapi	73.9	18.48	10.3	18.73	2.04	21.96
3. Kucukcekmece	53.0	13.25	7.3	13.28	1.45	15.60
4. Baltalimani	52.2	13.06	7.4	13.46	1.46	15.72
5. Tuzla	108.1	27.04	13.7	24.92	1.12	12.06
6. Buyukcekmece	12.5	3.13	1.7	3.10	0.33	3.55
7. Uskudar	10.7	2.68	1.5	2.74	0.30	3.23
8. Kucuksu	6.6	1.65	1.0	1.83	0.20	2.15
9. Tarabya	4.3	1.08	0.6	1.10	0.12	1.29
10. Adalar	3.2	0.80	0.5	0.91	0.09	0.97
11. Pasabahce	3.7	0.92	0.5	0.91	0.10	1.08
TOTAL	399.8	100	55	100	9.29	100

Table 2 : Characteristics of Discharges (Generated from Samsunlu, 1990).

Zone	Population Served	Area Served (Hectares)	Type of Treatment	Total Outfall Length (m)	Discharge Depth (m)	Discharge Into	Status
Buyukcekmece	630.000	13,750	Biological	1320	- 36	Marmara	Design Phase
Kucukcekmece	4.860.000	27,510	Biological	1770	- 34	Marmara	Under Construction
Yenikapi	2.000.000	10,000	Primary	1180	- 60	Bosphorous	In operation
Baltalimani	1.835.000	8,200	Primary	900	- 90	Bosphorous	Under Construction
Tarabya	180.000	5,200	Primary	3000	- 63	Bosphorous	Design Phase
Pasabahce	180.000	4,295	Primary	1600	- 75	Bosphorous	Design Phase
Kucuksu	770.000	3,100	Primary	800	- 71	Bosphorous	Design Phase
Uskudar	450.000	1,115	Primary	363	- 43	Marmara	Under Construction
Kadikoy	5.120.000	21,000	Primary	2800	- 55	Marmara	Under Construction
Tuzla	1.500.000	17,500	Biological	3446	- 47	Marmara	Under Construction
Adalar	87.000	754	Biological	300 - 500	-50 - 70	Marmara	In operation

coliforms (related to health risks), nutrients (in relation to eutrophication risks and depletion of oxygen, also ammonia toxicity), dissolved oxygen and number of species (for risks to aquatic organisms), and micropollutants (regarding public health risks and aquatic organisms). The specific conditions prevailing in terms of these water quality parameters in Istanbul may be summarized as follows.

3.1. Coliform and Other Public Health Concerns

The coliform group is a common indicator of pathogenic (disease causing) organisms. Within the context of this work, coliforms are significant in relation to swimming water quality. The swimming water standards adopted by the Turkish regulations are 1000 (MPN)/100 ml for total coliforms, and 200 (MPN)/100 ml for fecal coliforms.

In the figures 2, 3, and 4 the fecal coliform concentrations in surface and subhalocline waters are shown before and after the treatment plants were put into service (Yılmaz, et.al., 1990).

From figures 2 and 3, it can be seen that the 200/100 ml limit is exceeded in the Golden Horn and in some neighboring areas, both before and after the marine outfalls in the upper layer. The variation of coliforms with depth has a more marked effect after the sea disposal. Although before the sea outfalls, the concentration of fecal coliforms were not significant therein, the bottom layer has turned out to be richer in coliforms after the marine outfalls have started operation as may be seen from Figure 4 . This is because the discharges given into the sea at random is now collected and discharged into the sea at a particular position in the lower layer and a bilayer strafication with insignificant mixing exists in the Bosphorous. A comparison of Figures 2 and 3 shows that despite higher concentrations at the lower layer, the upper layer has improved in terms of coliforms which really is the area of concern for the swimming activity. However it is to be noted

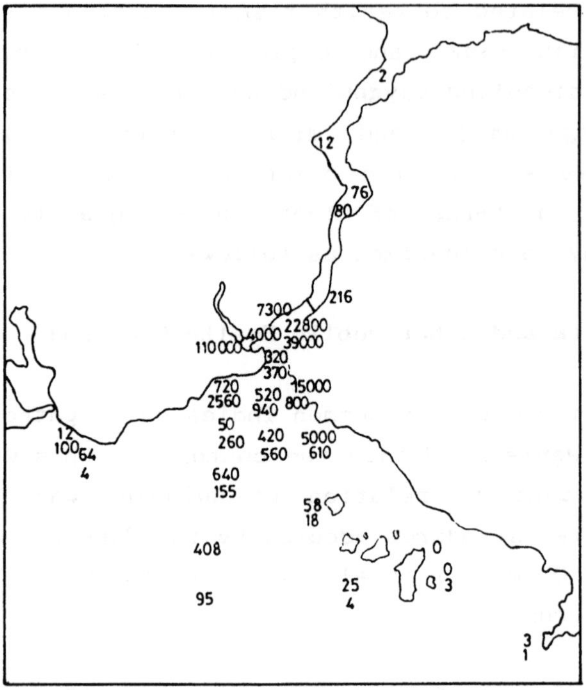

Figure 2. Coliform concentration in surface water in winter 1986 (Yilmaz et al., 1990)

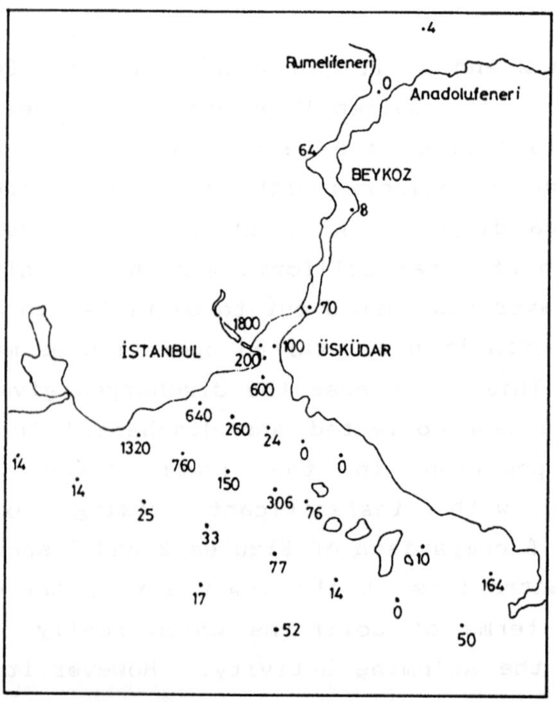

Figure 3. Coliform concentration in surface water in October 1989 (Yilmaz et al., 1990)

that the coliform concentration still exceeds the required limits at various stations in the upper layer.

The situation for the Golden Horn is that the Southern Golden Horn system is in operation at present. The improvement may be attributed to this system. However the still high concentration of coliforms is probably due to the fact that the Northern Golden Horn system is in the phase of construction at this time. The overall effect may be evaluated after its completion and it is expected that the limits will be complied with after the whole system goes into operation.

Except the risk of epidemics via pathogens, public health may be threatened by other, perhaps less obvious, direct and indirect risks. Among the direct risks is the possible ingestion of micropollutants. Consumption of fish or other water products that were previously contaminated by hazardous materials, like micropollutants and radioactive matter may be cited as an indirect risk. The latter is quite an important risk since some marine organisms have the ability to concentrate refractory matter within their tissues, although the concentration of such material is not high enough in the water to cause any direct problems. When those organisms are consumed by people for food, they may bear health risks.

3.2 Aquatic Species

In relatively unpolluted bodies of water, the number of species of organisms are at a maximum. In polluted waters however, only a few species can survive even though the amount of these few species may be large. Hence, the number of species present is a good indicator of water quality.

In 1960's about 125 species of fish were reported in the Marmara Sea. However this number decreased significantly to around 10 due to the constantly degrading water quality. Only the more tolerant species could survive, and actually grow in quantity. Currently the main species in abundance are, horse

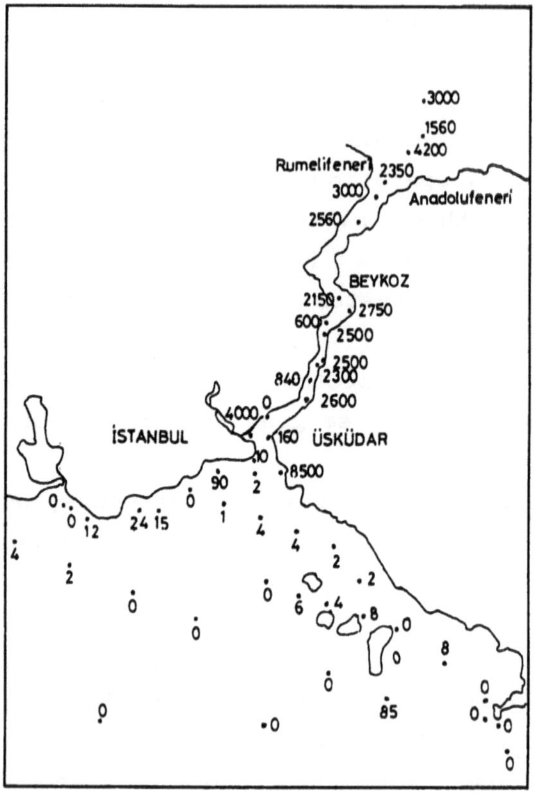

Figure 4. Coliform concentration in subhocline water in August 1989 (Yilmaz et al., 1990)

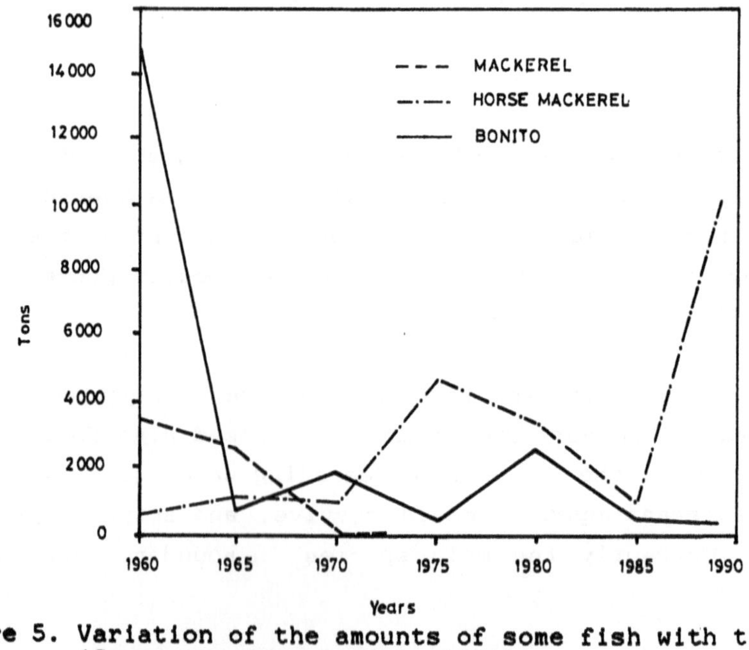

Figure 5. Variation of the amounts of some fish with time (Generated from Zambal, 1991)

mackerel(Trachurus trachurus), bluefish(Pomatomus saltarix), chub mackerel(Scomber colias), striped mullet(Mugilcephalus), whiting (Merlangus merlangus).

On the other hand some species have been completely gone. It was reported by Zambal (1991) that among thirteen species studied by Water Products Agency, four, namely lobsters (Homarus vulgaris), sword fish (Xiphias gladius), bonito (Pelamys sarda), mackerel(scomber scombrus), have almost completely disappeared while more tolerant ones have shown an increase in trend.

As an example, Figure 5 shows the variation of mackerel, horse mackerel, bonito from 1960 to 1989. Two of the more sensitive fish, namely mackarel and bonito have almost disappeared whereas horse mackerel which is a more tolerant fish has increased greatly in number.

The decrease in the number of species is indicative of risks brought about by various pollutants discharged to the sea. This may most likely be attributed to the oxygen deficit mostly due to organics and various micropollutants which include toxic materials like synthetic organics and heavy metals.

3.3 Dissolved Oxygen

Figure 6 depicts dissolved oxygen concentrations at the station E_2, the station on the Bosphorus across from the mouth of the Golden Horn before junction of the Bosphorous and the Marmara Sea . The solid curves are obtained by the minimums and maximums observed at each depth over the years before marine outfalls were put into service and the dashed curves after the discharges. Figure 6c shows two different measurements after the construction of sea outfalls.

An inspection of the Figure 6 shows that for winter and fall, the DO concentration before and after the sea outfall are similar. At the upper layer, it is a little better than the

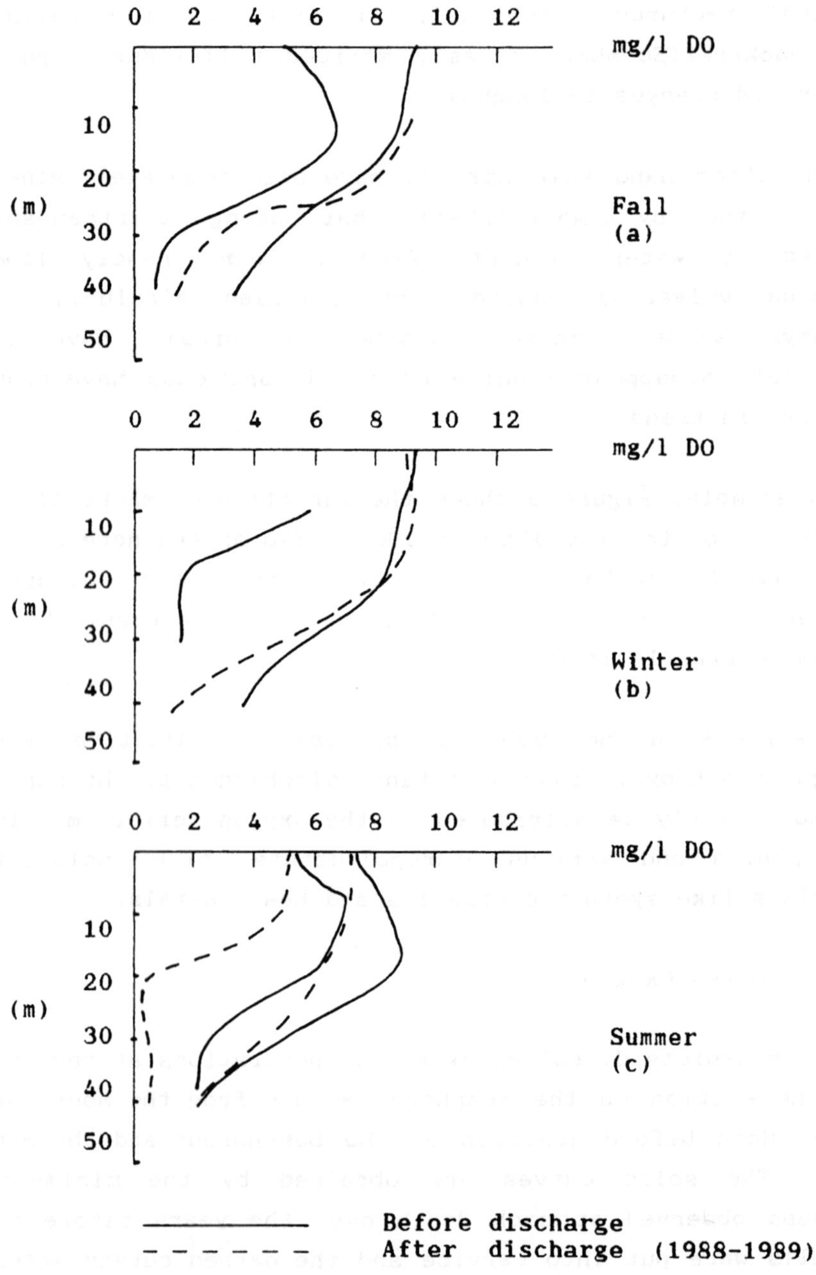

Figure 6. Dissolved Oxygen Concentrations at the Station E_2 (Zambal,1991).

maximum curve while it is some what worse than the maximum for the lower layer. However, both are better than the minimum curve at all times.

For the summer however, one of the measurements after discharges lie within the envelope bounded by the minimum and the maximum curves prior to dischages. However, the curve obtained from the second set of data is worse than even the minimum curve of the pre-discharge conditions. This suggests that thorough inspection of the situation is needed for the summer.

The strafication and negligible mixing are clearly obvious in that figure. The dissolved oxygen concentration deteriorates move in the lower layer. This is due to the fact that biodegredation of the organics occurs in the bottom layer at the expense of DO. Also the fact that penetration of oxygen into the lower layer is hindered due to insufficient mixing, plays an important role.

It can also be seen from the Figure 6 that the depth for 5 mg/l limit which is taken as a limit of biological life (Artuz, 1990), has shifted from approximately 20-25 meters to about 10 meters after the discharges in the worst case.

Figure 7 shows the distribution of DO in the Marmara Sea. In March there are stations where DO levels exceed 10 mg/l which is about the saturation concentration. Dissolved oxygen levels are lower (as low as 4.1 mg/l) near the shoreline but seem to rise to the south into the Marmara.

In July 1988, DO levels may seem to be deteriorated. However since the temperature has increased quite a lot in the summer, the solubility of oxygen decreases and the values around 9 are almost at the saturation concentration for the summer temperatures. It can be observed that the DO levels near the shoreline have improved significantly from about 4.1 to about 7.1 in the worst case after the discharges. However

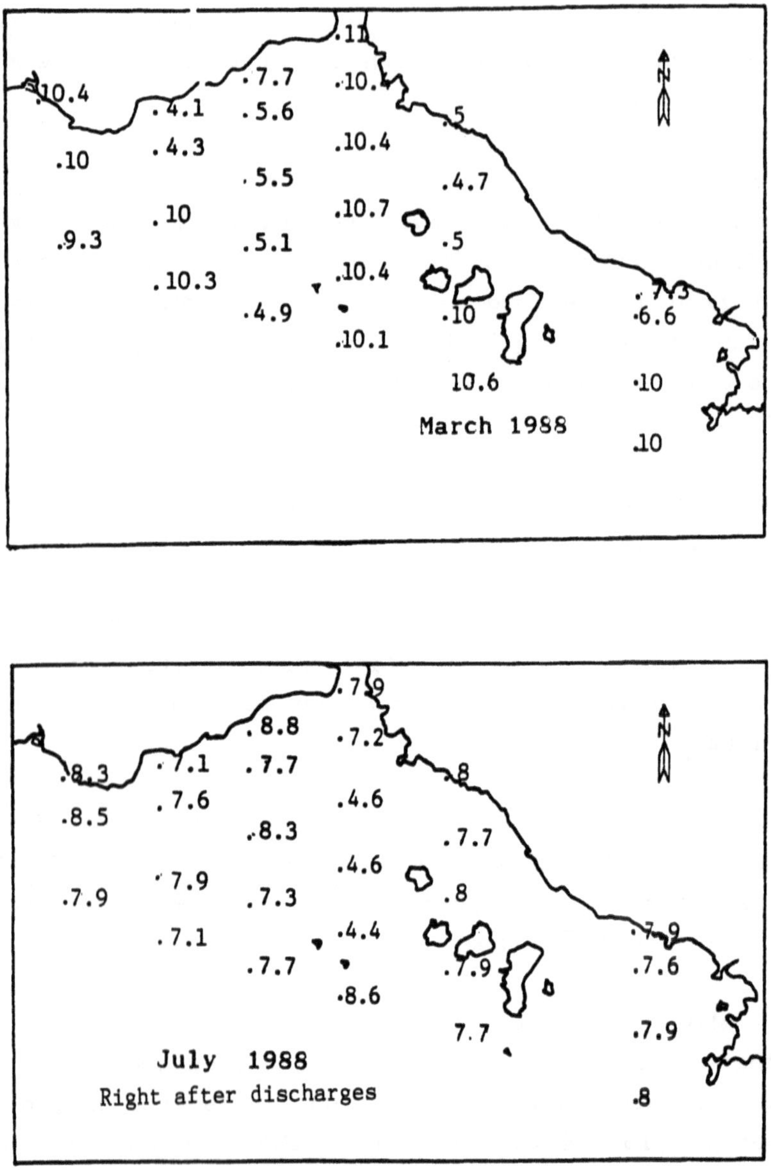

Figure 7 . Distrubution of Dissolved Oxygen at the Marmara Entrance of Bosphorous (mg/l) (Artuz, 1989)

there still is an area where the dissolved oxygen levels have dropped below 5 mg/l in the Marmara Sea.

Table 3 shows the variation of DO with depth for various years in the Marmara Sea (Artuz, 1990). Artuz (1990) confirms that the dissolved oxygen levels have dropped considerably, about -10 meters, following the marine discharges as opposed to -25 meters before the discharges in the Marmara.

Table 3 . The Variation of Dissolved Oxygen with depth for Various Years in the Marmara (Artuz,1990).

Years	0 m	10 m	25 m	50 m
1965	8.10	6.98	6.72	5.07
1971	8.24	6.27	7.79	5.89
1972	9.63	8.46	6.40	5.45
1973	8.25	7.95	6.18	4.32
1974	8.34	9.05	5.10	2.70
1977	8.94	8.17	6.29	2.81
1983	8.05	7.42	4.95	2.40
1984	8.30	7.95	4.54	2.80
1988	8.23	6.84	4.24	1.95

All the aforementioned evidence shows that although there is no significant additional risk for the marine organisms in the uppermost sections, those species which live below about -10 meters deep are subject to increased risks as far as dissolved oxygen is concerned.

3.4 Nutrients

Nutrients may be risky due to their potential for eutrophication which in simplest terms, is the fast aging of a surface water due to abnormal increases in productivity as a result of excessive amounts of nutrients. It is typically observed in stagnant bodies of water. Although connected to the Mediterranean and Black Sea by straights, the Marmara Sea can be considered as an inland sea, hence excessive quantities of nitrogen and phosphorus, may pose a eutrophication risk therein.

Table 4. Variation of NO_3+NO_2 (µg-at/l) (Meric, 1990)

Sta.		1987			1988			1989		
		Max.	Min.	Ave.	Max.	Min.	Ave.	Max.	Min.	Ave.
B1	U	12.20	0.90	2.90	3.35	0.82	1.79	3.13	2.45	2.79
	L	12.59	0.72	7.45	11.04	0.82	5.38	11.38	1.75	6.57
B2	U	4.13	0.13	1.35	2.96	0.77	1.80	2.70	0.37	1.74
	L	11.76	0.46	4.53	11.19	1.97	7.32	4.68	2.82	3.75
B11	U	6.21	0.04	1.35	3.05	0.09	0.87	2.15	0.19	1.01
	L	4.55	2.86	3.94	9.81	9.81	9.81	4.38	4.38	4.38
KO	U	5.24	1.59	3.81	3.22	0.07	0.81	2.08	0.28	0.93
	L	7.50	0.60	2.37	10.38	0.10	3.23	10.02	0.78	3.32
E2	U	1.71	0.07	1.61	2.76	0.55	1.64	4.48	0.05	1.87
	L	13.33	1.95	9.80	11.82	0.22	9.87	11.19	1.30	7.58
M3	U	14.18	0.07	4.50	4.71	0.23	1.78	4.02	0.05	3.12
	L	9.26	2.92	5.82	12.08	9.85	11.02	9.90	0.28	5.72
M8	U	4.70	0.07	1.13	9.68	0.07	2.62	4.27	0.21	2.04
	L	12.48	0.06	9.54	11.41	0.19	9.26	10.61	0.83	7.58

Tables 4 and 5 present nitrogen and phosphorus concentrations around Istanbul. From the figures in the tables it is difficult to critisize the effect of the marine discharges. The most important point is that the daily nutrient load discharged to the Marmara Sea from metropolitan Istanbul is 2 times the load of nitrogen and 4 times the load of phosphorus introduced from the Black Sea (Samsunlu,1990b) . This suggests that Istanbul's discharges may bear a significant risk for the eutrophication of the Marmara.

Table 5. Variation of PO_4-P (µg-at/l) (Meric, 1990)

		1987			1988			1989		
		Max.	Min.	Ave.	Max.	Min.	Ave.	Max.	Min.	Ave.
B1	U	5.70	0.07	0.80	1.97	0.24	0.50	1.12	0.10	0.65
	L	1.22	0.08	0.68	1.15	0.27	0.58	1.53	1.11	1.25
B2	U	0.97	0.03	0.33	0.21	0.06	0.14	6.37	0.31	0.34
	L	1.11	0.03	0.40	1.23	0.19	0.75	0.52	0.28	0.40
B11	U	0.56	0.02	0.32	0.11	0.04	0.08	0.31	0.12	0.23
	L	0.38	0.18	0.28	1.10	1.10	1.10	0.73	0.73	0.73
KO	U	0.28	0.04	0.13	0.12	0.05	0.08	0.37	0.08	0.20
	L	0.75	0.06	0.21	1.12	0.11	0.74	1.09	0.26	0.50
E2	U	0.16	0.05	0.10	0.49	0.04	0.22	0.52	0.15	0.39
	L	1.21	0.12	0.86	1.24	0.15	1.00	1.42	0.55	1.06
M3	U	1.27	0.02	0.33	0.47	0.09	0.26	0.51	0.19	0.34
	L	1.21	0.33	0.76	1.25	0.96	1.12	1.01	0.76	0.84
M8	U	1.11	0.10	0.24	0.96	0.03	0.20	0.54	0.19	0.34
	L	1.18	0.09	0.96	1.17	0.12	0.94	1.80	0.06	1.0

Zambal (1991) indicates that presently nitrogen and phosphorus concentrations in upper layer are below eutrophication limits. However, when all of Marmara is considered, the potential risk of eutrophication cannot be ignored, and this risk can be kept at a minimum by controlling nitrogen and phosphorus discharges from metropolitan Istanbul.

CONCLUDING REMARKS

The Metropolitan Sewerage Project was one of the first steps in reducing environmental risks in Istanbul. It is not completed yet. Only two plants out of eleven are in service.

Critisism from experts can be summarized as follows:

(1) In the year 2020 the population is estimated to be 20.000.000 corresponding to a wastewater flow of 5.200.000 m^3/day. This will exceed the self purifying capacity of the receiving waters. Hence the " primary treatment + marine outfalls " system should be extented to in favor of biological treatment.

(2) Micropollutants will accumulate in the receiving bodies and the ecological balance will be disturbed.

(3) The discharges go to the lower layer and practically consume all the oxygen. Anaerobic conditions may prevail in the future, which may pose problems for flora and fauna of the bottom layer, including complete disappearance of various species.

(4) Nitrogen and phosphorus loads discharged from Istanbul should be controlled to minimize the risk of eutrophication. Tertiary treatment for nitrogen and phosphorus removal should be employed.

(5) There are doubts about the existence of the halocline around junction of the Bosphorous and the Marmara, which would mean that the pollutants may be concentrated at the surface. This is an issue for further investigations.

All of these doubts probably deserve a second consideration. The first two are rational inferences. The risks of exceeding the natural assimilation capacity and accumulation of micropollutants prevail with marine outfalls. Present indications show that there are already some deteriorations in terms of dissolved oxygen in the lower layer. This situation may be even more pronounced, and may generate anaerobic conditions in the future. The nitrogen and phosphorus loads already exceed those coming from the Black Sea. Hence the risk of eutrophication cannot be underestimated. Although there is little solid evidence about the absence of the bilayer stratification around the junction of the Bosphorous and the Marmara, this is noted by the experts as a potential risk.

The most recently elected local administration has tried to review The Metropolitan Istanbul Project and they now consider biological treatment with nutrient removal as a more likely alternative. This will increase treatment costs from $10/person to $50/person for biological treatment and to $100/person for tertiary treatment. Tuzla and Kucukcekmece that have most recently been tendered will have tertiary treatment.

The potential risks and the costs associated to reduce them is a problem of optimization. Currently, the Metropolitan Istanbul Project is at its early stages. Only two plants out of eleven are in service and the combined effect and risk potential of all eleven cannot be put forth at this time. Nevertheless, Istanbul now has twice as long sewer lines as it used to have in early eighties. Most of the random discharges are and will be collected by sanitary sewers and discharged into the marine environment after precautions. The effect of currently operating plants can be seen in controlling coastal pollution both in terms of coliforms and in terms of dissolved oxygen in the upper layer. However, presently risks associated with coliforms, eutrophication and marine organisms seem to prevail.

It is the hope of the inhabitants of Istanbul to gradually attain a cleaner environment where the " green and the blue are saved " and environmental risks are minimized.

REFERENCES

Artuz I (1989) Historical evolution of pollution in the Marmara Sea and Bosphorous region and its present status. Proc. of the environmental problems in the Marmara Sea and Bosphorous Region symposium, Istanbul. Environmental and Woodland Society of Turkey.

Artuz I (1990) Historical evolution of pollution in the Marmara Sea and Bosphorous region and its present status. Proc.of the Environmental Problems of Istanbul Symp. p:281-291. Environmental and Woodland Society of Turkey.

CAMP-TEKSER (1975) Istanbul sewerage project master plan revision, Volume I, ISKI.

Damali A, Belek HT and Orhon D (1989) Water quality management in Istanbul: Creating a cleaner environment. Watershed 89. The future for water quality in Europe. Volume II, p: 505-514.

DAMOC (1971) Master plan and feasibility report for water supply and sewerage for the Istanbul region, Volume III. ISKI.

Kehr D (1962) Festschrift, Erschinen im eigenenverlage des Institutes für Siedlungswasser-wirtschaft Technische Hochschule , Hannover, Heft.9 S.936

Meric S (1990) Land based pollution in the Sea of Marmara and its consequences. M.Sc'thesis. ITU. Istanbul Technical University , Environmental Engineering Department.

Samsunlu A (1990a) The problems of sewerage and treatment systems of Istanbul. Proc.of the Environmental Problems of Istanbul Symp. p:145-160. Environmental and Woodland Society of Turkey.

Samsunlu A (1990b) Control of nitrogen sources and principles of treatment. North Atlantic Treaty Organization. Advanced Research Workshop.

Yilmaz A, Basturk O et al. (1990) Dynamical and environmental aspects of the Marmara Sea and Turkish Straits. Proc.of the Environmental Problems of Istanbul Symp. p:231-261. Environmental and Woodland Society of Turkey.

Zambal T (1991) Environmental impact evaluation of Istanbul wastewater treatment and marine disposal systems. M.Sc'thesis. ITU. Environmental and Woodland Society of Turkey.

DECISIONS WITH MULTIPLE ENVIRONMENTAL OBJECTIVES. THE SITING OF OIL DRILLING WELLS IN NORWAY

Knut L. Seip
Center for industrial research
Forskningsvn. 1, Oslo 3
NORWAY

ABSTRACT

This multiattribute analysis shows how "the seven steps of decision analysis" is applied to the siting of oil drilling wells in Northern Norway. The study includes an assessment of the frequency of accidents (oil blow-out), calculation of risks under uncertainty, (oil-drift modelling under variable sea-state conditions), consequence analysis (impacts on shores), and utility analysis, (preferences for natural resources). Results of the study show that exploratory oil drilling in the eastern off-shore waters north of the counties Troms and Finnmark (>200 km from the coast) gives the least expected oil pollution damage. Compared to this location, drilling in the western, off-shore parts would require an extra willingness to pay (WTP) of 70 mill NOK (10 mill USD). Drilling closer to the shore (≈20 km from the coast) requires an extra WTP of 400 to 500 mill NOK (80 mill USD).

1. INTRODUCTION

This paper describes "the seven steps of decision analysis" as they apply to the siting of oil drilling wells in Northern Norway. However, the approach is generic, and can be used for all decision problems which involve environmental decisions under uncertainty. In the study we calculate the expected and "worst case" damages to social, ecological and economic resources from oil pollution. Damages are calculated for ten alternative oil drilling sites in the Northern waters of Norway. The choice of the "safest sites" for oil drilling is one of the passive components of oil pollution combat preparedness. The objective of the present study is to find the safest oil drilling site from an oil pollution perspective, and to estimate the required willingness to pay (WTP) for moving from the safest site to alternative sites.

1. 1 Shore damages

Damages to shores are among the most "visual" pollution effects, and also among the damages which require the largest clean up efforts. For example, after the *Amoco Cadiz* grounding and oil spill in Brittany, France, in 1978, it took 300.000 man days and 300 mill francs (1982- value) to remove 150 000 tons of oil contaminated debris (7000 tons of oil) from the shores, (Congar 1982).

NATO ASI Series, Vol. G 29
Water Resources Engineering Risk Assessment
Edited by J. Ganoulis

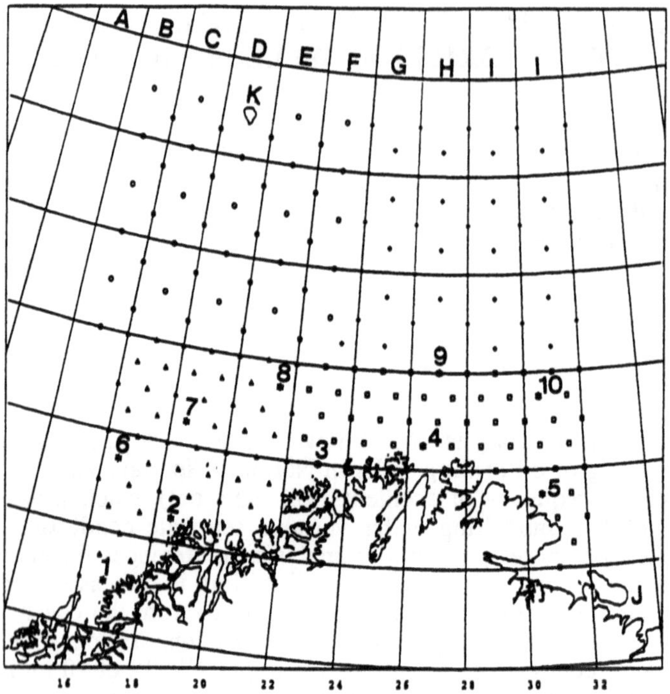

Figure 1. Possible drilling sites in Northern Norway. Numbers indicate possible drilling sites. Letters are used to identify coastal sections.

1.2 Study outline

We start with an assessment of the probability of oil pollution accidents, and with an inventory of oil susceptible resources along the coastal sections concerned. (Parts of the counties Troms and Finnmark in Northern Norway.) Thereafter we use models to calculate the amount of oil stranding. Next we calculate the effects of oil on the resources, and the losses in utility when the resources are damaged. Utility functions for the resources are defined by three groups of people which all have an interest in oil damage reduction and oil spill clean-up.

2. STUDY AREA

For this study we have selected 10 locations along the northernmost coast of Norway (70-71° North,15-31° East). The area is part of the Barents Sea, one of the largest ground water areas of the world, (100-400 m depth, 1 mill sq km). The area will be opened up for exploratory drilling operations in the near future. The shore area lies in the county of Troms and Finnmark and is estimated to be about 8500 km long (CBS 1983a; CBS 1983b) which is about 15 % of the total Norwegian coast. About 45 % of the coastline belongs to the main land and 55% belongs to islands.

3. METHOD

The problem is solved in the context of the multiattribute theory as described by for example Keeney and Raiffa (1976) and Seip et al. (1987). We therefore first present a short review of multiattribute utility methods as they appear when they are applied to the oil spill decision problem. Thereafter we discuss a few problems which are particular to decision situations which involve a high degree of stochastic events.

3.1 Multiattribute utility analysis

Multiattribute analysis can be divided into seven steps (Appendix A). Here we have applied the theory to the choice of oil drilling locations, but the steps are generally applicable

The general solution of the multiobjective problem may for $k \neq 0$ be stated as follows:

(1) $1 + kU_i(x) = P(1 + kk_j u_j(v_{ij}(x_j)))$

where

U_i aggregate utility value of the performance of the *i*th drilling site against $j=1...n$ attributes of the problem;

k_j preference weight associated with the *j* th attribute;

u_j utility value of the performance of the *j*th attribute, it assumes a value between 0 and 1

k scaling constant, or coefficient of synergy, which has a value different from 0 when $Sk_j \neq 1$;

$v_{ij}(x_j)$ direct value of performance of the *i*th siting alternative against the *j* th attribute (v is measured in whatever unit is most convenient)

Equation (1) is the multiplicative utility function of the problem. Since the attributes to be chosen for the present problem will contain some degree of synergy (as will be shown below), the multiplicative form is used in the present study. The aggregated utility U_i for each oil drilling site is used to rank the locations. The site with U_i closest to unity is the optimal site from the environmental perspective.

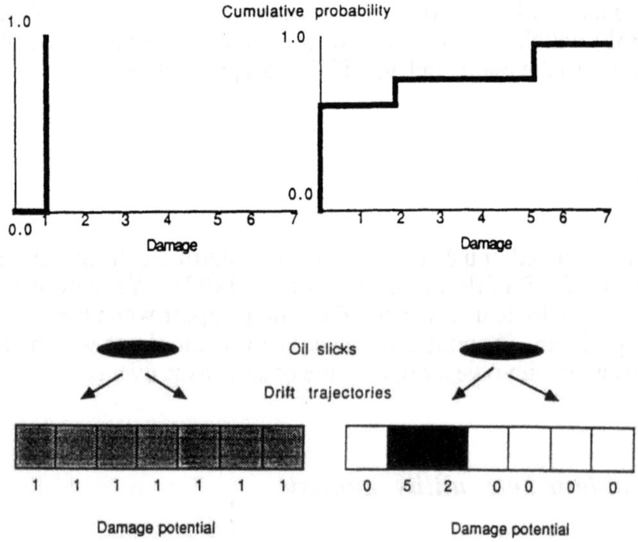

Figure 2

Damages to coastal resources as a function of the distribution of resources along the coast and the length of the oil slick. The oil slick to the left can maximally do damage to 2 resource units, but will do that wherever it hits the coast. The slick to the right may do damage up to 7 resource units, but only if it hits that part of the coast which contain the resources.

3.2 Damages to oil susceptible resources

In this study we define three levels of damages. The levels are: potential damage, expected damage and "Worst case" damages. The terms are defined more closely below. Some of the resources along the coast is extremely unevenly distributed. For example, many of the bird species for which we calculate probable oil spill damage live on bird cliffs. There may then be million of birds on an area of less than a square km. If an oil spill hits the cliff the damages will be very large (>1 million birds dead).

However, the chance to hit a bird cliff is small, because its extension is small. To be able to calculate resource damage estimates for such cases we have developed a separate calculation model. This model requires a certain spill extension (km) as input. It then first calculates the cumulative damage as a curve showing damages along the x-axis and cumulative probabilities (CP) along the y-axis. The expected damage is calculated by a five point interpolation in this curve, (Figure 2)

3.3. *Damage calculations*

The estimated damage from an oil spill caused by a blow out is calculated for each drilling site j. Oil will hit shore segments, i, with a certain distribution of oil susceptible resources. We start the calculations by estimating the potential damage, and modify this damage with factors related to:

- The amount of oil spilled and the size of the oil slick
- Characteristics of the oil, e.g., toxicity and viscosity
- Characteristics of the shore, e.g., slope, sediment size
- Distribution of oil susceptible resources on the coast
- Season, (month)

Details of the calculation are given in Appendix B

3.4 *Four calculation models*

We use four calculation models in this study. The first model, C1, (Appendix B) calculates the amount of oil which reaches nearshore waters under specified conditions of wind velocity and direction, wave heights and temperatures. It is implemented as the TKN - oil spill model, (Fredrikson et al. 1982). The computer model C2 is a model for stranding and subsequent run-off of oil from shores, (Hansen 1989, Seip et al. 1987). It is a very important model, because it describes a process which often is underestimated. Oil will hit the shores, but thereafter run off the shores and drift along the shore to inundate new stretches. The computer models C3 and C4 are damage models. There are different submodules for each damage category. These sub models will be described in more detail below.

4 RESULTS

4.1. *Bounding the problem (Step 1)*

There are three primary dimensions along which to bound the problem of siting oil drilling locations: ecological, economic and recreational. The economic boundaries may be divided into the economics of the private and the public sector. To identify the important attributes we consulted representatives for "The Norwegian council for environmental protection" (Norges Naturvernforbund), Norwegian Pollution Control Authorities (SFT), oil companies, and scientists at the University of Oslo.

In the present case the possible oil drilling sites was partly given by our assignment from the Norwegian Pollution Control Board and the Oil and Energy Department.

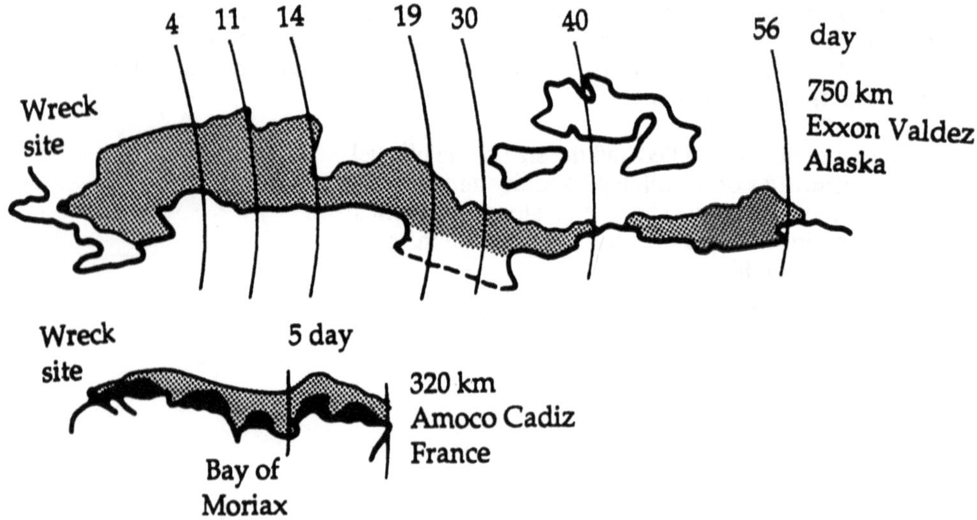

Figure 3
Comparison of the spreading of oil after the *Amoco Cadiz* and the *Exxon Valdes* groundings. The *Amoco Cadiz* spill on the coast of Brittany, France, 16 March 1978 lasted for 30 days and contaminated 320 km of shores with oil. The *Exxon Valdez* spill in the Gulf of Alaska 24 March 1989 lasted for 56 days and contaminated 750 km of shores. Dark shade shows the approximate extent of the oil slick.

However, based on previous knowledge of susceptible resources along the coast, and the influence of distances to the shores, we decided to define possible drilling sites along two trajectories parallel to the coast. We defined 5 sites at an outer trajectory about 200 km from the coast and 5 sites at an inner trajectory about 20 km from the coast. A survey of the drilling sites and their locations are given in Tables 1 and 2 and in Figure 1

4.3 Objectives and attributes (Step 3)

We have in an earlier study (Seip 1983) described oil damage to the open waters and shores by 8 attributes. These attributes are shown in Figure 4. Two attributes deal with damages to the shore. The first gives a measure of oiled shore (km), whereas the second expresses the ecological damage to the shores in terms of recovery time (years) for the shore to "normal" conditions. Four attributes expresses damages to birds. The first of these is concerned with the observation of dead birds irrespective of species (numbers of dead birds). The three next attributes are concerned with the damages to species groups of birds; e.g., auks, ducks and gulls, in terms of restitution time for these three groups (years). One attribute describes damages to recreation (numbers of recreation days lost), one to industry and property (monetary units) and one to fishing and fish farming (monetary units). The reason for having two attributes both expressed in monetary units is that fisheries are considered to be a special industry closely related to human settlement patterns along the coast. (Government regulations for this industry are extensive.)

Weber et al. (1988) have shown that the effect of splitting attributes, as we have done for shores and birds may increase the weight for these higher order categories relative to the weight on other categories on the same level. However, in this particular case we felt that we should separate among attributes expressing visual oil damages, e. g, number of dead birds, and attributes expressing ecological concerns, e.g., restitution times.

4.4 Impact calculations (Step 4)

The damages are calculated by the use of a damage calculation model, (Seip et al. 1987). The model consists of several sub models which each calculates damages to the different attributes. I give detailed results for the damage attribute "shore length" to illustrate typical results, but give only summary results for the other attributes.

By combining information of oil spill probabilities, oil drift and oil persistence, and information of the distribution of resources on the shore we obtain data on expected damages. "Worst case" damages are calculated as the damages which occur when oil actually drifts to the shores with the most susceptible resources, e.g., fish farms. Depending on the resource category the probability of such drift trajectories are one tenth to one hundred of the probability that the slick actually hit the shore.

Open - ocean oil drift model

We have assumed that all oil spills are caused by a blowout lasting for 15 days. The amount of spilled oil is 100,000 tons of crude oil. For crude oil about 20% of the oil will evaporate during the first 5 hours at 15°C water temperature and after 16 hours at 3°C, (Sveum and Bech 1990). Number of spill per ten years is 0.07 cf. (Dahle 1988, Westergaard 1987). Simulations have been carried out under two weather conditions. Alternative 1 corresponds to low temperature and quiet sea (sea state 4) and alternative 2 to high temperature and rough weather (sea state 1). The two combinations represent weather conditions approximately at each end of the weather "continuum" in the Barents Sea. Temperature means less for the fate of the oil than the sea state. The amount of stranded oil Oj* is found as:

(2) O*j = S Pr(t)*Ss(t) integrated from time zero to time 20 days

where Pr(t) is the probability that oil shall strand at time t; Ss(t) is the amount of oil stranded at time t

Table 1

Survey of drilling sites. Sites 1-5 is about 20 km from the coast. Probabilities for stranding are calculated by (Johansen 1988). The expected amount of stranded oil is calculated in the present work.

	Site				
	1	2	3	4	5
Coordinates (decimal)	16.75 E 69.50 N	18.50 E 70.25 N	23.00 E 71.00 N	26.50 E 71.25 N	30.50 E 70.75 N
Shortest drift time to shores (days)	0.25	0.25	0.25	0.25	0.25
Probability for stranding after 15 days, % -summer	58.1	69.9	86.1	75.8	61.0
Expected amount of stranded oil (Weather type 1) tons (1)	20,129	23,738	33,260	22,097	23,204
Expected amount of stranded oil (Weather type 2) tons (2)	8,217	9,542	14,000	9,012	9,297
Stranding location	Gryllefjord	Ringvassøya Tromsø Fugløy	Sørøysund Hjelmsøy	Mehamn	Vardø

(1) Weather type 1 occurs 25 % of the year. 0.0-0.5 m wave height
(2) Weather type 2 occurs 35 % of the year.> 0.5 m wave height

The probability for stranding increases with time (it is about 80 % for a drilling site 20 km from the coast after 20 days, Johansen 1988), but an increasing amount of the oil evaporates and mixes down into the water column as residence time on the water increases. Thus, the amount of oil left for stranding decreases.

Table 2

Survey of drilling sites. Sites 6-10 is about 200 km from the coast. Probabilities for stranding are calculated by (Johansen 1988). The expected amount of stranded oil is calculated in the present work.

	Site				
	6	7	8	9	10
Coordinates (decimal)	16.55 E 71.75 N	18.50 E 71.25 N	21.50 E 71.75 N	27.00 E 72.00 N	30.50 E 71.75 N
Shortest drift time to shores (days)	6.25	3.50	2.50	2.00	3.25
Probability for stranding after 15 days, % -summer	4.40	26.20	40.00	17.20	7.30
Expected amount of stranded oil (Weather type 1) tons (1)	290	3,000	5,000	2,000	700
Expected amount of stranded oil (Weather type 2) tons (2)	64	320	240	190	50
Stranding location	Fugløy Hammer-fest	Fugløy Skjøtings-berg	Fugløy Vardø	Hjelmsøya Vardø	Berlevåg Vardø

(1) Weather type 1 occurs 25 % of the year. 0.0-0.5 m wave height
(2) Weather type 2 occurs 35 % of the year.> 0.5 m wave height

For the locations closest to the shore (20 km) the largest amount of oil strands after 1 day and the amount stranding after 10 days is insignificant. (Stranded oil in location 1 to 5 is given in Table 1.) For the localities farthest from the shore (200 km) the first oil strands after about 5 days. A maximum of oil strands after 10 days, and there is still some oil (1/5 of the totally stranded oil) which strands after 15 days. The amount of oil stranding after 20 days is insignificant. (Stranded oil in locations 6 to 10 is given in Table 1).

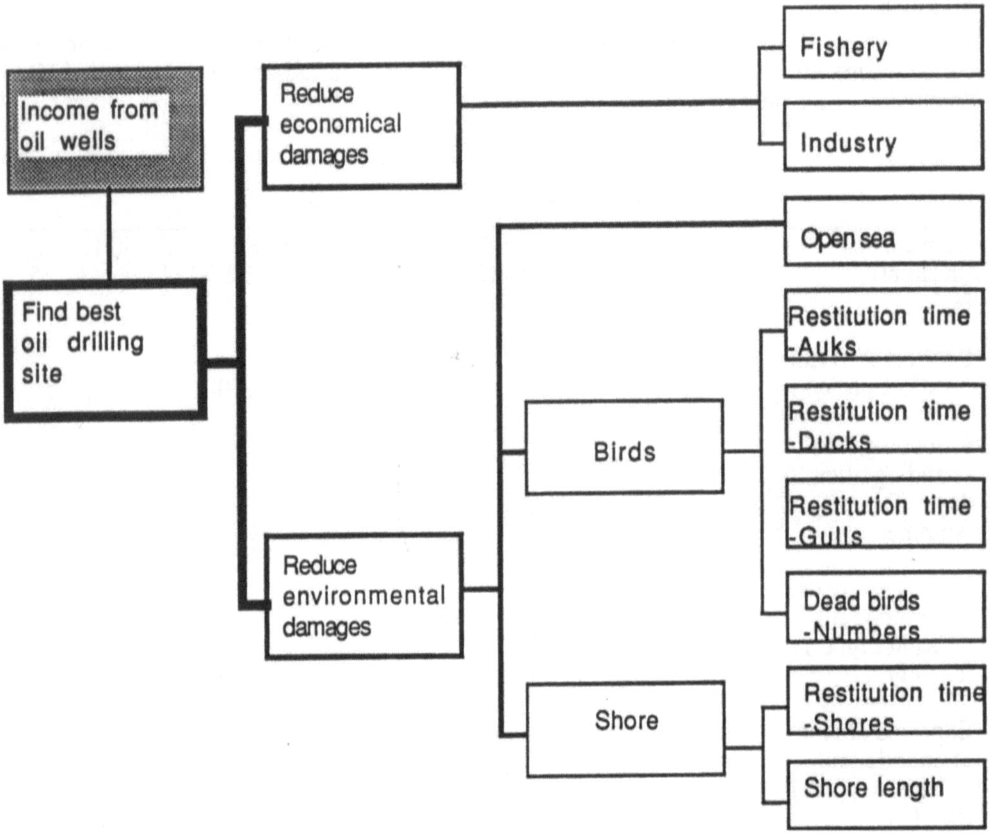

Figure 4. Attribute hierarchy for the siting of exploratory oil drilling. Primary attributes to the right, and increasing aggregation to the left. Income from oil wells is not calculated.

Oil-on-shore model

Shoreline contaminated is calculated by the carrying capacity method of Seip et al. (1987) and Gundlach and Reed (1985). Typical shore characteristics for the coast of Troms and Finnmark are as follows (Figures 5): Cliffs, rocky shores, gravel beaches, sand beaches, and mud flats have depths of about 2m, 3 to 10 m, 5 to 30 m, 5 to 100m, and 50 to 200 m respectively. (Average depths for shores along the Brittany coast where *Amoco Cadiz* grounded were about 100 m, (Gundlach et al. 1983). The length of shores contaminated from a 100,000 ton spill of crude oil from the different drilling locations were shown in Tables 1 and 2 .

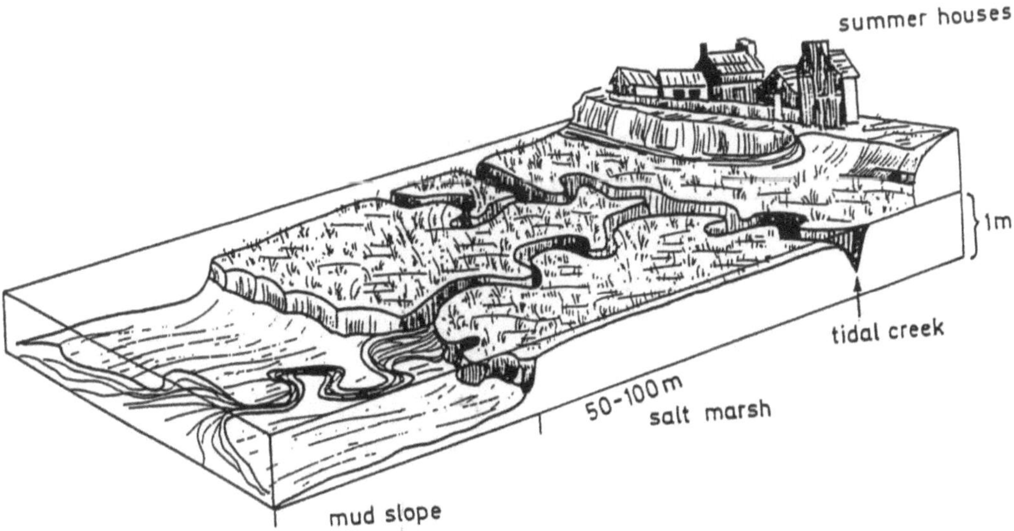

Figure 5
Section of a coastal profile (Partly after (Le Campion-Alsumard and Plante-Cuny 1982)

Shores restitution time

The rocky shores community comprises *Fucus* and *Ascophyllum* in the littoral zone and *Laminaria* in the sublittoral zone. Oil densities which may cause mortality to these species are reported to be 4 $kg.m^{-2}$ (Berne et al. 1980; Broman and Lindblad 1980). In terms of oil concentrations in the water Ganning and Billing (1974) show that bladder wrack may survive 1000 ppm of non emulsified oil. However, lower concentration of emulsified oil is toxic. Melzian and Lake (1987) show that crabs have a 96-h LC50 of 14 ppm, i.e., much lower than the benthic algae.

The impact of increased mortality on the shore vegetation and fauna was studied with a simulation model by Seip (1983). Shore restitution times may be 10 to 20 years, depending upon the definition of a restituted shore, e.g., the recolonization of non-abundant species. The gross expression for the shore restitution time is calculated as the weighted sum of the restitution times of rocky shores, sand shores and mudflats / marsh areas:

$$(3)\ T = (S\ (t_i^q\ .\ l_i)\ /\ Sl_i)^{1/q}$$

where

t_i is the restitution time for each shore type ; l_i is the length damaged of each shoretype; q is an integer (=4)

Table 3

Polluted shoreline (km), clean-up costs and number of "priority shores" (Elven and Johansen 1983). damaged in the study area (Maps, Ørja-Vardø). Simulated results.

	Shore length Spill occur	Expected 10 years	Worst case 10 years
Damage,			
-sites 1-5 (km)	72-126	7	11
-sites 6-190 (km)	15-55	3	11
Clean-up costs			
(mill NOK)	400	28	39
Priority shores			
-sites 1-10		1	5

We have chosen q = 4 to emphasis the pollution effect on shores with long restitution times. These are often mud flats and marsh areas with great ecological significance.

Birds

A characteristic feature of the Northern Norwegian coast is the dense populations of birds, particularly puffins and kitti wakes. On bird cliffs a million of pairs, more than two million of individuals (sic!, young immature individuals are not counted) may stay during the summer half year. However, in recent years the populations have declined.

Bird damage is calculated by dividing sea birds into three groups with increasing oil susceptibility. The number of birds dead is calculated by summing the number of dead birds in the three groups. Birds restitution time considers both the net fecundity of bird species and mobility factors which allows for recruitment from neighboring non- polluted shores. Almost all of the most susceptible birds' species will die if they are hit by an oil spill when they are most vulnerable. If the exact locations of the birds are not known, the model utilizes information of clustering behavior of the bird species considered. (E.g., auks are usually highly clustered, whereas sea gulls are not.) A certain fraction of the birds in the other two groups will die under the same conditions. The numbers obtained in this way are thereafter modified by applying a curvilinear modifier to take account of oil toxicity and the season of the accident. The numbers of dead birds relative to the total populations calculated in this manner are of the same order of magnitude as calculated by independent simulation models, (Ford et al. 1982, Seip et al. 1991).

Social damage and damages to economy

Social and economic attributes are calculated according to the formulae:

(4) $N = f(N).\ g\ (T, P, S)$
(5) $N = S\ N_i.DF_i$
(6) $DF_i = V_i.S_i.F_i$

where N_i is the number of oil susceptible units in resource category i
V_i is the value of each unit expressed in attribute denomination
DF_i is the fraction of resources which potentially (intrinsically) can be damaged by oil pollution.
T,P,S represent the factors temperature, oil persistence and time (month) of oil spill which are used as damage modifiers. The functional form of these modifiers is given as graphs.

Recreation

Recreation days lost is calculated by estimating the potential number of visits to the shore in Troms and Finnmark from Holiday surveys, (CBS 1983c; CBS 1985). We then estimating how many people will choose to stay away from the shore if it gets polluted by oil, (oil smell, oiling of recreational equipment, cf. Brown et al. (1983)) and then modifying this number according to season.

Property and business

Damages to industry include damages to harbors and vessels, and commercial recreation, but not to fisheries, which is treated separately. The cleaning of vessels constitutes the major damage category.

Fishing

Fishing, especially coastal and open sea fisheries, are a highly regulated primary industry in Norway, and is therefore treated separately from other industries. Damages to fisheries are dominated by potential damages to fish farming. Oil has been shown to induce panic reaction in fish pens inflicting damages to skin of the fish, (Reiersen and Berge 1985). Oil smell of cooked fish meat is also a problem, (Ogata et al. 1987). However experiments seem to indicate that the smell may disappear after some time in clean water. Recent experiences (the 1988 algae bloom) indicate that fish farms may be moved over fairly long distances if a sufficient warning time is given. However, we have here conservatively assumed that about two years of sales are lost if a farm is hit by an oil spill. This high value may be defended on the basis of the fish industry's high vulnerability to bad reputation.

Summary

A summary of damages from drilling locations 1 to 5 (the locations close to shore ≈ 20 km) and the locations 6 to 10 (the locations far from shore ≈ 200 km) is shown in Table 4.

5. WEIGHTING THE ATTRIBUTES (Step 5)

The fifth step in multiattribute analysis (Appendix A) is to construct the utility functions for each attribute and to make an assessment of the relative importance of the attributes in terms of the kjs and the k of equation (1). In the present case we presented the decision alternatives to three groups of people. The first group consisted of people employed at the Norwegian Pollution Control Authority. These people are responsible for oil combat operations in Norway. The second group consisted of people at the natural conservation society Bellona. The third group consisted of scientist employed at Center for industrial research and at the University of Oslo. The attribute utility assessment was made with a computerized, and interactive program developed at the Norwegian business administration school, BI. (Gottschalk and Wenstøp 1990). The decision makers were presented a fairly realistic decision scenario (Figure 6). They were told that a certain amount of oil booms available and was asked to position these booms to protect the resources they believed to be of highest priority.

The detailed results of the utility and preference study are given in Seip (1991). Here I only give the average values for all decision makers for the preference profile, i.e., the average weight of each attribute (Figure 7).

6. APPLYING THE DECISION RULE (Step 6)

The sixth and final step in multiattribute analysis, except for step 7 which is a post process to the analysis, is to calculate the single attribute utility and the aggregated utility for each drilling site. Since the attributes in this analysis include two attributes which are measured in monetary units, we can give the results both in "utiles" and directly in monetary units (NOK). To make the results more "robust," we combined the results for the ten drilling locations into four groups representing nearshore western locations: 1, 2, and 3, nearshore eastern locations: 4 and 5, open ocean western locations: 6, 7, and 8, and open ocean eastern locations: 9 and 10. The results are shown in Table 5

Table 4

Damage from oil spills by blow-out per ten years during the summer months. The amount of oil stranded is weighted with respect to the probability for different weather types. Expected damage is weighted with respect to probability for stranding locality and weather type. "Worst case" damages is ten to hundred times less probable than expected damage. It is here calculated as the damage, at the i) most unfavorable drilling location and ii) under the most unfavorable weather condition, and iii) assuming that the oil slick drifts towards the sites with the most concentrated resources. The probability for an expected damage is set to 0.007 per year and to 0.00035 per year for "worst case" damage.

Attribute	Drilling sites 1-5		Drilling sites 6-10	
	Expected	"Worst case"	Expected	"Worst case"
Spilled oil,tons	100,000	100,000	100,000	100,000
Stranded oil, tons	15,000	20,000	1,000	2,000
Probability for blow-out per ten years	0.07	0.0035	0.07	0.0035
Expected stranded oil (tons)	1,050	-	70	-
Shorelength polluted (km)	7	75	3	65
Shore clean-up costs (mill NOK)	35	70	4	8
Shore restitution time (years)	10-15	10-15	10-15	10-15
Priority shores damaged,(numbers)	1	5	1	3
Birds dead (numbers)	1,700	150,000	700	150,000
Auks restitution time (years)	10	20	4	16
Recreation (visits)	800	50,000	250	10,000
Recreation, costs (mill. NOK)	0.02	1.0	0.02	0.15
Property and business (mill NOK)	0.06	1.0	0.02	0.15
Fishery (mill NOK)	0.7	9.0	0.4	10.0
Cover, open ocean (km2) (1)	500 135	500 168	500 20	500 30

(1) The largest number corresponds to the oil slick size at the drilling location. The lowest number corresponds to the oil slick size just before stranding

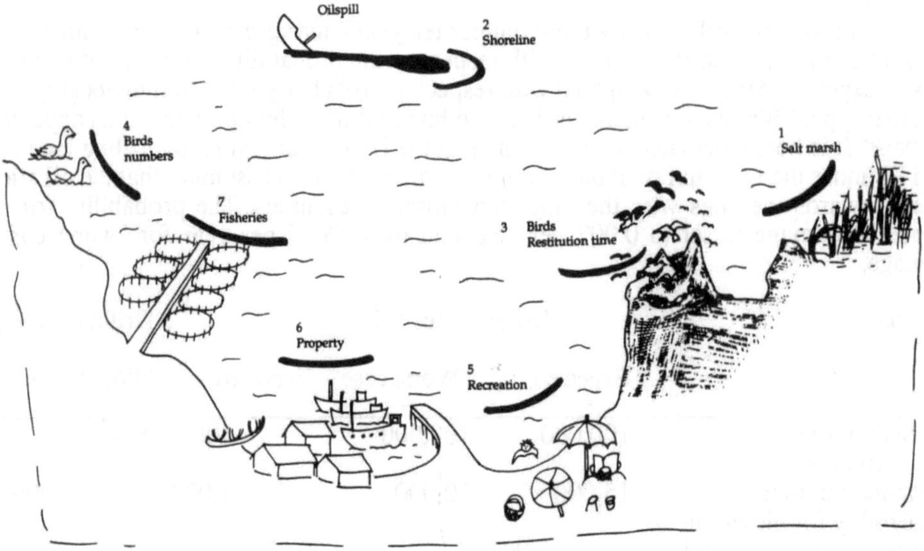

Figure 6.
Oil spill scenario presented to the decision maker during the sessions of preference acquisition. (1) Priority shores; (2) Shore line; (3) Restitution time for birds; (4) Number of dead birds; (5) Recreation; (6) Property, (7) Fishing.

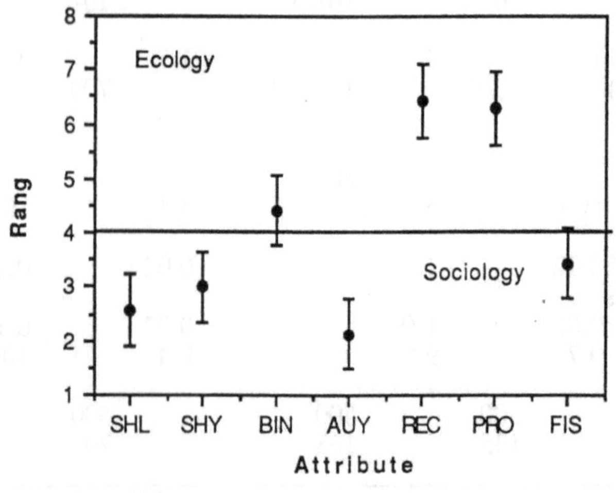

Figure 7
Preference profile for the siting of oil drilling location problem. Range reflects the importance of each attribute given the attributes domain of variation. "1" is most important. SHL is shore length (km); SHY is restitution time for shores (years); BIN is number of dead birds, irrespective of species (numbers); AUY is restitution time for populations of auks (years); REC is recreation (visits); PRO is property and business loss (NOK); FIS is fisheries losses (NOK). n= 7 for each attribute.

Table 5

Aggregated Utility values U for four drilling locations. Expected utility value and "worst case" utility value. "Extra" cost calculations are based on differences in the expected utility.

Drilling location	Expected utility 0.0 - 1.0	"Worst case" utility, 0.0 - 1.0	Extra cost, mill NOK
Off shore Western	0.927	0.472	67
Off shore Eastern	0.943	0.821	
Nearshore Western	0.848	0.529	397
Nearshore Eastern	0.821	0.861	510

The results show that the expected remaining utility after a blow -out will be largest for drilling sites in the off-shore eastern localities (0.972). The expected remaining utility will be lowest for near-shore eastern locations (0.821), i.e., this locality should be avoided for drilling operations. If we look at the "Worst case" results, the remaining utility after a blow-out is smallest on the off shore western locality and largest in near-shore eastern locations.

If we recalculate utilities to mill NOK and look at the annual extra cost of moving from the expected "best" location, i.e., the off shore eastern locations, to one of the other locations, we see that the extra cost in terms of environmental, social and economic impacts caused by blow-outs will be largest for the nearshore eastern locations (510 mill NOK). The cost will be of the same order when moving to the nearshore western locations.

7. DISCUSSION AND CONCLUSION (Step 7)

Our main conclusion is that to our best knowledge drilling in the off shore eastern waters will pose least damage to the environment, to social attributes and to economy. Moving closer to shore, either in nearshore western or eastern waters will increase the expected damage. The extra cost is of the order of 400 to 500 mil NOK. The "worst case" scenarios do express the costs if the "worst case" should occur, but these results do not make sense when we compare alternative drilling sites. The reason is probably that the "worst cases" differ ten fold in probabilities (ten to hundred times less probable than the "expected" damage) and are thus not comparable.

Sensitivity analysis

There are uncertainties in all the estimates. The uncertainty in the consequence estimates can partly be evaluated by examining the variations in damages calculated for drilling sites within the same region, i.e., sites 1 to 3 in the nearshore western region. For example the length of shore line polluted ranges from 72 to 126 km. However, all three results depend upon the use of the same model system. Thus, it would have been advantageous to know the goodness of the models. Unfortunately, oil pollution models are seldom extensively tested. The reason for this is mainly that too few data sets exist to do this properly. However, qualitative evaluation of the present model, and similar models have been reported, (Seip et al. 1987, Reed et al. 1989). I believe that experiences based on recent oil spills, e.g., *Amoco Cadiz* - 320 km, *Exxon Valdez* - 750 km suggests that the lengths of the polluted shores calculated by the models are reasonable.

The utility of marine resources was elicited from three groups of decision makers. The range of the utility of resources was similar, (Seip 1991), but the scientists obtained a higher remaining utility (0. 850 ±0.163, average over all sites, and over expected and "worst case") compared to the environmentalists (0.704±0.193). Also, values for units of single resources appear to be comparable to values obtained by questionnaire methods (Hem and Seip 1990) or reported in the literature (Bentcover, et al. 1986). Thus, I conclude that the results we have obtained a reasonable manner reflect the "cost" of oil pollutant from oil drilling in the Northern waters of Norway.

The multiattribute decision method has been critisized by Haimes (1991, this volume) because he claims that it is difficult, or impossible, to obtain "true" preferences by the Keeney and Raiffa (1976) method. However, I believe that the approach we have used may overcome some of his criticism. Also, the final decision in this type of analysis will always be taken by government authorities in Norway, so that the results presented here will only be part of the full decision support documentation. Other methods for decision support can be found in Woldt et al. (1991) on ground water remediation, Ganoulis (1991) on risk in coastal pollution, Plate (1991) on water quality in rivers, all in this volume. In these papers, refinements of the method we have used are presented, e.g., calculation of joint probability functions, a discussion of different distance measures, etc.

APPENDIX A. Steps in multiattribute utility studies

The seven steps in the application of multiattribute utility theory to the siting of oil drilling operations

Step 1 Bounding the problem. Important factors and constituents to be considered in the siting problem are identified. The appropriate region is defined.
Step 2 Identify feasible drilling sites.
Step 3 Choosing the objectives of the program. A comprehensive, mutually exclusive, and preferentially independent set of objectives is selected against which the feasible drilling sites will be judged. The attainment of each objective is represented by the value of an attribute representing that objective.
Step 4 Scoring the siting locations. Here we first calculate the amount of oil stranding on the shores from the ten different possible drilling sites. Then we calculate the impact on the attributes
Step 5 Setting the attribute weights. Here we establish the preference functions which describe decision makers weighting of different damage categories.
Step 6 Applying the decision rule. Scores and attribute weights for each possible drilling site are combined according to the predetermined decision rule and the sites are ranked according to scores on a utility scale. Sensitivity analysis is undertaken here.
Step 7 Resolving preference conflicts which persist after careful scrutiny of objectives and options.

APPENDIX B. Calculation models

This appendix describes the steps used to calculate expected and "worst case" damages.

A) Calculate oil in near shore areas O* from drilling site j as:

(2) $O^* = C1\ (Oj, AK, T, W_h)$

B) Calculate the amount of oil stranded, O_k, and the length of polluted shore L_k (the site index j is skipped during the following steps)

(3) $[L_k, O_k] = C2\ (O^*, A, v, V_h, S_k)$

C) Calculate potential damage $D_{k,i}$ when a certain fraction of the coast is polluted by oil

(4) $D_{k,i} = C3\ (X_{k,i}, f_{k,i}, Pr_i, L_k, O_k)$

D) Calculate realized damage $Dk_{,i}$ for different types of oil and different seasons, summer autumn, winter, and spring:

(5) $D^*_{k,i} = C4(D_{k,i}, g, v, M)$

E) Calculate expected damage:

(6) $E(x_i) = S\ f_j, f_h, f_l, f_k\ D^*_{j,k,h,l}\quad j,k,h,l$, sum over all alternatives

F) Calculate expected damage for the coast of Troms and Finnmark

(7) $E(x_i) = S\, f_k\, E(x_i)\, F^{-1}$

where

Oil characteristics

O_j is spilled oil; v is oil viscosity (scale 1-10); g is oil toxicity (scale(1-10); A is area of the near shore oil patch

Time of year

M is season, month; T is sea temperature; V_h is sea state, h=1,4

Computer model calculations

AK is a computer calculated estimation of the effects of an oil spill combat action

C1, C2, C3, C4 is represented by a computer model

Shore type characteristics

S_k is characteristics of the shore types which may be inundated, e.g., slopes, sand, rock etc.

$X_{k,i}$ is the amount of resources i within coastal section k

$f_{k,i}$ designates the distribution of resources within one coastal section

Probabilities

f_j is the probability of each drilling location being used. (Presently f_j = 1/N, N=10, all locations equally probable)

f_l is the probability that the spill reaches the coast

f_k is the conditional probability that the spill reaches coastal section k, if it reaches the coast

F is the proportion of the total spill probabilities which is represented by the 10 possible drilling locations

P_{ri} suggests if one wants to calculate expected value or a damage with lower or higher probability.

LITERATURE

Bentcover, J. D., V. T. Covello and J. Mumpower. 1986. Benefits assessments. The state of the art. Dordrecht, D. Reidel Publishin Company.
Berne, S., M. Marchand and L. DÓzonville. 1980. Pollution of sea-water and marine sediments in coastal areas. Ambio. **9**: 287-294.
Broman, D. and C. Lindblad. 1980. Biological effects of hot water cleaning. Effects on flora and fauna in the Stockholm arcipelag. University of Stockholm, Stockholm (In Swedish)
Brown, G. M., R. Congar and E. A. Wilman. 1983. Recreation: Tourist and residents. Assessing the social costs of oil spill. The Amoco Cadiz case study. Springfield, U.S. Department of commerce.
CBS. 1983a. County survey for Troms. Norwegian Bureau of Statistics, Oslo
CBS. 1983b. County survey of Finnmark. Norwegian Bureau of Statistics, Oslo
CBS. 1983c. Recreation in Norway. Norwegian Bureau of Statistics, Oslo
CBS. 1985. Outdoor life survey 1984. Norwegian Bureau of Statistics, Oslo
Congar, R. 1982. A chronological analysis of the cost of controlling pollution from the Amoco Cadiz oil spill. Finistere, 1978. The cost of oil spills. OECD, Paris pp.105- 111.
Dahle, E. A. 1988. Estimates for the frequencies and size distributions of various oil spills along the Norwegian coast in 1988 and 1990 with Sture and Mongstad developed. Det Norske Veritas. Høvik, Norway.
Elven, R. and V. Johansen. 1983. Shores in Finnmark- flora, vegetation and botanical value. Environmental directorate, Oslo.
Ford, R. G., J. A. Wiens, D. Heineman and G. L. Hunt. 1982. Modelling the sensitivity of colonial breeding marine birds to oil spills: Guillemot and kittiwake populations on the Pribilof islands, Beering sea. J.Applied Ecology. **19**: 1-31.
Fredrikson, G. W., H. Ibrekk, K. I. Johannessen, K. Kveseth, K. L. Seip, H. M. Seip and F. Wenstøp. 1982. Oil spill combat: Damage assessment using multiattribute utility analysis. Center for industrial research. Oslo.
Ganoulis, J. 1991. Risk in coastal pollution. NATO advanced study institute. Chalkidiki, Greece.
Ganning, B. and U. Billing. 1974. Effects on community metabolism of oil and chemically dispersed oil on Baltic bladder wrack Fucus vesiculosus. Ecological aspects of the toxicity testing of oils and dispersants. England, Appl. Sci. Publ. Ltd.
Gottschalk, P. and F. Wenstøp. 1990. Quantitative decision analysis for managers. Pro & Con. Norwegian school of management. Sandvika, Norway.
Gundlach, E. R., P. D. Boehm, M. Marchand, R. M. Atlas, D. M. Ward and D. A. Wolfe. 1983. The fate of Amoco Cadiz oil. Science. **221**: 122-129.
Gundlach, E. R. and M. Reed. 1985. Quantification of oil deposition and removal rates for a shoreline/oil spill interaction model. 6th Arctic marine oil spill program (AMOP) Technical seminar. pp. 65-76.
Haimes,Y.Y. 1991. Total risk management. Lecture notes. NATO advanced study institute. Chalkidiki, Greece.
Hansen, E. K. 1989. Matematisk modell for drift og ilandskylling av olje. University of Oslo. MA Thesis.
Hem, K.-G. and K. L. Seip. 1990. Valuation and ranging of pollution abatement efforts. VANN. **1**: 50-61.
Johansen, Ø. 1988. Oil drift in the Barents sea. Drift trajectories. OCEANOR, Trondheim.
Keeney, R. and H. Raiffa. 1976. Decisions with multiple objectives. New York, John Wiley.

Le Campion-Alsumard, T. and M. R. Plante-Cuny. 1982. Etydes micrcbiologiques et microphysiques dans les sediments des marais maritimes delÎle Grande a la suite de la pollution par lÁmoco Cadiz. Ecological study of the Amoco Cadiz oil spill. NOAA-CNEXO joint scientific commission. US Department of commerse.
Melzian, B. D. and J. Lake. 1987. Accumulation and retention of No.2 fuel oil compounds in the blue crab Callinecter sapidus Rathbun. Oil and Chemical Poll. **3**: 367-399.
Ogata, M., Y. Miyake and K. Fujisawa. 1987. Oil smell and oil components in fish flesh reared in seawater containing heavy oil. Oil and chemical pollution. **3**: 329-341.
Plate,E.J. 1991. Probabilistic modeling of water quality in rivers. NATO advanced study institute. Chalkidiki, Greece.(This volume)
Reed, M., D. French, T. Grigalunas and J. Opaluch. 1989. Overview of a natural resource damage assessment model system for coastal and marine environments. oil and chemical pollution. **5**: 85-97.
Seip, K. L. 1983. Mathematical models of the rocky shore ecosystem. Application of ecological modelling in environmental management. Part B. Elsevier.pp. 341-443.
Seip, K. L. 1991. Preferences for marine resources in Norway. Manuscript. Center for industrial research. Blindern, Oslo
Seip, K. L., K. A. Brekke, K. Kveseth and H. Ibrekk. 1987. Models for calculating oil spill damages to shores. Oil and Chemical pollution 3:69-81.
Seip, K. L., H. Ibrekk and F. Wenstøp. 1987. Multiattribute analysis of the impact on society of phosphorus abatement measures. Water Resources Research. **23**: 755-764.
Seip, K. L., E. Sandersen, F. Mehlum and J. Ryssdal. 1991. Damages to seabirds from oil spills: comparing simulation results and vulnerability indexes. Ecol Modelling. **52**: 39-59
Sveum, P. and C. Bech. 1990. Oil spill on Arctic shoreline sediments. SINTEF Group. Trondheim
Weber, E., F. Eisenführ and D. von Winterfeldt. 1988. The effects of splitting attributes on weights in multiattribute utility measurement. Management Science. **34**: 431-445.
Westergaard, R. 1987. All about blowout., Norwegian oil review. Oslo.
Woldt, W., Bogardi, I, and L. Duckstein. Consideration of reliability in system designs for groundwater remediation. NATO advanced study institute. Chalkidiki, Greece. (This volume).

NATO/ASI

RISK AND RELIABILITY IN WATER RESOURCES & ENVIRONMENTAL ENGINEERING

May 18-28, 1991 Porto Carras, Greece

ORGANIZING COMMITTEE

Jacques **GANOULIS**
ASI Director
Hydraulics & Environmental Engineering, School of Civil Engineering, Aristotle University of Thessaloniki
54006 Thessaloniki, GREECE

Francisco Nunes **CORREIA**
Laboratorio Nacional de Engenharia Civil
Av. Brasil, 101, P.1799 LISBOA
PORTUGAL

Lucien **DUCKSTEIN**
Systems & Industrial Eng. Dept.
University of Arizona
Tucson, AZ 85721
U.S.A.

Bernard **BOBEE**
Carrefour John Molson, INRS-Eau
2800, rue Einstein, bureau 105
QUEBEC, (Québec), G1X 4N8
CANADA

INVITED LECTURERS

Istvan **BOGARDI**
Department of Civil Engineering
University of Nebraska-Lincoln
W 358 Nebraska-Lincoln
Lincoln, Nebraska 68588-0531
U.S.A.

Louis **DE BACKER**
Unité de Génie Rural,
Place Croix du Sud 2, B[a] 2,
B-1348 Louvain-La-Neuve
BELGIUM

Yacov Y. **HAIMES**
Center for Risk Management of Eng. Systems, D111 Thornton Hall,
University of Virginia,
Charlottesville, VA 22903
U.S.A.

John A. **DRACUP**
Univ. of California, Los Angeles (UCLA), Civil Eng. Dept.
4532 Boelter Hall, CA 90024-1593
Los Angeles,
U.S.A.

Knut L. **SEIP**
Center for Industrial Research
P.O. Box. 124 Blindern,
0314 Oslo 3,
NORWAY

Eric **PARENT**
ENGREF
Applied Maths Dept.
19, Avenue du Maine
75732 PARIS CEDEX 15
FRANCE

Erich **PLATE**
Institut fur Hydrologie &
Wasserwirtschaft
Univ. Karlsruhe (TH)
Kaiserstrasse 12
D-7500 Karlsruhe 1
GERMANY

Jacques **BERNIER**
Lab. National d'Hydraulique
6, quai Watier, B.P. 49,
78401 CHATOU CEDEX
FRANCE

Ahmet **SAMSUNLU**
Istanbul Technical University
Environmental Engineering Dept.
80626 Maslak-Istanbul
TURKEY Fax: (1) 1766587

Bernard **CAUSSADE**
Inst. de Mécanique des Fluides
Avenue Camille Soula
31400-TOULOUSE
FRANCE

J. P. **GASTELLU-ETCHEGORRY**
Centre d' Etude Spatiale des
Rayonnements (CESR), B.P. 4346,
31029 TOULOUSE CEDEX
FRANCE

Paul **PILON**
Env. Canada, Inland Waters Direct.
Water Resources Branch,
8-th Floor, Place Vincent Massey,
Ottawa, Ontario, KIA OH3
CANADA

George **TSAKIRIS**
Lab. of Rural Technology
National Technical Univ. of Athens
Iroon Polytechniou 9
GR-15773, Athens
GREECE Fax./Tel.: 7774178

Eugene Z. **STAKHIV**
US. Army Corps of Engineers
Institute for Water Resources
Casey Building, Fort. Belvoir,
Virginia 22060-5586
U.S.A.

PARTICIPANTS

Nathalie **FANIEL**
Rue d' Andenne 85A
B-1060, BRUXELLES
Universite du Bruxelles
BELGIUM

Mr.Peter Funder **RASMUSSEN**
Inst. of Hydrodynamics and
HydraulicEngineering,Technical
University of Denmark
Building 115-DK-2800 Lyngby
DENMARK

Maria **TOUBEKTSI**
Faculté Polytechnique de Mons
rue de Houdain 9, 7000 MONS
BELGIQUE

Marie Georges **TOURNOUD**
Laboratoire Hydr.et Modelisation
USTL Place E.Bataillon
34095 MONTPELLIER Cedex 5
FRANCE

Claude **THIRRIOT**
ENSEEIHT
2 rue Charles Camichel
31071 TOULOUSE CEDEX
FRANCE

Pascal **BREIL**
CEMAGREF Div.Hydrologie-Hydraulique
3 bis, quai Chauveau
69336 LYON CEDEX 09
FRANCE

Marcus **HOEFKEN**
Cauerstrasse 4
D-8520 ERLANGEN
GERMANY

Franz **BISCHOF**
LSTM-ERLANGEN
Cauerstrasse 4
D-8520 ERLANGEN
GERMANY

Holger **MUSTER**
Inst. for Hydrology & Water
Resources, Univ. of Karlsruhe
D-7500 Karlsruhe 1
GERMANY

Marcus **DISSE**
(IHW) Univ. Karlsruhe
Kaiserstrasse 12, Postfach 69 80
D-7500 Karlsruhe 1
GERMANY

Bruno **MERZ**
Institut fur Hydrologie &
Wasserwirtschaft
University of Karlsruhe
Kaiserstr. 12
D-7500 Karlsruhe/GERMANY

Constantin **XANTHOPOULOS**
Inst. fur Siedlungswasserwirt-
schaft
University of Karlsruhe
Am Fasanengarten
D-7500 Karlsruhe 1/GERMANY

Mario **MAIOLO**
Via Gaetano Salvemini 16
87100 COSENZA
ITALIA

Leonard **LYE**
Memorial Univ. of Newfoundland
St. John' s, Newfoundland,
A1B 3X5
CANADA

Joseph **CORREA**
Ecole Polytechnique, Génie
Civil-Hydraulique, C.P. 6079,
succ. A, Montréal (Québec)
H3C 3A7
CANADA

Raafat **SAADE**
85 Rue ste-Catherine Quest,
Montréal, Québec, H2X 3P4
C A N A D A

Van-Thanh-Van **NGUYEN**
McGill University
Dept. de Génie Civil
817 Rue Sherbrooke Quest,
Montréal, Québec, H3A 2K6,
C A N A D A

Rene **ROY**
Service Previsons
75 Ouest Rene Levesque
9o Complexe Desjardins
Montreal, Quebec
C A N A D A H2Z 1A4

Luc **PERREAULT**
INRS-Eau, Carrefour Molson
2800, Rue Einstein, suite 105
Québec, G1V 4C7, Sainte-Foy
C A N A D A

Cristiane **MARCOUX**
Les Consultants BPR
4655 Boul. Hamel
Québec, GIP 2J7
C A N A D A

Ilrene **ABI-ZEID**
INRS-Eau, Carrefour Molson
2800, Rue Einstein, suite 105
Québec, G1V 4C7, Sainte-Foy
C A N A D A

Amvrossios **BAGTZOGLOU**
Department of Civil Engineering
University of California, Irvine
CA 92717 U.S.A.

Hans-Olaf **PFANNKUCH**
Dept. of Geology & Geophysics
University of Minnesota
310 Pillsbury Dr. SE
Minneapolis, MN. 55455-0219
U.S.A.

Mohamed **DAHAB**
University of Nebraska-Lincoln
Dept. of Civil Engineering
W348 Nebraska Hall
Lincoln, NE 68588-0531
U.S.A.

Scott **MILLER**
Dept.of Civil & Environmental Eng.
Michigan Technology University
1400 Townsend Drive
Houghton, MI 49931
U.S.A.

Richard **HEGGEN**
Civil Eng., Univ. of New Mexico
Albuquerque, NM. 87131,
U.S.A.

Yehouda **ENZEL**
Univ. of Arizona
Dept. of Geosciences
Tuscon, Arizona 85721
U.S.A.

Radwan **AL-MUBARAK**
2522 Hydrosystems Laboratory
205 N. Mathews Ave.
Urbana, IL 61801,
U.S.A.

Lisa **ELY**
University of Arizona
Tucson, Arizona 85721
U.S.A.

Bijaya **P. SHRESTHA**
Systems & Ind. Eng. Dept.,
Univ. of Arizona, Tucson, 85721
U.S.A.

Nabil **SHAFIKE**
Hydr. & Water Resources
Eng. Dept., Tucson,
AZ. 85722
U.S.A.

Aimée **BELLA**
Department of Systems & Ind.
Eng., Univ. of Arizona,
Tucson, AZ. 85721
U.S.A.

Donald **KENDALL**
Loyola Marymount Univ.
Loyola Blvd. at West 80th Str.
Los Angeles, CA 90045
U.S.A.

Pam **EMCH**
Univ.of California, Los Angeles
Civil Eng. Dept.
4532 Boelter Hall
Los Angeles, CA 90024-1593
U.S.A.

David **BELLUCK**
520 Lafayette Road, St. Paul
Minnesota 55155
U.S.A.

James **LAMBERT**
Dept. of Civil Eng.
Thornton Hall, Univ. of Virginia
Charlottesville, VA 22901
U.S.A.

Pericles **LATINOPOULOS**
Hydraulics & Environmental
Engineering, School of
Civil Engineering, Aristotle
University of Thessaloniki
54006 Thessaloniki, GREECE

Marios **VAFIADIS**
Hydraulics & Environmental
Engineering, School of
Civil Engineering, Aristotle
University of Thessaloniki
54006 Thessaloniki, GREECE

Yannis **MYLOPOULOS**
Hydraulics & Environmental
Engineering, School of
Civil Engineering, Aristotle
University of Thessaloniki
54006 Thessaloniki, GREECE

Nikos **THEODOSIOU**
Hydraulics & Environmental
Engineering, School of
Civil Engineering, Aristotle
University of Thessaloniki
54006 Thessaloniki, GREECE

Anastasios **STAMOU**
National Technical Univ.of Athens
Iroon Polytechniou 5
GR-15773, Athens
GREECE

Christos **STEFANIDIS**
Eng. of Hydraulique
Chalkidikis 6
Kozani-50100
GREECE

G. **MANOS**
DEI
Athens
GREECE

Panagiota **ANASTASSOPOULOU**
Ministry of Environment Town
Planning and Public Work
9, Fanarioton Street
10178 Athens GREECE

Kalliopie **ALEXOPOULOU**
Ministry of Environment Town
Planning & Public Work
9, Fanarioton Street
10178 Athens GREECE

Eleni **ALEXOPOULOU**
A. Zacharopoulos A.S.
Tositsa Street
10683 Athens
GREECE

Kalliopi **TZAVLAKI**
Municipal Enterprice for Water
and Sewage of Iraklion
1, Vironos Street
712 02 Irakrion-Kreta
GREECE

Eftihia **PAPALEXIOU**
Municipal Enterprice for Water
and Sewage of Kos
38, Grigoriou E' Street
85300 Kos
GREECE

Demetris **ARGIROPOULOS**
Ministry of the Environment of Greece
Division of Environmental Planning
11251 Athens
GREECE

Ilias **TRAKAKIS**
Municipal Enterprice for Water and
Sewage of Sparta
39, Kleobrotou Street
23100 Sparta, GREECE

Anna **MANTA**
Municipal Enterprice for Water and
Sewage of Nafplion
2, 25th Martiou Street
211 00 Nafplion, GREECE

S. **TZOVARIDIS**
Ministry of Public Works
Athens
GREECE

Sophia **GEROGIANNAKI**
48, Prasinou Lofou Str.
N. Iraklion
14121 Athens, GREECE

Maria-Luisa **BARATA DE TOVAR**
Projecto de Gestao Integrada
dos Recursos Hidricos do Norte
CCRN-PORTO-410o
PORTUGAL

Joao Paulo **LOBO-FERREIRA**
Lab. Nacional de Engenharia Civil
Av. do Brasil, 101
P-1799 LISBOA CODEX
PORTUGAL

K. F. **TURKMAN**
DEIOC, Faculdade de Ciencias
Bloco C/2-Campo Grande,
1700 Lisboa
PORTUGAL Fax: (3511)678308

José Gabriel **DE MENESES**
Dep. Ciencias Eng. do Ambiente
Quinta da Torre
28525 Monte Caparica, BRT.
PORTUGAL

Maria-Manuela **RAMOS DA SILVA**
Inst. Superior Técnico
Dept. of Civil Eng.
Av. Rovisco Pais
1096 LISBOA CODEX
PORTUGAL

Ant. Alex. **TRIGO TEIXEIRA**
Inst. Superior Técnico
Department of Civil Eng.
Av. Rovisco Pais
1096 LISBOA CODEX
PORTUGAL

Pedro Cunha **SERRA**
Lab. Nacional de Eng. Civil
Av. Brasil 101,
1799 LISBOA CODEX
PORTUGAL

Joao **ROCHA**
Lab. Nacional de Eng. Civil
Av. Brasil 101,
1799 LISBOA CODEX
PORTUGAL

Maria José **AMORIM**
Inst. Superior Técnico
Dept. of Civil Eng.
Av. Rovisco Pais
1096 LISBOA CODEX
PORTUGAL

Nilgun **HARMANCIOGLU**
Dokuz Eylul University
Faculty of Eng. & Architecture
Dept. of Civil Engineering
35100 Bornova-Izmir
TURKEY FAX: (51) 187864

Estela Goncalves **PEREIRA**
Environmental Systems Analysis
Group from UNINOVA
Inst. for the Development of New
Tecnologies-Quinta da Torre
2825 Monte da Caparica
PORTUGAL

Necdet **ALPASLAN**
Dokuz Eylul University
Faculty of Eng. & Architecture
Env. Eng. Dept.
35100 Bornova-Izmir
TURKEY Fax: (51) 187864

Bilsen **BELER BAYKAL**
Instanbul Tech. Univ. Env. Eng.Dept.
80626 Maslak-Istanbul
TURKEY

Gulseven **UBAY**
Istanbul Tech. Univ. Env. Eng.Dept.
80626, Maslak-Istanbul
TURKEY

Atil **BULU**
Technical Univ. of Istanbul
Civil Eng. Faculty, Dept. of
Hydraulics , 80626, Ayazaga,
Istanbul, TURKEY

Orhan **OZBAY**
Ataturk Universty Eng. Faculty
Head of the Envirionmental
Engineering Dept., 25240 Erzurum
TURKEY (Fax : (011) 17140

Dogan **ALTINBILEK**
Chairman, Civil Eng. Dept.,
Middle East Technical Univ.,
06531 Ankara-TURKEY

Ismail. **DURANYILDIZ**
Istanbul Teknik Univ.
Insaat Fakultesi, Hidrolik ABD.
Ayazaga 80626 Istanbul-TURKEY

Ferhat **TURKMAN**
Dokuz Eylul Univ., Faculty of
Eng. & Architecture, Dept. of
Civil Ehgineering,
Bornova 35100-IZMIR/TURKEY

Turkay **BARAN**
Civil Eng. Dept., Faculty of Eng. &
Architecture, Dokuz Eylul Univ.
Bornova 35100, IZMIR/TURKEY
Fax: (51) 187864

NON - NATO COUNTRIES

Alfir **AKHMETOV**
IMMS, Taymirskaya, 74
Inst.of Multiphase Meckanics
Systems of USSR Academy
62500 Tyumen GSP
USSR

Konstantin **FYODOROV**
IMMS, USSR Academy of Science
Taymirskaya, 74
62500 Tyumen GSP
USSR

Illka **LAMBOVA**
Dept. of Mathematics &
Informatics Univ. of Sofia
5, Anton Ivanov Avenue
BULGARIA

INDEX

NATO ASI Series G

Vol. 1: **Numerical Taxonomy.** Edited by J. Felsenstein. 644 pages. 1983. (out of print)

Vol. 2: **Immunotoxicology.** Edited by P.W. Mullen. 161 pages. 1984.

Vol. 3: **In Vitro Effects of Mineral Dusts.** Edited by E.G. Beck and J. Bignon. 548 pages. 1985.

Vol. 4: **Environmental Impact Assessment, Technology Assessment, and Risk Analysis.** Edited by V.T. Covello, J.L. Mumpower, P.J.M. Stallen, and V.R.R. Uppuluri. 1068 pages. 1985.

Vol. 5: **Genetic Differentiation and Dispersal in Plants.** Edited by P. Jacquard, G. Heim, and J. Antonovics. 452 pages. 1985.

Vol. 6: **Chemistry of Multiphase Atmospheric Systems.** Edited by W. Jaeschke. 773 pages. 1986.

Vol. 7: **The Role of Freshwater Outflow in Coastal Marine Ecosystems.** Edited by S. Skreslet. 453 pages. 1986.

Vol. 8: **Stratospheric Ozone Reduction, Solar Ultraviolet Radiation and Plant Life.** Edited by R.C. Worrest and M.M. Caldwell. 374 pages. 1986.

Vol. 9: **Strategies and Advanced Techniques for Marine Pollution Studies: Mediterranean Sea.** Edited by C.S. Giam and H.J.-M. Dou. 475 pages. 1986.

Vol. 10: **Urban Runoff Pollution.** Edited by H.C. Torno, J. Marsalek, and M. Desbordes. 893 pages. 1986.

Vol. 11: **Pest Control: Operations and Systems Analysis in Fruit Fly Management.** Edited by M. Mangel, J.R. Carey, and R.E. Plant. 465 pages. 1986.

Vol. 12: **Mediterranean Marine Avifauna: Population Studies and Conservation.** Edited by MEDMARAVIS and X. Monbailliu. 535 pages. 1986.

Vol. 13: **Taxonomy of Porifera from the N.E. Atlantic and Mediterranean Sea.** Edited by J. Vacelet and N. Boury-Esnault. 332 pages. 1987.

Vol. 14: **Developments in Numerical Ecology.** Edited by P. Legendre and L. Legendre. 585 pages. 1987.

Vol. 15: **Plant Response to Stress. Functional Analysis in Mediterranean Ecosystems.** Edited by J.D. Tenhunen, F.M. Catarino, O.L. Lange, and W.C. Oechel. 668 pages. 1987.

Vol. 16: **Effects of Atmospheric Pollutants on Forests, Wetlands and Agricultural Ecosystems.** Edited by T.C. Hutchinson and K.M. Meema. 652 pages. 1987.

Vol. 17: **Intelligence and Evolutionary Biology.** Edited by H.J. Jerison and I. Jerison. 481 pages. 1988.

Vol. 18: **Safety Assurance for Environmental Introductions of Genetically-Engineered Organisms.** Edited by J. Fiksel and V.T. Covello. 282 pages. 1988.

Vol. 19: **Environmental Stress in Plants. Biochemical and Physiological Mechanisms.** Edited by J.H. Cherry. 369 pages. 1989.

Vol. 20: **Behavioural Mechanisms of Food Selection.** Edited by R.N. Hughes. 886 pages. 1990.

Vol. 21: **Health Related Effects of Phyllosilicates.** Edited by J. Bignon. 462 pages. 1990.

NATO ASI Series G

Vol. 22: **Evolutionary Biogeography of the Marine Algae of the North Atlantic.** Edited by D. J. Garbary and G. R. South. 439 pages. 1990.

Vol. 23: **Metal Speciation in the Environment.** Edited by J. A. C. Broekaert, Ş. Güçer, and F. Adams. 655 pages. 1990.

Vol. 24: **Population Biology of Passerine Birds. An Integrated Approach.** Edited by J. Blondel, A. Gosler, J.-D. Lebreton, and R. McCleery. 513 pages. 1990.

Vol. 25: **Protozoa and Their Role in Marine Processes.** Edited by P. C. Reid, C. M. Turley, and P. H. Burkill. 516 pages. 1991.

Vol. 26: **Decision Support Systems.** Edited by D. P. Loucks and J. R. da Costa. 592 pages. 1991.

Vol. 27: **Particle Analysis in Oceanography.** Edited by S. Demers. 428 pages. 1991.

Vol. 28: **Seasonal Snowpacks. Processes of Compositional Change.** Edited by T. D. Davies, M. Tranter, and H. G. Jones. 484 pages. 1991.

Vol. 29: **Water Resources Engineering Risk Assessment.** Edited by J. Ganoulis. 551 pages. 1991.

MIX
Papier aus verantwortungsvollen Quellen
Paper from responsible sources
FSC® C105338
FSC
www.fsc.org

If you have any concerns about our products,
you can contact us on
ProductSafety@springernature.com

In case Publisher is established outside the EU,
the EU authorized representative is:
Springer Nature Customer Service Center GmbH
Europaplatz 3, 69115 Heidelberg, Germany

Printed by Libri Plureos GmbH
in Hamburg, Germany